# Die Kulturpflanze

Mitteilungen aus dem Zentralinstitut

für Genetik und Kulturpflanzenforschung

Gatersleben

der Akademie der Wissenschaften der DDR

**Band XXIX**

Herausgegeben von

H. Böhme, W. R. Müller-Stoll, K. Müntz,

R. Rieger, A. Rieth, F. Scholz, H. Stubbe

Schriftleitung: K. Müntz

*Mit 83 Abbildungen und 61 Tabellen*

THEORIA CUM PRAXI

AKADEMIE-VERLAG · BERLIN

1981

ISSN 0075-7209

Erschienen im Akademie-Verlag, DDR - 1080 Berlin, Leipziger Straße 3–4

Lizenznummer: 202 · 100/511/81
Gesamtherstellung: IV/2/14 VEB Druckerei »Gottfried Wilhelm Leibniz«,
4450 Gräfenhainichen · 5842
Umschlaggestaltung: Annemarie Wagner
Bestellnummer: 762 909 2 (2052/29) · LSV 1355
Printed in GDR
DDR 98,– M

*RUDOLF MANSFELD (1901–1960)*

*Wir erinnern uns in Dankbarkeit Professor Dr. RUDOLF MANSFELDS, des ersten Schriftleiters der „Kulturpflanze“ und langjährigen Leiters der früheren Abteilung Systematik und Sortiment. Seines 80. Geburtstages gedachten am 17. Januar 1981 seine ehemaligen Mitarbeiter und Kollegen.*

*Die Leitung des Zentralinstituts*

# Inhalt

**Sektion II:**
**Probleme der Taxonomie von Kulturpflanzen**

## Sektion III: Biosystematik von Kulturpflanzen und Züchtungsforschung

## II. Originalarbeiten aus dem Institut



## III. Das Institut im Jahre 1980



[1]) und alle unter I. erscheinenden Arbeiten englisch

# Content

**I. Symposium on Collection of European Land-races of Cultivated Plants and their Evaluation**

**Section II:**
**Problems of the Taxonomy of cultivated plants**

# Содержание

**II. Оригинальные работы института**

# I. Symposium über „Europäische Landsorten von Kulturpflanzen und ihre Nutzung“

Kulturpflanze XXIX · 1981 · S. 19–25

# Symposium on collection of European land-races of cultivated plants and their evaluation

## Introduction

Very much attention has been devoted to the genetic plant resources localized in the few classic gene centres of cultivated plants. Although not being of the opinion that these regions should not be investigated further on and much more intensively than hitherto we have had the impression that it is likewise worthwile to look into territories outside of these centres.

Here gene erosion has nearly always reached a much more advanced stage than in gene centres, land-races have almost completely been lost, and the considerable number of the cultivars of former decades has been replaced by few high-bred varieties which are dominating over large areas.

This situation is also true for Europe, land-races are only in a few regions still cultivated and there are only some groups of cultivated plants in which such indigenous material can be expected to be still in existence. It is an urgent necessity to collect and maintain this material, it may sometimes be more useful in current breeding work than gene resources from foreign regions.

Therefore we had the idea to organize a meeting which should give the opportunity to discuss these problems and to inform about the activities in different European countries referring to them.

At the suggestion of the chairman of the section 'Wild species and primitive forms' of EUCARPIA, Prof. Dr. J. G. HAWKES, Birmingham, the programme of the symposium had been extended to basic problems of taxonomy of cultivated plants especially the species concept and principles of infraspecific classifications as well as to biosystematic studies as an aid to breeding research.

Some colleagues had been invited to give main lectures on the selected topics and by the rather widespread consent that the first announcement of the meeting had found it was possible to establish a well-balanced scientific programme covering the three sections

1) Taxonomy, history, collection, maintenance and breeding value of European land-races,
2) Problems of the taxonomy of cultivated plants, and
3) Biosystematics of cultivated plants and breeding research.

The symposium was held at the Zentralinstitut für Genetik und Kulturpflanzenforschung in Gatersleben from July 1st to July 4th 1980. It started with opening addresses by Dr. M. ZACHARIAS, Deputy Director of the Gatersleben institute, and by Prof. Dr. J. G. HAWKES (delivered in German and English!) who gave also the concluding remarks. In all 73 scientists from 17 countries–more than 50% from

Fig. 1 The participants of the symposium in front of the auditory hall in Gatersleben
Foto G. TERPE

Fig. 2 F. KÜHN (left) and J. HAWKES
Foto G. TERPE

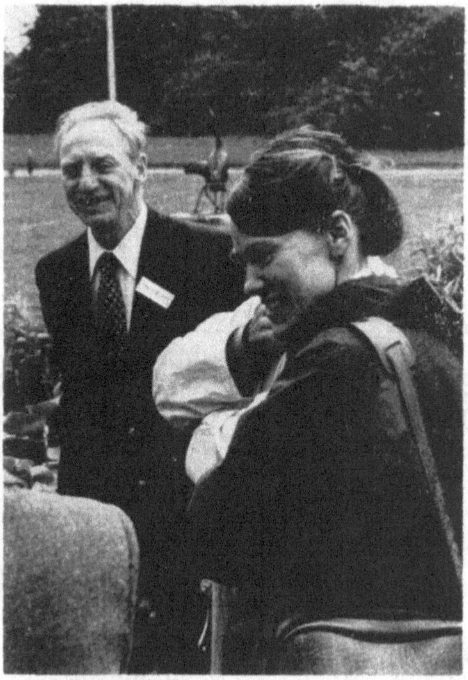

Fig. 3 J. M. J. DE WET (left) and B. PICKERSGILL
Foto H. KNÜPFFER

Fig. 4 R. SCHACHL (left) and H. KNÜPFFER Foto G. TERPE

Fig. 5 Visit of cereal land-races collection, foreground (from the left): P. HANELT, B. R. BAUM, J. MAC KEY, W. A. BRANDENBURG, K. PISTRICK and K. HAMMER, second row (from the left) R. FRITSCH, J. WEISS, B. PICKERSGILL, J. KRUSE, B. BOCHEV
Foto G. TERPE

abroad – participated. About 30 lectures and 10 posters had been presented, the scientific level of nearly all of these contributions and of the discussions having been impressively high according to the unanimous opinion of guests from abroad and of ourselves.

In the evening of July 1st a sight-seeing excursion through Quedlinburg, where the hotel accommodations could be provided for the participants, took place, the city's architect Mr. E. Bleicke having been our guide. In the 'Kunsthoken', a reconstructed building of the renaissance era, in Quedlinburg a farewell cocktail party have been given in the night of July 3rd. In the morning of July 4th our guests could have a look on the experimental fields and other facilities of the gene bank and the taxonomic group of the institute (long term seed storage building, herbarium, seed and spike collections).

In the morning of July 4th, the members of the EUCARPIA Gene Bank Committee and gene bank directors who participated in the symposium met with interested participators to the annual Extended EUCARPIA Gene Bank Committee meeting. Under the chairmanship of Professor Dr. J. G. Hawkes reports of the European gene bank directors and actual problems of their co-operation have been discussed.

The organization of the meeting could have been done only by the friendly assistance and discipline of all participants especially the speakers and with the enthusiastic support of the scientific and technical staff of the institute, all of them we have to give once more our best thanks.

We believe that the conference has given an insight into problems of collection and evaluation of land-races in some of the European countries and has contributed to a better understanding of systematics of cultivated plants and to a less controversial taxonomic treatment of their enormous variability. We hope that all participants have experienced the meeting as stimulating – as we have done it – and that they will remember the week in Gatersleben as a pleasant one, full of personal contacts and without serious shortcomings.

Gatersleben, November 1980

P. Hanelt Chr. O. Lehmann

## List of Participants

de Bakker, Gerrit
UNDP Office, Palais des Nations, CH-1211 Geneva 10, Switzerland

Bareš, Ivo
Research Institute of Plant Production, Drnovská 507, CS - 161 06 Praha-Ruzyně, ČSSR

Baum, Bernard R.
Biosystematics Research Institute, Agriculture Canada, William Saunders Building, C.E.F., Ottawa, Ontario K1A 0C6, Canada

Beridze, Rusudan K.
Institute of Botany of the Georgian Academy of Sciences, Kodjorskoe Road, SU - 380007 Tbilissi, USSR

Betzwar, Werner
Fa. Prochaska & Cie., Graben 14, A - 1010 Wien 1, Austria

BLIXT, Stig Gunnar
Cytogenetical Laboratory, Weibullsholm Plant Breeding Institute, S-261-10 Landskrona, Sweden
BOCHEV, Bogdan
Institute of Genetics, BG-1113 Sofia, Bulgaria
BRANDENBURG, Willem A.
Department of Taxonomy of Cultivated Plants, Agricultural University, Haagsteeg 3, NL-6708 PM Wageningen, The Netherlands
CUBERO, José-I.
Departamento de Genética, E.T.S. Ingenieros Agrónomos, Universidad de Córdoba, Spain
DEHNE, Jochen
Institut für Züchtungsforschung, Ethel-und-Julius-Rosenberg-Str., DDR-4300 Quedlinburg, GDR
ESQUINAS-ALCÁZAR, José T.
AGPE/IBPGR FAO, Via Terme di Caracalla, I-00 100 Roma, Italy
EVANS, Alice M.
Department of Applied Biology, University of Cambridge, Pembroke Street, Cambridge CB2 3DX, England, U.K.
FRANKE, Rolf
Institut für Züchtungsforschung, Ethel-und-Julius-Rosenberg-Str., DDR-4300 Quedlinburg, GDR
FRITSCH, Reinhard
Zentralinstitut für Genetik und Kulturpflanzenforschung, DDR-4325 Gatersleben, GDR
FURSA, Tatyana B.
N. I. VAVILOV Institute of Plant Industry, 44 Herzen Street, SU-190000 Leningrad, USSR
GEORGIEV, Sevdalin G.
Institute of Plant Introduction and Genetic Resources "K. MALKOV", BG-4122 Sadovo, Plovdiv district, Bulgaria
GIGOVA, M. Ljubenova
Institute of Genetics, BG-1113 Sofia, Bulgaria
GREBENŠČIKOV, Igor
Zentralinstitut für Genetik und Kulturpflanzenforschung, DDR-4325 Gatersleben, GDR
GRÖBER, Kurt
Zentralinstitut für Genetik und Kulturpflanzenforschung, DDR-4325 Gatersleben, GDR
HAMMER, Karl
Zentralinstitut für Genetik und Kulturpflanzenforschung, DDR-4325 Gatersleben, GDR
HANELT, Peter
Zentralinstitut für Genetik und Kulturpflanzenforschung, DDR-4325 Gatersleben, GDR
HAWKES, John G.
Department of Plant Biology, The University of Birmingham, P. O. Box 363, Birmingham B15 2TT, England, U.K.
HOLLY, László
Research Centre for Agrobotany, NIAVT, H-2766 Tápiószele, Hungary
HONDELMANN, Walter
Institut für Pflanzenbau und Pflanzenzüchtung der FAL, Braunschweig-Völkenrode, Bundesallee 50, D-3300 Braunschweig, F.R.G.
IVANJUKOVICH, Ludmila K.
N. I. VAVILOV Institute of Plant Industry, 44 Herzen Street, SU-190000 Leningrad, USSR
JAKUBIEC, J.
Instytut Hodowli i Aklimatyzacji Roślin, PL-05-870 Radzików, Poland
DE JONG, Johannes H.
Institute of Genetics, University of Amsterdam, Kruislaan 318, NL-1098 SM Amsterdam, The Netherlands
KÅHRE, Erik, L. J.
Swedish Seed Testing and Certification Institute, S-171 73 Solna, Sweden

Kjellqvist, Ebbe
Nordic Gene Bank, Fack, S-22101 Lund-1, Sweden
Knüpffer, Helmut
Zentralinstitut für Genetik und Kulturpflanzenforschung, DDR-4325 Gatersleben, GDR
Kostov, C.
Institute of Plant Introduction and Genetic Resources "K. Malkov", BG-4122 Sadovo, Plovdiv district, Bulgaria
Kruse, Joachim
Zentralinstitut für Genetik und Kulturpflanzenforschung, DDR-4325 Gatersleben, GDR
Kubiczek, Roman
Botanical Garden of Polish Academy of Sciences, ul. Prawdziwka 2, PL-00-979 Warszawa p-84, Poland
Kühn, František
Vysoká Škola Zemědělská v Brne, Katedra obecné a speciální botaniky zemědělské, Zemědělská 1, CS-662 65 Brno, ČSSR
Kulpa, Władysław
Instytut Hodowli i Aklimatyzacji Roślin, PL-05-870 Radzików, Poland
Kummer, Manfred
Institut für Züchtungsforschung, Ethel-und-Julius-Rosenberg-Str., DDR-4300 Quedlinburg, GDR
Lafiandra, Domenico
Istituto del Germoplasma, C.N.R., Via G. Amendola 165/A, I-70 126 Bari, Italy
Lamberts, Hendrik
Stichting voor Plantenveredeling, Postbus 117, Droevendaalsesteeg 1, NL-6700 AC Wageningen, The Netherlands
Lange, Wouter
Stichting voor Plantenveredeling, Postbus 117, Droevendaalsesteeg 1, NL-6700 AC Wageningen, The Netherlands
Lehmann, Christian
Zentralinstitut für Genetik und Kulturpflanzenforschung, DDR-4325 Gatersleben, GDR
Maass, Helga
Zentralinstitut für Genetik und Kulturpflanzenforschung, DDR-4325 Gatersleben, GDR
Mac Key, James
Department of Plant Breeding, Swedish University of Agricultural Sciences, S-75007 Uppsala-7, Sweden
Meyer, Doris
Zentralinstitut für Genetik und Kulturpflanzenforschung, DDR-4325 Gatersleben, GDR
Meyer, Horst
Institut für Getreideforschung, DDR-3234 Hadmersleben, GDR
Molski, Boguslaw
Botanical Garden of Polish Academy of Sciences, ul. Prawdziwka 2, PL-00-979 Warszawa p-84, Poland
Müller, K.
Institut für Pflanzenzüchtung, DDR-1701 Petkus, GDR
Näther, Jochen
Institut für Getreideforschung, DDR-4350 Bernburg, GDR
Ohle, Harald
Zentralinstitut für Genetik und Kulturpflanzenforschung, DDR-4325 Gatersleben, GDR
Perrino, Pietro
Istituto del Germoplasma, C. N. R., Via G. Amendola 165/A, I-70 126 Bari, Italy
Pickersgill, Barbara
Department of Agricultural Botany, University of Reading, Whiteknights, Reading, RG6 2AS, England, U.K.
Pignone, Domenico
Istituto del Germoplasma, C.N.R., Via G. Amendola 165/A, I-70 126 Bari, Italy
Pistrick, Klaus
Zentralinstitut für Genetik und Kulturpflanzenforschung, DDR-4325 Gatersleben, GDR

PORCEDDU, Enrico
Istituto del Germoplasma, C.N.R., Via G. Amendola 165/A, I - 70 126 Bari, Italy
PUCHALSKI, Jerzy
Botanical Garden of Polish Academy of Sciences, ul. Prawdziwka 2, PL - 00-979 Warszawa p-84, Poland
RIEMANN, Karl-Heinz
Institut für Pflanzenzüchtung, DDR - 2601 Gülzow-Güstrow, GDR
ROELOFSEN, Hendrika J.
Institute of Horticultural Plant Breeding, Mansholtlaan 15, NL - 6700 AA Wageningen, The Netherlands
SCHACHL, Rudolf M.
Landwirtschaftlich-chemische Bundesversuchsanstalt in Linz, Wieningerstr. 8, A - 4025 Linz, Austria
SCHOLZ, Friedrich
Zentralinstitut für Genetik und Kulturpflanzenforschung, DDR - 4325 Gatersleben, GDR
SCHULTZE-MOTEL, Jürgen
Zentralinstitut für Genetik und Kulturpflanzenforschung, DDR - 4325 Gatersleben, GDR
SEHNALOVÁ, Jana
Research Institute of Plant Production, Drnovská 507, CS - 161 06 Praha-Ruzyně, ČSSR
van SLOTEN, D.
AGPE/IBPGR FAO, Via Terme di Caracalla, I - 00 100 Roma, Italy
SMITH, Roger Drummond
Royal Botanic Gardens, Wakehurst Place, Ardingly, Haywardsheath, W. Sussex, RH 17 6TN, U. K.
STREITZ, Doris
VEG-Saatzucht Zierpflanzen, Gorkistr. 9, Postfach 209, DDR - 5010 Erfurt, GDR
SZABÓ, Attila T.
Institutul Agronomic, Gradina Agrobotanica, 3400 Cluj-Napoca, R. S. Romania
TABBERT, E.
VEB Datenverarbeitungszentrum, Gadebuscher Str. 74, DDR - 2700 Schwerin, GDR
TIEMANN, Horst
Institut für Kartoffelforschung, DDR - 2551 Groß-Lüsewitz, GDR
TITTEL, Claus
Zentralinstitut für Genetik und Kulturpflanzenforschung, DDR - 4325 Gaterseben, GDR
TODOROVA, P.
Institute of Plant Introduction and Genetic Resources "K. MALKOV", BG - 4122 Sadovo, Plovdiv district, Bulgaria
WEISS, Jürgen
Zentralinstitut für Genetik und Kulturpflanzenforschung, DDR - 4325 Gatersleben, GDR
de WET, Johannes, M. J.
Crop Evolution Laboratory, Department of Agronomy, University of Illinois at Urbana-Champaign, N 214 Turner Hall, Urbana, Illinois 61801, USA
WOLTERSTORFF, Bernd
Zentralinstitut für Genetik und Kulturpflanzenforschung, DDR - 4325 Gatersleben, GDR
ZACHARIAS, Martin
Zentralinstitut für Genetik und Kulturpflanzenforschung, DDR - 4325 Gatersleben, GDR
ZEVEN, Anton C.
Instituut voor Plantenveredeling, Postbus 386, Lawickse Allee 166, Wageningen, The Netherlands

# Section 1:
# Taxonomy, history, maintenance and breeding value of European land-races

Chairmen:

J. Hawkes
E. Porceddu
B. Molski
S. Blixt

Kulturpflanze XXIX . 1981 . S. 29–40

# Collecting European land-races and development of European gene banks–historical remarks

CHRISTIAN O. LEHMANN (Gatersleben, GDR)


## Summary

Plant breeding in Europe began in the 19th century by selection from land-races. It was recognized very early that land-races in Europe vanished from cultivation both by improved farming techniques and spread of new cultivars particularly after rediscovery of MENDEL's laws. E. von PROSKOWETZ and F. SCHINDLER were probably the first to draw attention, in 1890, to the valuable characters of land-races and the dangers for future plant breeding following from their disappearance.

Unfortunately, the following repeated appeals for the preservation of European land-races had only little results.

The first description of "gene bank activities" was given by E. von PROSKOWETZ. The concept of a gene bank was realized for the first time by N. I. VAVILOV at the then Institute of Plant Industry in Leningrad.

Gene banks in Socialist European countries were established mainly in the years after World War II for the benefit of plant breeding and its activities stimulated by N. I. VAVILOV's ideas.

In other European countries gene banks started their activities mainly during the past ten/fifteen years as a result of wide discussions and proposals of EUCARPIA since 1960. Some of them are related to programmes of the International Board for Plant Genetic Resources (IBPGR), e.g., to its Mediterranean Germplasm Programme.

At present, efforts are being made by the EUCARPIA Gene Bank Committee and the FAO/UNDP/IBPGR European Cooperative Programme for Crop Genetic Resources to establish relations of close co-operation between the European national (and regional) gene banks and their networks.


In the European countries systematic plant breeding on the most important food plants began in the course of the 19th century. Breeding methods were selection from local populations of land-races based upon their phenotypic variation and planned hybridization.

At that time varieties were considered to be land-races "which are cultivated in this region from time immemorial of which they are named" (RÜMKER 1908). Included was the existence of distinct morphological features and physiological characters in these forms or mixtures of forms (e.g., early or late maturing, spring- or winter-form) (FRUWIRTH 1910).

According to our present knowledge (cf. HARLAN 1975) land-races populations show a high phenotypic variation but they are each more or less identifiable.

Their components are adapted to local soil types, climatic conditions, diseases, pests, as well as to cultural practices.

Genetic and phenotypic variations correspond to each other. This is the result of local selection for many generations. The genotypes are adapted both to natural and man-made environments as well as to each other.

The level of their yield is low but the genetic variability provides a certain kind of yield stability and, with it, of insurance against crop failures.

Both the intensification of agriculture and the development of plant breeding led to a diminishing and later to a complete replacement of land-races. Probably the first who drew attention to the dangers for plant breeding caused by the diminishing of land-race populations were EMANUEL von PROSKOWETZ, plant breeder and big landowner at Kwassitz, now ČSSR, and FRANZ SCHINDLER, at that time Professor of Agronomy in Riga.

E. von PROSKOWETZ had begun line selection in spring barleys of the Hana-region in Moravia in the seventies of the last century. His "Proskowetz-Hanna-Pedigree" barley superseded "Chevallier"-barleys which in those days were leading in Europe. The "Hanna"-barleys were widespread not only in Europe but also in overseas.

In 1890 E. von PROSKOWETZ gave to the International Agricultural and Forestry Congress in Vienna a report on the value of land- or natural races in comparison to so-called bred races. He had realized that the value of land-races lies in their "relative immunity" to fungi and insects. He drew attention to the gradual disappearance and demanded that the land-races should be collected, exactly described, and their description compiled in a catalogue of land-races. He proposed to establish stations to carry out there systematic studies of different characters of land-races, for the time being only from the scientific point of view. F. SCHINDLER endorsed, in a second paper, E. von PROSKOWETZ's opinion.

By these proposals E. von PROSKOWETZ has outlined the objectives of the present gene bank activities: collection, exploration, documentation, and conservation.

Stimulated by rising combination breeding after the rediscovery of MENDEL's laws reference was made repeatedly, in the first decades of our century, to the value of land-races as raw material for plant breeding and to the urgency of their conservation (NILSSON-EHLE 1911, 1924, BAUR 1914, TSCHERMAK 1915, SCHINDLER 1918).

Two new topics were introduced into the discussion by BAUR (1914): (1) He went beyond the collection of European land-races and drew attention to the vanashing cultivated primitive races especially of wheat and barley in the Orient, Asia, North-Africa, and Ethiopia. They should be collected and preserved together with wild-growing relatives of our crop plants. (2) Collection and conservation should be done at a special centre since not every plant breeder can set up a big collection of old and primitive varieties.

Two resolutions shall be quoted as examples of the continuation of discussions on the preservation of land-races.

In 1927, the Section Genetics of the Scientific Advisory Board of the International Agricultural Institute in Rome studied the problem of land-races and stated: (1) A centre for the conservation of land-races should be established in every member country; (2) land-races should be maintained possibly in the region of their origin and (3) without any selection.

The General Assembly of the International Association of Plant Breeders held in Berlin, in 1931, passed a resolution calling upon the governments to take care that (1) land-races, including those of fruits and vegetables, growing in their countries should be collected and maintained; (2) the maintenance should be performed in primary regions and by traditional cultivation methods, and (3) at the same time as many morphological and physiological lines as possible should be isolated which then were to be maintained in respective institutes.

These different and repeated appeals hardly received desired notice at that time.

What could have been the possible motives for the lack of resonance to these numerous requests for collecting and preserving European land-races?

In no European country except in the Soviet Union existed an institute for planned collection and preservation of land-races. Land-races, lines selected from them and cultivars were preserved in the collections of private breeders and of state or university institutes with plant breeding activities.

The interest in indigenous land-races decreased as agricultural practices improved. Cultivars developed by selection from land-races responded to improved methods of cultivation scarcely with higher yields. Breeders had to go early the more complicated and longer way of crossing to raise the yielding capacity of their cultivars. In selecting parental material for crossings they preferred cultivars with an already higher yielding capacity (Dorst 1958, Mayr 1951, Roemer 1953, Wienhues und Giessen 1957).

Lines, resistant to spreading fungal diseases, could be selected only exceptionally from European land-races. Therefore, Roemer (1942) denied European land-races any value as raw material in breeding for disease resistance.

One of these exceptions were the mildew and stripe rust resistant barleys descending from South-German land-races, namely the mildew resistant 'Pflugs Intensiv' and the stripe rust resistant 'Ackermanns Bavaria', 'Heils Franken', 'Fuchs Pfälzer', etc. The mildew resistance derived from lines of the Dalmatian local barley Ragusa determined this resistance in European barleys for many years. This population was collected by G. Aufhammer in the market at Ragusa and in the fields of this region in 1929 (Honecker 1942).

Occasionally the view was heard that the useful characters would be preserved in the progeny because all parents used in earlier crossings were selected from land-races.

The first description of present gene bank activities was probably given by E. von Proskowetz in 1890, the botanic gardens, however, can be regarded as the oldest forerunners. Initially they served in teaching of medicine but gradually they developed into centres of plant introduction. Many extra-European plant species were introduced, through botanic gardens, into horticulture and agriculture.

The present concept of a gene bank, viz. collection, exploration, utilization and long-term conservation, was first realized at the N. I. Vavilov Institute of Plant Industry in Leningrad (Frankel 1975). It took its origin from the Bureau of Applied Botany established in then Petersburg, in 1894. The collection, exploration and description of Russian land-races of major crop plants and the introduction of less known or new species of cultivated plants were a part of its responsibilities. From 1904 till his death in 1920, ROBERT REGEL was the director of this institution. He became well-known by his comprehensive treatment of Russian barleys based on the large barley collection. In his report on the activities of the Bureau over the first twenty years (1894–1914) he wrote that the collections included approximately 9000 seed and spike samples of cereals and 900 seed samples of industrial plants. They came mainly from the provinces of the Russian Empire (REGEL 1915).

In 1920 NIKOLAI IVANOVITCH VAVILOV succeeded R. REGEL in the directorship. In 1924 the Bureau of Applied Botany was reorganized in Allunion Institute of Applied Botany and New Crops. It received the more known name of Allunion Institute of Plant Industry (VIR) in 1930, and finally, on the occasion of the 80th anniversary of N. I. VAVILOV's birthday in 1967 the present name.

Under the direction of N. I. VAVILOV the staff of the institute devoted themselves to collecting and explorating the crop genetic resources of the entire world with the aim of utilizing this material in developing new cultivars for Soviet agriculture and horticulture.

Between 1920 and 1940, 140 expeditions in the U.S.S.R. and 40 expeditions to 64 foreign countries were organized (BREŽNEV 1970). The collections increased rapidly and had in 1927 140000 and 1940 more than 200000 accessions (PISSAREV 1929, BREŽNEV 1969).

VAVILOV's theories on the origin of crops, their evolution and their variability in space and time formed the scientific basis for the collection, exploration and preservation of "as wide a range as possible of genetic variability to form a continued basic supply for breeding new cultivars" (HAWKES 1978).

After the World War II so-called Collections of Cultivated Plants, i.e., national gene banks, were established in the Socialist countries. Their establishment was closely related to the recovery of plant breeding and met with universal support. Their tasks are the collection, preservation and exploration of the genetic resources with a view to using them in plant breeding at present and in future.

At that time the basis of the collections were mostly landraces, old and recent cultivars of the respective country as far as still existing in breeding stations or institutes. Later, this material was supplemented and extended both by exchange with other collections abroad and by collections in own and/or foreign countries.

The concept of these collections and their activities goes back to the Allunion Institute of Plant Industry in Leningrad and to the work of N. I. VAVILOV and his collaborators.

A short survey of the gene banks in Socialist countries in a roughly chronological order of their establishment shall demonstrate this development.

In the GDR, the gene bank is maintained at the Central Institute for Genetics and Cultivated Plant Research, Academy of Sciences of the GDR, at Gatersleben, the potato collection at the Institute of Potato Research at Groß Lüsewitz, Academy of Agricultural Sciences of the GDR.

Stimulated by the ideas of N. I. VAVILOV, as early as in the midthirties considerations were underway to establish an Institute for Cultivated Plant Research. On the basis of extensive collections of wild and primitive forms, land-races and cultivars problems of cultivated plants should be studied from various points of view, including evolution and taxonomy, genetics, physiology and biochemistry. The institute was established in Vienna, in 1943, under the directorship of H. STUBBE but it was only after World War II that it could be built up as an institute of the Academy of Sciences at Gatersleben (STUBBE 1970).

In the first years various collections used in breeding or research studies at different institutes and stations were brought together for further preservation in the gene bank. These were collections of wild and primitive forms, land-races, old or more recent cultivars of cereals, vegetables, medicinal herbs and industrial plants. Moreover, it was to maintain the material collected by H. STUBBE and his collaborators in the Eastern European parts of the Mediterranean Centre of Diversity, mainly in Albania and Greece, in 1941 and 1942.

In the following years the collections were extended by an extensive seed exchange with gene banks, botanic gardens, plant research and breeding institutes and by expeditions, e.g., to South Italy, China, Mongolia, Cuba, ČSSR, Poland and Spain (LEHMANN 1972, 1979, LEHMANN und HAMMER 1979).

In the People's Republic of Bulgaria the gene bank is now attached to the K. MALKOV Institute for Introduction and Plant Resources at Sadovo near Plovdiv. This institute originated from the department of Plant Resources of the Institute of Genetics and Plant Breeding of the Bulgarian Academy of Sciences. Since the foundation of the institute in 1947 its staff collected very extensively the wealth of land-races in Bulgaria.

In the ČSSR planned efforts at collecting, preserving, explorating and utilizing the genetic resources began in 1951 when various collections scattered on breeding stations and agricultural research institutes were concentrated under the supervision of the Division of Genetics and Plant Breeding Methods of the Research Institute of Crop Production in Prague-Ruzyně. These collections were established and utilized in breeding, to a certain extent, since the beginning of this century.

At present, the collections are maintained mainly in institutes specializing in individual groups of crops and exploited in their breeding programmes (BAREŠ 1979).

In the Hungarian People's Republic the utilization of land-races and ecotypes in plant breeding has a long-standing tradition. After the World War II A. JÁNOSSY did a great deal for the exploration, collection and preservation of still existing land-races, old cultivars and other locally adapted plant populations. In 1959 the Institute of Agrobotany (now Research Centre for Agrobotany) was established at Tápiószele under the directorship of A. JÁNOSSY where the maintenance and exploration of genetic resources is now concentrated.

The Institute of Plant Breeding and Acclimatization at Radzików near Warsaw is both the coordination centre for the various activities in genetic resources in the Polish People's Republic and the place of the main collection (mostly cereals and some pulses). The other collections are maintained at institutes and breeding stations specializing in individual groups of agricultural and horticultural crop plants.

In the Botanic Garden of the Polish Academy of Sciences in Warsaw, established in 1974, a special centre was developed for the collection and exploration of genetic resources of rye.

In the Socialist Republic of Romania the collections of crop genetic resources are scattered on institutes and stations specializing in breeding individual groups of crops (cf. SZABÓ 1981). At the Agrobotanic Garden of the Agronomy Institute at Cluj-Napoca during the last ten years efforts have been made to develop a centre of the exploration and collection of land-races cultivated in Transylvania.

In the Soviet Union the work on genetic resources is continued on a large scale and with great enthusiasm by N. I. VAVILOV's successors. Their activities cover a wide scope ranging from world-wide exploration to extensive evaluation and utilization. The Soviet colleagues, in a selfless manner, make available their rich experiences to the development of gene banks in the member countries of the Council for Mutual Economic Assistance.

The gene bank activities of these countries are furthered by a Technical and Scientific Council for Genetic Resources established in 1973 and headed by the director of the N. I. VAVILOV Institute of Plant Industry. Its tasks include mutual information on projects implemented, the organization of experts meetings to discuss current problems, e.g., on standardized documentation for computerized data management (storage, retrieval and exchange), long-term storage and evaluation.

A world-wide movement for the collection and preservation of the genetic resources was started in the 1960s.

At that time it became evident that the land-race populations are threatened and increasingly displaced by new cultivars and improved farming techniques in the Third World countries where the centres of their diversity were most concentrated. A process of "genetic erosion" had begun and its annual advance required actions to prevent complete extinction of genetic resources.

New impulses were given by the first FAO Technical Meeting on Plant Exploration and Introduction held in 1961 which reviewed the situation and by the FAO/IBP Technical Conference on the Exploration, Utilization and Conservation of Plant Genetic Resources in 1967.

In 1964 the International Biological Programme (IBP) was established by the International Council of Scientific Unions. The subject of the IBP was defined as "Biological Basis of Productivity and Human Welfare". In the section concerned with Use and Management of Biological Resources a so-called "Gene Pool" Subcommittee was set up. The collaboration between this committee and the FAO became very close.

By the activities of the IBP attention of a broader public of scientists, administrators and politicians was focussed on the crop genetic resources both on their significance for breeding better cultivars and on the danger of their disappearance.

The International Board for Plant Genetic Resources (IBPGR) of the Consultative Group on International Agricultural Research (CGIAR) was established in 1974. The main objectives of this autonomous, international scientific organization are to promote a world-wide network of genetic resources centres, i.e., of gene banks, and to further the collection, conservation, documentation and evaluation of crop genetic resources.

These developments had also stimulating effects on the situation in Europe. Some of the European gene banks are involved, e.g., in the "action orientated" programmes of the IBPGR.

In Europe, since 1960, the EUCARPIA – European Association for Research in

Plant Breeding – has been continuously concerned with the problems of genetic resources conservation.

Discussions in 1968 resulted in a proposal to establish a regional network of gene banks for Europe instead of costly national gene banks in every country. Some already existing gene banks as for instance in the Socialist countries and some gene banks to be newly established in Western Europe should form this network (HAWKES and LAMBERTS 1977).

The various discussions, resolutions and proposals of EUCARPIA stimulated the foundation of gene banks at Bari, Italy, and Braunschweig, FRG, and more recently at Lund, Sweden, for the Nordic countries, as well as similar activities in other countries.

The National Research Council in Italy decided to establish a germplasm laboratory (now Istituto del Germoplasma) at the end of 1970 at Bari (SCARASCIA-MUGNOZZA and PORCEDDU 1973). Collecting activities of its staff are mainly concentrated on the countries of the Mediterranean Basin. A close collaboration exists with FAO and IBPGR. The gene bank is, e.g., the centre for the IBPGR Mediterranean Germplasm Programme and acts, following an invitation from IBPGR, as one of the four centres for the preservation of base wheat collections and related species (PORCEDDU 1979).

In the Federal Republic of Germany the gene bank of the Institute of Crop Sciences and Plant Breeding of the Federal Centre of Agriculture at Braunschweig started its activities in 1971. The German-Netherlands Potato Gene Bank established in 1974 by an official government agreement is part of this bank. At present the cereal collections are most important, in future, however, grain legumes, potatoes and Beta-roots for which the gene bank already acts as a main conservation centre following a proposal of the IBPGR will become the main collections (HONDELMANN 1979).

From the very outset the gene bank work has been closely linked with the development of a system for data documentation. Standardization of descriptors necessary both for a computer-aided data management and exchange of informations on plant genetic resources resulted in the edition of a "Thesaurus for the International Standardization of Gene Bank Documentation" (SEIDEWITZ 1973–1976, DAMBROTH 1979).

The Nordic Gene Bank for Agricultural and Horticultural Plants at Lund, Sweden, started its activities in 1979. It is a joint undertaking of Denmark, Finland, Iceland, Norway and Sweden and is funded by the Nordic Council of Ministers. At present, material of the seed-propagated species is being transferred from the individual collections in these countries to the gene bank, material of vegetatively propagated species is being included later. In all five countries landraces and ecotypes are still available for collection, e.g., from grasses and fruit trees. The Nordic Gene Bank will preserve one of the major collections of peas according to a proposal made by the IBPGR (pers. comm. by L. KÅHRE 1979).

In the Netherlands various collections of crop plants are being maintained in the breeding institutes or through other organizations, including also the German-Netherlands Potato Collection in the gene bank at Braunschweig, FRG. Since 1976 a working group has been preparing the establishment of a Netherlands gene bank. Its tasks will comprise the further maintenance and documentation, etc. of the already existing collections of the institutes, as well as the enlargement and improvement of these collections through international exchange and through collection and conservation of still remaining indigenous land-races and other material which is of interest for current breeding or research programmes (ROELOFSEN 1979). Recently ZEVEN (1979) has reported on the latest collecting activities and on some land-races and old cultivar collections in the Netherlands, e.g., of potatoes, vegetables, fodder crops and ornamentals.

In recent years, extensive collecting activities were started in Spain and Portugal stimulated mainly by the IBPGR to preserve the threatened genetic resources on the Iberian Peninsula (cereals, maize, grain legumes, vegetables). In Spain the gene bank has been completed now near Alcalá de Henares. Legumes from the Mediterranean region and national crops will probably be its major responsibilities. In Portugal a gene bank is being constructed at Braga. It may have regional responsibility for maize and for national crops as well (pers. comm. J. T. ESQUINAS-ALCÁZAR 1980).

In the United Kingdom the care for crop plant genetic resources is likely to remain under the responsibility of the main plant breeding or research institutes. Each institute will be responsible for the crops used in their breeding or research programmes (pers. comm. R. WATKINS 1979). The vegetable gene bank, e.g., was set up in 1980 at the Vegetable Research Station of the UK Agricultural Research Council at Wellesbourne, Warwickshire. Its aim is to collect and preserve seeds of old and wild varieties of mainly temperate vegetables (ANONYMUS 1980). An unique gene bank (seed bank) for wild species threatened by extinction was set up by the Royal Botanic Gardens at Kew and Wakehurst Place.

Two organisations make efforts to establish co-operation between the European gene banks, i.e., the EUCARPIA through its Gene Bank Committee by carrying out scientific advisory activities and disseminating experiences on genetic resources work and the FAO, IBPGR and UNDP with its recently started "European Co-operative Programme for the Conservation and Exchange of Crop Genetic Resources for Plant Breeding" by strengthening the co-operation between the gene banks through establishment of inter-Governmental links.

The rather provisional Gene Bank Committee of EUCARPIA acquired an official status at the 6th EUCARPIA Congress at Cambridge in 1971 and both the committee and the section "Wild Species and Primitive Forms" were linked together under the chairmanship of J. G. HAWKES. Due to increasing contacts between the European gene banks, the Gene Bank Committee was remodelled and enlarged at the 8th Congress in Madrid in 1977. The gene banks in Western Europe and in the Socialist European countries are now equally represented on the committee.

The annual meetings of the so-called Extended EUCARPIA Gene Bank Committee to which the regular committee members and the heads of all European gene banks are invited have a catalytic effect on the wide range of co-operation between gene banks throughout Europe.

In Geneva, at a Government Consultation in December 1979, the "European Co-operative Programme for Crop Genetic Resources" was considered and its activities were approved. The scientific and technical collaboration of the national (and regional) gene banks should be furthered by a government agreement to promote an optimum utilization of genetic resources in plant breding in Europe and also in Third World countries.

It is high time now to continue as fast as possible the collection and preservation of land-races in their niches, especially of those which have come to occur very sparsely, and to continue closer co-operation of the European gene banks to use their facilities for mutual benefit.

Acknowledgement.–I wish to thank the colleagues at the European gene banks for numerous helpful informations.

## Zusammenfassung

Die Sammlung europäischer Landsorten und die Entwicklung der Genbanken in Europa – historische Bemerkungen

Die Pflanzenzüchtung begann im 19. Jahrhundert in Europa mit der Auslese aus Landsorten. Sehr früh wurde in Europa erkannt, daß die Landsorten infolge der Verbesserung der Anbaumethoden und der Ausbreitung neuer Sorten, besonders nach der Wiederentdeckung der MENDEL'schen Regeln, aus dem Anbau verdrängt werden. E. von PROSKOWETZ und F. SCHINDLER lenkten wahrscheinlich als erste bereits im Jahre 1890 die Aufmerksamkeit auf die wertvollen Merkmale der Landsorten und auf die Gefahren, die sich für die zukünftige Pflanzenzüchtung ergeben, wenn sie verloren gehen.

Von E. von PROSKOWETZ stammt die erste Beschreibung der Tätigkeit einer Genbank. Das Konzept einer Genbank wurde von N. I. VAVILOV zum ersten Mal mit dem Institut für Pflanzenbau in Leningrad verwirklicht.

In den sozialistischen Ländern Europas entstanden, angeregt durch die Ideen N. I. VAVILOV's, nach dem 2. Weltkrieg Genbanken. In anderen europäischen Ländern wurden sie im Verlauf der zurückliegenden 10–15 Jahre eingerichtet. Dazu gaben die seit dem Jahre 1960 geführten Diskussionen und zahlreiche Vorschläge der EUCARPIA einen wesentlichen Anstoß.

Gegenwärtig gibt es Bemühungen, die Beziehungen zwischen den europäischen nationalen (und regionalen) Genbanken sowie ihren regionalen Organisationen zu vertiefen.

## Краткое содержание

Сборы европейских местных сортов и возникновение генных банков в Европе (Исторические замечания)

Селекция растений началась в 19 в. в Европе с отбора из местных сортов. Уже очень давно в Европе стало ясно, что, вследствие улучшения приёмов возделывания и внедрения новых сортов – особенно после переоткрытия законов Менделя – местные сорта будут вытеснены из употребления. Вероятно, Э. фон Просковец и Ф. Шиндлер были первыми, обратившими внимание уже в 1890 г. на ценные признаки местных сортов и на опасности, которые ожидают селекцию растений, если эти признаки окажутся утерянными.

Э. фон Просковец первый описал возможную деятельность генного банка. Концепция генного банка была впервые осуществлена Н. И. Вавиловым в ленинградском Институте растениеводства.

В социалистических странах Европы генные банки, под влиянием идей Н. И. Вавилова, возникли после Второй мировой войны. В других европейских странах они начали возникать лет 10–15 тому назад. Значительный вклад в дело их организации внесли дискуссии и многочисленные предложения организации ЭУКАРПИА.

В настоящее время наблюдается тенденция углубить связи между европейскими национальными (и региональными) генными банками и их региональными организациями.

## Literature

Anonymus, 1980: Temperate vegetable seed bank set up. – Nature **283**, 514.

Bareš, I., 1979: Study of genetic resources of crops in Czechoslovakia. – In: A. C. Zeven and A. M. van Harten (Edits.), Proceed. of the Conference Broadening the Genetic Base of Crops, Wageningen 3–7 July 1978, pp. 145–149. – Pudoc, Wageningen.

Baur, E., 1914: Die Bedeutung der primitiven Kulturrassen und der wilden Verwandten unserer Kulturpflanzen für die Pflanzenzüchtung. – Jb. Deut. Landw.-Ges. **29**, 104–109.

Brežnev, D. D., 1969: Vsesojuznomu ordena Lenina naučno-issledovatel'skomu institutu rastenievodstva imenu N. I. Vavilova – 75 let. – Trudy prikl. bot., genet. i selekc. **41** (**1**), 5–30.

–, 1970: Mobilization, conservation and utilization of plant resources at the N. I. Vavilov All-Union Institute of Plant Industry, Leningrad. – In: O. H. Frankel and E. Bennett (Edits.), Genetic Resources in Plants – their Exploration and Conservation. IBP Handbook No. 11, pp. 533–538. – Blackwell Scientific Publications, Oxford and Edinburgh.

Dambroth, M., 1979: The Institute of Crop Science and Seed Research and its dedication to the preservation of plant genetically valuable material for plant breeding. – In: Proceed. of a Colloquium Technical Aspects of Information Management and Means of Communication in Plant Genetic Resources Work for a Future Utilization of Genetic Material in Plant Breeding, Braunschweig 7th and 8th November 1978, pp. 17–22.

Dorst, J. C., 1958: A quarter of a century of plant breeding in The Netherlands. – Euphytica **7**, 9–20.

Frankel, O. H., 1975: Genetic resources centres – a co-operative global network. – In: O. H. Frankel and J. G. Hawkes (Edits.), Crop genetic resources for today and tomorrow. International Biological Programme 2, pp. 473–481. – Cambridge University Press, Cambridge, London, New York, Melbourne.

Fruwirth, C., 1910: Über die Vielförmigkeit der Landsorten. – Monatshefte f. Landw. **3**, 48–62.

Harlan, J. R., 1975: Crops and man. – American Society of Agronomy, Madison, Wisconsin.

Hawkes, J. G., 1978: The taxonomist's role in the conservation of genetic diversity. – In: Street, H. E. (Edit.), Essays in Plant Taxonomy, pp. 125–142. – Academic Press, London, New York, San Francisco.

–, and H. Lamberts, 1977: EUCARPIA's fifteen years of activities in genetic resources. – Euphytica **26**, 1–3.

Hondelmann, W., 1979: Gene bank activities in the Federal Republic of Germany. – In: A. C. Zeven and A. M. van Harten (Edits.), Proceed. of the Conference Broadening the Genetic Base of Crops, Wageningen 3–7 July 1978, pp. 117–122. – Pudoc, Wageningen.

HONECKER, L., 1942: Erbanalytische Untersuchungen über das Verhalten der Gerste gegenüber verschiedenen physiologischen Rassen des Mehltaues (*Erysiphe graminis hordei* MARCHAL). – Z. Pflanzenz. **24**, 429–506.

LEHMANN, CHR. O., 1972: Das Kulturpflanzenweltsortiment des Zentralinstituts für Genetik und Kulturpflanzenforschung in Gatersleben, ein Genreservoir für die Pflanzenzüchtung in der DDR. – Arch. Züchtungsforsch. **2**, 109–127.

–, 1979: The Gatersleben gene bank. – In: A. C. ZEVEN and A. M. VAN HARTEN (Edits.), Proceed. of the Conference Broadening the Genetic Base of Crops, Wageningen 3–7 July 1978, pp. 111 –116. – Pudoc, Wageningen.

–, und K. HAMMER, 1979: Bericht über die Reise nach Südwest-Spanien zur Sammlung kultivierter und wild wachsender Leguminosen im Jahre 1978. – Kulturpflanze **27**, 97–108.

MAYR, E., 1951: Ergebnisse der österreichischen Landsortenforschung bei Getreide in den letzten zwei Jahrzehnten. – Z. Pflanzenz. **30**, 434–444.

NILSSON-EHLE, H., 1911 (1913): Mendélisme et acclimatation. – Compte rendu de la IVième conférence internationale de génétique à Paris 1911 (1913), pp. 136–155.

–, 1924: Versuche über die Möglichkeit, Frühreife mit hoher Ertragsfähigkeit durch Kreuzungen zu kombinieren. – In: Beiträge zum landwirtschaftlichen Pflanzenbau insbesondere Getreidebau. Festschrift zum 70. Geburtstag Professor Dr. F. SCHINDLER, pp. 239–263. – P. PAREY, Berlin.

PISSAREV, V., 1929: Fortschritte auf dem Gebiete der angewandten Botanik und Pflanzenzüchtung in der Union der Sozialistischen Sowjetrepubliken während der letzten Jahre. – Z. Pflanzenz. **14**, 175–232.

PORCEDDU, E., 1979: Gene bank activities in Italy. – In: A. C. ZEVEN and A. M. VAN HARTEN (Edits.), Proceed. of the Conference Broadening the Gentic base of Crops, Wageningen 3–7 July 1978, pp. 123–125. – Pudoc, Wageningen.

PROSKOWETZ, E. von, 1890: Welches Werthverhältniss besteht zwischen den Landrassen landwirthschaftlicher Culturpflanzen und den sogenannten Züchtungsrassen? – Internationaler land- und forstwirthschaftlicher Congress zu Wien 1890. Section I: Landwirthschaft. Subsection: Pflanzenbau. Frage 5. Heft **13**, S. 3–18.

REGEL, R., 1915: L'organisation et les travaux du bureau de botanique appliquée pendant vingt ans de son fonctionnement. – Bull. appl. bot. **8**, 328–723, 1465–1637.

ROELOFSEN, H. J., 1979: The Wageningen collections. – In: A. C. ZEVEN and A. M. VAN HARTEN (Edits.), Proceed. of the Conference Broadening the Genetic Base of Crops, Wageningen 3–7 July 1978, pp. 133–137. – Pudoc, Wageningen.

ROEMER, TH., 1942: Ausgangsmaterial für die Resistenzzüchtung bei Getreide. Ergebnisse 20jähriger Arbeit der Pflanzenzuchtstation Halle a. S. – Z. Pflanzenz. **24**, 304–332.

–, 1953: Getreide der gemäßigten Zone. – In: TH. ROEMER, A. SCHEIBE, J. SCHMIDT und E. WOERMANN (Herausgeb.), Handbuch der Landwirtschaft, Bd. 2, Pflanzenbaulehre, pp. 1–85. – P. PAREY, Berlin und Hamburg.

RÜMKER, K. VON, 1908: Die systematische Einteilung und Benennung der Getreidesorten für praktische Zwecke. – Jb. Deut. Landw.-Ges. **23**, 137–160.

SCARASCIA-MUGNOZZA, G. T., and E. PORCEDDU, 1973: Activities and programmes of the Italian germ plasm laboratory. – Bari, 16 pp.

SCHINDLER, F., 1890: Welches Werthverhältniss besteht zwischen den Landrassen landwirthschaftlicher Culturpflanzen und den sogenannten Züchtungsrassen? – Internationaler land- und forstwirthschaftlicher Congress zu Wien 1890. Section I: Landwirthschaft. Subsection: Pflanzenbau. Frage 5. Heft **13**, S. 19–24.

–, 1918: Einige Bemerkungen über die züchterische und wirtschaftliche Bedeutung der Landrassen unserer Kulturpflanzen. – Deut. Landw. Presse **46**, No. 25, 155–156.

SEIDEWITZ, L., 1973–1976: Thesaurus for the international standardization of the gene bank documentation. Part I (1973), part II (1974), part III (1974), part IV (1975), part V A (1976). – Braunschweig.

STUBBE, H., 1970: Das Institut für Kulturpflanzenforschung Gatersleben (Aufgaben, Ergebnisse, Probleme) 1943–1968. – Kulturpflanze, Beiheft **6**, 29–52.

SZABÓ, A. T., 1981: Problems of genetic erosion in Transylvania, Romania. – Kulturpflanze **29**, 47–62.

TSCHERMAK, E. von, 1915: Über die Notwendigkeit der Sammlung und Erhaltung unserer bewährten, noch unveredelten Getreidelandrassen. – Wiener Landw. Zeitg. No. **104**, 750–761.
WIENHUES, F., und J.-E. GIESSEN, 1957: Die Abstammung europäischer Weizensorten. – Z. Pflanzenz. **37**, 217–230.
ZEVEN, A. C., 1979: Collecting genetic resources in highly industrialized Europe, especially The Netherlands. – In: A. C. ZEVEN and A. M. VAN HARTEN (Edits.), Proceed. of the Conference Broadening the Genetic Base of Crops, Wageningen 3–7 July 1978, pp. 49–58. – Pudoc, Wageningen.

Dr. CHR. O. LEHMANN
Zentralinstitut für Genetik
und Kulturpflanzenforschung
DDR - 4325 Gatersleben

Kulturpflanze XXIX · 1981 · S. 41–45

# Some notes on the collection of land-races of cultivated plants indigenous to the territory of the Federal Republic of Germany

MANFRED DAMBROTH and WALTER HONDELMANN (Braunschweig, F. R. G.)

## Summary

The occurrence of old land-races indigenous to the territory of the Federal Republic of Germany under special consideration of their availability is being described. Some typical features of the findings are explained.

Land-races of cultivated plants are to be considered "complex evolutionary populations" (SCHNELL 1980). From this it follows that special attention has to be given to them within the field of conservation of plant genetic resources. This should include not only exotic provenances but also be extended to the indigenous land-races, even in a highly industrialized country like the Federal Republic of Germany. Therefore from the very beginning of our gene bank activities in the early seventies we took this into consideration.

Two different approaches to obtain this type of plant genetic material seemed to be possible:

1) to search in existing collections;
   this should refer especially to crops which had been collected in former decades serving as working collections for breeding purposes
2) to start collection work now;
   for those crops, which for whatever reasons were either not available in those collections or only on a small scale.

It is rather easily understandable that no. 1) is mainly applying to major crop plants like the cereals especially the inbreeding species wheat, barley and oats; no. 2) then should take in most of the vegetables, oil crops and others.

Actually, as far as the inbreeding cereals were concerned, we could go back to older collections. As an outstanding example a comprehensive collection of *Triticum aestivum* var. *spelta* land-races of South-western Germany should be mentioned, which formerly was maintained at the Eidgenössische Forschungsanstalt für landwirtschaftlichen Pflanzenbau Zürich-Reckenholz (Switzerland) and curated by Dr. WAGNER.

These land-races were collected from 1933 to 1937 in Badenia and Württemberg under special consideration of the southern Black Forest area (Schwarzwald), the Swabian uplands (Schwäbische Alb) as well as from the Bauland, a region neighbouring southeast the Odenwald forest and even from the Eifel mountains. These regions in former times were wellknown for their *spelta*-cultivation partly due to their inclement climatic and soil conditions. So these accessions represent a well-defined, genetically differentiated and rather complete collection, which in its general importance cannot be overestimated, even though *T. aestivum* var. *spelta* nowadays is not given any priority (see also KUCKUCK 1964). (Details were given by slide). This at the same time is a good example for classical collection work in Europe in order to safeguard plant germplasm. Altogether we now have more than 250 accessions of indigenous *spelta*-wheat, which are preserved using modern long-term storage facilities.

As to the barley (*Hordeum vulgare* L.) and oats (*Avena sativa* L.) the indigenous specimens are more scattered throughout the country ranging from northern to southern areas. Though both crops in their figures are not comparable to that of the *spelta*-wheat, barley comes to 98 and oats to 34 accessions, they do represent samples of remarkable genetic diversity as shown by their ancestry and other available information. With the exception of some smaller numbers of bean (*Phaseolus vulgaris* L.) and pea (*Pisum sativum* L.) accessions of no other species were available in existing collections.

Therefore the second possibility had to be followed. Encouraged by the findings of HAMMER, HANELT and TITTEL (1977) who showed that even in an industrialized country numerous and different species of old land-races can be found, DAMBROTH in 1979 started a first attempt for this purpose, following

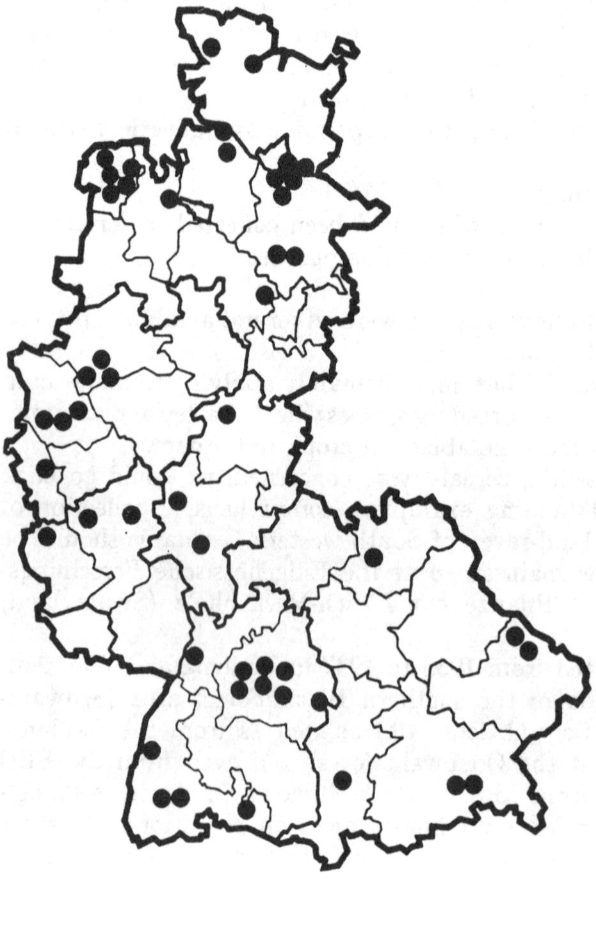

Fig. 1 Provenances of accessions in 1979

about the same procedure. In general the results correspond to those of the aforementioned authors. The response came from all over the country, especially for various vegetable crops. Figure 1 shows the collection sites of the accessions.

Some features of the experiment might be of general interest:
1) the crops concerned and 2) their distribution.

*To 1) Crop Species:*

Vegetables rank first followed by oil crops, special crops (tobacco and others of that kind) and forage crops.

The fact that vegetables come in first presumably depends mainly on the traditional vegetable growings that are wide-spread among amateur gardeners. This also might be the reason for the old origin of some specimens. For example among the *Phaseolus* beans are such remarkable types like 'Ostfriesische Bunte' (originated about 1930), 'Wollbohne' (originated about 1920) 'Napoleons Bohnen' (originated before 1900), 'Rotbunte Suppenbohne' (originated in the 1920?), unnamed climbing types more than 100 years old, or 'Samersberger Buschbohne' also from rather old origin. Or among the peas especially the 'Kesselheimer Zuckererbse' (originated before 1880), an old Friesian type of "Zuckererbse" as well as land-races simply regionally named as yellow or green peas (originated before 1930) could be mentioned. Then among the onions such types as the viviparous "Luftzwiebel" or "Etagenzwiebel" (originated about 1920) as well as "Blaue Schalotten" (originated about 1900) are of special interest, as well as the regionally limited sources of tobacco land-races like 'Badische Geudertheimer', 'Goundie', 'Friedrichstaler' or 'Rheinzabern'.

Coming to the forage crops 'Altfränkische Landsorte' of alfalfa, 'Altwürttembergische Landsorte' of red clover, the 'Nocherner Futterrübe' (*Beta vulgaris* L.), the 'Alb-Schwedenklee' (*Trifolium hybridum* L.) or even potatoes named 'Rotblaue' and 'Bamberger Hörnle', all of them very old in origin, are good examples of land-races which still are grown in special areas or localities. This is also valid for the oilcrops where even species, which were grown more generally in former times, like *Camelina sativa* (L.) Cr., *Eruca sativa* Mill. or *Madia sativa* Molina besides the Brassicas and linseed (*Linum usitatissimum* L.) could be obtained, not to forget the "Schüttmohn", an old variety of the cultivated poppy (*Papaver somniferum* L.).

All the aformentioned species in general are considered as typical examples for a less developed agriculture than we have nowadays in Western Europe. But as we see the conservation of certain primitive forms of land-races even under highly developed agronomic or horticultural conditions is still continuing.

*To 2) Distribution of the crops:*

Even though their distribution is all over the country as stated before, a thorough analysis does result in some typical features. Regarding the agricultural crops it can be said, that the occurrence of land-races and primitive forms is not so much found in large continuous growing areas of more or less modern origin but rather in smaller scattered ones with a restricted cultivation from former times. This does not only refer to remote or hidden areas. Examples are given by the potato 'Bamberger Hörnle' in the surroundings of Bamberg, the various clover species in the Eifel uplands as well as in some parts of Southern Bavaria, alfalfa from Franconia or the 'Nochener Futterrübe' near Koblenz (Rhine-Palatine).

As to the horticultural crops, especially the vegetables, one has to notice a quite different feature. Firstly, land-races of vegetables are to be found in the fringes or in the surroundings of old vegetable growing areas; secondly in regions with traditional amateur gardening including the „Schreber"- or small-gardening districts of larger cities and thirdly in the backyard gardens of farmers.

Examples to be mentioned are: the wellknown horticultural area of Vierlande (Hamburg) where a good number of old vegetable land-races still is available, the Filder plain, south of Stuttgart, also with good vegetable specimens, many private gardens especially for beans, peas and onions but also for cucumber, radish, salads, celery. (The overall situation was illustrated by means of a slide showing the land uses of the Federal Republic of Germany.)

From this we would like to make the following conclusions:

1) Recognizing the situation of distribution a more systematic approach for exploration seems to be possible;
2) The results obtained so far are encouraging enough to establish future efforts.

## Zusammenfassung

Einige Bemerkungen zur Sammlung von für das Gebiet der Bundesrepublik Deutschland indigenen Landsorten von Kulturpflanzen

Das Vorkommen alter Landsorten, die für das Gebiet der Bundesrepublik Deutschland indigen sind, wird unter besonderer Berücksichtigung ihrer Verfügbarkeit beschrieben. Einige charakteristische Besonderheiten der Nachweise werden erläutert.

## Краткое содержание

Некоторые замечания по поводу сборов местных автохтонных сортов на территории Федеративной Республики Германии

При описании старых местных сортов, которые являются автохтонными для территории ФРГ, особое внимание уделяется их возделыванию и сохранению. Даётся объяснение некоторым характерным особенностям их выявления.

## Literature

HAMMER, K., P. HANELT and C. TITTEL, 1977: Sammlung autochthoner Kulturpflanzen auf dem Gebiet der DDR. – Kulturpflanze **XXV**, 89–99.

KUCKUCK, H., 1964: Experimentelle Untersuchungen zur Entstehung der Kulturweizen. I. Die Variation des iranischen Spelzweizens und seine genetische Beziehung zu *Triticum aestivum* ssp. *vulgare* (VILL., HOST) MACKEY, ssp. *spelta* (L.) THELL. und ssp. *macha* (DEK. et MEN.) MACKEY mit einem Beitrag zur Genetik des *Spelta*-Komplexes. – Z. Pflanzenzüchtung **51**, 97–140.

SCHNELL, F. W., 1980: Aspekte der genetischen Diversität im Problemkreis der Pflanzenzüchtung: In: Biochemische Bestimmung und züchterische Nutzung genetischer Diversität. 4. Göttinger Pflanzenzüchterseminar, 11. und 12. Februar 1980.

Prof. Dr. M. DAMBROTH, and Prof. Dr. W. HONDELMANN
Institut für Pflanzenbau und Pflanzenzüchtung
Bundesforschungsanstalt für Landwirtschaft Braunschweig-Völkenrode (FAL)
Bundesallee 50, D - 3300 Braunschweig

Kulturpflanze XXIX · 1981 · S. 47–62

# Problems of genetic erosion in Transylvania, Romania

ATTILA T. SZABÓ (Cluj-Napoca, Romania)

## Summary

Started from the basic principles of IBP and MAB programmes, genetic resources of agricultural plant species cultivated traditionally in Transylvania have been studied. Land-races from this territory deserve special attention, as Transylvania is situated on the north-eastern border of the Mediterranian centre of diversity, between Near-East and Western Europe, has a relatively isolated position surrounded by the eastern range of the Carpathian Mountains. Traditional agriculture practiced in many of the intramountain hollows is favourable for the preservation of ancient land races and the territory is also promising in respect of spontaneous germplasm resources. In the Agrobotanical Garden of the Institute of Agronomy at Cluj-Napoca seed samples of economically important plant species – cultivated land-races and wild relatives – have been collected with the beginning of 1970. Based on this collection the threatened variability of agricultural plant species have been grouped in four major categories. Trends and peculiarities of genetic erosion acting in these groups have been exemplified with the cases of *Trifolium pratense* L., *Phaseolus vulgaris* L. and *Triticum monococcum* L.

## Introduction

During the evolution of a cultivated species in the periods of transition from types with lower productivity to more productive ones, or from a general purpose ("primitive") land race to a more specialized ("advanced") cultivar, narrowing of the genetic base of the crop follows, due to consequent lost of previous types. This phenomenon is usually reffered to as "genetic erosion". The ultimate aim and purpose of the collection and preservation of land races all over the world is to create artificial reservoirs, where threatened germplasm may be stored for the future. In this connection genetic erosion is a central problem of the crop plant evolution and it deserves a special attention. In spite of that, due perhaps the lack of definitions also (FRANKEL and BENNETT 1970, RIEGER et al. 1975), or because some kinds of confusions, as "eroded" germplasm generally is not deposited elsewhere but definitely lost, the technical term has been avoided by authors like HARLAN 1970, BENNET 1970, FRANKEL 1970 in their papers, but it proved to be useful for indexing (FRANKEL and BENNETT 1970) and for conceptual approaches (HAWKES 1977 ap. VARGA et al. 1979).

The force acting against genetic erosion is genetic conservation. This conservation in its unconscious form (e.g. tabus) may be considered as the most ancient of all conservationist aspirations. Domestication – at least in the case of grain crops – represented the beginning of conscious conservation of seeds produced in nature. In this first step common stocks for human food and for consequent agricultural practices (sowing, planting) have been created. Breeding postulates the differentiation of this common stocks and reserves separated as reserves for food and reserves for the reproduction of the most convenient types. As breeding becomes more sophisticated differential conservation evolved, assortments and experimental fields arrived. In the perspective of genetic manipulations the necessity of the comprehensive conservation and evidence of the variability may be regarded as the next step in the historical evolution process of this ancient human activity. In our times, in the age of comprehensive genetic conservation, food reserves, reserves for further cultivation, reserves for direct breeding purposes and long term reserves valuable perhaps only in the perspective of a more distant future must be separated (Tab. 1).

Table 1

The differentiation of reserve stocks during the history of cultivated plants

| Stage | Type of conservation | Differentiation of reserve stocks | Observations |
|---|---|---|---|
| A | Only in nature | Only in wild | only plant-gathering |
| B | Primitive | Reserves for food and designated for further multiplication not differentiated | domestication |
| C | Differential | Stocks designated to further multiplication are selected and separated from food reserves | breeding |
| D | Comprehensive | Reserves further differentiate (food, reproduction, short term and long term breeding reserves) | breeding evolves in the direction of genetical manipulation("engineering") |

In Romania genetic conservation followed the outlined general pattern. Collections regarding land races of different species have been organized in the 20th century in the case of some crop plants, e.g. *Zea mays* L. (Velican 1934, Cristea 1962, 1975, 1979, Cristea et al. 1973; Roman et. al. 1973, Homorodeanu 1974, Şuba 1967, 1973, Grecu 1962, Căbulea et al. 1975), *Triticum aestivum* L. em. Fior. et Paol. (Ceapoiu et al. 1975), *Hordeum vulgare* L. s. l. (Drăghici et al. 1975), *Helianthus annuus* L. (Vrinceanu 1974), *Secale cereale* L. (Gaspar 1977), in the case of fruit trees (Cociu 1979), grape Vine (Popescu et al. 1979) and forest trees (Enescu 1979, Stănescu et al. 1979). Researches have been initiated in respect of genetic resources of spontaneous and cultivated forage species (Kovács and Breazu 1980). The collection, study and conservation of genetic resources of agricultural, silvicultural and other economically important plant species gained also special importance and interest during the past decade. (Ceapoiu 1975, Giosan 1979, Cristea and Gaspar 1979, Kellner and Giura 1979, Szabó 1978, Varga et al. 1979, Vrinceanu et al. 1979).

As only few of the above mentioned authors studied land races collected from

Transylvania, with the beginning of 1970 research have been started in the Agrobotanical Garden of the Agronomy Institute Cluj-Napoca in order to complete our knowledge about the situation of land races cultivated in Transylvania.

## Material and Methods

Research have been started from a multidisciplinary basis, trying to reconciliate useful principles of different scientific fields. For territorial analysis and representation the chorological system proposed by NYÁRÁDY and VICOL 1968, 1973 have been adopted. In the first step the teritory was divided in an extensively and an intensively studied area. (Fig. 1). The extensively studied area includes

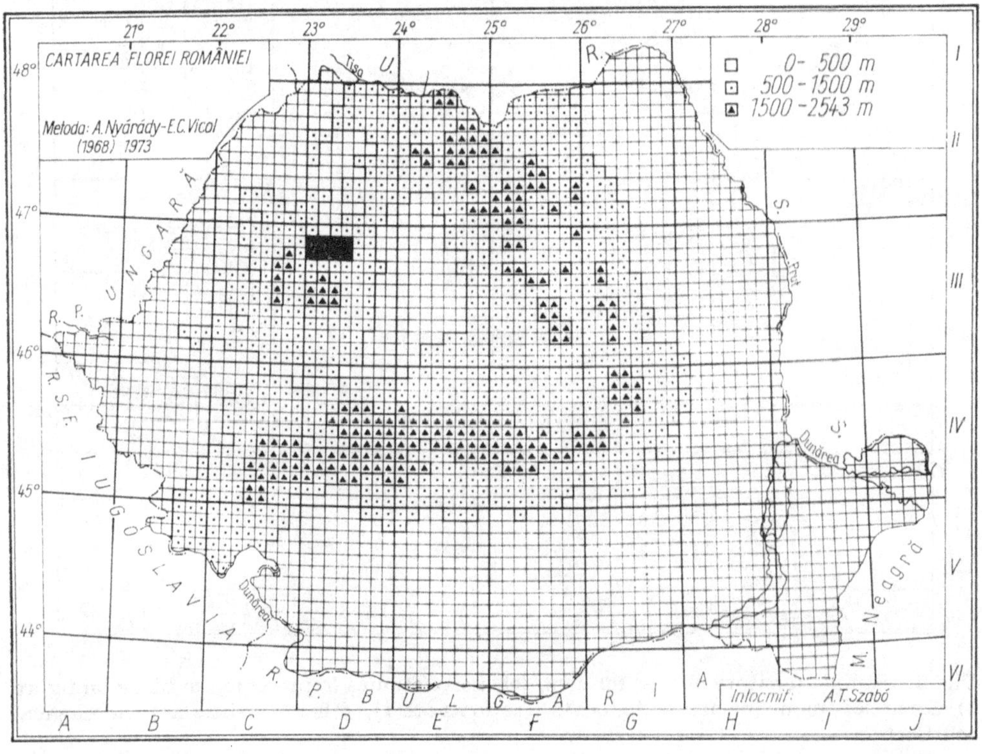


**Fig. 1** Chorological map used for ethnobotanical research and territorial survey of land-races in Transylvania. Sample territory chosen for comprehensive study (in black) is represented in detail on figure 2

a territory bordered by a conventional line traced along the highest peaks of Southern- and Eastern Charpatians and Munții Apuseni respectively (maximum hight above see level: Moldoveanu Peak – 2543 m – in Mt. Făgăraș). This mountains embrace the Transylvanian Highland. The medium hight above see level of the different geomorphological units of this highland is between 403 and

468 m. The great number of different intramountain hollows cultivated by villagers belonging to different ethnic groups, the great diversity of strong and ancient (agri)cultural traditions are undoubtedly favourable for preservation of crop plant diversity. Collections of orientative value have been also organized on this territory mostly with the help of uncompelled collectors (Péntek and Szabó 1976b, Szabó and Péntek 1976). Taking into account the relatively high number of endemic species (Morariu and Beldie 1976), collections regarding spontaneous germplasm resources are also promising.

In the first period a sample area including 61 settlements has been chosen on the montaneous and hilly area of the northwestern side of the Transylvanian Highland for intensive research (Fig. 1, Fig. 2). Researches of botanical, agro-

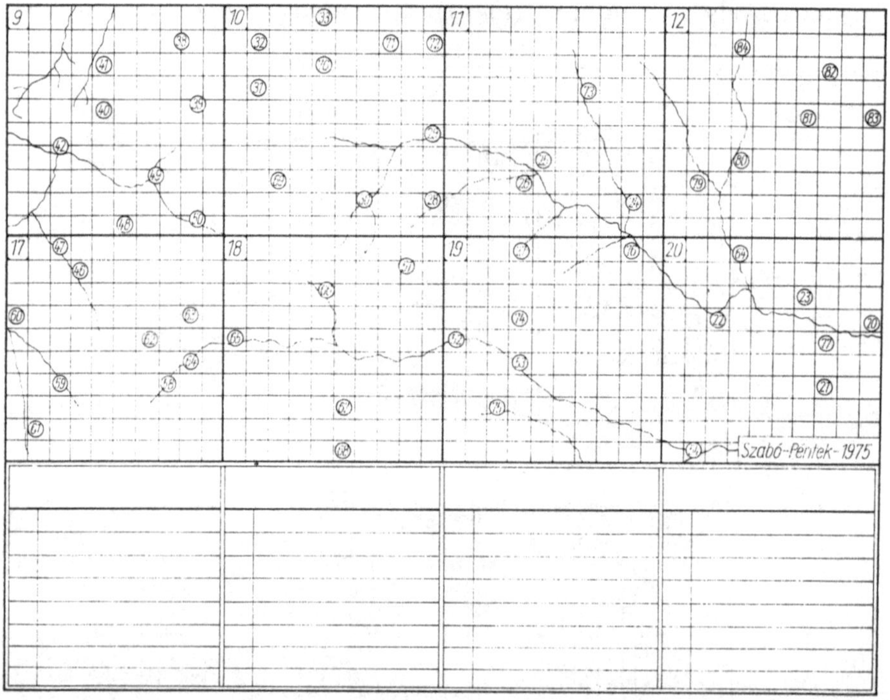


Fig. 2 Sample territory chosen for comprehensive studies (represented in black on figure 1). Encircled numbers are settlements (see footnote 1), other numbers are chorological codifications

1 The list of settlements from the sample territory/Fig. 2, encircled numbers/: 20 – Baciu, 21 – Suceacu, 22 – Viştea, 23 – Mera, 24 – Turea, 25 – Băgara, 26 – Doroltu, 27 – Macău, 28 – Inucu, 29 – Aghireş, 30 – Leghia, 31 – Stîna, 32 – Petrinzel, 33 – Petrindu, 38 – Sfăraş, 39 – Jebuc, 40 – Bicalat, 41 – Tetişu, 42 – Huedin, 46 – Horlacea, 47 – Domoşu, 48 – Nearsova, 49 – Şaula, 50 – Izvorul Crişului, 51 – Dumbrava, 52 – Căpuşul Mic, 53 – Căpuşul Mare, 54 – Gilău, 58 – Mănăstireni, 59 – Văleni, 60 – Călata, 61 – Călăţele, 62 – Bica, 63 – Dretea, 64 – Ardeova, 65 – Bedeciu, 66 – Păniceni, 67 – Bălceşti, 68 – Dingău, 69 – Nadăş, 70 – Gălăşeni, 71 – Dincu, 72 – Ticu, 73 – Corneşti, 74 – Straja, 75 – Agîrbiciu, 76 – Gîrbău and Rădaia (77), 78 – Nădăşel, 79 – Şardu, 80 – Sînpaul, 81 – Sumurducu, 82 – Săliştea Veche, 83 – Săliştea Nouă, 84 – Mihăieşti. Lacking numbers are settlements outside the frame.

botanical and ethnobotanical character have been fulfilled on this territory regarding all spontaneous and cultivated plant species (Szabó and Péntek 1976, 1980, Péntek and Szabó 1976b). Historical approach with respect to the study of written records about plant cultivation have been also attempted (Szabó in Szabó T., 1975, 1978).

The agrobotanical results regarding research, and the collection of seed samples of land races and spontaneous species where systematised in seed lists (Puia et al. 1975–1976, 1977–1978), specimens collected in the field and cultivatad later in collections of the agrobotanical garden are preserved as herbarium samples in the Herbarium of the Institute of Agronomy Cluj-Napoca.

Variability studies have been fulfilled in limited extent till now, only in the case of some leguminous and grass species (Puia and Szabó 1978, Szabó 1978, Szabó and Dankanits 1976, Szabó 1979).

In the nomenclature and systematics of the cultivated plants the solutions adopted by Böhme et al. 1979 where mainily followed; author names omitted from reason of space may be completed accordingly.

## Results

At this relatively early stage of research the situation of the main cultivated plant species in respect of genetic erosion may be outlined for Transylvanian Highland only approximately. (Tab. 2). According to the available data, the spontaneous and the relatively newly domesticated species, used mostly for forage, biomass or secondary metabolic substances (drugs) are not directly and seriously endangered (Tab. 2.). In the case of plants cultivated in small scale

Table 2

Plants used traditionaly for food or other purposes in Transylvania, grouped according the effects of genetic erosion acting against them

1. Spontaneous species of economic importance, wild relatives, relatively newly domesticated species:
   *Achillea millefolium, Agropyron intermedium, Anthriscus cerefolium, Astragalus cicer, Atropa belladonna, Brassica campestris, B. nigra, B. juncea, Bromus erectus, B. inermis, Bunias orientalis, Carum carvi, Castanea sativa, Chaerophyllum bulbosum, Cichorium intybus, Cornus mas, Coronilla varia, Corylus avellana, Crambe tataria, Dactylis glomerata, Daucus carota* ssp. *carota, Festuca arundinacea, F. pratensis, Helianthus tuberosus, Hyoscyamus niger, Hypericum perforatum, Humulus lupulus, Lactuca serriola, Lolium perenne, Lotus corniculatus, Medicago falcata, M. varia, Melilotus officinalis, Nasturtium officinale, Onobrychis viciaefolia, Phleum pratense, Portulaca oleracea, Prunus spinosa, Pyrus pyraster, Raphanus raphanistrum, Rubus caesius, Rubus idaeus, Secale montanum, S. sylvestre, Solanum dulcamara, S. nigrum, Taraxacum officinale, Tragopogon pratensis, Trifolium fragiferum, T. hybridum, T. pratense, T. repens, Tussilago farfara, Urtica dioica, Valeriana officinalis, Vicia* spp., *Vitis sylvestris.*
2. Cultivated only in small scale, represented in flora only by land-races and spontaneous populations:
   *Althaea officinalis, A. rosea, Artemisia abrotanum, A. dracunculus, Anethum graveolens, Apium graveolens, Armoracia rusticana, Atriplex hortensis, Beta vulgaris* var. *cicla, Carthamus tinctorius, Cornus mas, Cucurbita ficifolia, Dracocephalum moldavica,*

Fortsetzung Tabelle 2

*Lagenaria siceraria, Lathyrus* spp., *Lens culinaris, Levisticum officinale, Majorana hortensis, Nicotiana rustica, Nigella sativa, Ocimum basilicum, Pastinaca sativa, Phaseolus coccineus, Ph. vulgaris* convar. *vulgaris, Phytolacca* spp., *Rheum* spp., *Salvia officinalis, S. sclarea, Silybum marianum, Triticum monococcum, Vicia faba.*

3. Cultivated in a large scale, mostly as advanced cultivars:
*Allium* spp., *Avena sativa, Beta vulgaris, Brassica oleracea, Cannabis sativa, Cucumis sativus, Cucurbita maxima, C. pepo, Cydonia oblonga, Daucus carota, Helianthus annuus, Hordeum vulgare, Humulus lupulus, Lactuca sativa, Linum usitatissimum, Lupinus* spp., *Mentha* spp., *Papaver somniferum, Petroselinum crispum, Phaseolus vulgaris* (mostly convar. *nanus*), *Pisum sativum, Prunus armeniaca, P. avium, P. domestica, P. persica,* etc., *Pyrus communis, Raphanus sativus, Ribes rubrum, R. grossularia, R. nigrum, Rubus caesius, Rumex rugosus, Satureja hortensis, Secale cereale, Spinacia oleracea, Solanum tuberosum, Trifolium pratense, Triticum aestivum, Vitis vinifera, Zea mays* etc.
4. Historical record of cultivation only:
*Cicer arietinum, Cnicus benedictus, Coriandrum sativum, Crocus sativus, Cynara* spp., *Dipsacus fullonum, Fagopyrum* spp., *Panicum miliaceum, Hyssopus officinalis, Isatis tinctoria, Rubia tinctorum, Ruta graveolens, Sorbus domestica, Triticum dicoccon, T. spelta* etc.

in home gardens, maintenance is strongly dependent on the economic motivations and the traditions. Species with low space requirements and high nutritive value (e. g. *Phaseolus vulgaris*) or species used traditionally in condimentary purposes (e. g. *Levisticum officinale, Satureja hortensis*) are not seriously endangered, while others (e. g. *Lens culinaris, Vicia faba*) are highly affected, and on the intensively studied territory no more land-races in cultivation have been found.

With the species cultivated in large scale by agricultural cooperatives and state departments the erosion of land-races is in an advanced stage, or have been already completely finished, especially on regions situated on lower altitudes.

A fairly large number of cultivated plants are no more accessible at all for variability studies, and the ancient cultivation is documented only by archeological and historical (written) records.

Some examples for the indicated cases would be presented in the following part.

1. *Trifolium pratense* L. ssp. *sativum* (SCHREBER) SCHÜBL. et MART. have been cultivated in Transylvania till the XVIIIth century. During the XIXth century, due to intense cultivation, some land-races of this species slowly emerged here. These land-races and types gained appreciation in XXth century under the name of "Transylvanian Red Clover", having some morphological and ecophysiological characters intermediate between the northern and southern group of red clover cultivars (provar. *sativum* SCHREBER and provar. *hispanicum* GAUD.). Based on this differences a new taxonomic category have been recommended – provar. *intermedium*. This new category permits the circumscription of a specific group of ecotypes representing a very valuable germplasm resource within the species (PUIA and SZABÓ 1978). The germplasm of provar. *intermedium* was used already in the breeding of more efficient selected varieties (PANFIL et al. 1976, SAVATTI 1976). As far as the original variability pattern

of the land-races concerned, due to centralized seed production and menagement practices a homogenisation process occured, with consequent loss of dinstinct populations of ancient types.

2. Garden bean (*Phaseolus vulgaris* L.) is highly appreciated in human nutrition on the examined settlements. The variability pattern of land-races in the case of this autogamous introduced species is entirely different from that of *Trifolium pratense*. The data collected till now have been examined according to agrobotanical (SZABÓ and DANKANITS 1978) and ethnobotanical (PÉNTEK 1978) criteria. Results regarding the territorial variability of certain morphological characters (growth type, seed shape, seed size and colour) are represented in Table 3.

Table 3

The occurence (frequency %) of the main seed characters by *Phaseolus vulgaris* L. in Transylvania

| Seed shape | Seed size | | | | Seedcoat colour | | | | | | | |
|---|---|---|---|---|---|---|---|---|---|---|---|---|
| Provar. | Total % | Mi | Mo | Ma | W | Y | B | R | Vt | Bl | Dm | Vr |
| | I. convar. *vulgaris*, 549 samples, 70,7 % | | | | | | | | | | | |
| *sphaericus* | 31.7 | 8.2 | 20.1 | 5.6 | 2.5 | 3.9 | 4.3 | 1.8 | – | 0.5 | 11.3 | 10.2 |
| *ellipticus* | 21.5 | 8.2 | 7.7 | 4.9 | 2.7 | – | 6.3 | – | 1.3 | 2.7 | 1.8 | 9.1 |
| *compressus* | 15.1 | 0.8 | 14.2 | 0.5 | 6.3 | – | 2.3 | – | – | 0.1 | 0.4 | 0.9 |
| *oblongus* | 2.4 | 0.3 | 1.4 | – | 0.1 | – | – | – | – | 0.5 | 0.1 | 0.1 |
| | II. convar. *nanus*, 226 samples, 29,4% | | | | | | | | | | | |
| *sphaericus* | 1.7 | 0.3 | 1.2 | 0.3 | 0.5 | 0.5 | – | – | – | – | 0.8 | – |
| *ellipticus* | 19.6 | 0.8 | 9.9 | 9.5 | 2.5 | 0.5 | 2,7 | 0.8 | – | 2.3 | 0.5 | 10.1 |
| *compressus* | 2.0 | – | 1.7 | 0.3 | – | 1.7 | – | – | – | – | – | – |
| *oblongus* | 6·1 | 2.2 | 3.1 | 0.8 | 0.8 | 1.6 | – | – | – | – | 0.5 | 4.4 |

Mi = microspermus, Mo = mesospermus, Ma = macrospermus, W = white, Y = yellowish, B = brownish, R = redish, Vt = bluish and violet, Bl = black, Dm = dimidiatus, Vr = variegatus.

The general frequency of different growth types – convar. *vulgaris* and convar. *nanus* (JUSL.) ASCHERS. – in Transylvanian land-races was strikingly similar to that recorded by BÖHME et al. 1975 for the *Phaseolus* world germplasm collection preserved in Gatersleben. Based on the variability registered in the field and that found in historical (written) documents it seems highly probable, that "primitive" types, introduced in Transylvania during XVIII–XIXth centuries are still present in the traditional agriculture, especially in that based on the simultaneous cultivation of some crop plants of American origin (*Zea mays*, *Phaseolus vulgaris*, *Cucurbita pepo*) completed sometimes and in some places with *Beta vulgaris*. Such mixed cultures of "indian type" are often preferred here on small scale (private fields) because they produce a high and diversified total yield. Different land-races of *Phaseolus vulgaris* are in cultivation here usually not as pure seed, but as mixtures of more or less distinct morphological types (Fig. 3). It seems, that in the territory *Phaseolus* variability is in

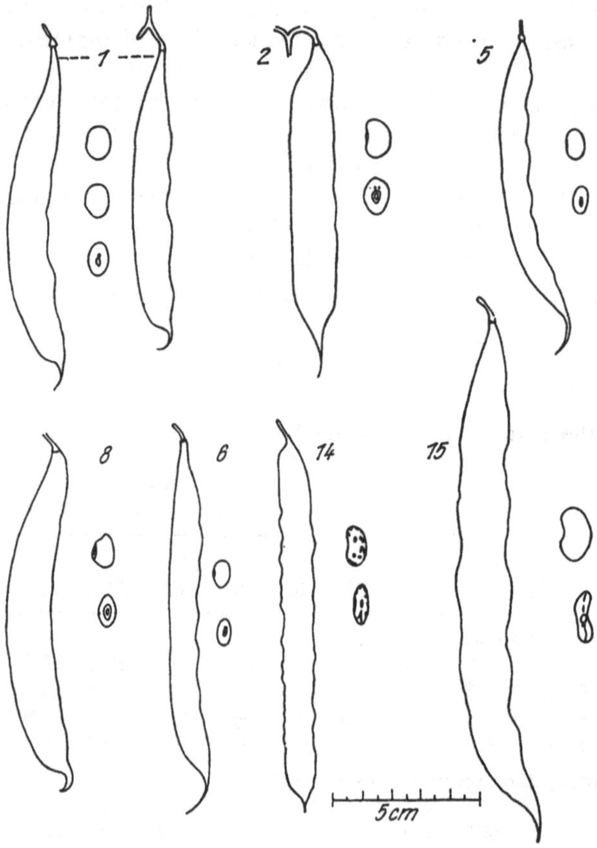


Fig. 3 Seed and pod characters of some *Phaseolus vulgaris* L. convar. *vulgaris* land-races collected from the sample territory. 1. 'Măzăruică albă mare' from Agîrbiciu; 2. 'Mazăre de Jebuc' from Jebucu; 5. 'Măzăruică albă mică' from Văleni; 6. 'Ouşoare albe' from Căpuşul Mic; 8. 'Măzăruică galbenă' from Văleni; 14. A hybrid sample from Izvorul Crişului; 15. 'Auriu' from Babiu

the state of steady enrichment. Older types still persist; newly introduced land-races of different origin steadyly appear and new cultivars are experimented yearly in many gardens mostly by woman farmers. This phenomenon is surely related with the increasing importance of bean protein in rural nutrition.

Periods of territorial enrichments of the genetic variability are characteristic for the evolution of most crop plants and may lead to the appearance of new, secondary centres of diversity. For the designation of the process the technical term "genetic sedimentation" seems to be appropriate and useful.

3. The opposite phenomenon of introduction and sedimentation is genetic erosion and ultimately the final extinction of the crop on a given territory. During the human history a definite loss of the genetical reserve of a crop cultivated through millennia was perhaps a rare event. In our days such events may occur much more commonly in a given geographical region. Strangely enough, the thorough analysis of the phenómenon is lacking in the agrobotanical literature consulted by the author of this paper. In our sample territory the analysis of the process was possible in the case of a number of species, as *Lens culinaris, Vicia faba, Triticum monococcum*. The case of *Triticum monococcum* will be presented in the following.

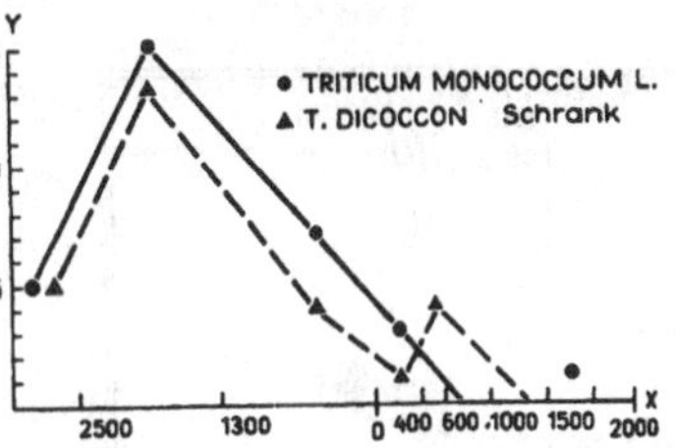


Fig. 4 Cultivation of *Triticum monococcum* L. and T. *dicoccon* Schrank in Pannonian Basin as reflected in archeological records compiled by HARTYÁNYI and NOVÁKI 1975. X = time scale in years; Y = number of records per period

The history of the Einkorn cultivation in the Carpathian Basin may be reconstructed according to the paleobotanical evidences reviewed by HARTYÁNYI and NOVÁKI 1975. The first archeological records are dated from the 3rd millennium B. C., the species reached its greatest importance as a grain crop at the beginning of the second millennium B. C. Afterwards paleobotanical records provide evidences for a gradual erosion process (Fig. 4). The crop disappeared in the Pannonian Basin during the 17th century A. D. In Transylvania written records related with einkorn ("alakor") cultivation prove the persistence of the species till 20th century (SZABÓ T. 1975). A thorough revue of the situation of *Triticum monococcum* cultivation for the period before World War II have been presented by BORZA 1945. As the Transylvanian data are poor represented

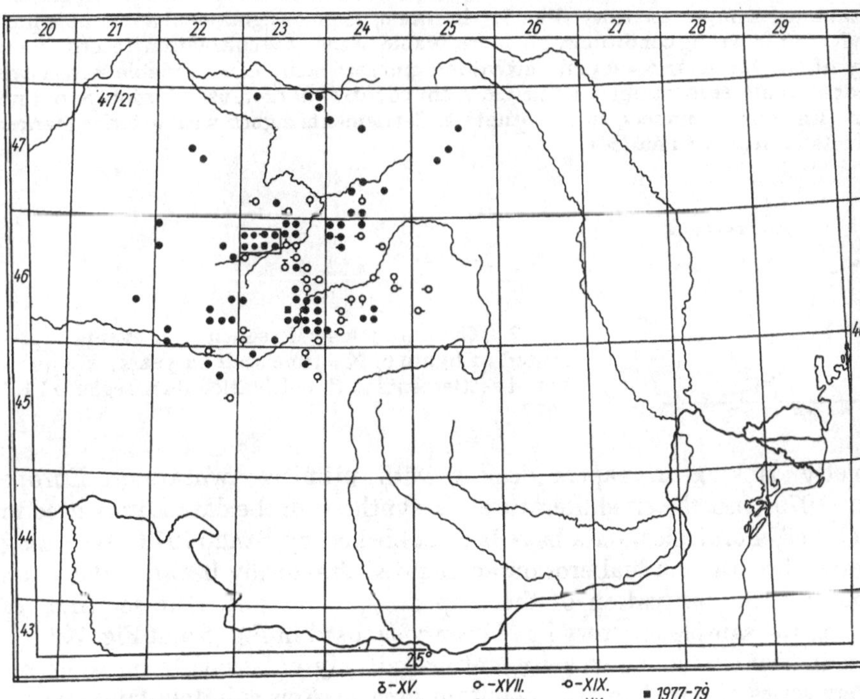


Fig. 5 The territorial dispersal of data regarding *Triticum monococcum* cultivation in Transylvania during the XVth-XXth centuries. The actual situation on the sample territory (in frame) is presented on figure 6

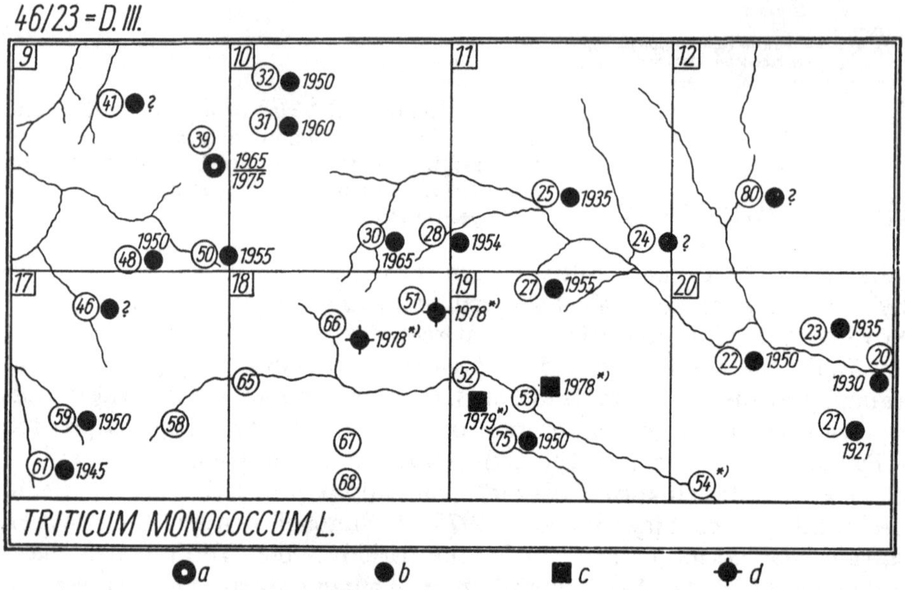


Fig. 6 The dynamics of the genetic erosion in the case of *Triticum monococcum* as registred for the examined sample territory (Fig. 5, in frame). Sign designations: *a* = sample preserved by farmer in wrong conditions, no more viable seeds; *b* = cultivation records only in memory of the farmers; *c* = actual cultivation and seed samples with viable seeds have been found; *d* = only seed samples aquisited from country market, used for forage but no cultivation. Encircled numbers as in footnote 1. Settlements signed with asterics market their products mostly in Gilău (54)

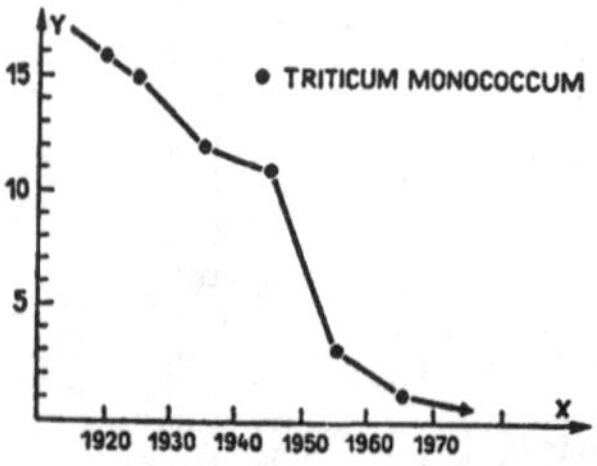


Fig. 7 The chronological sequence of events represented in figure 6. X = time scale in years; Y = number of settlements with cultivation data registred till 1977

or completly lacking in papers dealing with primitive wheats in Europe (Kuckuck 1970, and the cited literature), a synthesis of the data is presented in Fig. 5. Detailed chorological data have been published by Szabó 1978. According to our researches the gradual erosion accelerated drastically between the years 1950–1960 and the cultivation of the crop dropped sharply. The sequence of the events on the sample territory have been visualised in Fig. 6 and Fig. 7.

At present *Triticum monococcum* is cultivated only occasionally on some restricted territories situated in the mountain area of Apuseni Mountains, where germplasm reserves of the species still may be collected. Some cultivation—using mostly seeds originated from this mountain area and aquisited from country markets by local farmers – occured also till 1978–1979 in the sample territory (Fig. 8 and 9). The crop was sown in marginal land as forage in pure cul-

Fig. 8 A couple of traditional farmers from Căpuşul Mare, preserving *Triticum monococcum* germplasm.

Fig. 9 Forage mixture containing *Triticum monococcum. Avena sativa, Secale cereale* and other seeds in the hands of the farmer from fig. 8

tures or as a component of forage mixtures with *Avena sativa* L., *Hordeum vulgare* L., *Secale cereale* L., *Triticum aestivum* L., *Vicia sativa* L. and other, mostly weedy species (Szabó and Márton 1980a, b).

## Conclusions

1. The study of the land-races is a typical interdisciplinary field of research, where botanical (taxonomical, chorological ethnobotanical), genetical, historical and lingvistic approches are promising and team work is useful (see also Hanelt 1972, Zeven 1979).

2. "Genetic erosion" and "genetic sedimentation" are two technical terms which, once defined exactly, may be useful for the description of different stages in the evolution of the variability of a crop in a given territory.

3. Based on agrobotanical, chorological, ethnobotanical and historical methods both phenomena can be analysed numerically, as demonstrated in the cases of *Phaseolus vulgaris* and *Triticum monococcum*.

4. In respect of genetic erosion acting in Transylvania, in the present stage of research four main groups of cultivated species may be outlined:

a. Spontaneously occuring, or relatively newly domesticated species, as forage crops or medical plants are not usually very endangered, but certain loss of variability regularly occurs.

b. In the case of plants cultivated on a small scale and represented only or mostly by land-races genetic erosion is strongly dependent from the motivation of the farmer. If economic or other motivation is strong genetic reserves are normally preserved or are still increasing and genetic sedimentation may occur (e. g. by *Phaseolus vulgaris* L.). If motivation is of traditional character the process of erosion is dependent from the changes occuring in traditional peasant life (e. g. by *Levisticum officinale* L.). Land-races which require more land and labour for cultivation and economical or traditional motivation is weak, may be lost definitely (e. g. *Lens culinaris* L.).

c. In the case of the species cultivated on a large scale by cooperative or state departments, erosion of land-races is very advanced, the original variability was impoverished or lost.

d. Some of the cultivated species are no more available for research in any form on the examined territory, and the ancient cultivation my be proved only with historical evidences.

**Zusammenfassung**

Probleme der Generosion in Transylvanien, Rumänien

Ausgehend von den Grundprinzipien des IBP und MAB-Programms wurden die genetischen Ressourcen landwirtschaftlicher Pflanzenarten, die traditionell in Transylvanien angebaut werden untersucht. Landsorten aus dieser Gegend verdienen besondere Beachtung, da Transylvanien an der nordöstlichen Grenze des mediterranen Genzentrums zwischen nahem Osten und Westeuropa liegt und eine relativ isolierte Lage, umgeben von der Ostkette der Karpaten besitzt. Traditioneller Ackerbau, in vielen Gebirgsgegenden noch praktiziert, begünstigt die Erhaltung einstiger Landsorten und das Gebiet ist vielversprechend in Bezug auf das Vorkommen genetischer Ressourcen. Im Agrobotanischen Garten des Instituts für Ackerbau in Cluj-Napoca werden Samenproben von ökonomisch bedeutenden Pflanzenarten, Landsorten und verwandte Wildarten, beginnend seit 1970 gesammelt. Auf dieser Sammlung aufbauend wurden die in Bezug auf ihre Variabilität gefährdeten landwirtschaftlichen Pflanzenarten in Vier Hauptgruppen eingeteilt. Entwicklung und Besonderheiten der Generosion, die sich in diesen Gruppen vollzieht, werden am Beispiel von *Trifolium pratense* L., *Phaseolus vulgaris* L. und *Triticum monococcum* L. dargestellt.

**Краткое содержание**

Проблемы генной эрозии в Трансильвании (С. Р. Румыния)

Принимая во внимание основные принципы МБП и программы «Человек и биосфера», рассматриваются генетические ресурсы видов сельскохозяйственных растений, издавна возделываемых в Трансильвании. Местные сорта этой области заслуживают особого внимания, так как Трансильвания лежит на северо восточной границе Средиземноморского генного центра, между Ближним Востоком и Западной Европой, занимая относительно изолированное положение за Восточными Карпатами. Земледелие, ведущееся по старым традициям, которое можно встретить ещё во многих гористых районах, способствует сохранению местных сортов и в этой области можно ожидать нахождения интересных генетических ресурсов. Для агроботанического сада Института земледелия в Клуже (Напока) с 1970 г. начали собирать образцы семян видов, имеющих экономическую ценность, а также местные сорта и родственные им дикорастущие виды. В этой коллекции виды, находящиеся под угрозой потери их изменчивости, в зависимости от степени генной эрозии, делятся на четыре главные группы. Развитие и особенности генной эрозии, протекающей в этих группах, представлены на примерах *Trifolium pratense* L., *Phaseolus vulgaris* L. и *Triticum monococcum* L.

## Literature

Böhme, H., Chr. Lehmann, K. Hammer, P. Hanelt, H. Knüpffer, C. Tittel et J. Weiss, 1979: Index seminum (Zentralinstitut für Genetik und Kulturpflanzenforschung Gatersleben). Akademie-Verlag, Berlin.

Borza, A., 1945: Alacul (*Triticum monococcum*) la Români. [L'alac (*Triticum monococim* ] chèz les Roumains]. – Bul. Grăd. Bot. Cluj **25** (1–2), 93–119.

Căbulea, I., L. Roman, V. Tătaru şi C. Grecu, 1975: Germoplasma locală de porumb din Transilvania şi utilizarea ei în lucrările de ameliorare. (Local maize germplasm from Transylvania and its utilisation in maize breeding). – Probl. Genet. Teor. Apl. **7**, 57–90.

Ceapoiu N., 1975: Cuvînt înainte la Simpozionul "Conservarea fondului genetic vegetal şi animal". (Opening Address on the Symposium on Conservation of Plant and Animal Genetical Resources). – Probl. Genet. Teor. Apl. **7**, 1.

–, N. Eustatiu, N. N. Săulescu, Gh. Ittu, A. Giura şi Z. Tapu, 1975: Colecţia de grîu de la I. C. C. P. T. Fundulea ca sursá de gene valoroase pentru procesul de ameliorare. (The wheat germplasm collection of the Research Institute for Cereals and Industrial Crops Fundulea as a source of valuable genes for breeding). – Probl. Genet. Teor. Apl. **7**, 25–41.

Cociu, V., 1979: Colectarea, conservarea şi utilizarea raţională a fondului de germoplasmă pomicolă din România. (Collection, conservation and utilization of fruit tree germplasm in Romania). – Probl. Genet. Teor. Apl. **9**, 117–128.

Cristea, M., 1975: Germoplasma la porumb. – Editura Ceres, Bucureşti.

–, şi I. Gaspar, 1979: Explorarea şi colectarea resurselor genetice vegetale. (Exploration and collection of plant genetic resources). – Probl. Genet. Teor. Apl. **9**, 65–75.

Cristea, M., S. Reichenbuch şi D. Funduianu, 1979: Resurse genetice de porumb evaluate şi conservate la staţiunea de cercetări agricole Suceava. (Evaluation and conservation of maize genetic resources at the Suceava Agricultural Research Station). – Probl. Genet. Teor. Apl. **9**, 345–356.

DRĂGHICI, L., A. L. BUDE şi C. TUSA, 1975: Valorificarea resurselor mondiale de germoplasmă la lucrările de ameliorarea orzului. (Utilization of the world germplasm sources in barley breeding). – Probl. Genet. Teor. Apl. **7**, 42–56.

ENESCU, V., 1979: – Probleme actuale ale conservării resurselor genetice forestiere. (Present problems of forest resources conservation in Romania). – Probl. Genet. Teor. Apl. **9**, 129–136.

FRANKEL, H. O. and E. BENNETT (Ed.), 1970: – Genetic Resources in Plants – their Exploration and Conservation. – Blackwell Sci. Publ., Oxford and Edinburgh.

GIOSAN, N., 1979: Cuvînt de deschidere la cel de al VI-lea Simpozion naţional de genetică cu tema "Colectarea, studiul, conservarea şi valorificarea în lucrările de ameliorare a germoplasmei agricole şi silvice". (Opening Address on the Proceedings of the 6th National Genetics Symposium on "Collection, study conservation and utilisation of genetic resources for the breeding of agricultural and silvicultural plants"). – Probl. Genet. Teor. Apl. **9**, 1–2.

GRECU, C., 1962: Observaţii asupra cîtorva populaţii locale de porumb din Transilvania. (Observations regarding some land-races of maiz from Transylvania). – Analele I. C. C. P. T. Fundulea, Ser. C. 30.

HARLAN, J. R., 1970: Evolution of Cultivated Plants, in FRANKEL, H. O. and BENNETT, E. (Ed.), Genetic Resources in Plants – their Exploration and Conservation. Blackwell Sci. Publ. Oxford, Edinburgh.

HANELT, P., 1972: Zur Geschichte des Anbaues von Vicia faba L. und ihrer verschiedenen Formen. – Kulturpflanze **20**, 209–223.

HARTYÁNYI, P. B. und GY. NOVÁKI, 1975: Samen und Fruchtfunde in Ungarn von der Neuzeit bis zum 18. Jahrhundert. – Agrártört. Szemle, Hist. Rer. Rust. **17**, Suppl.,1–89.

HOMORODEANU, S., 1974: Comportarea populaţiilor locale de porumb colectate din judeţele Alba şi Hunedoara în condiţiile staţiunii experimentale Geoagiu. (Local maize populations from Alba and Hunedoara districts cultivated under the conditions of the Agricultural Experiment Station Geoagiu). – Probl. Genet. Teor. Apl. **6**, 214–237.

KELLNER, E., P. VARGA, C. GĂLAN şi M. GEORGESCU, 1969: Studiul colecţiei de graminee perene ca material iniţial pentru ameliorare (The study of an assortment of perennial grasses as basic material for breeding). – Analele I. C. C. P. T. **35**, 579–593.

–, şi A. GIURA, 1979: Studiul colecţiilor de plante cultivate şi cooperarea internaţională în domeniul resurselor vegetale (The study of plant collections and international cooperation in genetic resources). – Probl. Genet. Teor. Apl. **9**, 77–95.

KÜHN, F., K. HAMMER und P. HANELT, 1976: Botanische Ergebnisse einer Reise in die ČSSR im Jahre 1974 zur Sammlung autochtoner Landsorten von Kulturpflanzen. – Die Kulturpflanze **24**, 207–215.

KUCKUCK, H., 1970: Primitive Wheats. In: FRANKEL, O. H. and E. BENNETT, op. cit., 249–266.

KOVÁCS, A. şi I. BREAZU, 1980: Resursele genetice şi ecologice autohtone la gramineele perene de pajişti (Autochtonous genetical and ecological resources by the forgae grasses). – Prob. Genet. Teor. Apl. **3**, (in press).

MORARIU, I. şi A. BELDIE, 1976: Endemismele din flora R. S. România. In: Flora R. S. România, vol. XIII/Ed. pr. T. SĂVULESCU). – Ed. Acad. R. S. România, Bucureşti, 97–105.

NYÁRÁDY, A. şi E. C. VICOL, 1973: O nouă contribuţie metodologică la cartarea florei Romaniei. (A new methodological contribution to the mapping of the Romanian flora). – Not. Bot. Hort. Agrobot. Cluj. **7**, 35–37.

OPREA, ST., 1968: Studiul însuşirilor agrobiologice ale unor soiuri de viţă de vie în principalele podgorii şi centre viticole din nordul Transilvaniei (The study of agrobiological characters of some grape vine cultivars in main viticultural zones from Northern Transylvania). – An. I. C. V. V. **1**, 141–152.

PANFIL, C., M. SAVATTI und M. RÜHRIG, 1976: Die Rotklee-Sorte "Napoca-Tetra". – Not. Bot. Hort. Agrobot. Cluj. **8**, 45–46.

PÉNTEK, J., 1978: Rendszerszerűség és produktivitás a *Phaseolus* népi terminológiájában. (System and productivity in the popular terminology regarding the land-races of Phaseolus). – Nyelv- és Irodalmoud. Közl. **22**, 175–190.

Péntek, J. 1978, és A. Szabó, 1976a: Tájékoztató mutatvány egy népi növényismereti gyüjtésből. (Samples from a collection regarding popular botanical knowledge). – Nyelv- és Irodalomtud. Közl. **20**, 52–60.
–, –, 1976b: Egy háromszéki falu népi növényismerete. (Popular botanical knowledge in a village of Háromszék, judeţ Covasna, Romania). – Ethnographia **87**, 203–225.
–, –, 1980: A régi növényvilág és változásai a kalotaszegi földrajzi nevek tükrében (Changes in the plant life of Calata district as reflected in geografical names). in Nyelvészeti tanulmányok, Bucureşti 131–172.
Pintér, I., Szabó I., Kóczián, G., Gál M. és Szabó, L., 1976: Kultúrnövény tájfajták, vad növényfajok és etnobotanikai adatok gyujtése a Kászoni Medencéből. [Exploration of cultivated land-races, wild species and ethnobotanical data in the Kászon-Bazin (Transylvania, Roumania)]. – Agrobotanika 1974 **16**, 123–138.
Popescu, Gh., M. Iacob, M. Badistean, I. Ionita, M. Stoian şi M. Petrescu, 1979: Rezerva de germoplasmă a viţei de vie şi variabilitatea principalelor caractere ale soiurilor aflate în colecţiile ampelografice din Romania. (Germplasm reserves and variability of characteristics of the varieties preserved in grape-vine collections in Romania). – Probl. Genet. Teor. Apl. **11**, 97–116.
Puia, I., D. Pázmány, A. T. Szabó, et I. Petridean, 1976, 1978: Index Seminum Horti Agrobotanici Cluj-Napoca 1975–1976; Idem 1977–1978; Ateliere Didactice Inst. Agr., Cluj-Napoca.
–, and A. T. Szabó, 1978: On the systematic position of Red Clover (*Trifolium pratense* L.) cultivated in Romania. – Not. Bot. Hort. Agrobot. Cluj. **9**, 47–50; On the cultivated Red Clover (*Trifolium pratense* ssp. *sativum*) taxonomy. Bulet. Inst. Agr. Cluj-Napoca **32**, 9–14.
Rieger, R., A. Michaelis and M. M. Green, 1976: Glossary of Genetics and Cytogenetics. Classical and Molecular. Fourth edition. VEB Gustav Fischer Verlag, Jena.
Roman, L., V. Tătaru şi C. Grecu, 1973: Imbunătăţirea valorii biologice a provenienţelor locale de porumb prin crearea de populaţii sintetice. (Improving the biological value of the maize land-races through sinthetic populations). – Contribuţii ale cercetării ştiinţifice la dezvoltarea agriculturii. Ed. Dacia, Cluj-Napoca.
Savatti, M., L. Buda şi I. Kain, 1961: Contribuţii la ameliorarea trifoiului roşu de Transilvania. (Contribution à l'amelioration du trèfle rouge de Transylvanie). – Analele Inst. Cercet. Agr., Ser. C., **28**, 239–249.
–, 1976: Die Rotklee-Sorte 'Select-1'. – Not. Bot. Hort. Agrobot. Cluj. **8**, 47–48.
Stanescu, V., J. Damian, D. Parascan şi F. Negrutiu, 1979: Surse de gene de răşinoase din Carpaţii Orientali. Consideraţii genetic-evolutive. (Hardwood genetic sources in the Oriental Carpathians. Genetic-evolutive considerations). – Probl. Genet. Teor. Apl. **9**, 137–146.
Şuba, T., 1973: Valoarea populaţiilor de porumb din vestul ţării ca material iniţial în lucrările de ameliorare. (The value of the maize populations from the western part of Romania in respect of initial material for breeding work). – Doctoral dissertation, Inst. Agr. Timişoara.
Szabó, T. A., 1975: Erdélyi Magyar Szótörténeti Tár. (Wortgeschichtlicher Thesaurus der Siebenbürgisch-Ungarischen Sprache). **I.** (A–C); Idem, **II**, (CS-Elsz), 1978. Kriterion Publ. House, Bucureşti.
–, 1977: Studii asupra plasticităţii fenotipice la plante I. Analize privind statura speciilor din etajul alpin al Republicii Socialiste Romania. [studies on the phenotypic plasticity of plants. I. Analysis of growth data (stature) of the alpin plants from Romania]. – Comunicări de Botanică ţinute la a X-a Consfătuire Naţională de Geobotanică – Mt. Făgăraş 22–27 iulie 1974. Bucureşti. 253–266.
–, 1978: Eroziunea genetică şi protecţia resurselor autohtone la plante din ecosistemele artificializate. (Genetic erosion and the protection of autohtonous plant resources from the agroecosystems). – In Ecosistemele artificiale şi însemnătatea lor pentru omenire (Ed. V. Preda), **I**, Academia R. S. Romania, Cluj-Napoca, 116–142.
–, 1979: Wheatgrass variability (*Agropyron*, Sect. *Elytrigia*) in a native collection from Transylvania, Romania. I. *Agropyron intermedium* Complex. – Not. Bot. Hort. Agrobot. Cluj. **10**, 89–99.

Szabó, T. A., and V. Dankanits, 1978: Regional variability in *Phaseolus vulgaris* L. – Not. Bot. Hort. Agrobot. Cluj. **9**, 73–78.
–, és A. Márton, 1980: Az alkor nyomában. Egy ősi búzafaj a korszerű kutatás fényében. (Auf den Spuren des Einkorns. Eine uralte Weizenart im Lichte der modernen Forschung) I, II. – Művelődés (Bukarest) **33**, 1, 35–38; 2, 44–46.
–, és J. Péntek, 1976: Ezerjófű. Etnobotanikai utmutató. (Centaurium. A guide-book in ethnobotany). Kriterion Publ. House, Bukarest.
Varga, P., I. Moga., E. Kellner, C. Nalan şi M. Ionescu, 1973: Lucerna. Ed. Ceres, Bucureşti.
–, M. Tudose şi E. Alionte, 1979: Conservarea resurselor genetice. (Conservation of genetic resources). – Probl. Genet. Teor. Apl., **11**, 147–159.
Velican, V., 1934: Experienţe cu soiuri de porumb în Transilvania 1929–1932. Analele I. C. A. R. Impr. Nat. Bucureşti.
Vrânceanu, A. V., 1974: Floarea soarelui. Ed. Acad. R. S. Romania, Bucureşti.
–, M. Fl. Stoenescu şi M. Tudose, 1979: Centre genetice, de origine şi de diversificare ale plantelor cultivate. (Gene centres, centres of origin and diversity in cultivated plants). – Probl. Genet. Teor. Apl. **11**, 39–64.
Zeven, A. C., 1979: Collecting genetic resources in highly industrialized Europe, especially in Netherlands. – Proc. Conf. Broadening Genet. Base Crops, Wageningen, 1978, Pudoc, Wageningen, 49–58.

Dr. A. T. Szabó,
Institutul Agronomic "Dr. P. Groza"
Grădina Agrobotanică
3400 Cluj-Napoca, Str. Mănăştur 3
R. S. România

Kulturpflanze XXIX · 1981 · S. 63–65

# Preservation of Hungarian land-races as genetic resources

LÁSZLÓ HOLLY and JÁNOS UNK (Tápiószele, Hungary)

## Summary

The history of collecting Hungarian land-races and their present preservation techniques and system are briefly summarized in this paper.

## Introduction

Some parts of Hungary belong to the SE-European secondary gene centre. According to this, local types and land-races of certain agricultural and horticultural crops show a considerable amount of diversity. Local plant materials have been playing an important role as raw material for scientifically based plant breeding programmes. Although their systematic exploration and collection started only at the end of the forty's in Hungary, under the leadership and supervision of late Professor Dr. A. JÁNOSSY.

## History

At the beginning of the work on land-race collections, techniques used by breeders in their working collections were mainly adapted. Some later, when a centralised plant collection was established at Tápiószele in 1957–59, research on methods and techniques of the preservation of plant collections received much more attention. In addition to the experimental fields at Tápiószele, three further Experimental Stations served the maintainance of crop collections which required different ecological conditions. An important improvement was achieved at the late sixty's, when a so called "back garden system" was established for isoclimatic regeneration of land-races and some outpollinated crops, and also when an experimental cold seed-store was put into operation in 1972 (JÁNOSSY 1974).

During the last few years, collecting of still existing land-races and ecotypes has been receiving high attention and priority in our work (UNK 1979). An increasing number of samples have been collected in these years, most of which are vegetables and grain legumes but a number of maize land-races, clover and grass ecotypes have also been explored and collected (Table 1).

As a consequence of our increasing collecting activity, collections of Hungarian material increased considerably, and they now consist of more than 5000 accessions comparing with their number of 1629 in 1974.

Table 1

Hungarian land-races and ecotypes collected in 1977–1979

| Year | No. of species | No. of accessions | Main crops collected |
|---|---|---|---|
| 1977 | 42 | 631 | Onion, paprika, poppy, bean, pea, clovers, grasses, maize, etc. |
| 1978 | 60 | 806 | Onion, paprika, tomato, lettuce, bean, grasses, etc. |
| 1979 | 56 | 801 | Grasses, cucurbits, poppy, sunflower, vetches, paprika, onion, maize, bean, etc. |

## Present system of preservation

In parallel with the collecting work, an urgent need arised for the isoclimatic rejuvenation of Hungarian land-races which had been collected in earlier years. Therefore, our "back garden system" has been considerably extended and improved, and it includes 87 contributors in 1980. The system is subdivided into 9 districts of gardens directed by 8 district supervisors who are retired research workers, biologists or teachers. 4 of the districts situate on the Transdanubian area, 4 others on the Hungarian Great Plain, and the last one includes the villages around Tápiószele. Using this system, we now are able to rejuvenate 500–600 accessions each year, and this capacity seems to be sufficient for the systematic regeneration of Hungarian ecotypes and land-races (Table 2).

Table 2

Rejuvenation of land-races in the "back garden system"

| Year | No. of districts | No. of land-races | Main crops/Genera/ rejuvenated |
|---|---|---|---|
| 1977 | 4 | 239 | *Allium, Citrullus, Cucurbita,* |
| 1978 | 8 | 593 | *Cucumis, Daucus, Medicago,* |
| 1979 | 8 | 508 | *Pastinaca, Raphanus, Rumex,* |
| 1980 | 9 | 582 | *Sorgum, Spinacia, Trifolium,* |
| | | | *Zea,* seed legumes. etc. |

The collected or rejuvenated seed samples are treated in a way similar to the treatment of any other kind of material (e. g. foreign material from seed exchange, material multiplied on the experimental fields, etc.): Seed samples are dried down to an equillibrium with air of 15% relative humidity at room temperature (20–25 °C) in an air tight chamber using a Dutch-made EA-100-type dryer machine and put into and sealed in air tight glass jars. Seed samples of land-races and ecotypes usually divided into at least two parts: one subsample (4000–10.000 seeds/sample) is for long term storage at –20 °C, and the remaining part serves for medium term storage at +4 °C. At present (1 st June,

1980), 1012 subsamples at −20 °C and 3428 subsamples at +4 °C are being stored from our Hungarian collections.

Subsamples stored at +4 °C, serve different purposes (e. g. seed exchange, distribution, preliminary and more detailed evaluation, etc.). Accessions published in our Index Seminum have a surplus material which stored separately in a subdivided state to ensure a quick and easier response to seed requests.

During the last few years, most efforts has been payed for collecting and preserving land-races and ecotypes, and therefore much less capacity was left for a more detailed evaluation of these genetically highly variable populations. Usually small samples of the newly collected material are being used for preliminary evaluation. The main purpose of this work is to compare them with materials of similar types, for obtaining sufficient data for differentiation and identification purposes. Some parts of the land-races collections have already been evaluated in more details, and evaluation has been recently initiated in certain districts of "back garden system" too, where the necessary supervision could have been arranged for.

## Zusammenfassung

Die Erhaltung ungarischer Landsorten als genetische Ressourcen

Die Geschichte der Sammlung ungarischer Landsorten und die gegenwärtige Technik ihrer Erhaltung wird kurz zusammengefaßt dargestellt.

### Краткое содержание

О сохранении венгерских местных сортов как генетических ресурсов

Вкратце описывается история сборов венгерских местных сортов и современная техника их сохранения.

## Literature

JÁNOSSY, A., 1974: Az Agrobotanikai Intézet génbank munkája. (Gene-bank work of the Institute of Agrobotany) Agrobotanika **XV**, 5–9.

UNK, J., 1979: A tápiószelei növényigénbank-munka eredményei. (Work performed at the Plant Gene Bank of Tápiószele) Fajtakisérletezés **XXVIII**, 23–35.

Dr. L. HOLLY and Dr. J. UNK
Research Centre for Agrobotany, NIAVT H–2766 Tápiószele, Hungary

Kulturpflanze XXIX · 1981 · S. 67–77

# Preservation of land-races of cultivated plants in Czechoslovakia

Ivo Bareš and Jana Sehnalová (Praha-Ruzyně, ČSSR)

## Summary

At present, the collection of cultivated plants comprises in Czechoslovakia 40883 cultivars of field, vegetables and fruits, vine grape and ornamental plants. The collection has been located in the Research Institute of Crop Production, Praha–Ruzyně, which also renders the central coordination, and in 17 specialized Research Institutes and Breeding Stations which are responsible for investigation, maintenance, exploration of the collections of special crops. Continual research has assumed since 1951 when former collections of different research stations (amounting only 6000 cultivars) were united.

The first efforts to collect cultivars by the different research and breeding stations were started in the ČSSR already at the beginning of this century. A part of the then collections coming from different research and breeding stations has been conserved up to present time and represents, together with later collected cultivars, a valuable gene pool of 2000 cultivars of Czechoslovak origin. The original Czechoslovak cultivars are represented by selections, lines or clones from land-races (wheat, barley, oat, legumes, flax, hop, self-pollinated fodders and vegetables), or populations (maize, oil plants, fodders vegetables and ornamental plants). The major part of the collection is represented by Czechoslovak cultivars coming from combination crossing.

The old cultivars of vine grape, fruits and ornamental plants have been maintained since the beginning of this century in old plantations of the research institutes or they are registred by them. Also cultivars of wheat (Czech red wheats, Czech alternate wheats, South-Moravian awnless and awned wheats, Slovak awned wheats) and of barley (spring malting Haná barley, Haná is territory in middle Moravia) come from the same period. The oldest native as well as foreign cultivars of other agricultural crops originate from thirties being conserved at different degree of breeding intensity at the time of their cultivation.

## Introduction

The collection of useful plants gene recources in the Czechoslovakia amounts 40833 cultivars or forms (the status on December 31.1979, see Table 1). It has been maintained in the Research Institute of Crop Production (RICP), Praha–Ruzyně, which has been also the co-ordinating center, and in 17 other research and breeding workplaces. These research institutes and breeding stations being

Table 1

Survey on the amounts of crop cultivars (on December 31. 1979) in Czechoslovak collections

| Group of crops | Number of samples |
|---|---|
| Cereals | 10638 |
| Maize | 1472 |
| Legumes | 2653 |
| Oil plants | 394 |
| Fibre crops | 1707 |
| Technical crops | 157 |
| Potatoes | 672 |
| Fodders | 2566 |
| Hop (inclusive of wild forms) | 587 |
| Vegetables and aromatic plants | 10385 |
| Vine grape | 1556 |
| Fruits | 2248 |
| Ornamental plants | 5248 |
| The collection of the Central Institute for Control and Testing in Agriculture (Department of Cultivars Testing) | 600 |
| Total | 40883 |

specialized to breeding as well as growing techniques in individual crops, are also responsible for their study, maintenance, investigation and exploration the collection of gene resources. This decentralized organisation makes possible to employ a relatively great amount of research workers (about 120 persons), to evaluate profoundly the most important biological crop character (on in average 70 ha of experiments yearly), as well as to exploit the resources effectively in the breeding program of breeding stations or, eventually, to introduce the cultivars immediately into practice (as cultivars registred for the import).

The program of complementing the collection by desirable resources has been realized mainly by the central activity of the RICP Praha–Ruzyně which imports yearly 3000–4000 cultivars. In the RICP also the central system of registering the resources by computer is being developed and the project of gene bank is under preparation. A part of the collection is being offered for exchange by editions of Index seminum (Sehnalová et al. 1976, 1979).

A systematic investigation of the collections has been carried out since 1951

Table 2

The amount of collected cultivars of cultivated plants at years 1951–1979 in Czechoslovak collections

| Years | Numbers of cultivars |
|---|---|
| 1951 | 6000 |
| 1954 | 9300 |
| 1960 | 18600 |
| 1964 | 29942 |
| 1970 | 35636 |
| 1974 | 40129 |
| 1979 | 40883 |

when formerly dispersed collections of different breeding stations and institutes were united. The status of the united collection amounted then 6000 cultivars of forms (the increasing amounts see Table 2).

## The beginning of cultivars collecting

Already since the beginning of this century, different research and breeding stations have evolved the activity in collecting the cultivars. Collections with a broader set of crops were found then in three institutions.

1. In the Agricultural and Botanical Research Station at Tábor (established in 1880, abolished in 1919) experiments with cultivars of barley had been carried out already since 1899, with wheat and other crops since 1903 (ERBEN, JIRSA and VILIKOVSKÝ 1916).
2. The Research Station for Chemistry and Physiology of the Technical University at Jeneč near Praha (established in 1898) examined different species and cultivars of crops. In 1920, the experiments were transfered to the Experimental Field at Uhříněves which became a constituent part of the newly established State Research Agricultural Institutes in Praha. These Institutes assumed later also the responsibility for the collection of the former Agricultural and Botanical Station at Tábor. In 1948 the united collection was transferred to the Experimental Field at Doksany and finally, in 1952, to the RICP Praha–Ruzyně. At that time, the collection amounted 2847 cultivars of cereals, legumes and oil plants and partly of fodders.
3. In 1919, Moravian National Research Agricultural Institutes were established in Brno (CHMELAŘ 1920). Likewise the State Research Agricultural Institutes in Praha, they developed intensive efforts in collecting as well as exploring of land-races (they organized for instance different public inquiries in thirties). In 1951–1954 the collections of these institutes were transferred to the newly established workplaces of the present Research and Breeding Institute of Cereals at Kroměříž, Research and Breeding Institute of Fodders at Troubsko and Research Station of Grasses at Rožnov pod Radhoštěm (Moravia).

In the collection of mentioned institutes land-races as well as improved cultivars from the beginning of this century have been conserved. Also old cultivars of vine grape, fruits and ornamental plants from the same period have been conserved in old plantations registered now by the research institutes.

From thirties there have been maintained also original native and foreign cultivars of other agricultural crops which had been collected by pre-war research and breeding stations. They are at a different stage of breeding intensity as it corresponds to the period of their cultivation.

At present the oldest original cultivars in many crops are represented in the form of selections (lines, clones) from land-races (wheat, barley, oat, legumes, flax, hops, self-pollinated fodders and vegetables) or their populations (maize, oil plants, fodders, vegetables, ornamental plants). The majority of Czechoslovak cultivars is represented by cultivars coming from combinative crossing. The present amount of cultivars of Czechoslowak provenance in the collection attains 2000 (BAREŠ 1974, KOLEKTIV 1955–1980).

## The characteristic of the oldest Czechoslovak cultivars

(Only cultivars are included which have been conserved in the collection).

*Cereals*

In the **wheat** collection (*Triticum aestivum* L. emend. Fiori et Paoletti) there is a great number of Czech alternate wheats and Czech red wheats coming from populations of land-races. They are characterized by a high adaptibility and winter resistance as conditioned by a special developmental characteritics, namely a long lasting photoperiodical response. Alternate wheats have a very short vernalization stage (3 days); red wheats have a short vernalization stage (14–18 days). Some Czech alternate wheats come from the beginning of this century, as follows: cv. 'Jinonická', cv. 'Přesívka Červený Újezd', cv. 'Česká Přesívka' and cv. 'Dobrovická Česká Červená Přesívka P 2' ( a synthetic cultivar from lines of Czech alternate wheat), cv. 'Postoloprtská Přesívka', cv. 'Kaštická Přesívka' come from twenties. Red wheats are as follows: cv. 'Chlumecká 12' (=cv. 'Dregerova 12'), cv. 'Dregerova G 10', cv. 'Dobrovická Červená B', cv. 'Liblická', cv. 'Dregerova B I/22' and cv. 'Dregerova 126'.

Among Czech original cultivars the following land wheats are the oldest ones: cv. 'Bělá od Dukovan', cv. 'Kostomlatská Sametka', cv. 'Rokycanská Sametka' (all coming from the end of 19th century) and cv. 'Vouska z Třemošnice' (with a pubescent spike). The oldest Moravian cultivars (awnless and awned land-races) are as follows: cv. 'Pavlovická Vouska', cv. 'Hanácká Osinatka', cv. 'Kelčanská Osinatka', cv. 'Hanácká Bělka' and cv. 'Kelčanská Holice'.

Considerably earlier than in wheat, the process of selecting was started in the old land-races of **spring barley** from Haná (E. Proskowetz at village Kvasice in 1872–1875). All cultivars developed from those land-races were two-row types with a nutant spike (convar. *distichon* var. *nutans*) and high malting quality. The oldest ones among them are as follows (Hana pedigreé, developed 1884): cv. 'Hana', cv. 'Staroveský', cv. 'Jarohněvický', cv. 'Selekční Netčický'', cv. 'Stupický Hanácký', cv. 'Proskovcův Hanácký', cv. 'Selekční Hanácký', cv. 'Hanácký Staroveský', cv. 'Pudmerický Pivovarský' and cv. 'Terrasol Pivovarský'. From Czech land-races the cultivars cv. 'Dobrovický Staročeský, cv. 'Hořický', cv. 'Nohelův Staročeský', cv. 'Selecty Staročechý R', cv. 'Kaštický 91' and cv. 'Kaštický 132' were developed. The Czechoslovak regional barley cultivars which had been later developed from them, produced a considerable influence on improving the malting quality of barley all over the world (Lekeš 1961).

The present Czechoslovak cultivars which originated from Haná cultivars and exploited the short-straw mutation of the cv. 'Valtický' prove an increased productivity as connected with the resistance to lodging and high malting quality (cv. 'Favorit', cv. 'Spartan', cv. 'Korál', cv. 'Safír').

Most Czechoslovak cultivars of **winter rye** (*Secale cereale* L.) originate from the German breeding program. To land-races there belong the cultivars cv. 'Stupické S II', cv. 'Víglašské', spring rye cv. 'Těšovské' and a number of rye populations called "křibice" (*Secale cereale* f. *multicaule* Metzger) coming from Krkonoše and Beskydy mountains. This type rye has been represented by the only selected cultivar, namely 'Dobřenické Krmné'.

In **oat** (*Avena sativa* L.) most cultivars originate from Swedish or German breeding program; of land origin there come only the cvs. 'Doupovský', 'Hořický', 'Jindřichovský Bílý', 'Klatovský Bílý', 'Radošinsky', 'Rychlík', 'Stupický', 'Studnický', 'Valečovský', 'Zlaťák' and other land-races from the Žďár region.

In the collection of **maize** (*Zea mays* L.) there are conserved the original land populations of convar. *mays*: cvs. 'Adamcová Budynská', 'Slovenská Velkozrnná', 'Florentinka', 'Valticka' and the cultivars developed by crossing of domestic cultivars with the foreign ones: cv. 'Stupická Raná', cv. 'Čelechovická ADQ', cv. 'Hodonínská Florentinka', cv. 'Bučianská Žlutá', cv. 'Valtická (kmen C)'.

Breeding work in maize was influenced mainly by the American and Hungarian resources, especially in the conv. *dentiformis*. At the beginning of sixties, the cultivar populations were replaced by hybrids; the collection contains also Czechoslovak lines developed from mentioned land-races.

*Legumes*

**Common pea** (*Pisum sativum* L. subsp. *sativum* convar. *sativum*). Since the start of pea breeding up to 1930, tall pea cultivars were being developed by means of selecting from land-races (of unknown, probably German origin) which had been grown here for a longtime. At that time following cultivars were developed: cvs. 'Dětenický Žlutý Velkozrnný', 'Slovenský Viktoria 75', 'Klatovský Zelený', 'Slovenský Expres', 'Stupický Zelený', 'Milion Zelený'. Later, the cultivars with a small yellow grain originated by crossing with land-races, as for instance cvs.: 'Liblický Bastard', 'Milion Žlutý'. All present cultivars ('Meteor', 'Borek', 'Dlaibor', 'Inter', 'Jupiter', 'Proteus', 'Dukát', 'Juran', 'Smaragd') are of short growth; they come from highly productive Hollandish cultivars.

From land-races have come also the **field pea** (*Pisum sativum* L. subsp. *sativum* convar. *speciosum* (Dierbach) Alef.) cultivars 'Stupicka Jarní' and 'Přerovská Krajová'. The cultivars 'Klatovská Jarní' and 'Klatovská Ozimá' have resulted from Russian cultivar which had been acclimatized under conditions of Šumava region.

Most **lentil** (*Lens culinaris* Medik.) cultivars cultivated up to fifties came from land-races 'Valtická Haléřová', 'Hrotovická Velkozrnná', 'Písárecká Velkozrnná', 'Moravská Drobnozrnná Krajová', 'Slovenská Krajová' and 'Trebišovská'.

Following **common bean** (*Phaseolus vulgaris* L.) cultivars originated from land-races: cv. 'Jihomoravská Bílá Krajová', cv. 'Slovenská Ledvinkovitá Bílá Krajová', cv. 'Slovenská Perlová Bielá Krajová', cv. 'Slovenská Sírově Žlutá Krajová'. Of Slovak land origin are many others, especially half-climbing bean forms with coloured grains (in the collection of the RICP Piešťany).

In **horse bean** (*Vicia faba* L., var. *minuta* (Alef.) Mansfeld) the cultivars 'Přerovský' and 'Tatranský' have originated from land-races while cv. 'Chlumecký' (registered 1926) and cv. 'Selecty' were developed by many years repeated selections in stands of unknow origin or Russian origin (cv. 'Zborovický').

*Oil and fiber plants*

Oil plants. A very valuable source of **winter rape** (*Brassica napus* var. *napus* f. *biennis* (Schübler et Mart.) Thellung) is being represented by the land-races 'Třebíčská' which shows a relatively high content of erucid acid, high adaptability and winter resistance; also the cvs. 'Slapská', 'Přerovská', 'Slovenská Krajová' and the spring cv. 'Česká Krajová' are of land origin.

The collection of **poppy** (*Papaver somniferum* L.) contains a relatively extensive set of cultivars developed from land-races: cv. 'Azur', cv. 'Hanácký Modrý' and cv. 'Dubského Stříbrošedý' as the most productive among them, further on cv. 'Dětenický Modrý', cv. 'Elity', cv. 'Freundkův', cv. 'Karlův 10/45', cv. 'Selecty Stříbrošedý', cv. 'Táborský', cv. 'Stupický Bělosemenný', cv. 'Sušický Červenosemenný', cv. 'Hnědosemenný'.

The collection of **common flax** (*Linum usitatissimun* L.) for fibres comprises also the oldest flax cultivar 'Krajový-Staré Město' which was firstly under cultivation in 1881; also other land-races were designated according to region of their cultivation as follows: cv. 'Krkonošský', cv. 'Táborský', cv. 'Orlický', cv. 'Českomoravský' etc.; in the present breeding program they are explored as a source of high earliness. Already in fifties cultivars were grown which had been developed by selecting or combination crossing of Polish, German and Russian flax cultivars. From pre-war period the cultivars 'Domanínský Výnosný' and 'Keřkovský Diplomat' are in the collection. The present cultivars prove a very high efficiency in fibre of high quality.

*Root crops*

Breeding work in **sugar beet** (*Beta vulgaris* L. convar. *crassa* ALEF. s. l. var. *altissima* DÖLL) was influenced mainly by the German and French cultivars. Developed cv. 'Dobrovická N' (into cultivation since 1898, registered 1915) created base for all Czechoslovak breeding work from which most yielding diploid cultivar 'Dobrovická A' is till this day registered in ČSSR.

*Tuber crops*

*Potatoes*

Breeding **potatoes** (*Solanum tuberosum* L.) has a long tradition in this country. Already in 1877 the station at Valečov maintained a collection of 63 cultivars; in 1922 the collection of domestic and foreign cultivars amounted 366 accessions. Due to virus attack the cultivars of that period have been conserved only at a limited extent; they were, however, fully explored in the breeding work. In the ČSSR there exist no land-races of potatoes; all original Czechoslovak cultivars originated by crossing of Europian cultivars. The present extensive collection of about 100 cultivars represents a valuable source for breeding. The most important of them are as follows: **Early**: cv. 'Český Erstling', cv. 'Keřkovské Rohlíčky', cv. 'Jara', cv. 'Hera'; **Semi-early**: cv. 'Krasava', cv. 'Český Triumf', cv. 'Kardinal', cv. 'Karmen', cv. 'Čajka', cv. 'Jiskra', cv. 'Jizera', cv. 'Daria', cv. 'Radka', cv. 'Lučnica', cv. 'Breza'; **Industrial**: cv. 'Kotnov', cv. 'Blaník', cv. 'Limba', cv. 'Renata'.

*Fodder plants*

Breeding **alfalfa** (*Medicago sativa* L.) started in thirties. In fourties and fifties, besides land-races the seed of which was purchased according to regions, also improved cultivars developed from land-races were grown. Czechoslovak cultivars prove a good adaptibility to environment conditions, persistance, medium up to high green mass productivity as well as a good seed productivity. The collection comprises land-races from different regions of this country as for instance: Česká Krajová, Moravská Krajová, Slovenská Podunajská Krajová, Slovenská Trebišovská Krajová and from land-races developed cultivars: cv. 'Přerovská', cv. 'Brňenská', cv. 'Hodonínská', cv. 'Stupická', cv. 'Kaštická', cv. 'Nitranka'. Also the most productive present cultivars which originated by crossing with Czechoslovak land-races are in the collection, as for instance: cv. 'Palava', cv. 'Hodonínska', cv. 'Bobrava'.

A similar situation was in breeding and seed production in **red clover** (*Trifolium pratense* L.). Until fifties, land-races were grown bearing biological differences according to regions of their cultivation (Český Krajový, Litomyšlský Krajový, Moravský Krajový, Slovenský Podtatransky Krajový) and cultivars improved by means of selection: cv. 'Chlumecký', cv. 'Jíčinský', cv. 'Keřkovský', cv. 'Přerovský', cv. 'Horal', cv. 'Táborský', cv. 'Třebíčský', cv. Víglašský'. The remarkably productive clover cv. 'Start' was developed by crossing of some land-recas.

The extent areas of promontory and mountain meadows and pastures as well as wooded areas of the ČSSR produce suitable conditions for giving rise and conserving the natural localities of **wild grasses**. In the region of Beskydy mountains (Moravia) the Research Station of Grasses was established in 1920 which has payed a great attention to grass seed production. Exploring the collected wild grass forms and selecting the productive types among them resulted in native improved cultivars designated "Rožnovský" in most important grass species. Similarly the cultivars were later developed of land-races in Breeding Station Větrov (named "Větrovský") and in Breeding Station Levočské Lúky (Slovakia) (named "Levočský"). The Research Station of Grasses at Rožnov pod Radhoštěm evolves permanent collecting activy in wild grass species a part of which has formed a living gene bank comprising 1000 ecotypes in the natural mountain area near Rožnov pod Radhoštěm.

Similarly the Research and Breeding Institute of Fodder Plants at Troubsko (near town Brno) is organizing collecting and evaluating of wild species of grasses and clovers for breeding aimed at new fodders.

Only in the last decade the collection of grass cultivars was complemented by new cultivars coming of crossing with one of the Czech landrace parent and also of polyploidization.

The **hop** (*Humulus lupulus* L.) collection comprises the oldest Czechoslovak land-races since 1931: 'Maštýřovický', 'Starožatecký', 'Staroúštěcký', 'Dubský Zeleňák', 'Úštěcký Smetaňák', 'Hřebčí', 'Semšův'; further selected land modifications: 'Lučan', 'Blato', 'Roudnický Krajový', 'Tršický Krajový' and clones coming from the Czech semiearly red hops (Osvalds clones 31, 72, 114) as well as the registred selections for present from populations as follows: 'Aromat', 'Siřem' (='Aurea'), 'Zlatan'.

*Vegetables*

The species and cultivars of cultivated vegetables in the ČSSR is relatively narrow. In thirties, the composition of cultivars was influenced by seed import; the native breeding program explored the sources of West-European countries. However, in some species land-races were traditionally cultivated; they were improved by selection in fifties, or employed for crossing with imported sources.

An example of a long tradition is the breeding of **kohlrabi** (*Brassica oleracea* L. convar. *acephala* (DC.) ALEF. var. *gongylodes* L.) noted by extreme earliness and high inner quality. Following white kohlrabi cultivars are of land origin: 'Dvorského Pražská Bílá Raná', 'Selecty Bílá Raná', 'Stupická Polní Raná', 'Boháčkova Bílá'. The cultivars 'Moravia' and 'Libochovická Bílá Raná' were developed by crossing land-races with foreign cultivars. Blue kohlrabis have come from the cultivars 'Pražská Modrá', 'Matouškova Pražská Modrá Raná', 'Kozmanova Modrá' and finally, the cv. 'Dětenická Modrá' originated from a blue late land-race.

The **cabbage** (*Brassica oleracea* L. convar. *capitata* (L.) ALEF. var. *capitata*) white land-races named 'Křimické' and 'Klokotské' prove a high inner quality. The cv. 'Dobrovodské Polopozdní' and 'Dobrovodské Pozdní', 'Táborské' and 'Pourovo Červené' were developed from land-races while 'Pourovo Zimní Bílé' originated from crossing of a land-race with a foreign source.

**Celery** (*Apium graveolens* L.) proves a high productivity and keeping quality. From the land-race 'Pražský' the cultivars 'Pražský Obrovský' and 'Nerez' were selected.

The most popular land-race of **horse radish** (*Armoracia rusticana* GAERTNER, MEYER et SCHERB.) has been called cv. 'Malínský'.

From land-races of **parsley** (*Petroselinum crispum* (MILLER) NYMAN convar. *radicosum* (ALEF.) DANERT var. *radicosum*) the cv. Přerovská Polodlouhá, Hanácká and Olomoucká Dlouhá have been developed.

The **radish** (*Raphanus sativus* L. var. *niger* KERNER) cultivar 'Kulatá Černá' is also of land origin.

What concerns the **lettuce** (*Lactuca sativa* L.) for forcing, land-races of the rocky type (named Kamenáč) were cultivated; the present cv. 'Stupický Kamenáč' and 'Kamýk' (crossing of Kamenáč with foreign source) were developed from them. From early land-races named "rychlík" the cv. 'Domažlický Rychlík' and 'Prostějovský Rychlík' were selected.

Within the group of common **bean** (*Phaseolus vulgaris* L.) cultivars only the climbing cv. 'Ostravská' is of land origin.

In South Slovakia the land-races gave origin to the **watermelon** (*Citrullus lanatus* (THUNBERG) MATS. et NAKAI) cv. 'Dunaj' as well as **pumpkin** (*Cucurbita pepo* L.) cv. 'Kveta'.

Before hybrids entered the practice following Czechoslovak **cucumber** (*Cucumis sativus* L.) for pickling of land origin had been very popular: 'Bílské', 'Mělnické', 'Znojemské', Also the slicing cv. 'Mladoboleslavské salátnice' was developed from land-race, namely 'Černavské Salátnice'.

In South Moravia and Slovakia **green pepper** (*Capsicum annuum* L.) has been traditionally grown and selected. The origin of land-races can be followed in the vegetable cv. 'PCR' and in the condiment cv. 'Hodonínská Raná Vzpřímená', 'Hodonínská Ostrá', 'Tesa', 'Žitavská'. 'Tenkostěnná Krajová is a pure land-race.

In the mild territories of Bohemia and Moravia **true onion** (*Allium cepa* L.) cultivars were developed from land-races. The most popular of them were as follows: the round-shaped 'Všetatská' and flat-shaped 'Žitavská', 'Kvetoslavovská', 'Obrovská Žlutá', 'Moravanka'.

Also **garlic** (*Allium sativum* L. var. *sativum*) cultivars are of land origin. There are the flowering types called "paličák": 'Bzenecký Paličák', 'Zeiorský'; then broadleaved non-flowering cvs. 'Přerovský', 'Táborský'; and finally, narrowleaved non-flowering cvs. 'Japo' and 'Polaban'. In **chives** (*Allium schoenoprasum* L.) the cvs.' 'Pražská Krajová' and 'Jemná' were selected from land-races.

Within the group of staked **tomato** (*Lycopersicon esculentum* MILLER) only the cv. 'Průhonické' has been of land origin.

The **asparagus** (*Asparagus officinalis* L.) cv. 'Ivančický Krajový' was also selected from a land-race.

*Fruits*

The well-known land-races of **apple** (*Malus sylvestris* MILLER subsp. *sylvestris* var. *domestica* (BORKH.) MANSF.) are as follows: 'Čistecké Lahůdkové' (a mutation of the cv. 'Princess Luise' found at Čistá near town Rakovník). 'Chodske' (from surroundings of town Domažlice), 'Jaderníčka Moravská' (origin from Valašsko and Kravaře – territory in Moravia), 'Holovouské Malinové' (from surroundings of town Hořice; the race was awarded in Vienna in 1888), 'Smiřické Vzácné' (originated in 1870 at village Smiřice), 'Sudetská Reneta' (appeared at the end of the last century at Budov in North Moravia), 'Žďárské Úrodné' (originated at the end of the last century at Veselý Žďár near town Havlíčkův Brod).

The **common pear** (*Pyrus domestica* MEDIK.) land-race named Solanka originated at Solany in North Bohemia and Koporečka at Koporeč near town Most, namely at the end of 18th century.

Among **sweet cherry** (*Prunus avium* L.) cultivars 'Karešova', 'Kaštánka' and 'Těchlovická' are of land origin similarly as the **sour cherry** (*Prunus cerasus* L.) cv. 'Vackova'.

In Czechoslovakia no original **vine grape** (*Vitis vinifera* L.) cultivars were cultivated; since 1970 the only cv. 'Palava' of native origin has been registred. There are grown here predominantly European cultivars which had been introduced since Middle Ages. There were mainly French Burgundy wines, Sauvignon, Semillion – white, blue and gray –, Italian wines of Veltlin type, German Sylvan and Müller Thurgau and Hungarian Furmint, Lipovina, Leanka, Ezerjö and – at the beginning of this century – also Hungarian table cultivars. Many years lasting influence of Czechoslovak ecological conditions produced some effect on biological characters of these cultivars differring now according to regions of cultivation.

*Ornamental plants*

All well-known species of ornamental **woody plants** which are found in parks have been registred. There exist a relatively extensive collection of Czechoslovak selections of rhododendron. In **annual** as well as **perenial ornamental plants** no land-races have been conserved. Already in thirties improved foreign cultivars or Czechoslovak cultivars developed from them appeared in cultivation. The collection of ornamental plants of Czechoslovak origin amounts 300 cultivars and forms. Relatively extensive collections exist in following ornamental species: *Callistephus chinensis* (L.) NEES *Matthiola incana* (L.) R. BROWN., *Begonia* L. sp., *Petunia* JUSS. sp., *Dahlia pinnata* CAV., *Chrysanthemum* L. sp. and *Pelargonium* BURMAN sp.

## Zusammenfassung

Die Erhaltung der Landsorten von Kulturpflanzen in der ČSSR

Gegenwärtig umfaßt die Kollektion in der ČSSR 40883 Sorten von Feldfrüchten, Gemüse, Obst, Weinreben und Zierpflanzen. Die Kollektion befindet sich im Forschungsinstitut für Pflanzenbau Prag-Ruzyně, das für die zentrale Koordinierung zuständig ist, sowie in 17 spezialisierten Forschungsinstituten und Zuchtstationen, die für die Untersuchung und Erhaltung der Kollektionen spezieller Kulturpflanzen verantwortlich sind. Kontinuierliche wissenschaftliche Untersuchungen begannen im Jahr 1951, als die früheren Kollektionen verschiedener Forschungsstationen (nur etwa 6000 Sorten) vereinigt wurden.

In der ČSSR bemühten sich bereits zu Beginn dieses Jahrhunderts verschiedene Forschungs- und Zuchtstationen um die Sammlung von Sorten. Ein Teil der damaligen, aus verschiedenen Forschungs- und Zuchtstationen stammenden Sammlungen wurde bis zur Gegenwart erhalten und stellt, zusammen mit später gesammelten Sorten, einen wertvollen Genpool von 2000 Sorten tschechoslowakischer Herkunft dar. Die tschechoslowakischen Originalsorten sind durch Linien oder Klone von Landsorten (Weizen, Gerste, Hafer, Leguminosen, Lein, Hopfen, selbstbefruchtende Futterpflanzen und Gemüse) oder Populationen (Mais, Ölfrüchte, Futterpflanzen, Gemüse und Zierpflanzen) vertreten. Den größten Teil der Kollektion bilden tschechoslowakische Sorten, die durch Kombinationszüchtung entstanden sind.

Die alten Sorten von Weinrebe, Obst und Zierpflanzen werden seit Beginn, dieses Jahrhunderts in alten Anpflanzungen von wissenschaftlichen Instituten erhalten und durch diese registriert. Auch Sorten von Weizen (böhmische Rotweizen, böhmische Wechselweizen, südmährische grannenlose und begrannte Weizen, slowakische begrannte Weizen) und Gerste (Sommerbraugerste 'Hanna'; die Hanna ist ein Gebiet in Mittelmähren) stammen aus dem gleichen Zeitraum. Die ältesten einheimischen und ausländischen Sorten anderer landwirtschaftlicher Kulturpflanzen, die aus den dreißiger Jahren stammen, werden entsprechend dem Stand der Züchtung z. Z. ihres Anbaues erhalten.

## Краткое содержание

Сохранение местных сортов культурных растений в ЧССР

В настоящее время коллекция в ЧССР насчитывает 40883 сорта полевых культур, овощей, плодовых, винограда и декоративных растений. Она находится в НИ Институте растениеводства в Праге-Рузине, который является координационным центром, и в 17 специализированных институтах и на селекционных станциях, ответственных за исследования и сохранение коллекций специальных культур. Постоянные научные исследования начались в 1951 г., после объединения разрозненных коллекций различных опытных станций (всего около 6000 сортов). Эти станции начали собирать сорта в Чехословакии уже в начале нашего века. Часть тогдашних коллекций сохранилась до настоящего времени и, вместе с последовавшими сборами, представляют собой ценный генофонд в виде 2000 сортов чехословацкого происхождения. Оригинальные чехословацкие формы представлены линиями или клонами местных сортов (пшеница, ячмень, овёс, бобовые, лён, хмель, самоопыляющиеся кормовые и овощи) или популяциями (кукуруза, масличные, кормовые, овощи и декоративные). Большая часть коллекции состоит из чехословацких сортов, полученных путём комбинационной селекции. Старые сорта винограда, плодовых и декоративных растений возделываются с начала этого века на старых местах и регистрируются научными учреждениями. Сорта пшеницы (чешские красные, чешские пшеницы-двуручки, южно-моравские безостые и остистые формы, словацкие остистые сорта) и ячменя (яровой пивоваренный «Ханачкие» – название происходит от названия местности в средней Моравии) были выведены в том же периоде времени. Наиболее старые автохтонные сорта и сорта заграничного происхождения прочих сельскохозяйственных культур, возникшие в тридцатых годах, сохраняются на селекционном уровне их тогдашнего возделования.

## Literature

Bareš, I., 1974: Genové zdroje ve šlechtění rostlin. Sborník vědeckých prací z celostátní konference "Genetické zdroje ve šlechtění rostlin", Sbírka ÚGŠ – Genové zdroje **2**, 31–49.

Chmelař, F., 1920: Úkoly a organizace sortovního pokusnictví na Moravě – Zprávy Moravského výzkumného ústavu zemědělského v Brně č. 68.

Erben, B., Jirsa, S., Vilikovský, V., 1916: Novější sorty plodin hospodářských pěstované na polích zkušebních v letech 1914, 1915. – Zpráva hospodářsko botanické stanice výzkumné v Táboře č. 83.

Kolektiv, 1974: Genetické zdroje ve šlechtění rostlin – Sborník vědeckých prací z celostátní konference; Praha, Sbírka ÚGŠ – Genové zdroje **2**, Praha, 1–702.

–, 1955–80: Výsledky studia kolekcí jednotlivých druhů kulturních rostlin. – Závěrečné zprávy z průběhu řešení.

Lekeš, J., 1961: Původ československých odrůd sladovnického ječmene a jejich vliv na šlechtění předních světových sladovnických odrůd v zahraničí. – Slezský ústav ČSAV, Opava.

Schmidt, J., a Kolektiv, 1964: Povolené odrůdy zelenin, SZN, Praha, 1–175.

–, 1978: Odrůdová agrotechnika polních plodin, SZN, Praha, 1–452.

Sehnalová, J., a Kolektiv, 1976: Index seminum 1976–1978, SZN Praha, 1–221.

–, 1979: Index seminum 1979–1982, SZN Praha, 1–230.

Sobotka, M., a Kolektiv, 1957: Stručný popis československých povolených a rajonovaných odrůd zemědělských rostlin (1956–57), SZN Praha, 1–192.

Dr. I. Bareš and Dr. J. Sehnalová
Research Institute of Plant Production
Drnovská 507
CS - 161 06 Praha-Ruzyně, ČSSR

Kulturpflanze XXIX · 1981 · S. 79–80

# Study of the specific and varietal diversity of field crops in Bulgaria[1]

GEORGI STAJKOV (Plovdiv, PR Bulgaria)

## (Abstract)

1. The particular geographical situation of Bulgaria, its natural connection between Asia and Europe, the various landscapes, the diverse soil and climatical conditions have been and are favourable conditions for the development of a diverse vegetation.

2. The flora of the country consists of more than 5000 species; among them 19 species, 35 subspecies, 115 varieties and 60 forms have been newly described in recent times.

3. With a large specific and varietal diversity the group of field crops is represented by wheat, barley, oat, rye, chickpea, *Lathyrus*, lucerne, clover, rose and flax.

4. As a result of the systematic study of 22 species of field crops the following taxa have been described as new ones:

*Triticum pseudopolonicum* POP. (POPOV 1949/50); *Tr. boeoticum* BOISS. var. *bulgaricum* STRANSKI; *Tr. monococcum* L. var. *sofianum* STRANSKI (STRANSKI 1929)

*Secale anatolicum* BOISS. var. *rhodopeum* (DELIP.) GAN.; *S. montanum* GUSS. var. *balcanum* GAN. (JORDANOV 1963)

*Cicer arietinum* L. var. *afghanico-aurantiacum* KOIN., var. *turcico-aurantiacum* KOIN., var. *bohemico-atrum* KOIN. and var. *hispanico-aurantiacum* KOIN. (KOJNOV 1968)

*Lens esculenta* MOENCH var. *roseum* STAIK. and var. *rubrum* STAIK. (STAJKOV 1963)

*Trifolium incarnatum* L. ssp. *incarnatum* var. *intermedium* KOŽ. and var. *minor* KOŽ.; *Tr. pratense* L. var. *stojanovii* KOŽ. and var. *stranskii* STOJ. et STEF. f. *velenowskyana* KOŽ., *Tr. alexandrinum* L. var. *madarensae* KOŽ. (JORDANOV 1976).

5. The rich specific and varietal diversity of field crops gives the opportunity of carrying out an intensive and successful breeding work.

## Literature

JORDANOV, D., (ed.), 1963: Flora reipublicae popularis Bulgaricae, vol. 1. – Izd. B'lgarsk. Akad. Nauk., Sofija.

1 Paper not delivered during the symposium

JORDANOV, D., 1976: Flora reipublicae popularis Bulgaricae, vol. **6**. – Izd. B'lgarsk. Akad. Nauk., Sofija.
KOJNOV, G., 1968: Nachut. *Cicer arietinum* L. – Izd. B'lgarsk. Akad. Nauk., Sofija.
POPOV, A., 1949/50: Edin nov vid pšenica. – God. sel'skostopansk. akad. Sof. agr. fak. **28**.
STAJKOV, G., 1963: Dva novi varieteta lešča. – Priroda, Sofija.
STRÁNSKI, I., 1929: Za limeca v B'lgarija. – Izv. bot. družestvo **3**, p. 202–208.

Prof. Dr. G. STAJKOV,
Institute of Plant Production, Agricultural University "V. KOLAROV",
Plovdiv, Bulgaria

Kulturpflanze XXIX · 1981 · S. 81–90

# Activities regarding collection and evaluation of Polish land-races

Władysław Kulpa (Radzików, Poland) and Peter Hanelt (Gatersleben, GDR)


## Summary

Polish plant breeders had already recognized the importance of the conservation of land-races at the beginning of this century. But systematic approachments for collecting, maintenance and evaluation began only after World War II, mainly in the mountaineous districts south of Kraków.

In the meantime serious gene erosion has taken place and indigenous land-races are to be found at present only sporadically in some parts of the country. To analyse the actual situation the Radzików institute (IHAR) being responsible for coordinating plant genetic resource work in Poland has undertaken 5 collecting trips from 1976 to 1979 partly in cooperation with the Gatersleben institute. Refugial areas of land-race cultivation could be confirmed for different parts of the Beskids and other mountain chains in South Poland although the decrease of land-races has been accelerated here too considerably since the last decades. Minor areas of land-race cultivation were observed in E and SE Poland. Always only a small number of farmers still is using these indigenous races which may be superior to modern cultivars under special ecological conditions and regarding to some qualitative characters.

Altogether 445 seed samples have been collected, mainly cereals and leguminous crops, many of them being highly variable in respect to morphological characters. Land-race cultivation is often combined with the cultivation of rare relic crops on fields and especially in gardens, some formerly cultivated primitive races of several cultigens nowadays occur merely as weeds.

Roughly calculated one can assume that within the next 10–15 years also the recently still existing land-races will disappear completely from the agricultural cultivation. First observations proved the high value of the collections for breeding purposes which has been outlined in some detail for several cereals.


As in other European countries also in Poland the replacement of land-races of cultivated plants has been a gradual process. Its recent phase since the last 2 to 3 decades is characterized by a rapid extension into regions which are in agricultural respect marginal ones and by a complete disappearance of such forms even very often in these sites. The decrease of the cultivation of land-races is always correlated with and partly caused by the development of a plant breeding operating consciously with scientific methods, thus the beginning of this process can be temporally compared with the beginning of the modern breeding. In Poland this has happened more than one century ago when between 1860 and 1885 breeders and breeding firms as WŁ. PEPŁOWSKI (Sarnowo near Płock), K. BIELAWSKI (in Wysokie Litewskie near Brest, now Belorussia), A. JANASZ (in Dańków near Warsaw) and the State Agricultural Research Centre (1862 transferred from Warsaw to Puławy) produced the first Polish wheat cultivars. They represented mostly improved selections of land-races widely cultivated at that time, ancestral material having been populations of the 3 great variable groups of land-races predominating in Poland up to the sixties of the last century ('Gołka Biała' = White Awnless, 'Sandomierka' = from Sandomierz, 'Ostka Czerwona' = Red Awn) but also local wheat races from different parts of Russia. Thus the first Polish winter wheat variety 'Sarnowska' by WŁ. PEPŁOWSKI was bred from a local race of the region of Płock belonging to 'Gołka Biała'; 'Wysokolitewka' of K. BIELAWSKI (still nowadays used as cross parent because of its white kernels) was selected from 'Gołka Bialoziarnista', a race from the Eastern part of Podlasie, and 'Puławka' from a Kostroma land-race of spring wheat (!), the latter cultivar from Puławy stood later on also at the beginning of the famous series of 'Dankowskie'-varieties. Further well known Polish breeder firms of the last decades of the 19th century up to the time of World War I and afterwards have been K. BUSZCZYNSKI and ŁAZYŃSKI (originally situated in Podolia, later on established near Krakow, and at other places) who released oat and sugar beet varieties e. g., and E. KOSTECKI who created–besides well-known oat, beet and clover cultivars – 'Eka' (1922), a winter wheat variety which is still nowadays propagated and cultivated frequently by farmers in Southern Poland.

Very early Polish breeders have been recognized the great practical value of the old land-races and have pointed to their maintenance and further evaluation. Such opinions have already been expressed by A. SEMPOŁOWSKI, a leading plant breeder from the end of the last and the beginning of our century who had worked at Poznan, Warsaw and Puławy. Detailed proposals regarding such investigations for Poland were made for the first time by K. MICZYŃSKI sen. in 1907 and since 1918 L. KAZNOWSKI began in Puławy to collect and evaluate local races of peas, broad beans, clovers and oats (KAZNOWSKI 1923). Later on these activities have been concentrated mainly to the mountaineous regions and their forelands in Southern Poland, where combined efforts have been made to collect and to describe the land-races, to improve them and to breed varieties adapted to the special local environmental conditions using the indigenous plant material. These intentions were initiated by J. RYX who studied before 1920 cereal land-races from the Podhale region and continued by K. MICZYŃSKI jun. before World War II and the first years afterwards who collected in some regions of the mountaineous parts of the Kraków district. On a large scale this collection work has been proceeded by J. SAWICKI between 1946 and 1963 in the Western Beskids of the districts of Nowy Sącz, Limanowa and Nowy Targ. His collection which is still being evaluated and used for breeding purposes in Kraków consists of more than 500 local forms of spring wheat, spring barley, oat and rye grouped into different agroecotypes (SAWICKI 1975).

Since in 1954 during a session of the Polish Academy of Sciences the importance of such agroecotype studies has been officially stressed similar investigations have been encouraged in different centres of the country, e. g. in Lublin where KAZNOWSKI's work regarding forage legumes and grasses could be continued, in Olsztyn, Bydgoszcz, etc; nowadays all these activities are coordinated by the Institute for Plant Improvement and Acclimatisation (IHAR) in Radzików.

For a study of the actual situation this institute initiated a series of uptillnow 5 collecting trips from 1976 to 1979; the covered area is to be seen from the

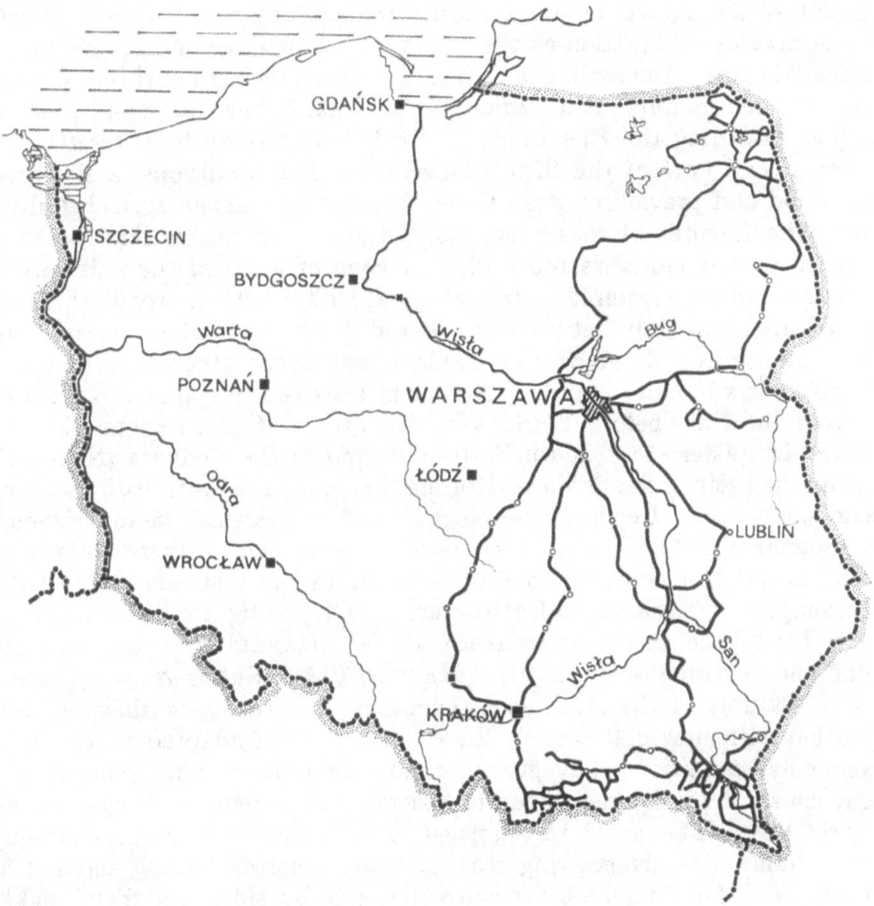


**Fig. Routes of the collecting trips initiated by IHAR between 1976 and 1979**

travel route in fig. 1; 2 of these trips (1976, 1978) have been undertaken in cooperation with the Gatersleben institute.

As a result of these activities the following general remarks can be made: During the last decades the replacement of old land-races has been rapidly increased. This can be correlated with the general raising of the level of agricultural production methods, with changes in the infrastructure of agriculture, with officially strengthened recommendations for seed exchanges favouring modern cultivars, with the development of industrial complexes and increase of tourism in regions with non-optimal agricultural conditions. The speed of this process has been already earlier indicated (HANELT 1979) and it may be once more demonstrated by the fact that in regions of the Beskids where according to SAWICKI (1958, 1973) 20 years ago still the cultivation of indigenous spring barleys or oats called 'Jarzec' and 'Górski' respectively were dominating, nowadays (1978) in the villages only few farmers cultivate these types on small fields to be discovered only after careful search. Also in regions where climatic-orographic conditions or special soil factors inhibited hitherto an intensification of agriculture the old varieties are mostly replaced by modern ones. However just these regions do function as the only refugial areas for indigenous material of cultivated plants. The main one is situated in the Southern part of the country and includes the mountaineous region of the Beskids, the Tatra and their forelands (but excluding the Bieszczady in SE-Poland having been resettled only since the second part of the fifties); severe climatic conditions, a short vegetation period and prevailing steep slopes difficult to manage agriculturally are characteristic features of these districts. Although the arable land being predominated by new cultivars too a distinct concentration of the cultivation of local races could be registered in the marginal parts of the Lower Beskids (Beskid Niski) approximately between Sanok and Nowy Sącz where in some villages for instance near Grybów and Dukla a significant percentage of the farmers still prefers to cultivate land-races or at least self-propagated old varieties often from the time before World War II. Minor refugial regions have been discovered in Eastern and South-Eastern Poland in the Podlasie (Nizina Podlaska) and the basin of Sandomierz (Kotlina Sandomierska) with badly drainaged or poor sandy soils. Geographic-ecological and sociological factors favouring the maintenance of land-races in the agricultural system of these regions (geographical isolation, unsuitable conditions for industrial methods of agricultural production, the agricultural infrastructure and a strong tradition among the farmers) have been mentioned already earlier (HANELT and HAMMER 1977, HANELT and SCHULTZE-MOTEL 1979, HANELT 1979). But it must be stressed that also features of the land-races themselves contribute to their successful competition with new cultivars in these regions: Well adapted to the special environmental conditions they guarantee also in unfavourable years a rather low but constant yield when modern varieties will experience serious or even total yield losses. This could be observed in 1978 when a late and cold spring caused an unusual short growing season: Modern spring barleys did not ripe at all but the local 'Jarzec'-forms cultivated side by side with them could be harvested normally at the beginning of September; late frosts in the spring of this year caused shooting of a considerable percentage of beet plants in stands of new cultivars while local beet races proved to be highly shooting resistant. Also in respect to characters of quality land-races (or old cultivars) are often superior to modern ones mainly selected for high yield as it is commonly known (and having been heavily stressed by the farmers) regarding baking quality of wheat varieties, the same may be due to characters of resistance against diseases or winter-hardiness.

The mentioned regions are likewise characterized by the cultivation of some relic or old traditional crops; these may be agricultural ones as two-rowed and – very seldom – six-rowed naked barley (Northern slopes of the Lower Beskids),

German sesame (*Camelina sativa*, near Dukla), oil radish (*Raphanus sativus* var. *oleiformis*, basin of Sandomierz) and millet (*Panicum miliaceum*), land-race populations very polymorphous regarding to characters of the panicle and the grain colour have been collected in the Sandomierz district too (but this species is still sporadically distributed easternwards in the country). Sometimes for these crops a probabely more recent shift of the use was observed, e. g. grains of naked barley are nowadays no more used for human consumption but mostly for feeding horses and the oil of German sesame is needed only for veterinary purposes! Also the backyard gardens in these districts are distinguished by traditional cultivation of relic garden crops e. g. of some medicinal plants as *Malva crispa*, *Artemisia abrotanum* or different hybridogenic taxa of *Mentha* in which mostly *M. arvensis* is involved as one parent; likewise old varieties of peas, *Phaseolus* beans and large-grained *faba* beans are often to be met with.

In some cases within the described regions primitive races of cultivated species apparently cultivated here earlier have been descended to a weedy stage, for instance *Vicia sativa* s. str. and field peas (*Pisum sativum* convar. *speciosum*) with unusual small flowers, pods and grains are presently regular components of the weed flora in spring cereal fields and have been used by plant sociologists for the characterization of segetal communities typical for parts of the Beskid mountains. Likewise small-flowered variants of poppy (*Papaver somniferum*) with shattering capsules observed in beet fields at different localities can be regarded merely as a weed, the capsules of which are sometimes collected. But otherwise also the weed flora is exploited as a genetic resource: In 2 villages farmers had selected cultivated forage strains of hairy vetch (*Vicia dasycarpa*) from local weedy types of this species!

Returning to the recent phase of land-race cultivation in Poland it must be said once more that its descrease has been proceeded rapidly comparing the time of SAWICKI's collecting work 20 years ago, is still proceeding considerably even within few years and having in mind planned economic changes and structural alterations within the agricultural system to be foreseen one may calculate that within the next 10–15 years also the nowadays still existing cultivation of indigenous land-races of field crops will completely disappear and remains of it may persist only for some years on abandoned fields, field margins etc.

Thus these very years offer the last opportunity to collect and maintain this material and discussions are under way to continue the collecting work, to extend it to not yet explored districts of the country, e. g. to the region East and Southeast of Lublin (Wyżyna Lubelska), to the Holy Cross Mountains (Góry Świętokryskie) in Central Poland, to the Westernmost part of the Beskids and to intensify exploration work in the already known refugial centres. The collection of the 5 mentioned travels consists of 445 samples, more the half of it represented by cereals. For details of the material compare table 1. Samples especially of wheat, oat, millet, peas, broad beans and *Phaseolus* beans contain up to ten morphologically different types so that the number of available strains after separation of these samples is much higher and the collection thus will probabely be enlarged up to more than 1000 strains to be maintained in Radzików and partly in Gatersleben. A likewise important genetic resource is represented by

Table 1

Collection of land-races from Poland (1976–1979)

| species | no. of samples | |
|---|---|---|
| *Triticum aestivum* | 65 | |
| *Secale cereale* | 55 | |
| *Hordeum vulgare* | 54 | |
| *Avena* spp. | 53 | |
| *Panicum miliaceum* | 6 | |
| *Fagopyrum esculentum* | 5 | |
| cereals total | | 238 |
| *Pisum sativum* | 28 | |
| *Vicia faba* | 28 | |
| *Phaseolus* spp. | 26 | |
| small seeded legumes | 26 | |
| leguminous crops total | | 108 |
| *Papaver somniferum* | 9 | |
| *Brassica* spp. | 8 | |
| further oil, vegetable and other cultigens | 43 | |
| weedy species | 39 | |
| total | | 445 |

the collection of forage grasses of the Botanical Garden in Bydgoszcz brought together mainly by the activities of Dr. B. OSIŃSKI. It consists at present of more than 4.600 ecotypes of economic important grass species (especially *Dactylis glomerata, Festuca pratensis, Lolium perenne, Phleum pratense* and *Poa pratensis*) which have been collected mostly in North and Northwestern Poland from permanent grassland sites and which has been partly already incorporated into recent breeding work.

It is not possible here to go into details of the collections of the different crops, some will be covered more thoroughly by following papers (see HAMMER and HAMMER et al. in this volume). It should be stressed however that as well as in botanical and in economical respect the following crop collections are especially noteworthy (see also HAMMER and HANELT 1979): 1. Wheat (*Triticum aestivum*), land-races with a high intra- and interpopulational variability, often curious mixtures of indigenous material with modern cultivars; 2. Barley (*Hordeum vulgare*), almost exclusively spring forms, mostly of the two-rowed *nutans*-type, some six-rowed races and also a few already mentioned naked barleys (var. *nudum* and var. *coeleste*); 3. Oat (*Avena sativa*), land-races with a broad varietal spectrum, including some flag oats with contracted panicles, often mixtures of indigenous dark-glumed or awned types with recent cultivars, some samples of the diploid *A. nuda* (*A. strigosa*) and introgressions from *A. fatua* into *A. sativa*; 4. Rye (*Secale cereale*), variable land-race material (also spring types) adapted to short growing periods, some of them, e. g. one collected as a self-sown remnant of a former cultivation in Eastern Mazowie superior to modern standard cultivars in respect to some yield characters; the search for the perennial var. *multicaule* still collected 20 years ago (SAWICKI 1975) has been in vain;

5. Broad beans (*Vicia faba*), most remarkable samples of ssp. *faba* with medium- to large-sized grains used as vegetable, often rather heterogenous samples from gardens or different crop mixtures; 6. Peas (*Pisum sativum*), especially variable field pea samples (convar. *speciosum*, see above), vegetable and dry pea races from backyard garden however being more uniform and representing apparently old varieties from the first decades of this century. – Although being represented only by few samples the importance of the minor cereals millet and buckwheat (*Panicum miliaceum*, *Fagopyrum esculentum*) should not be underestimated; breeding work has been ceased some years ago in Poland but these crops may gain in future once more an economic value because of their nutritional and dietic value and the collected variable land-races of them may be very useful for the breeders in future.

Table 2
Breeding value of land-race material

| cropa | observed characters | Origin of samples carrying these characters |
|---|---|---|
| winter wheat | resistance against stripe rust | basin of Sącz, Bieszczad Mts. and foreland |
| | logging resistance | Bieszczad Mts. and foreland |
| | high winterhardiness | W-Pienine and Bieszczad Mts., Beskid Sądecki |
| | high TKM | Bieszczad Mts. and foreland |
| winter rye[1] | very high winterhardiness | W-Pienine Mts., Lower Beskids, basin of Sącz |
| | improved mildew resistance | widespread |
| | logging resistance | basin of Sącz |
| | high grain number/spike | Small Pienine Mts. |
| | high grain weight/spike | basins of Sącz and Sandomierz, E Mazowie |
| | high TKM | E Mazowie |
| oat | earliness | basins of Sácz and Sandomierz |
| | logging resistance | basin of Sandomierz, Gorce and W-Pienine Mts., NW Bieszczad Mts., E Mazowie |
| | resistance against mildew and crown rust | Podhale region and other mountaineous districts |
| | high TKM | W-Pienine and Bieszczad Mts. |
| | low glume percentage | Lower Beskids |
| | high protein content of the grain | Lower Beskids, W-Pienine Mts. |
| spring barley | earliness | W-Pienine Mts., Podhale region |
| | short straw | Lower Beskids, basin of Sącz |
| | resistance against stripe rust | widespread in the mountains |
| | resistance against leaf rust | Lower Beskids, Bieszczad Mts., basin of Sącz, Beskid Sądecki |
| | resistance against mildew | Podhale region |
| | resistance against logging | W-Pienine and Bieszczad Mts., Lower Beskids |
| | high TKM | W-Pienine and Bieszczad Mts. |

[1] in the mentioned characters significantly better than the standard cultivar 'Dańkowskie Złote'

The importance of land-race material for breeding needs not to be underlined generally, this concerns of course also the Polish material. It should be remembered only that already long ago Polish wheats of the old 'Sandomierka' group were successfully used also by wheat breeders outside of Poland or – more recently – that the rapeseed cultivar 'Bronowski' selected from a Polish land-race (its exact origin being unknown) got importance as cross parent for breeding eruca-acid free cultivars also abroad. SAWICKI had repeatedly stressed the economic usefulness of his land-race collections (1958, 1960, 1973, 1975), he referred to their general low demands regarding different soil factors, to earliness and winterhardiness and could detect characters as resistance against loose smut and logging within the local spring wheats, disease and logging resistance and unusual frosthardiness among his rye populations, high TKM, high grain yield per panicle, high protein and oil content of the grains and low glume percentage within the oat collection etc. and could prove the superiority of land-race selections against compared standard cultivars many times in many respects.

Similar results could be obtained for our material (see for details KULPA and GÓRSKI as well as KULPA and JASTRZĘBSKI, in press). Several selected characters of high value for breeding purposes which have been observed and confirmed in Radzików during 2 to 3 vegetation periods are listed for some cereal crops in table 2 which is demonstrating very well the importance of such land-race material for plant breeding especially as genetic source of characters of quality.

Besides the authors the following colleagues were participants of the collecting trips: from Radzików Ing. W. DOBOSZYŃSKI, Mag. Ing. M. GÓRSKI, Mag. A. JASTRZĘBSKI, Mag. H. UTRATA, from Gatersleben Dr. K. HAMMER and Dr. J. SCHULTZE-MOTEL.

## Zusammenfassung

Arbeiten zur Sammlung und Evaluierung polnischer Landsorten

Polnische Pflanzenzüchter haben bereits zu Beginn unseres Jahrhunderts die Bedeutung der Erhaltung von Landsorten erkannt, systematische Sammlungen begannen jedoch erst nach dem 2. Weltkrieg (vor allem in den gebirgigen Gebieten im Süden des Landes). Inzwischen hat die Generosion auch hier zu einem weitgehenden Verschwinden der indigenen Landsorten geführt. Vom IHAR in Radzików, das in Polen alle Aktivitäten hinsichtlich pflanzlicher genetischer Ressourcen koordiniert, sind von 1976 bis 1979 z. T. in Zusammenarbeit mit dem Institut in Gatersleben 5 Sammelreisen unternommen worden, auf denen Reliktareale des Anbaus von Landsorten in verschiedenen Teilen der Beskiden und anderer Gebirgszüge in Südpolen festgestellt wurden. Jedoch werden hier wie in weiteren kleinen Gebieten im Osten und Südosten des Landes nur noch von wenigen Bauern die alten Landsorten kultiviert, die jedoch unter bestimmten ökologischen Verhältnissen und in bezug auf manche Qualitätsmerkmale modernen Sorten überlegen sein können. Es wurden insgesamt 445 z. T. stark variable Samenproben (meist Getreide und Körnerleguminosen) gesammelt. Verbunden mit dem Landsorten-Anbau ist oft die Kultur seltener traditioneller Kulturarten, vor allem in Gärten, sowie das Vorkommen früherer Kulturformen als Unkraut.

Erste Beobachtungen erwiesen den hohen züchterischen Wert der Kollektionen polnischer Landsorten, für die ein völliges Verschwinden aus dem landwirtschaftlichen Anbau innerhalb der nächsten 10–15 Jahre angenommen werden kann.

**Краткое содержание**

Работы по сбору и оценке польских местных сортов

Ещё в начале нашего столетия польские селекционеры поняли значение сохранения старых местных сортов, однако систематические сборы их начались только после второй мировой войны (главным образом в гористых областях на юге страны). К этому времени генная эрозия привела уже и здесь к исчезновению многочисленных автохтонных местных сортов. Отделение IHAR в Раджикове, которое координирует всю деятельность по сохранению растительных ресурсов, предприняло, частью совместно с институтом в Гатерслебене, 5 экспедиций (с 1976 по 1979 гг.), которыми были установлены реликтовые ареалы возделывания местных сортов в различных участках Бескид и других гористых местностях южной Польши. Но и здесь, как и в некоторых местах на востоке и юго-востоке страны, только очень ограниченное число крестьян возделывают старые местные сорта, которые, однако, в определённых экологических условиях и при наличии некоторых качественных признаков, могут оказаться ценнее современных сортов. Всего было собрано 445 образцов семян, часто очень изменчивых, главным образом зерновых (злаков и бобовых). С возделыванием местных сортов нередко связана и культура редких традиционных культурных видов, обыкновенно в садах и огородах, как и встречаемость «бывших» культурных форм в виде сорняков.

Первые этапы обработки материала выявили высокую ценность коллекций польских местных сортов, однако, их полное исчезновение с сельскохозяйственных участков следует ожидать уже в течение следующих 10—15 лет.

## Literature

Hammer, K., and P. Hanelt, 1979: Botanische Ergebnisse einer Reise in die VR Polen 1976 zur Sammlung autochthoner Landsorten von Kulturpflanzen. – Kulturpflanze **27**, 109–149.

Hanelt, P., 1979: Collecting in Czechoslovakia and Poland. In: Zeven, A. C., and A. M. van Harten (edit.), Proc. Conf. Broadening the genetic base of crops, 59–66. – Wageningen, Pudoc.

–, und K. Hammer, 1977: Bericht über eine Reise nach der VR Polen 1976 zur Sammlung autochthoner Sippen von Kulturpflanzen. – Kulturpflanze **25**, 33–44.

–, und J. Schultze-Motel, 1979: Bericht über die Reise in die VR Polen zur Sammlung autochthoner Sippen von Kulturpflanzen im Jahre 1978. – Kulturpflanze **27**, 151–163.

Kaznowski, L., 1923: O potrzebie organizacji ogrodu botaniczno-rolniczego. – Roczniki nauk rolniczych **10** (1), 213–218.

Kulpa, W., and M. Górski,: Wyniki eksploracji miejscowych form roślin uprawnych w północno-wschodniej Polsce w latach 1977 i 1979; in press.

Kulpa, W., and A. Jastrzębski,: Miejscowe formy roślin uprawnych na Płaskowyżu kolbuszowskim i Pogórzu Karpat w latach 1976 i 1978; in press.

Sawicki, J., 1958: Studia nad miejscowymi odmianami zbóż z rejonu Karpat. Cz. I. Jęczmień jary – Hordeum vulgare L. s. I. – Acta Agrobot. **8**, 67–150.

–, 1960: Studia nad miejscowymi odmianami zbóż z rejonu Karpat. Cz. II. Pszenice jare z powiatu limanowskiego i nowotarskiego. – Acta Agrobot. **9**, 5–62.

– 1973: Studia nad agroekotypami owsa z rejonu Karpat. Cz. I. Wyniki badań fizjograficznych i wstępna ocena zebranego materiału. – Komitet Zagospodarowania Ziem Górskich **12**, 5–34.

– 1975: Znaczenie agroekotypów zbóż z rejonu Karpat dla uprawy i hodowli. – Postępy nauk rolniczych nr. **1/150** (1975), 37–70.

Dr. W. Kulpa,
Instytut Hodowli i Aklimatyzacji Roślin (IHAR),
Pl - 05-870 Radzików p. Błonie, PR Poland
Dr. P. Hanelt,
Zentralinstitut für Genetik und Kulturpflanzenforschung,
DDR - 4325 Gatersleben, GDR

Kulturpflanze XXIX · 1981 · S. 91–97

# Variability of wheat land-races from Czechoslovakia and Poland

K. HAMMER (Gatersleben, GDR), M. GÓRSKI (Radzików, PR Poland), P. HANELT (Gatersleben, GDR), F. KÜHN (Brno, ČSSR), W. KULPA (Radzików, PR Poland), and J. SCHULTZE-MOTEL (Gatersleben, GDR)

## Summary

The wheat material of collecting missions to Czechoslovakia (in 1974 and 1977) and Poland (in 1976 and 1978) has been analyzed botanically and in respect to its possible breeding value by studying the original samples as well as their progenies grown up in the following years.

The material represents populations of land-races and some old cultivars. Though there is a rapid genetic erosion in the collecting areas, the populations are very variable containing up to 6 botanical varieties. Some of the most variable land-races are described. In the populations of *Triticum aestivum* land-races altogether 11 botanical varieties could be found, all with red grains. Exceptional findings are var. *nigroferrugineum* from the mountainous parts of Eastern Slovakia in 1977, and emmer wheat, *Triticum dicoccon*, from southern parts of Slovakia in 1974. The most promising characters for breeding purposes are short straw, winterhardiness, earliness, and field resistance against some diseases.

In Gatersleben the land-races are separated into morphologically distinguishable types and maintained for their use in future breeding programmes.

Some years ago a search for land-races in some promising parts of central Europe was initiated and collecting missions to Czechoslovakia in 1974 and 1977 (KÜHN et al. 1976, 1980) and to Poland in 1976 and 1978 (HAMMER and HANELT 1979, HANELT and SCHULTZE-MOTEL 1979) turned out to be very successful especially with regard to cereals. Some results from these missions, concerning the important crop wheat, will be discussed in this paper. The collecting missions went to mountainous regions of Moravia and Slovakia in Czechoslovakia and to the south-eastern mountainous parts of Poland, because these regions provide good refuges for land-races. But also in these areas land-races are increasingly displaced by newer cultivars and often they are found as more or less frequent admixtures in fields of newer varieties only.

Sometimes it was difficult to judge whether there was an autochthonous land-race or an old cultivar. Consequently, old cultivars have been included into our collecting project. In the gene-banks there is often a lack of such old cultivars, too.

The composition of morphological types in some extremely variable fields will be described briefly. During our mission to Czechoslovakia in 1974 we found in one field of the village of Málinec-Hámor (450 m above sea level) in Slovakia near the border to Hungary a mixture of wheat and rye. *Agrostemma githago* and *Bromus secalinus* were typical weeds as indicators for land-races, here. The wheat population contained var. *lutescens*, var. *erythrospermum*, var. *milturum*, var. *ferrugineum*, and var. *suberythrospermum*. (In this and the following examples the morphological races are given in the order of their frequencies). In two other fields of that village also var. *subferrugineum* could be detected. Thus, in the village the whole of the botanical varieties collected during that trip could be found and most of them in a single population.

In Eastern Slovakia in 1977 near the village Starina in the Beskid Mts., in a region with relics of shifting cultivation detected only recently (KÜHN and HAMMER 1979), in one field the following population was present: var. *lutescens*, var. *erythrospermum*, var. *milturum*, var. *ferrugineum*, var. *subferrugineum*, and var. *nigroferrugineum*. Only forms with medium dense spikes and square head types, found otherwhere during that trip, did not grow in this field. Forms with medium dense spikes occurred rather seldom in Czechoslovakia whereas in Poland they were present to a more or less extent in almost every field of land-races.

Extremely variable was a land-race collected on our expedition to Poland in 1976 in the village Czarny Potok (basin of Nowy Sącz at the northern slope of the Beskid Mts.). The owner of the field apparently very much liked a mixture of different types and thus favoured the variable population containing var. *lutescens*, var. *ferrugineum*, var. *erythrospermum*, var. *milturum*, var. *ferrugineum-compactoides*, and var. *lutescens-compactoides* during the whole of his life. In all of the botanical varieties there were types with relatively short and long spikes vs. straw, different tinges of colours etc. In the same basin a land-race was cultivated called 'Bawarka', possibly stemming from Bavaria, containing the botanical varieties var. *ferrugineum*, var. *erythrospermum*, and var. *lutescens*. There was a difference from field to field in the frequencies of their components and therefore in the colour of the stands. Another old land-race, 'Gołka Biała', from the village of Ropa in the forelands of the Beskid Mts., on the other hand, showed a great uniformity (only var. *suberythrospermum*).

A land-race collected in Poland 1978 from the village of Kopcie in the basin of Sandomierz, cultivated in mixture with rye, contained var. *ferrugineum-compactoides*, var. *lutescens-compactoides*, var. *erythrospermum-compactoides*, and var. *milturum*, characterizing the frequency or sometimes, as in this example, predominance of types, with spikes of medium density.

Usually, in our area of collecting the fields in the plains contained beardless, white-glumed cultivars of the var. *lutescens* whereas farther to the mountains the number of bearded and/or red-glumed types increased. On our expeditions 11 botanical varieties have been collected (table 1). All of them have red grains. Land-races with white grains, formerly cultivated to a little extent in southern Poland (BARBACKI et al. 1937, GIŻBERTÓWNA 1938, SAWICKI 1960) and central Moravia, could be found no more. From table 1 the frequencies for the different morphological types can be estimated only to a limited extent because the rare types have been collected preferently. Generally, white-glumed wheats are more frequent than the red-glumed ones. Sometimes land-races with compact spikes could be found and very seldom, only in the surroundings of the little East-

Table 1

Components of wheat land-races collected in Czechoslovakia and Poland

| | Czechoslovakia 1974 | Czechoslovakia 1977 | Poland 1976 | Poland 1978 | sum |
|---|---|---|---|---|---|
| var. *ferrugineum* | 8 | 5 | 24 | 15 | 52 |
| var. *ferrugineum-compactoides* | – | – | 13 | 3 | 16 |
| var. *erythrospermum* | 20 | 18 | 21 | 17 | 76 |
| var. *erythrospermum-compactoides* | – | 1 | 3 | 2 | 6 |
| var. *nigroferrugineum* | – | 5 | – | – | 5 |
| var. *subferrugineum* | 2 | – | 1 | 1 | 4 |
| var. *suberythrospermum* | 9 | – | – | 3 | 12 |
| var. *milturum* | 15 | 16 | 11 | 6 | 48 |
| (var. *milturum-compactoides*) | – | – | – | 1 | 1 |
| var. *lutescens* | 32 | 25 | 17 | 8 | 82 |
| var. *lutescens-compactoides* | – | 3 | 5 | 4 | 12 |
| sum | 86 | 73 | 95 | 60 | 314 |
| number of samples collected | 27 | 20 | 28 | 31 | 106 |

Slovakian town of Snina, we met the var. *nigroferrugineum* with black tinged glumes as an exceptional finding amongst the white- and red-glumed wheats. (In 1976 Kühn has found var. *melanopogon* (Chiov.) Mansf. in western Moravia).

Within *T. aestivum* only 11 recent botanical varieties could be found as compared with more than 400 formally described varieties listed in Mansfeld (1951). It is quite sure that formerly there were more different races. As already mentioned in Poland and Moravia wheat land-races with white grains have been grown and Kühn (1974) has found races with hairy glumes in Czechoslovakia some years ago. Thus the reduction of morphological types in wheat land-races can be demonstrated. But, surprisingly, in the existing pure stands of land-races or even in intermixed few plants of these land-races in modern cultivars there can be observed a marked variability until now and some examples are demonstated with the following character states. As colours in races with white glumes could be observed yellowish-white, pale yellow, and yellow and for the red-glumed races from brownish-yellow to dark brown. The density of the spikes also shows a great variability, e. g. in the forms with lax spikes the density reached from medium to very lax with very long (19 cm) *spelta*-like spikes. Great variability could be observed also in other morphological characters such as plant height, length of awns, number of tillers per plant, and number of grains per spike.

The collected land-races have been segregated to their morphological constituents in Gatersleben prior to propagation. This is a precaution for the maintaining of rare morphological types which may get lost because of occasional incautious handling and selection pressures at the fields in Gatersleben differing from those at the collection sites, promoting some types whereas others are suppressed and may get lost during propagation. Therefore, every morphological form will be propagated as a single unit. In table 1 these units have been computed for the separated spike samples and for the grain samples separated into different types when growing in the Gatersleben fields for the first time. The numbers are summed up and there are altogether 314 types, the sum of originally collected samples beeing 106. This is about a threefold higher number of separated types than the number of original samples. This is only the first step to separate the material into segregates. New morphological types appearing in the following two generations are selected as well and propagated separately. Thus, the amount of material may increase considerably. The question arises, which are the causes for these segregations in a strongly autogamous crop as wheat? Possible answers are: Some glume colorations are not to be seen in every year. During the trip to Czechoslovakia in 1977 e. g. there were rather cool and wet weather conditions. Often the glumes have been darkened by fungi and even the var. *nigroferrugineum* collected at that time could be detected only when growing in the fields at Gatersleben. Moreover, this character depends on the intensity of sun radiation very much and the colour may reach from light brown in some years to a distinct black in others. On the other hand there may be a remarkable degree of outcrossings in years with conditions favouring cross pollinations. The mixtures of different wheat types usually grown in the fields may, therefore, facilitate outcrossings considerably and consequently segregations occur in autogamous wheat. Foreign pollinations are apparently responsible for new character combinations usually not met with in cultivars. During the trip through Poland in 1978 we found a red-glumed beardless wheat with spikes of medium density (basin of Sandomierz – village of Kopcie). This combination is not mentioned in the already cited paper of MANSFELD (1951) but using those principles it would be called var. *milturum-compactoides*. The glumes are only brownish yellow and this may suggest an introgression of red-glumed types into the frequently met with var. *lutescens-compactoides*. As already pointed out, hairy types could not be found during our expeditions. But a form from Slovakia 1974 (village of Dolná Strehová-Prieloh) showed somewhat pubescent glumes possibly indicating an earlier introgression from types with hairy glumes. Another example for introgressions may be types with somewhat inflated glumes (Czechoslovakia 1974 – collecting numbers 50, 147a; Poland 1976 – collecting number 106a). Thus, in the Polish and Czechoslovakian fields of land-races or even in stands of modern cultivars containing portions of land-races, hybridization appears to be an acting evolutionary factor.

So far only winter wheats have been mentioned. By far, they constitute the majority of the collected material. Only 13 of the 106 collected samples were spring forms. The predominance of winter forms may derive from the low elevations of the fields in the collection area (not more than about 700 m above sea level) with relatively mild winters. Spring forms show a less variability as compared to winter wheats and are characterized by long spikes, generally. In the village of Hutisko Solanec in the Moravian – Silesian – Beskids of Czechoslovakia (1974) we found a special spring wheat with short straw (about 1 m plant height) which was sown in mixture with naked barley (tends to lodging considerably) to improve its lodging resistance.

A rather spectacular result of our collecting missions is the evidence of the recently cultivated emmer wheat, *Triticum dicoccon* SCHRANK, in Slovakia in 1974 (villages of Sobotiště and Tisovské Lazy) with white and red glumes. Five samples have been gathered, but only three of them have been collected from more or less pure stands in little fields of no more than 1/2 ha. Two samples have been drawn from oat fields where the emmer wheat was only an occasional admixture. Its grain is used for a special type of sausages and for porridges or soups. Emmer wheat is said to have the best quality for these purposes and this fact may be the reason of its relatively long lasting culture as a relic crop in Slovakia.

The collected wheat land-races from Czechoslovakia and Poland consist of a great amount of morphological forms and some of the morphological traits, e. g. short straw, may be of interest for plant breeders. But furthermore there will be physiological and other characters also usable in plant breeding. First results of our investigations in this direction are shown in table 2. In the land-

Table 2

Material from wheat land-races with possible value for breeding

| characters observed | origin of promising samples |
|---|---|
| logging resistance mainly conditioned by short straw | Eastern Beskids, Vsetinské Vrchy Mts., Bieszczad Mts. and foreland |
| weight of 1000 grains high | Eastern Beskids and foreland, mountainous parts of Moravia, Bieszczad Mts. and foreland |
| many grains per spike | Slovakian Ore Mts., basin of Nowy Sącz |
| good winterhardiness | widespread |
| earliness | Vtáčnik Mts., Slovakian Ore Mts. |
| field resistance against | |
| yellow-rust | Eastern Beskids, basin of Nowy Sącz, Bieszczad Mts. and foreland |
| brown-rust | Vsetinské Vrchy Mts., foreland of Bieszczad Mts. |
| mildew | Eastern Beskids |

races no vertical resistance has been found but some of them show a good field resistance. Special interest deserve also quality characters as protein-content and -quality. Köck (1973) and Schachl (1973) tested Austrian land-races concerning these characters and found them often to be superior to modern cultivars. As has been pointed out, peasants sowing land-races often adduce this fact to the quality of their wheats. Further arguments are that the land-races are better adapted to the conditions of the growing place and a long tradition for special utilizations (e. g. *T. dicoccon* and special forms of spring wheats). For these reasons land-races will be grown yet for some time. But the progress in the distribution of new cultivars will at least diminish their variability and finally eradicate them in their last mountainous refuges, too. Our collections have saved some Czechoslovakian and Polish land-races for a future use in plant breeding.

**Zusammenfassung**

Die Variabilität von Weizenlandsorten aus der ČSSR und Polen

Das Weizenmaterial der Sammelreisen in die ČSSR (1974 und 1977) und nach Polen (1976 und 1978) wurde botanisch und im Hinblick auf seinen möglichen züchterischen Wert anhand der Untersuchungen der Originalproben sowie ihres Nachbaus analysiert.

Das Material besteht aus Landsortenpopulationen und einigen alten Sorten. Obwohl eine rasche Generosion in den Sammelgebieten zu verzeichnen ist, sind die Populationen sehr variabel und enthalten bis zu 6 Varietäten. Einige der variabelsten Landsorten werden beschrieben. In den Populationen von *Triticum aestivum*-Landsorten konnten insgesamt 11 Varietäten gefunden werden; alle mit braunem Korn. Bemerkenswerte Funde sind var. *nigroferrugineum* aus den gebirgigen Regionen der östlichen Slowakei 1977 und Emmer, *Triticum dicoccon*, aus südlichen Gebieten der Slowakei 1974.

Für züchterische Zwecke besonders aussichtsreiche Merkmale sind kurzer Halm, Winterfestigkeit, Frühreife und Feldresistenz gegenüber einigen Krankheiten.

In Gatersleben werden die Landsorten in morphologische Typen aufgetrennt und für künftige Züchtungsaufgaben erhalten.

**Краткое содержание**

Изменчивость местных сортов пшеницы из Чехословакии и Польши

Образцы пшенцы, собранные во время экспедиции в Чехословакию (1974 и 1977 гг.) и Польшу (1976 и 1978 гг.) анализировались как ботанически, так и с точки зрения их ценности для селекционных целей, причём во внимание принимались как оригинальные образцы, так и растения, выращенные из собранных семян.

Анализированный материал состоит из популяций местных сортов и некоторых сортов старой селекции. Хотя в местах сборов и наблюдается интенсивная генная эрозия, популяции оказались очень изменчивыми и имеют в своём составе до 6 разновидностей. Описываются некоторые из наиболее изменчивых местных сортов. В популяциях местных сортов *Triticum aestivum* были найдены, в общей сложности, 11 разновидностей; все они имели коричневые зёрна. Стоит обратить внимание на var. *nigroferrugineum* из горных областей восточной Словакии в 1977 г. и эммер, *Triticum dicoccon*, из южных частей Словакии в 1974 г.

Для селекционных целей особенно перспективными признаками являются: короткий стебель, зимостойкость, раннезрелость и полевая резистентность к некоторым болезням.

В Гатерслебене предполагается расчленить местные сорта на отдельные морфологические типы и сохранить таковые для будущих селекционных заданий.

## Literature

BARBACKI, ST., ST. LEWICKI, K. MICZYŃSKI sen. i A. SŁABOŃSKI, 1937: Pszenice Polskie. – Biblioteka Puławska Nr. 15, 163pp.
GIŻBERTÓWNA, W., 1938: Polskie odmiany pszenicy. – Kraków, 172pp.
HAMMER, K. und P. HANELT, 1979: Botanische Ergebnisse einer Reise in die VR Polen 1976 zur Sammlung autochthoner Landsorten von Kulturpflanzen. – Kulturpflanze **27**, 109–149.
HANELT, P. und J. SCHULTZE-MOTEL, 1979: Bericht über die Reise in die VR Polen zur Sammlung autochthoner Sippen von Kulturpflanzen im Jahre 1978. – Kulturpflanze **27**, 151–163.
KÖCK, L., 1973: Die Landforschung im alpinen Raum. – Ber. Arbeitstag. 1973 Saatzuchtleiter Gumpenstein, 143–157.
KÜHN, F., 1974: Genové zdroje mistnich plodin v Čechoslovensku. – In: Genetické zdroje v šlechtěni rostlin, 685–694. – Praha.
– und K. HAMMER, 1979: Das Ausklingen der Brandrodungskultur in Zentraleuropa. – Kulturpflanze **27**, 165–173.
–, K. HAMMER und P. HANELT, 1976: Botanische Ergebnisse einer Reise in die ČSSR 1974 zur Sammlung autochthoner Landsorten von Kulturpflanzen. – Kulturpflanze **24**, 283–347.
–, – und –, 1980: Botanische Ergebnisse einer Reise in die ČSSR 1977 zur Sammlung autochthoner Landsorten von Kulturpflanzen. – Kulturpflanze **28**, 183–226.
MANSFELD, R., 1951: Das morphologische System des Saatweizens, *Triticum aestivum* L. s. l. – Züchter **21**, 41–60.
SAWICKI, J., 1960: Studia nad miejscowymi odmianami zbóż z rejonu Karpat. Cz. II. Pszenice jare z powiatu limanowskiego i nowotarskiego. – Acta Agrobot. **9**, 5–62.
SCHACHL, R., 1973: Sammlung und Erhaltung alter Sorten und Herkünfte. – Ber. Arbeitstag. 1973 Saatzuchtleiter Gumpenstein, 127–142.

Dr. K. HAMMER, Zentralinstitut für Genetik
und Kulturpflanzenforschung
DDR - 4325 Gatersleben
Corrensstraße 3

Kulturpflanze XXIX ·1981 ·S. 99–110

# Cereal land-races from Austria and their utilization in plant breeding

RUDOLF SCHACHL (Linz, Austria)

## 1. Summary

Land-races are somewhat younger than usually mentioned. Also land-races are not abandoned populations but influenced by a simple continuous, equalizing human selection, as well as by natural evolution. Thus the phenotypical uniformity will be high, while the genetical variability is larger than in commercial bred varieties, but much less than in polulations grown in a gene centre. Many effects like resistance, adaptation and continuous annual yield are resulting from the land-race's multiline character, which can be seen as biological model for surviving of agricultural populations. The breeder's effort to isolate certain genes will be troublesome and often less effective because of this multiline symbiosis and genes approaching demand and botanical features of a gone period.

### *2.1. Introduction*

Research on land-races in Austria has a very old tradition. In the museum in Linz we can find one of the oldest herbariums in the world, which dates from the year 1599. We have very exact informations, also about agricultural plants, made by SAILER in the year 1841; BRITTINGER and RITZBERGER during the second half of last century, and finally in the beginning of our century by WERNECK, BURGGASSER and MAYR. While MAYR studied in particular the Alpine-region, collecting and maintaining all the land-races there, WERNECK and BURGGASSER concentrated on the Subalpine region. Of special interest are the studies of BURGGASSER, who gives us exact descriptions of the land-race's features during 1920 and than, in the same area for 1950.

Unfortunately WERNECK's suggestion to collect and maintain the land-races in the Subalpine region for further breeding-efforts was not realized further more, which he than collected has now lost its viability. Nevertheless his book "Die naturgesetzlichen Grundlagen in Oberösterreich" (the naturalhistorical base in Upper-Austria) is of paramount importance for us, because it was written during a periode when breeding commercial varieties was in its infancy. He fixed the exact date, when the wheat-land-races, and we must expect all other land-races too, were substituted by commercial varieties. WERNECK mentioned, "that in the years 1927 and 1928 the land-race of the Sipbachzeller-wheat has been extinguished totally by heavy hail-storms".

*2.2. Collection and studies of our material*

When we started in our institute, the Landw. chem. Bundesversuchsanstalt Linz to collect land-races, following Werneck's appeal in 1922, the period of the widespread use of the land-races had already finished. Nevertheless, in the last 12 years we have been able to get a good impression of the features of the old land-races and we could reconstruct their populations, with certain reservations.

Our studies have been made with a knowledge of modern genetics and evolution; so our results show some contradictions to Werneck (1935) or Mayr (1928), but also explain some phenomena, observed by these authors. Concentrating on a few small and isolated ecological niches, where land-races grew, has resulted in a clear summary.

Is it legitimate to ask if the materials we studied were land-races in the original sense and if the original genetical variability has still been preserved? To do this it is necessary to explain how Austrian plant breeding began, with the work of Pammer and Ranninger (1928). Both rejected the idea of a pure line variety, but proposed a mixture of phenotypically uniform lines. In the special case of the winter-wheat variety 'Ritzlhofer', selected from the Sipbachzeller-land-race, the breeding activities were started three times, with totaly new material, collected at each time around the village Sipbachzell (the centre of spreading for this land-race). By this method, as well as by the incomplete exchange of seed, the different strains of land-races could survive, and the variability of the Sipbachzeller land-race and other land-races could be preserved, that this variability, mentioned by Werneck (1931), can be still found. Unfortunately I have to report, that the rest of these land-races disappeared almost completely during the last five years.

The influence of breeding activities have been much less effective on rye, barley and oats than in wheat. On the other hand, especially after the second world war, the competition from commercial oat and barley varieties increased rapidly and they replaced the old land-races. In the case of oat, most land-races were lost by a reduction in the acreage sown. Amazingly one oat land-race survives – it is the landrace of the 'Kaltenberger'–oat, which still survives in the area, where it was originally grown, together with another untouched rye-land-race ('Kaltenberger' rye). The last winter wheat land-race, the 'Alpine–beard-wheat' (Alpiner Bart-Weizen) which was never really influenced by breeding, can still be found on the Haunsberg-mountain, near Salzburg, though in the last two years its area of survival is being drastically reduced, especially along the Upper Austrian-Salzburgian lake area. The last niches of the Sipbachzeller land-race were lost in the rainy summers of the years 1975 and 1976 when lodging induced a bad harvest.

## 3. The spreading of land-races

### *3.1 The theory of varietal replacement*

The survey of wheat varieties in the Upper-Austrian-Salzburgian Subalpine region has shown, that in the most intensively cultivated central area, between Linz-Wels-Enns, where also the previously mentioned village Sipbachzell is located, the most modern commercial varieties (e. g. 'Önus', 'Lentia', 'Linzerbraun' and 'Danubius') can be found. In this area also the exchange of seed has been most intensive. Around this central area in a concentric ring the older and oldest varieties have been grown. They have been 'Christian', 'Hubertus' and finally 'Tassilo'. In the next step the land varieties of 'Ritzlhofer', 'Achleitner' and 'Otterbacher', selected from the naturally grown land-race of the Sipbachzeller wheat, have been cultivated, followed by the really unbred and untouched land-race of the Alpine beard-wheat, like the Haunsberg-wheat.

We can also try to reconstruct the situation on land-races of barley, during the 1920's and we will get a similar story to that of wheat (SCHACHL 1975).

Summarizing, these observation we must say that with the introduction of a new variety, the old ones are replaced from the intensive to the extensive agricultural area and from the centre to the perimeter, as in a ring formation, like the waves in a pool. The changing over from one varietal ring to the next is not abrupt but marked by a mixing belt, consisting of two, three or sometimes more varieties, mixed together in one and the same field, though one variety may be dominant. The varietal rings are more or less conform with steptracts of land and specialized natural selection is found there. On the periphery the rings become smaller, while the varietal mixture and the gene introgression increases, until the varieties in the surf-zone of the mountains are shored together like breakes filling geographical niches or being eliminated under the increasing selection pressure of altitude.

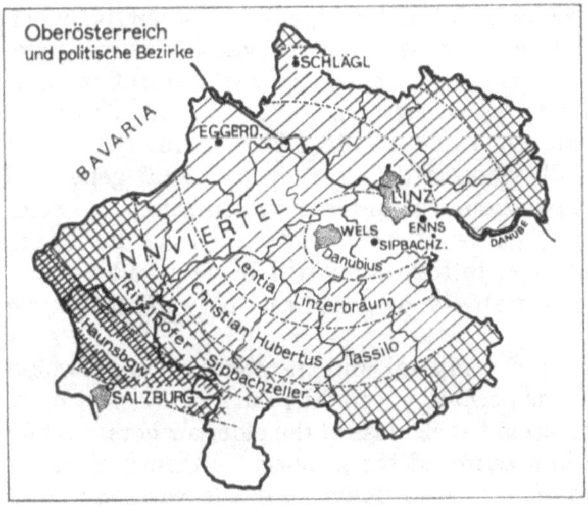


Fig. 1 Theoretical ring formed replacement of wheat varieties in Upper Austria

### 3.2. *Valley isolation*

The rings of varietal replacement may be broken, when isolation in valleys occurs. The natural selection will became stronger in these circumstances, the population more uniform, and new evolutionary factors can be expected. The population will be cut off from the common varietal gene pool and causual factors, like gene-drift may transform the individual varieties. This may explain why we can find in the Alpine valleys different land-races distinguished by some special features, but having resulted finally from one and the same land-race group.

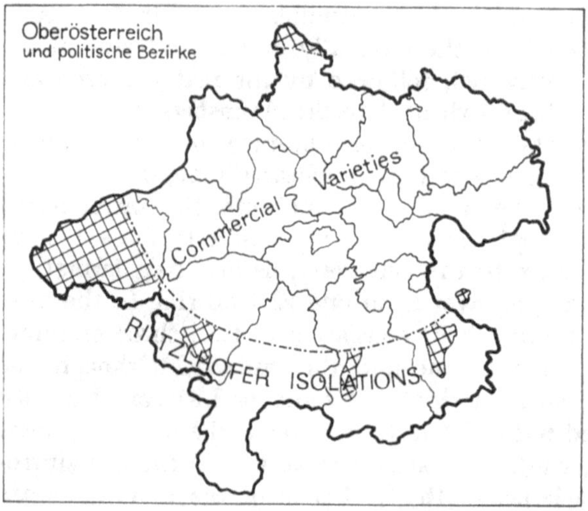


Fig. 2 Isolations of 'Ritzlhofer'-wheat: The ring is broken by valley isolation and the population cut off from the common gene pool

### *3.3. Genetical results from varietal replacement*

The theory of the varietal replacement is interesting from two points of view: First, it explains where and why we can still find land races, and secondly, which and how many land-races can be expected, as well as their variability and genetical structure. The result of the continuous ring formed replacement is, that we can expect within one and the same varietal ring – even when this ring is broken by valley isolation – one and the same land-race is surviving.

For collecting activities especially those which survey the total gene pool for future breeding efforts, the population of land-races still surviving in ecological isolation in the intensive areas will be of greater interest. Here due to the nonreplacement the mixture and gene introgression will be less and because of reduced natural selection, the variability of genes can be expected to be larger.

The varietal replacement seen as a temporal and historical process, would indicate, that varieties or races on the periphery would be the most ancient ones.

In this case we will get a contradiction to the ideas of the older authors, namely WERNECK (1935), who gives the impression of the gene-centre-theory of variation "cycles" in VAVILOV's sense. There is no doubt that the variability has been overestimated and also environmental influences have been mistaken for genetical ones. Nevertheless the studies of WERNECK (1935) and BURGGASSER (1956) in particular would give room for the interpretation, that these land-races, which have been found on the periphery, of the agricultural used area, and have been mentioned as original grown there, are infact replaced groups of land-races.

Agricultural used crops of larger economical interest will show this theoretical ring formation in any way. Horticultural crops, like poppy used as vegetable in the East part of Austria, between the Danube and the Chechoslovakian border – will survive in a summary of disconnected ecological niches around the farm houses.

In Bangladesh for instance spread of jute seems to be influenced by these two phenomenons resulting from its divergent use: In the ecological uniform central area, the Gangetic flood plains, few, relatively uniform jute-varieties are grown for fibre. Here we can expect a similar ringformation, than shown above. On the other hand in the South-East, the peninsula of Teknaf, where we have probable a gene-centre for jute, jute is grown as vegetable with this typical aspects of hortical cultivation in a large number of more or less isolated housegardens.

## 4. Evolution and genetics

### *4.1. The model of lines*

For cross-pollinated populations the rule of HARDY-WEINBERG is applicable, if the population is big enough and kept free from any influences.

On the other hand self-pollinated land-races consist of various numbers of genetically different lines which are relatively uniform due to human and natural selection. The phenotype of a population is the product of genetical and environmental influences and can be graphically expressed by a group of distribution-curves. Each of these curves represents a number of individuals and a certain phenotype.

### *4.1.1 The equilibrium of lines*

Following the HARDY-WEINBERG rule the lines must be in an equilibrium too, if there is no disturbing factors. This equilibrium can be expressed by

$$P^2+Q^2+R^2+\ldots\ldots N^2=1$$

where $P^2$, $Q^2$, e.t.c. is the multiplication rate of lines. In each selfpollinated population some spontaneous crossing may also occur. But the resultant hybrids will be of little influence, due to retrogression towards the parental forms. Nevertheless, the hybrids will be the basis of gene-introgression and the origin of new lines if they have any selection advantage.

Each line must have its own selection-factor. It provides the increasing of one line's individuals beyond the equilibrium; in other words, in any population still in equilibrium, these special selection factors seems to be the same.

$$s_pP^2+s_2Q^2\ldots\ldots s_nN^2=1$$

### *4.2. Disturbing the equilibrium*

### *4.2.1 Natural selection*

The equilibrium of lines will be disturbed, when the line-specific selection-factors become effective again.

$$s_pP^2+S_2Q^2+s_rR^2\ldots s_nN^2\neq1$$

It is possible, that a new equilibrium will finally occur and that after this or another line may disappear. Two natural factors, disturbing the line's equilibrium are of special interest, studying the land-races:

a) Varietal replacement: Each changing of the environment will activate line-specific selection factors and land-races will change commensurative with environmental change. In the case of replacement from an intensive into an extensive area, line after line will be extinguished by progressive stronger selection. This explains the phenomenon, already observed by Mayr (1928), that the land-races become more uniform with altitude.

b) Varietal mixture: The mixture of lines within one and the same variety must be seen as stabilizing and the equilibrium continuing factor, due to the supply of genes from the common gene-pool. On the other hand, intervarietal mixtures which can be seen in the mixing belts will disturbe the equilibrium by creating new competition between the lines of these diverse varieties. But

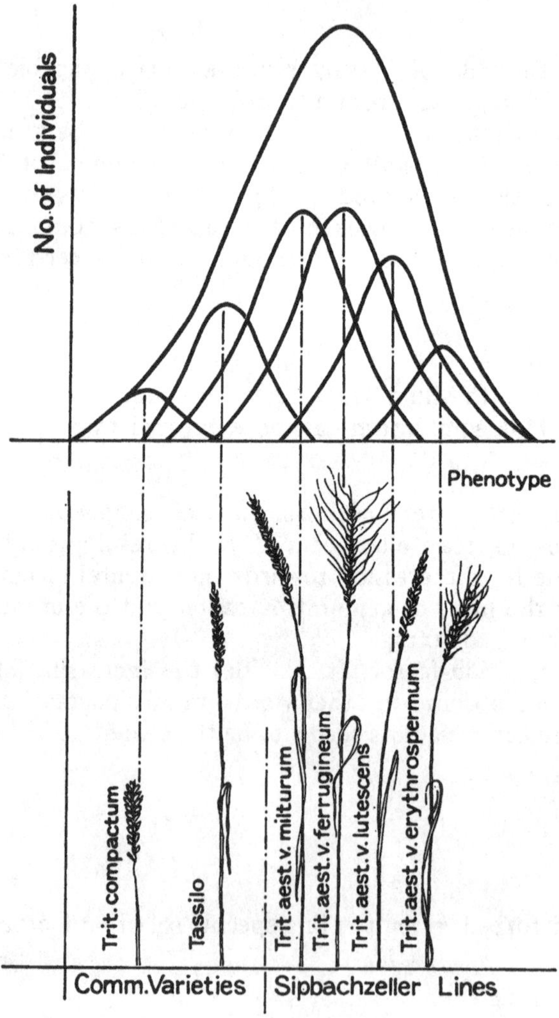


Fig. 3 The model of lines: introgressed lines of *T. compactum* und 'Tassilo' as an integral part of a 'Sipbachzeller' population

mixture of varieties is a very common practice during seed exchange, though it is difficult to say, if it is the farmer's carelesness, or may be intention. However, when seed is renewed especially in the extensive areas, usually 2/3 of the seed will be new and the rest mixed with the previous years' seed.

By this means not only whole lines, but also genes of an other variety can introgress into land-races. Also by this means hybrids can become individual lines. For instance during the 1920's, the land-races of two rows barley are reported as A-type: nevertheless in the two-row land-barley of today, the C-type is prevalent. That may be easily unterstood, considering, that the C-type may carry a vernalization-gene for frost resistance which will bring an advantage for the land-race replaced in hilly areas, where late frosts regularly occur.

Introgressed genes will become an integral part of the land-races, as will also introgressed lines. An introgressed *compactum*-type, as well as some 'Tassilo' – lines become infact elements of today's replaced 'Sipbachzeller'-populations. These introgression lines seem to be in an equilibrium with the rest of the original 'Sipbachzeller' lines. We can find a similar case in many 'Haunsberg' populations, mixed with some unindentified brownish awned and awnless lines.

Some types of *Avena strigosa*, growing in the whole Kaltenberger-oat-population are most probably such introgression lines. It's not originally a weed, as usually assumed, but about the year 1870 *Avena strigosa* was introduced from Northern Germany to the rough areas, north of the Danube, where it was than extensively cultivated.

### *4.2.2 Human selection*

Attention should be paid also to the effect of breeding activities on land-races. This is commonly mass-selection of easily distinguishable phenotypical features but the continuous, equalizing selection has influenced the characteristic of the land-races to a high degree. In connection with the heavy hail storms, extinguishing the land-races, WERNECK (1935) continues with the important observation, that the farmers didn't pay attention, to reconstruct their land-races, as they did previously, but they substituted them by commercial varieties.

By such and similar breeding activities the farmers altered their old land-races obviously unknowingly or may be knowingly. Thus the long awnless, pyramidal, red-brownish ear of the 'Sipbachzeller'-wheat and its red grain, which was assumed of value for high baking quality, became selected. Wether this process was based on conscious selection or is just a mode-trend, is not known. However, white eared lines have by a similar process been eliminated. Another example is the total loss of non-lodging oat types, because of the demand for fine-strawed plants, suitable for feeding animals, which all lodge.

### *4.2.3 Trade and multiplication*

An important fact is that not all farmers selected their own cultivated land-race. Still today we can find the traditional seed growers, breeding land-races once – surely today mostly commercial varieties – by this simple mass-selection and multiplying it beyond their own needs. By this means specific locations become famous for certain land-races and became the supplier for the rest of the nearbye region. This occured for oat and rye in the districts north-west of the Danube, with the monastery Schlägl as centre. For barley the land around the village Eggerding was noted and the area around the village Sipbachzell for wheat. Werneck (1935) mentioned the traditional trade in this better land-race, coming from the Upper-Austrian central area, to west, into the Innviertel-district, close to the Bavarian border, and even into Lower-Bavaria. When Werneck (1935) is speaking of his own, "variability-circle" in this West-Upper-Austrian district, he concides, that this group of land-wheats would be at least "descendants" of the Sipbachzeller-land-race.

Not only commercial interests have played a role in the spread of land-races. A barley-land-race, still under cultivation, was traditionally presented as a dowry with the bride 50 years ago! In many cases the farmer's curiosity may be the reason for spreading varieties. Even in countries with extremely bad communications, like Bangladesh a jute-land-race could be found transfered over a distance of 70 km from Maskali-Island up to the Chitagong Hill tracts.

In comparsion to the commercial or public breeding institutes of today, the number of farmers, engaged with "breeding" and multiplying land-races was much larger. Beyond the general line of breeding objectives, the individual skill of each farmer preserved a large genetical variability. On the other hand, the interval between exchanging seed or purchasing a new one from the traditional land-race's growers, was larger than it is today, and so, some gene – combinations, being reduced during the selective procedure, got a chance for propagation again.

### 4.3. *The limitations of population*

It is important to remember, that land-races are agriculturally used populations, which will be reduced drastically and accidentally. Considering that only 180–200 kg out of a good land-races's yield of 4000 kg, that are 4 to 5 per cent of the yield only are used for further sowing. So it's easy to understand, that it will be a lucky chance, if hybrids or mutations are sown again, and that there must be an enormous advantage in the selection, if they are to survive. This process will always result in a higher uniformity of the race, than we can expect in really abandoned, wild populations, e. g. in a gene centre.

In testing these some land-races today, however, we cannot approach the yield levels obtained from them in the 1930's. The explanation will be that landraces run wild, since the demand for commercial varieties was increasing. The farmers breeding activities became less and less, finally petered out. Nevertheless farmers 8 years ago once again startet to select their own landrace of four-row spring barley, due to few bad harvests from commercial varieties. (Schachl and Weisz 1972, 1973, 1974)

However, we usually find land-races today, very far from intensive agricultural areas, where the best conditions prevail, and here factors other than yield and quality decide their survival. Not always the lines with the best adaptability give the highest yield.

### 4.4. *Yields and resistance*

Although land-races show a high resistance to different diseases, the plant breeders efforts to isolate such resistant genes had not often been successful. The answer is given in the line-construction of the land-races, where we can find:
a) individual lines being differentially infected by deseases,
b) temporal resistance in which the lines are infected in the same manner, but due to the stage of each line's development being different, than the infection will be at different stages too and a field resistance simulated. The entire land-race often therefore looks sounder and is healthier too, due to this more difficult plant to plant infection.

Another phenomenon seems to be that the germs of diseases have been changed with the genetical construction of the commercial varieties. The catastrophic mildew epidemic, which extinguished almost the total South-German barley land-races in the second half of last century is a clear example, how less resistant genes may be in a land-race.

This line construction of land-races solves the problem of ecological adaptability and continuous yield. No each line in a larger mixture of lines is ever stressed. There is a gene-pool of still latently existing genes, offering a large facility for evolution. Thus, land-races have much higher reserves of genes and as a consequence greater adaptability than bred, commercial varieties.

Stability of annual yield can be also seen as a special form of this adaptability. This is because the lower yield of one line will be compensated by a higher yield of another one. However these annual fluctuations will hardly disturb the lines equilibrium, if there is not extremely bad weather.

## 5. Conclusions

### 5.1. *Spreading of land-races*

As shown by the varietal replacement, land-races are usually somewhat younger than noted by early authors. When we transform this Upper-Austrian replacement into a larger scale, say on the total South-German Alpine and Subalpine region, we will get a contradiction to Mayr (1934), who supposed, that land-races were originally there (autochton) or grown over a very long time (autochtogen). Similar suggestions are made by Werneck (1949), looking for continuous evolution since the Stone-age. In every case we must not forget, however, that even during our own historical period the Austrian Subalpine region has been sometimes depopulated and that the climatic and botanical base would be unsuitable for the survival of cultivated cereals. Furthermore, the examples of the 'Sipbachzeller' wheat and 'Himalaya' barley have shown, that well adapted "ancient" land-races have been in this area less than 120 to 150 years.

Also land-races are rarely many distinct races, but a part of a larger group of local races representing a common gene-pool, dispersed over an ecologically uniform area. In this connection it must be noticed, that the botanical features of the 'Sipbachzeller' lines are identical in the whole Austrian and Bavarian

Subalpine region and seem to be related with those land-races, collected from South-Styria. In the same kind the local races of the Alpine beard wheat are more or less identical in the whole Alps.

### *5.2. The multiline character as biological model*

Land-races consist of a various number of lines. This multiline construction may be seen as biological model for survival and adaptation under most difficult and different conditions. There will be a higher value may be in the studies of this biological and natural base, than in the trial to isolate certain genes, which will be more or less effective only in the totality of all latently existing genes. Thus the isolation of a special desired gene will be more a lucky chance.

Also we have seen, that not all land-races survived but we must expect, that many of them have been given up after a short experimental sowing or have been lost due to stronger selection during replacement. However, lacking the commercial interest of plant breeding stations, having limited communications, and a less hectic life, the land-races at past had greater facilities for spreading, adaptation, evolution and finally survival, than today's commercial varieties.

### *5.3. The origin of land-races*

The origin of land-races is not a single process, but will be repeated every time, varieties start to run wild due to carelessness of farmers and plant breeders and when there is scope for natural evolution, like hybridization, selection and adaptation. We can still observe such processes in the very extensive areas where the last land-races have already been displaced by commercial varieties, left to many years of multiplication by farmers. The wheat varieties 'Tassilo' and 'Jubilar' the oat variety 'Austria Weiß' and the two barleys 'Union' and 'Martha' are typical examples of this. Studies on different multiplication stages of 'Union' have been not only an example of gene introgression, and adaptation by natural selection, but has also shown, that this "now land race" has become more adapted and suitable for some special areas, than the original, certified seed, and even, that no deterioration yield-wise has occured (SCHACHL and WEISZ 1975).

Finally we may say, land-races in the Subalpine region – and I would believe in most Central Europe – which have been found during the last two hundred years, are not abandoned populations, but populations approaching the yield, demand and also botanical features of their gone period. As a consequence there is no hope for breeding varieties by selection only (KÖCK 1973), and by hybridization of commercial varieties with land-races always unsuitable genes will be transfered too, already eliminated by plant breeder's success. These facts will become more effective as older the land-races will be. In these circumstances plant breeders – especially breeding under commercial aspects – will hardly use land-races, even not for hybridization, if they have a larger source of modern commercial varieties. Thus at present land-races will be more a still existing resource of genes for tomorrow's demand, which we can't realize in the moment, and warn us against the evolutionary "one way" of pure line varieties.

## Zusammenfassung

Getreide-Landsorten aus Österreich und ihre Verwendung in der Züchtung

Landsorten sind etwas jünger als allgemein angenommen. Landsorten sind auch keine zufälligen Populationen, sondern sie werden sowohl von der ständigen egalisierenden menschlichen Selektion als auch von der natürlichen Evolution beeinflußt. Sie sind deshalb phänotypisch sehr einheitlich. Ihre genetische Variabilität ist zwar höher als bei Zuchtsorten, jedoch nicht so groß wie bei Populationen in einem Genzentrum. Viele Eigenschaften wie Resistenz, Anpassung und kontinuierliche Jahreserträge lassen sich auf den Multilinien-Charakter der Landsorten zurückführen, der wiederum als biologisches Modell für den Fortbestand landwirtschaftlicher Populationen gelten kann. Infolge dieses Multilinien-Charakters der Landsorten ist es für den Züchter schwierig und oft wenig effektiv, bestimmte Gene zu isolieren.

## Краткое содержание

Местные сорта зерновых злаков из Австрии и их применение в селекции

Местные сорта, на самом деле – моложе, чем это принято считать. Они не являются случайными популяциями, а подвержены как постоянному уравнивающему влиянию человеческой селекции, так и естественной эволюции. Поэтому они фенотипически – очень выровнены. Их генетическая изменчивость, правда, выше, чем у селекционных сортов, но не так высока, как у популяций в каком-нибудь генном центре. Многие свойства, как напр. резистентность, приспособляемость и равномерность урожаев, можно объяснить «многолинейностью» местных сортов, которую можно принять и как биологическую модель для поддержания и сохранения сельскохозяйственных популяций. Из за большого числа линий, входящих в состав местных сортов, для селекционера оказывается не лёгким, а часто и малоэффективным, делом изолировать определённые гены.

## Literature


Becker Dillingen, J., 1927: Handbuch des Getreidebaues. – Paul Parey, Berlin **1**, 242–259, 359–376, 460–467.

Brittinger, C., 1862: Flora von Oberösterreich. – C. Überreiter-Verlag, Wien **1**, 41–47, 97–108.

Burggasser, E., 1956: Getreidebau und Getreidevarietäten von Oberösterreich. Tätigkeitsbericht 1953–1955 der Landw. chem. Bundesversuchsanstalt Linz. – Eigenverlag, Linz, 121–133.

Köck, L., 1973: Die Landsorten im alpinen Raum. – Bericht über die Arbeitstagung der Arbeitsgemeinschaft der Saatzuchtleiter, Gumpenstein, 143–157.

Mayr, E., 1928: Die Getreidelandsorten und der Getreidebau im Salzachtal und seinen Nebentälern. – Eigenverlag der Bundesanstalt für Pflanzenbau und Samenprüfung Wien, Wien, 15–65.

Mayr, E., 1935: Über wissenschaftliche und praktische Ergebnisse der alpinen Landsortenforschung an Getreide. – Forschung und Fortschritte **11**, 376–378.

Pammer, G., und R. Ranninger, 1928: Der rationelle Getreidebau. – Springer-Verlag, Wien, 104–109, 127–188.

Ritzberger, E., 1904: Prodromus einer Flora von Oberösterreich. – J. Wimmer, Linz, **1**, 41–47, 97–108.

Sailer, F. S., 1841: Die Flora Oberösterreichs. – Eigenverlag, Linz, **1**, 48–51, 61–63; 2, 285–286.

Schachl, R., 1974: Die Entwicklung des oberösterr. Getreidesortiments. – In: Land- und Forstwirtschaftl. Forschung in Österreich. Agrarverlag, Wien. **6**, 241–289.

– 1975: Das oberösterr.-salzburgische Getreidesortiment (Landsorten). – In: Festschrift der Landw. chem. Bundesversuchsanstalt Linz, Landesverlag Linz, Linz. **10**, 63–78.

–, 1976: Refugia of Austrian Land-Varieties in the Subalpine Region. – In: Barley Genetics **III**. Proc. III. Intern. Barley Genetics Sympos. Garching 1975, pp. 70–75. Karl Thiemig Verlag, München.

–, und F. Weisz, 1972: Ergebnisse aus Feldversuchen. Bericht V 46/72. – Eigenverlag der Landw. chem. Bundesversuchsanstalt Linz, 70–73.

– und –, 1973: Ergebnisse aus Feldversuchen. Bericht V 48/73. – Eigenverlag der Landw. chem. Bundesversuchsanstalt Linz, 41, 42.

– und –, 1974: Ergebnisse aus Feldversuchen. Bericht V 49/74. – Eigenverlag der Landw. chem. Bundesversuchsanstalt Linz, 85, 86, 91, 92.

– und –, 1975: Ergebnisse aus Feldversuchen. Bericht V. 50/75. – Eigenverlag der Landw. chem. Bundesversuchsanstalt Linz, Sonderversuche.

Werneck, H. L., 1930: Die 2. oberösterreichische Gerstenschau. Herausgeber: O. Ö. Landeskulturrat Linz, 1–12.

– 1931: Oberösterreichische Weizensorten. Herausgeber: O. Ö. Landeskulturrat Linz, 1–24.

– 1935: Die naturgesetzlichen Grundlagen der Land- und Forstwirtschaft in Oberösterreich. – In: Jahrbuch des oberösterr. Musealvereines pp. 270–288. Feichtingers Erben, Linz.

– 1949: Kultur- und Nutzpflanzen in den Ostalpen und am Rande des Böhmerwaldes. – O. Ö. Landesverlag, Wels, 69, 79–98, 115–122, 129–138, 143–145.

Dipl.-Ing. Dr. Rudolf Schachl. Landw. chem. Bundesversuchsanstalt Linz,
4025 Linz, Wieninger Str. 8

Kulturpflanze XXIX · 1981 · S. 111–114

# Use of Hungarian land-races in breeding

JÁNOS UNK and LÁSZLÓ HOLLY (Tápiószele, Hungary)

## Summary

The history of land-race utilization in Hungarian plant breeding programmes is briefly summarized with special emphasis on the potential use of Hungarian land-race collections in future breeding activities.

## Introduction

A feature of Hungarian climate results from the special geographical situation of Hungary. She is situated in the Carpathian Basin surrounded by the Carpathians, Alps and Dinarids in an area affected by the Atlantic, Mediterranean and Continental climates. This situation gives rise to a rather fluctuating climate, but which is relatively well buffered against rapid changes because of the basin situation. The fluctuating climate requires a relatively wide adaptability from cultivars, and perhaps it is one of the reasons why Hungarian land-races are still playing a relatively important role as sources of general and special adaptability in recent breeding programmes in our country.

## History

During the second part of the last century, when scientifically based plant breeding started to take place, land-races were mainly used as raw material in Hungary in the same way as it happened in other parts of Europe. Since the end of the last century a number of breeding stations had been established, most of which belonged to large private farms and worked on several crops according to the interest of the certain farm. At the first stage, mass selection technique was most frequently applied resulting in so called "farm varieties" which were maintained in somewhat similar way as it happened to most of land-races, but on a larger scale. After the First World War, foreign introductions started to play an increasing role in Hungarian breeding. Although, this kind of material were first of all used as parent stock in crosses with locally adapted material, the danger of extinction of land-races due to the rapid spread of new cultivars was early recognised in Hungary (RAPAICS 1934, 1943; VILLAX 1938). Large working collections were maintained by most breeders in which local land-races were also well represented. Several successful new cultivars were developed and released in this period, for example the series of Bánkúti wheats, Bánkúti early spring barley, FLEISCHMANN maize open pollinated cultivars and variety hybrid, etc. which had been cultivated until the sixty's (KAPÁS 1969).

After the Second World War, it was essential to recollect and renew breeders' working collections in parallel with the rebuilding of Breeding Stations both of which were almost completely destroyed during the war. This heroic work, which was directed by Professor Dr. A. JÁNOSSY, resulted in the reconstruction of most of the earlier bred varieties and in the collection of most of the existing field crop land-races. These achivements played an outstanding role in the reestablishment of breeding activity in Hungary and also initiated a more centralised work on plant raw material collections.

In 1961, still more than 50% of the commercialised varieties of field and vegetable crops were land-race selections or such cultivars of which at least one of the parents was Hungarian land-race or other local population. Among the 163 registered field crop varieties, for example, 63 were land-race selection and 37 further cultivars had land-race or other locally adapted material as one of their parents (e. g. wheat 10, maize 15, flax 5, tobacco 6, vetches 5, lucerne 5, red clover 3, etc.) (KAPÁS 1961).

For present time, the proportion of direct foreign introductions and cultivars produced from foreign basic material in Hungary have increased considerably. The main reason for this is that modern farming techniques spread all over Hungary very quickly after small private farms had been reorganized into large cooperatives at the early sixty's. The modern farming system required intensive type cultivars, and meeting this quickly arised requirement was only possible by using new introductions. But there are a few crops (e. g. grasses, forage legumes, grain legumes, some vegetables, etc.) in which cases land-race or ecotype selections are still widely grown (KAPÁS et al. 1977).

## Use of land-race collections

During the last few years, an increasing interest of Hungarian breeders in genetically adapted and diverse raw material can be detected, which is mainly due to the increasing need for quality factors, specific adaptability, and yield stability. Although a high number of seed requests have been received by our Centre, it is hardly possible to estimate the actual use of these materials. The main reason for this is that land-races as any other genetically variable raw materials usually can not be used directly in breeding programmes. In most of the cases, an extensive selection programme is needed to obtain strains or lines with the desirable characters, but the success of this kind of work is hardly if at all predictable in large land-race collections using available evaluation techniques. We would really need relatively simple methods to estimate genetic potentialities for certain important characters, and perhaps by the same way a more extensive "prebreeding" activity should be conducted in active collections to produce breeding materials which are immediately applicable in breeding programmes.

Of course, this kind of activity should be strictly separated from the preservation function, but it could certainly contribute to a better characterisation and utilization of locally adapted, genetically variable populations.

Although our land-race collections have not yet evaluated in details, according to our experiences, Hungarian land-races of wheat, maize, lucerne, clovers, tomato, green and red paprika, onion, poppy, cucurbits, etc. may represent valuable gene resources for improving the following characters: general and special adaptation, resistance to pests, different growth habits, certain quality factors, yield stability, etc. (Jánossy 1971, Unk 1979).

There are also a few new examples for the successful utilization of land-races in recent breeding programmes in Hungary. For example, several inbred lines are being used by maize breeders which have been produced from land-races at Tápiószele, tetraploid red clover cultivars ('Hungaropoly', 'Tápiói poly') were produced from inland ecotypes and land-races, and most of the dry bean, green paprika and grass cultivars grown in Hungary are of land-race or ecotype origin.

It seems clear that potentialities of land-races in breeding programmes are not yet fully utilized, and more detailed study should be done on this field in the future, in parallel with the exploration and preservation of still existing land-races and ecotypes.

## Zusammenfassung

Die Nutzung ungarischer Landsorten in der Züchtung

Die Geschichte der Verwendung von Landsorten in ungarischen Pflanzenzuchtprogrammen wird kurz zusammengefaßt. Besondere Betonung wird dabei auf den möglichen Nutzen der ungarischen Landsortenkollektion bei zukünftigen züchterischen Vorhaben gelegt.

## Краткое содержание

Использование местных венгерских сортов в селекции

Даётся краткий обзор истории применения венгерских местных сортов в программах селекции растений. Особо подчёркивается возможность использования коллекций местных сортов в будущих селекционных работах.

## Literature

Jánossy, A., 1971: A növénynemesitési alapanyag-kutatás Magyarországon. (Research on plant breeding basic material in Hungary) Agrártudományi Közlemények, **30**.

Kapás, S. (ed), 1961: Államilag minősitett növényfajtak jegyzéke. (National list of qualified plant varieties) Mezőgazdasági Kiadó, Budapest.

–, 1969: Magyar Növénynemesités. (Hungarian Plant Breeding) Akadémiai Kiadó, Budapest.

KAPÁS, S., 1977: Szántóföldi növényfajták. (Field crop cultivars) MÉM Információs Központja, Budapest.

RAPAICS, R., 1934: A kenyér és táplálékot szolgáló növényeink története. (History of our bread and food crops) Természettudományi Társulat, Budapest.

–, 1943: Termesztett növényeink eredete. (Origin of our cultivated plants) Földmüvelési Minisztérium, Budapest.

UNK, J., 1979: A tápiószelei növényigénbank-munka eredményei. (Work performed at the Plant Gene Bank of Tápiószele) Fajtakisérletezés, **XXVIII**, 23–35.

VILLAX, Ö., 1938: A magyar növénynemesités múltja, jelene és jövő feladatai. (Hungarian plant improvement, the past, present and future tasks) Mezőgazdasági Közlöny, **2**.

Dr. J. UNK and Dr. L. HOLLY
Research Centre for Agrobotany, NIAVT
H - 2766 Tápiószele, Hungary

Kulturpflanze XXIX . 1981 . S. 115–127

# Genetic variability for protein content and S-aminoacids in broad-beans (*Vicia faba* L.)

DOMENICO LAFIANDRA, Gian Battista POLIGNANO, ANDREA FILIPPETTI and ENRICO PORCEDDU (Bari, Italy)

## Summary

Approximately 600 different samples of the broad bean collection have been evaluated for crude protein content of the seeds.

111 samples were also evaluated for S-aminoacids. The values ranged from 19.5 to 34.0 for protein content and 0.97 to 1.45 for half-cystine and 0.62 to 0.99 for methionine. Protein content was negatively correlated with seed weights, highest values been found in the small seeded samples of Ethiopia and Afghanistan. Mediterranean samples exibited quite large variability. S-aminoacid per protein unit were negatively correlated with the protein content and determination coefficients had intermediate values. Sulphur aminoacids expressed as % of dry seed weight were positively correlated with protein content, but determination values were rather low. Analyses of protein and sulphur aminoacids in seeds from a 10×10 diallel cross scheme gave information on genetic mechanisms involved in the character's control; additive and non additive components of variance were significant, indicating difficulties in practical breeding programmes.

## Introduction

More than two-thirds of the world's protein supply is from plant sources. Of these, approximately 67% come from cereals to which seed legumes are an important partner in diets. In the Mediterranean area, *Vicia faba* provides one of the most important sources of pulse pıotein, besides playing an important role as animal feed.

Several workers (SMITH and ALDRICH 1967, EDEN 1968, BOND and TOYNBEE-CLARKE 1968, CLARKE 1970, BOND 1970, BHATTY 1974, GRIFFITHS 1977) have reported results of experiments on protein determination in *Vicia faba* cultivars. Generally, little variation has been found.

On the other hand, the analysis of large germplasm collections has revealed a range of variation from 19 to 34%, (POLIGNANO et al. 1980) from 17 to 38% (FILIPPETTI 1979) and from 23 to 40% (PICARD 1978).

A considerable degree of variation has also been found most cultivated varieties (BOND 1970, SJÖDIN 1971, 1974; MUNCK et al. 1975). An analysis carried out by GRIFFITHS and LAWES (1978) on progenies of the highest and lowest selections in the same variety revealed a range of approximately 20 units between 19.5 and 39.3%, also, it showed broad sense heritability values equal to 70% thus indicating the possibility of fairly positive responses to selection.

Diallel analysis among six spring inbreds, having different nitrogen contens, showed that the greater part of variance was additive and the narrow sense heritability estimate greater than 50%. Also the non-additive portion of variance was significant though not all in the same direction (BOND 1978).

As far as the quality of protein is concerned, *Vicia faba,* in common with all pulses, is known to have high lysine contens but is poor in sulphur aminoacids, cystine and methionine.

BOLTON (1967), WARING and SHANNON (1969), PALMER and THOMPSON (1975), SJÖDIN and MARTENSON (1979) LAFIANDRA et al. (1979) reported broadly similar aminoacid patterns for all the materials examined. However, in a collection of 500 accessions, HANELT et al. (1978) found a range between 0.61 and 0.99 (g/16g N) for methionine, and 1.10–1.73 (g/16g N) for cystine. In addition, studies on variation in protein content and aminoacid composition showed that protein content and sulphur aminoacid were negatively correlated, while cystine was positevely correlated to methionine (FRÖLICH et al. 1974, HANELT et al. 1978, LAFIANDRA et al. 1979).

The purpose of the present study was to asses the magnitude of variability for total protein and sulphur aminoacids in a collection of *Vicia faba* of various origins and the nature of the gene action involved in the control of these characters.

## Materials and methods

Three sets of materials were studied. The first set was made up of about 600 accessions, from different countries, as in table 1; the material was multiplied at the same location – Bari – in 1977–1978, on a typical "terrarossa" soil. Seed weight (g) and crude protein content (Microkjeldahl N×6.25) were determined on seeds from 30 plants.

The second set of material included 111 seed specimens as in table 1; the plants

Table 1

Number of accessions from different countries analyzed for seed weight and protein (left), protein and sulphur aminoacids (right)

| Country | Seed weight protein | Protein S-aminoacids |
|---|---|---|
| Italy | 163 | 16 |
| Ethiopia | 143 | – |
| Afghanistan | 62 | – |
| Egypt | 51 | 25 |
| Turkey | 51 | 18 |
| Iraq | 28 | 10 |
| Algeria | 28 | 12 |
| Spain | 25 | 10 |
| Lebanon | 19 | 10 |
| Tunisia | – | 10 |

were grown at Metaponto, on the Ionian sea, in 1977–1978, on a clay-silt soil. Seeds from 10 plants were analyzed for cystine (half), methionine and crude protein content; sulphur aminoacids were determined by means of the performic acid oxidation method (MOORE 1963), and reported as protein percentage.

Finally, seeds from $F_1$ plants of a 10×10 diallel cross scheme performed among lines derived from Italian varieties, and grown in Bari in 1977–1978 at Bari, were evaluated for protein content and sulphur aminoacids, using the above mentioned methodology.

## Results

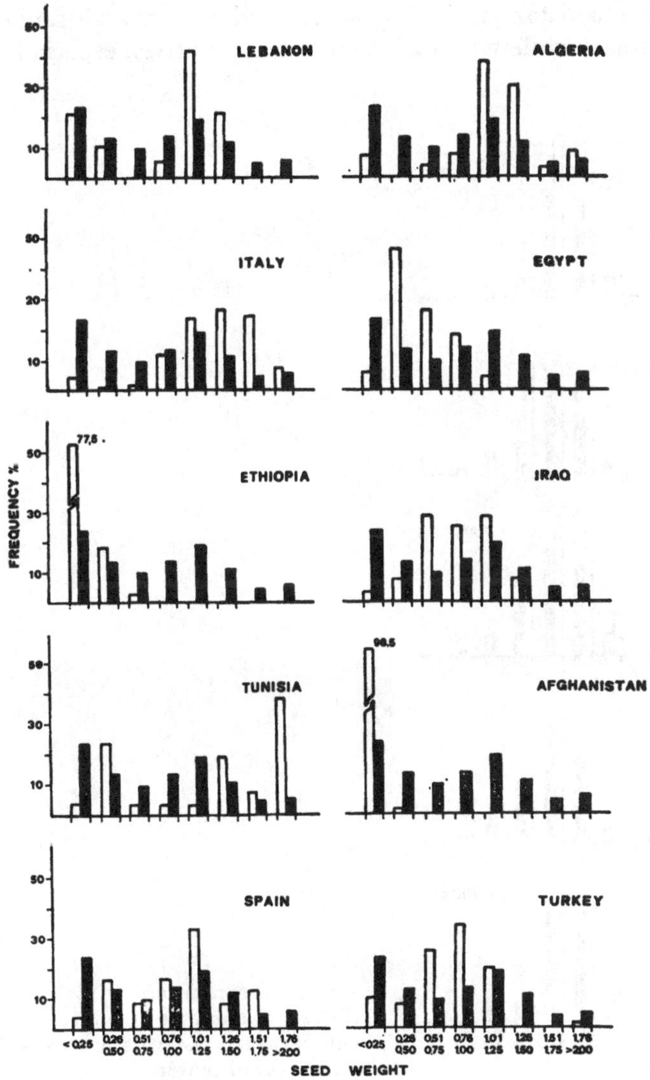


Fig. 1 Distribution of seed weight in different countries (white) in comparison to mean values (dark) from overall material

Results of seed weight determinations are shown in fig. 1. Seed weight average was equal to 0.77 with extreme values ranging from 0.13 to 2.03; the most frequent types were those included in classes 0.13–0.25 and 1.00–1.25. Small seed types mostly originated from Afghanistan and Ethiopia, while accessions with large seeds were derived from Lebanon, Tunisia and Italy. Protein content determinations gave the results reported in fig. 2. Values for the overall material ranged from 19 to 34%, with a concentration around 23–26%.

All countries possess material with a similar range, with the exception of Lebanon and Iraq, which lack the lowest and the highest class, respectively.

Low protein values were predominant in material from Iraq, Turkey, Spain and Tunisia, whereas high values characterize Afghan material and to some extent the Ethiopian one too.

Fig. 3 shows a negative correlation between seed weight and protein content. Small seeded types from Afghanistan and Ethiopia showed a very high range of variation for protein content, while materials from other countries, especially Italy, were well spread.

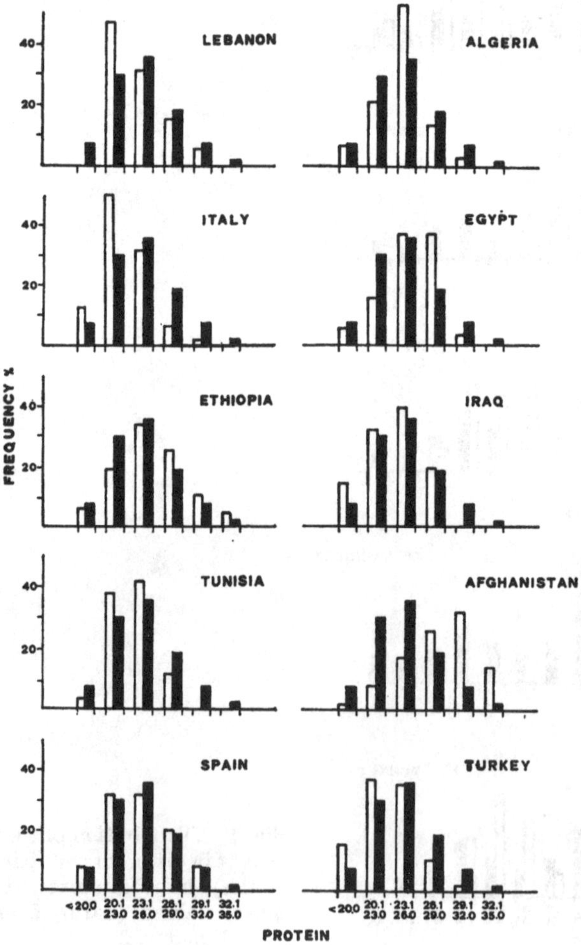


Fig. 2 Distribution of protein in different countries (white) in comparison to mean values (dark) from overall material

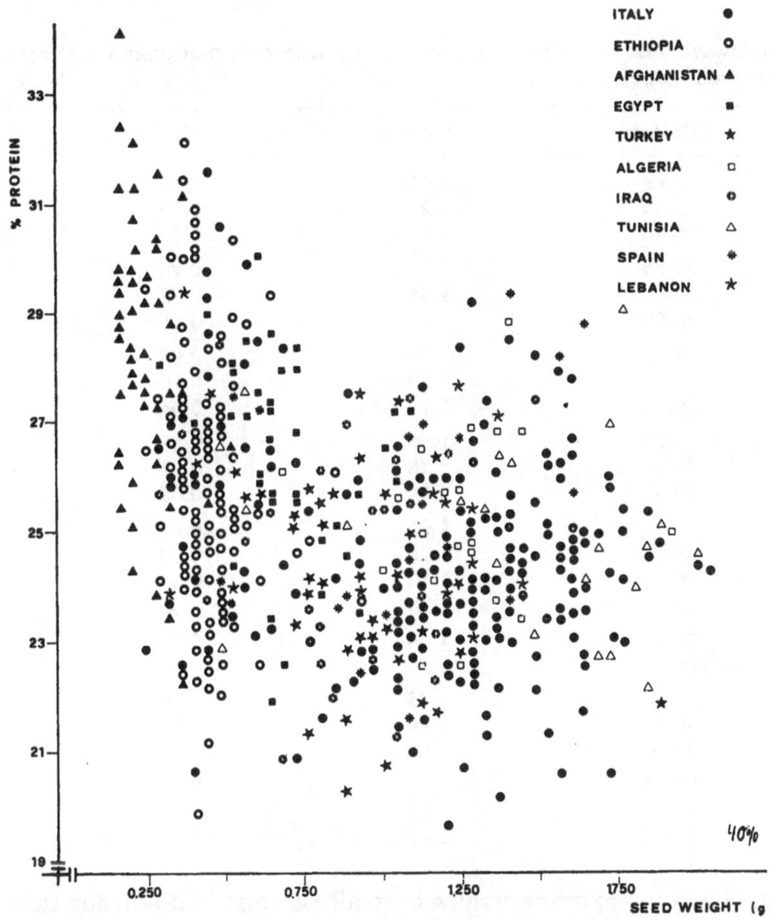


Fig. 3 Relationship between protein ($N \times 6.25$) and seed weight from overall material

Results of the analysis on the second set of material are given in table 2. Spain exhibited the highest mean value for protein content, and Italy the lowest; however, in absolute terms, the highest value was observed in the Egyptian material (35.43) and the lowest in the one from Iraq (24.03). Cystine mean values espressed as g/16g N, ranged from 1.29 for Italy, to 1.16 for Spain, but the highest value was detected in the Algerian (1.46) and the lowest in the Egyptian material (0.97). Methionine mean values were concentrated between 0.70 (Spain) and 0.77 (Algeria), with the highest value in the Italian (0.98) and the lowest in the Tunisian material (0.57). In general both sulphur aminoacids were negatively related with protein content (fig. 4), the correlation coefficients being equal to $-0.68$ for cystine and $-0.42$ for methionine. The relation however, is intermediate. About 46% of the variation in cystine and 18% in methionine could be attributed to variation in protein content.

Both the cystine and the methionine contents, expressed as percentage of grain dry weight, are positively correlated with protein content. About 14% of the variation in cystine and methionine could be attributed to variation in the protein content of the grain (fig. 5).

Table 2

Highest, mean and lowest values for protein content, cystine and methionine exibited by the material from each country

| Country | | Protein | Cystine | Methionine |
|---|---|---|---|---|
| Egypt | H | 35.43 | 1.41 | 0.93 |
| | x̄ | 30.25 | 1.21 | 0.72 |
| | L | 26.06 | 0.97 | 0.62 |
| Turkey | H | 31.56 | 1.37 | 0.79 |
| | x̄ | 28.84 | 1.25 | 0.75 |
| | L | 26.58 | 1.11 | 0.66 |
| Italy | H | 34.43 | 1.43 | 0.98 |
| | x̄ | 27.98 | 1.29 | 0.76 |
| | L | 24.40 | 1.18 | 0.64 |
| Algeria | H | 33.69 | 1.46 | 0.84 |
| | x̄ | 29.38 | 1.28 | 0.77 |
| | L | 25.53 | 1.11 | 0.68 |
| Iraq | H | 31.82 | 1.45 | 0.87 |
| | x̄ | 28.69 | 1.25 | 0.73 |
| | L | 24.03 | 1.10 | 0.63 |
| Lebanon | H | 32.47 | 1.33 | 0.86 |
| | x̄ | 30.23 | 1.24 | 0.76 |
| | L | 28.03 | 1.10 | 0.66 |
| Tunisia | H | 35.28 | 1.33 | 0.86 |
| | x̄ | 30.60 | 1.19 | 0.73 |
| | L | 25.66 | 1.07 | 0.56 |
| Spain | H | 34.22 | 1.24 | 0.80 |
| | x̄ | 31.16 | 1.16 | 0.70 |
| | L | 29.63 | 1.07 | 0.63 |

Characteristic features for extreme values over all the material and for each country are reported in table 3.

Cystine and methionine values in the accessions with the highest and lowest protein content were equal to 0.98 and 0.66 and to 1.45 and 0.76, respectively. Samples with extreme cystine values showed protein contents of 24.03 and 34.81 and methionine values equal to 0.84 and 0.62.

Finally, using methionine content as a selection criteria, samples with 1.43 (cystine) and 24.40% (protein) for highest methionine values and 1.10 (cystine) and 32.84% (protein) for lowest methionine values were obtained.

Table 3 also shows the extreme values of protein, cystine and methionine in materials from each country, together with values from a selected number of samples, indicated by a digit in fig. 4.

Results for material from diallel crosses are reported in table 4; the significance of the effects were estimated by adopting the random model. General combining ability mean squares were 5 to 8 times higher than the effects of specific combining ability and reciprocals. No significant difference were found between specific combining ability and reciprocal effects.

The latters were due mainly to maternal effects for protein, while they had about the same value for the two sulphur aminoacids.

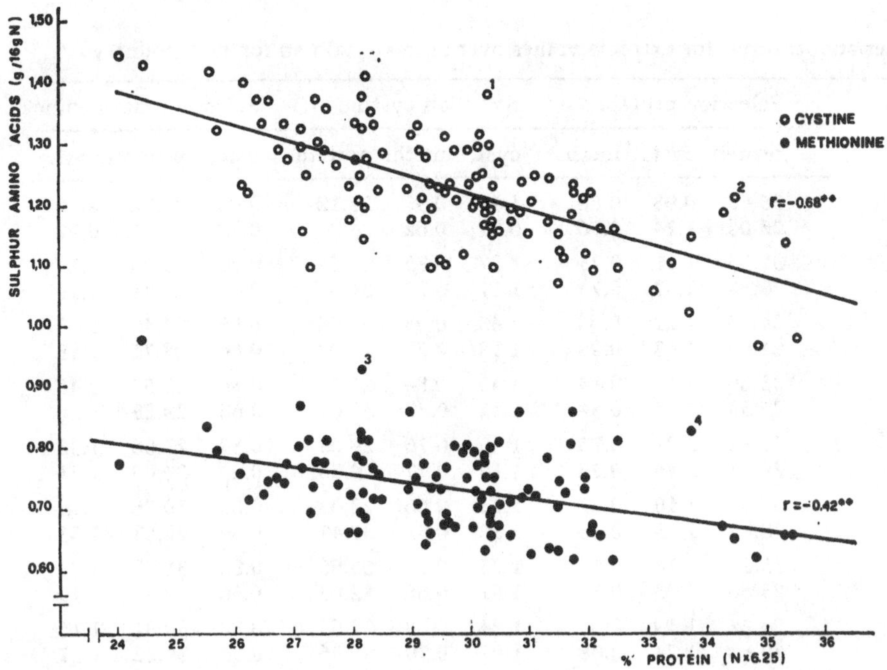


Fig. 4 Relationship between protein ($N \times 6.25$) and sulphur aminoacids expressed as g/16g $N$, in 111 samples of different origins

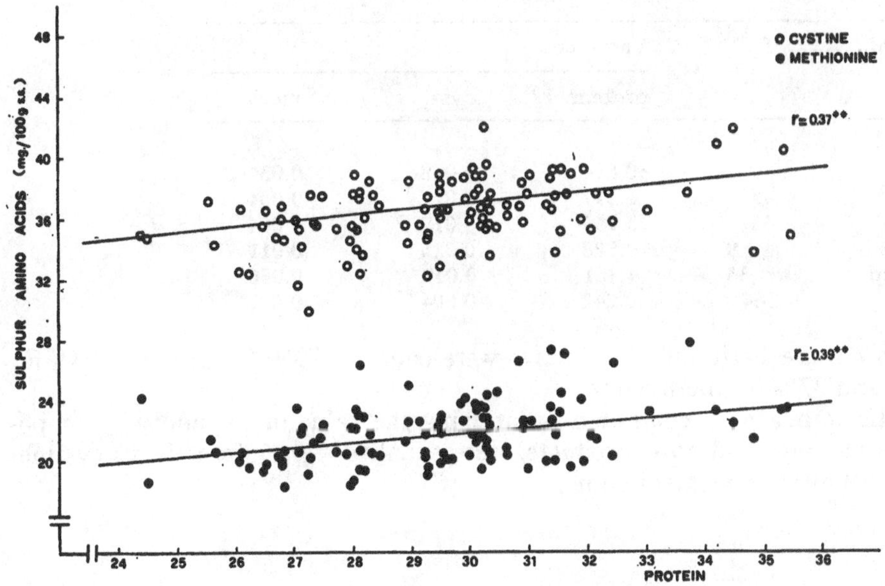


Fig. 5 Relationship between protein ($N \times 6.25$) and sulphur aminoacids expressed as percentage of dry seed

Table 3

Characteristic features, for extreme values over all material and for each country

| Country | Selection protein | | | Selection cystine | | | Selection methionine | | |
|---|---|---|---|---|---|---|---|---|---|
| | protein | cyst. | meth. | cyst. | meth. | protein | meth. | protein | cyst. |
| Egypt | 35.43 | 0.98 | 0.66 | 1.41 | 0.93 | 28.12 | 0.93 | 28.12 | 1.41 |
| | 26.06 | 1.24 | 0.76 | 0.97 | 0.62 | 34.81 | 0.62 | 34.81 | 0.97 |
| Turkey | 31.56 | 1.11 | 0.73 | 1.37 | 0.75 | 26.58 | 0.79 | 30.03 | 1.21 |
| | 26.58 | 1.37 | 0.75 | 1.11 | 0.73 | 31.56 | 0.66 | 30.31 | 1.19 |
| Italy | 34.43 | 1.21 | 0.65 | 1.43 | 0.98 | 24.40 | 0.98 | 24.40 | 1.43 |
| | 24.40 | 1.43 | 0.98 | 1.18 | 0.77 | 28.97 | 0.64 | 28.71 | 1.18 |
| Algeria | 33.69 | 1.11 | 0.83 | 1.45 | 0.84 | 25.53 | 0.84 | 25.53 | 1.45 |
| | 25.53 | 1.45 | 0.84 | 1.11 | 0.83 | 33.69 | 0.68 | 29.25 | 1.25 |
| Iraq | 31.82 | 1.21 | 0.73 | 1.45 | 0.76 | 24.03 | 0.87 | 27.06 | 1.33 |
| | 24.03 | 1.45 | 0.76 | 1.10 | 0.77 | 29.53 | 0.63 | 31.00 | 1.21 |
| Lebanon | 32.47 | 1.10 | 0.81 | 1.33 | 0.66 | 28.03 | 0.86 | 30.78 | 1.19 |
| | 28.03 | 1.33 | 0.66 | 1.10 | 0.81 | 32.47 | 0.66 | 28.03 | 1.33 |
| Tunisia | 35.28 | 1.14 | 0.66 | 1.33 | 0.80 | 25.66 | 0.86 | 31.69 | 1.23 |
| | 25.66 | 1.33 | 0.80 | 1.07 | 0.66 | 32.03 | 0.56 | 32.84 | 1.10 |
| Spain | 34.22 | 1.19 | 0.67 | 1.24 | 0.68 | 29.63 | 0.80 | 30.31 | 1.15 |
| | 29.63 | 1.24 | 0.68 | 1.07 | 0.70 | 33.06 | 0.63 | 30.22 | 1.17 |
| Over | 35.43 | 0.98 | 0.66 | 1.45 | 0.76 | 24.03 | 0.98 | 24.40 | 1.43 |
| Countries | 24.03 | 1.45 | 0.76 | 0.97 | 0.62 | 34.81 | 0.67 | 32.84 | 1.10 |

Table 4

Results of analysis of variance on diallel dates

| Source of variation | D. F. | Variances | | |
|---|---|---|---|---|
| | | protein | cyst. | meth. |
| Reps | 1 | – | – | – |
| Gca | 9 | 30.614 | 0.078 | 0.034 |
| Sca | 45 | 6.377 | 0.017 | 0.004 |
| Recp | 45 | 5.750 | 0.017 | 0.004 |
| – Mat | 9 | 10.828 | 0.015 | 0.011 |
| – Recp | 36 | 4.481 | 0.017 | 0.006 |
| Error | 99 | 0.242 | 0.004 | 0.001 |

Narrow sense heritability estimates were equal to 29% for protein, 24% for cystine and 37% for methionine.

Genetic correlations, computed by utilizing the value of the additive components of variance and cross products, were equal to −0.37 (protein vs cystine) and −0.89 (protein vs methionine).

## Discussion

Seed weight represents one of the most important characters for intraspecific classification in *Vicia faba*. Observed values have almost the same range as found by Bianco et al. (1979) in a collection of broad beans from many countries.

Light seed forms come from Afghanistan, the probable primary centre of diversification of the minor group, which is considered to be more primitive (Hanelt 1972) in the philogenetic sense. Small seeds are also characteristic of wild types and primitive forms of *Pisum* (Blixt 1979).

Light seeds are also common in the Ethiopian material, while accessions from other countries have heavier seeds. This pattern supports Hanelt's (l.c.) hypothesis on the existence of two main geographical groups; the first ranging from Afghanistan to Ethiopia and Egypt, and the second from the Iranian-Afghan area to the western Mediterranean countries.

Since minor types are almost absent in S. W. Asia (Lebanon, Iraq, etc.) accessions, one could hypothesize that they migrated from the Indian-Afghan region to Ethiopia across the sea, and then, through the Nile valley, to the Mediterranean area, where they spread over limited land areas.

Heavier seed types possibly originated a little more westward, hence spread in the Mediterranean region, where they were utilized for human consumption in addition to animal feeds.

This pattern seems to be also supported by the fact that the Ethiopian word for broad beans is same as the Indian one and that the same Latin word panis (Italian pane, French pain, English bread) originated from *Πυανοδ*, the Greek word for bean which in the turn produced the word *Πυανέ Θια*, to designate an outdoor festival dedicated to Apollo and Diana, during which broad beans used to be eaten (Scarascia Mugnozza and Marzi 1979).

Rather interesting is the relationship between seed weight and protein content since both light and heavy seeds show a rather high range of variation.

Although protein content did not reach the highest values of the Afghan material, accessions from the Italian and Iberian peninsulas and the Magreb region, mainly represented by heavy seeds, show a large range of variation both for seed weight and protein content, confirming Vavilov's (1951) hypothesis that the Mediterranean region may have been a centre of diversification for large bean forms.

*Vicia faba* is the most important pulse crop in the Mediterranean area, as wheat is for cereals; its cultivation cannot compare with the amount of land surface grown to the cereal; the breeding system is intermediate, about 50% of flowers are cross pollinated (Porceddu et al. 1980); interspecific and, in some cases, unilateral intraspecific (Abdalla 1977) incompatibility is present, with single genome controlling the plant.

In spite of these facts the range of variability, both for seed weight and protein content, is larger than that in wheat, supporting the hypothesis of a species still in active avolution (Hanelt 1972), and potentially favourable for breeding work.

Protein content and sulphur aminoacids are negatively correlated, as it was already found for limiting aminoacids in other species, e. g. lysine in cereals. Lysine is negatively correlated to protein content in *Vicia faba* also (BOND 1974; LAFIANDRA et al. 1979). Unfortunatly studies on this species have not progressed as much as those on cereals, and further research is needed to determine exactly which protein fractions have increased in high protein types.

Genetical correlation values confirm the sound basis of this correlation. The tendency for an increased protein content to be associated with slight reductions in protein quality clearly represents a problem in practical breeding work.

Additional problems derive from the fact that dominance and reciprocal effects are significant, along with additive ones.

These facts call for more large collections to be analyzed and for more extensive research work to ascertain the process of accumulation of storage protein and its genetical control.

## Zusammenfassung

Genetische Variabilität des Gehalts an Rohprotein und schwefelhaltigen Aminosäuren bei Ackerbohnen (*Vicia faba* L.)

Fast 600 Sortimentsnummern der Ackerbohnenkollektion wurden auf den Rohproteingehalt der Samen untersucht, 111 auch auf den auf Gehalt an schwefelhaltigen Aminosäuren. Der Rohproteingehalt schwankt von 19,5 bis 34,0%, der Gehalt an Cystein von 0,97 bis 1,45% und der an Methionin von 0,62 bis 0,99%. Der Rohproteingehalt war negativ mit dem Samengewicht korreliert, die höchsten Werte wurden bei den kleinsamigen Sippen aus Äthiopien und Afghanistan gefunden. Die Sippen aus dem Mittelmeergebiet zeigen eine sehr große Variabilität. Der relative Gehalt an S-haltigen Aminosäuren war negativ mit dem Rohproteingehalt korreliert und das Bestimmtheitsmaß hatte mittlere Werte. Der absolute Gehalt an S-haltigen Aminosäuren war positiv mit dem Rohproteingehalt korreliert, aber das Bestimmtheitsmaß war niedriger.

Die Analysen des Rohproteingehaltes sowie des Gehaltes an schwefelhaltigen Aminosäuren der Samen (von $F_1$-Pflanzen) aus einem 10×10 Diallel-Kreuzungsschema ergaben Informationen über genetische Mechanismen in der Kontrolle der Merkmale; die additive und die nicht additive Varianzkomponente waren signifikant und offenbaren die Schwierigkeiten in den praktischen Züchtungsprogrammen.

## Краткое содержание

Генетическая изменчивость содержания сырого протеина и аминокислот, содержащих серу, у конских бобов (*Vicia faba*)

Около 600 образцов коллекции конских бобов исследовалось на содержание сырого протеина в семенах.

111 образцов анализировались и на содержание аминокислот, содержащих

серу (С-аминокислоты). Содержание сырого протеина колебалось от 19,5 до 34,0%; содержание цистеина – от 0,97 до 1,45%, а метионина – от 0,62 до 0,99%. Содержание протеина показало отрицательную корреляцию с весом семян; самое высокое содержание было отмечено у мелкосемянных форм из Эфиопии и Афганистана. Средиземноморские формы оказались очень изменчивыми. Относительное содержание С-аминокислот показало отрицательную корреляцию с содержанием сырого протеина, хотя коэффициент детерминации был невысоким. Однако абсолютное содержание С-аминокислот имело позитивную корреляцию с содержанием сырого протеина, но коэффициент детерминации был ниже.

Анализы содержания сырого протеина и содержания С-аминокислот в семенах растений $F_1$ (по схеме диаллельных скрещиваний $10 \times 10$) дали информацию о генетических механизмах, контролирующих эти признаки; аддитивная и неаддитивная компоненты варианcы были статистически достоверны, что говорит о сложности проведения селекционных программ.

## Literature

ABDALLA, M. M. F., 1977: Intraspecific unilateral incompatibility in *Vicia faba* L. – Theor. Appl. Genet. **50**, 227–233.

BHATTY, R. S., 1974: Chemical composition of some faba bean cultivars. – Canad. J. Plant Sci. **54**, 413–421.

BIANCO, V. V., G. DAMATO, V. MICCOLIS, G. B. POLIGNANO, E. PORCEDDU and G. SCIPPA, 1979: Variation in a collection of *Vicia faba* L. and correlations among agronomically important characters. – In: Some current research on *Vicia faba* in Western Europe (Eds.: D. A. BOND, G. T. SCARASCIA-MUGNOZZA and M. H. POULSEN) pp. 217–250. Commission of the European Communities. Brussels, Luxemburg.

BLIXT, S., 1979: Natural and induced variability for seed protein in temperate legumes. – In: Seed protein improvement in cereals and grain legumes pp. 3–21. Proc. Internat. Symp. FAO/IAEA. Neuherberg, 1978. IAEA, Vienna

BOLTON, W., 1967: Poultry nutrition. Bullettin No. 174 (3rd edition). – H. M. S. O., London.

BOND, D. A., 1970: The development of field beans as a crop in Britain. – Proc. Nutr. Soc. **29**, 74–79.

–, 1974: Die Züchtung von Hybrid- und synthetischen Sorten in Cambridge, England. – In: Ernährungsqualität und Züchtung von Ackerbohnen (*Vicia faba minor*) pp. 39–62, Göttinger Pflanzenzüchter-Seminar 2. Institut für Pflanzenbau und Pflanzenzüchtung, Universität Göttingen.

–, 1978: Breeding for zero-tannin and protein yield in field beans (*Vicia faba* L.). – In: Protein quality from leguminous crops pp. 348–359. A seminar in the EEC programme organized by the INRS Station d'Amélioration des plantes, Dijon, France.

– and G. TOYNBEE-CLARKE, 1968: Protein content of spring and winter varieties of field beans (*Vicia faba* L.) sown and harvested on the same dates. – J. Agric. Sci., Camb. **70**, 403–404.

CLARKE, H. E., 1970: The evaluation of the field bean (*Vicia faba* L.) in animal nutrition. – Proc. Nutr. Soc. **29**, 64–73.

EDEN, A., 1968: A survey of the analytical composition of field beans (*Vicia faba* L.). – L. Agric. Sci., Camb. **70**, 229–301.

FILIPPETTI, A., 1979: Breeding projects and work for the improvement of broad beans (*Vicia faba*) in Puglia. In: Some current research on *Vicia faba* in Western Europe (Eds.: D. A. BOND, G. T. SCARASCIA-MUGNOZZA and M. H. POULSEN) pp. 168–188. Commission of the European Communities. Brussels, Luxembourg.

FRÖLICH, W. G., W. G. POLLMER and W. CHRIST, 1974: Variation of the contents of protein and of methionine and cystine in *Vicia faba* L. – Z. Pflanzenzüchtung **72**, 160–165.

GRIFFITHS, D. W., 1977: Genetic variation in the chemical composition of field beans (*Vicia faba* L.). – Ph. D. Thesis, Univ. of Wales. Aberystwyth.

– and D. A. LAWES, 1978: Variation in the crude protein content of field beans (*Vicia faba* L.) in relation to the possible improvement of the protein content of the crop. – Euphytica **27**, 487–495.

HANELT, P., 1972: The intraspecific variability of *Vicia faba* L. and its classification. – Kulturpflanze **20**, 75–128.

–, A. RUDOLPH, K. HAMMER, H.-W. JANK, K. MÜNTZ and F. SCHOLZ, 1978: Protein screening of the Gatersleben collection of cereals and pulses. 3. Protein and SH-amino contents of field beans (*Vicia faba* L.). – Kulturpflanze **26**, 183–212.

JEROCH, H. and H. HENNING, 1964: Field beans for poultry. – Tierzucht **18**, 486–488.

KALDY, M. S. and R. KASTING, 1974: Amino acid composition and protein quality of eight faba bean cultivars. – Can. J. Pl. Sci. **54**, 869–871.

LAFIANDRA, D., G. B. POLIGNANO and G. COLAPRICO, 1979: Protein content and amino acid composition of seed in varieties of *Vicia faba* L. – Z. Pflanzenzüchtung. **83**, 308–314.

MOORE, S., 1963: On the determination of cystine as cysteic acid. – J. Biol. Chem. **238**, 235–237.

MUNCK, L., E. HOLMBERG, K. E. KARLSSON, B. LOFQVIST and J. SJÖDIN, 1975: Breeding for protein quantity and quality in field beans (*Vicia faba* L.) and barley (*Hordeum vulgare*). In: Nutritional improvements of food legumes by breeding. Ed.: M. MILNER) pp. 197–204. – Wiley Interscience, New York.

PALMER, R. and R. THOMPSON, 1975: A comparison of the protein nutritive value and composition of four cultivars of faba beans. – J. Sci. Fd. Agric. **26**, 1577–1583.

PICARD, J., 1976: Some results dealing with breeding for protein content in *Vicia faba* L., pp. 339–347. – In: Protein quality from leguminous crops. A seminar in the EEC programme organized by the INRS Station d'Amélioration des plantes, Dijon, France.

POLIGNANO, G. B., D. LAFIANDRA, V. MICCOLIS and G. TARANTO, 1980: Variabilità di caratteri agronomici e del contenuto proteico in una collezione di *Vicia faba* L. – Atti incontro seminario sul frumento duro e leguminose da granella 27–28 ottobre 1979 (in press).

PORCEDDU, E., L. M. MONTI, L. FRUSCIANTE and N. VOLPE, 1980: Analysis of cross-poollination in *Vicia faba* L. – Z. Pflanzenzüchtung **84**, 313–322.

SCARASCIA-MUGNOZZA, G. T. and V. MARZI, 1979: Retrospective and prospective views for the *Vicia faba* crop in southern Italy. – In: Some current research on *Vicia faba* in Western Europe (Eds.: D. A. BOND, G. T. SCARASCIA-MUGNOZZA and M. H. POULSEN) pp. 7–22. Commission of the European Communities. Brussels, Luxembourg.

SJÖDIN, J., 1971: Induced morphological variation in *Vicia faba* L. – Hereditas **67**, 155–180.

–, 1974: Induzierte Variabilität von Qualitätseigenschaften der Ackerbohne. – In: Ernährungsqualität und Züchtung von Ackerbohnen (*Vicia faba minor*) pp. 101–105. – Göttinger Pflanzenzüchter-Seminar 2. Institut für Pflanzenbau und Pflanzenzüchtung, Universität Göttingen.

–, and P. MÅRTENSSON, 1979: Problems of quality breeding in *Vicia faba*. – In: Some current research on *Vicia faba* in Western Europe (Eds.: D. A. BOND, G. T. SCARASCIA-MUGNOZZA and M. H. POULSEN) pp. 313–323. Commission of the European Communities. Brüssels, Luxembourg.

SMITH, B. F. and D. T. H. ALDRICH, 1967: Spring bean variety trials, 1954–1964. – F. natn. Inst. Agric. Bot. **11**, 114–133.

VAVILOV, N., 1951: Phytogeographic basis of plant breeding. The origin, variation immunity and breeding of cultivated plants. – Chronica Bot. **13**, 1–366.

WARING, J. J. and D. W. F. SHANNON, 1969: Studies on the energy and protein values of soya bean meals and two varieties of field beans using colostomised laying hens. – Br. Poult. Sci. **10**, 331–336.

WILSON, B. J., 1972: The nutritive value of field beans (*Vicia faba* L.) for poultry. – Ph. D. Thesis, University of Edinburgh. Edinburgh, Scotland.

Dr. DOMENICO LAFIANDRA
Dr. Gian Battista POLIGNANO
Prof. Dr. ENRICO PORCEDDU
Germplasm Institute
Via Amendola, 165/A
70126 Bari

ANDREA FILIPPETTI
Institute of Plant Breeding
University of Bari
Via G. Amendola, 165/A
70126 Bari

Kulturpflanze XXIX · 1981 · S. 129–136

# Rye genetic resources evaluation in the Botanical Garden of the Polish Academy of Sciences in Warsaw

BOGUSŁAW MOLSKI, ROMAN KUBICZEK, JERZY PUCHALSKI (Warszawa, Poland)

## Summary

The new Botanical Garden of the Polish Academy of Sciences was established in 1974 as an institution dedicated to collecting, conserving, and evaluating crop genetic resources and their wild progenitors. The main object of this research is rye (*Secale* L. spp.) – the cereal of great importance to Poland. The genetic resources of rye are preserved in living field collections, as well as in a seed stock bank. The living collection contains 1.193 taxa, including 12 species of *Secale* and 788 different rye cultivars. The seed bank stores 773 taxa of rye kept in glass containers (in vacuum, neutral gases or with silica gel) with optimum low moisture content in the seeds. Biochemical, physiological and genetical investigations are carried out for the estimation of safe and suitable conditions of seed dehydration and long term storage. These genetic resources of rye serve as material for taxonomic and phylogenetic studies, as well as for the evaluation of their selected breeding properties, especially seed protein content and its nutritional value. The taxonomic studies are performed using the isoenzyme technique and numerical methods. The enzymatic variation within the rye species, cultivars and inbred lines, are analysed. For the protein resources evaluation, different techniques are applied: the histochemical staining of seed slices. scanning electron microscope observations, and the chemical analyses of the entire amino acid composition of seed proteins. These results are used to select the most valuable taxa, which are then proposed to be used in the rye breeding programme.

## Introduction

The Botanical Garden of the Polish Academy of Sciences was established in 1974 as an institution obliged to perform research on plant genetic resources conservation. The main task is the investigation of crop plants and their wild progenitors, especially in the materials of important features for breeding cereal cultivars in Poland. Rye is the most important cereal crop in Poland because of the soil and climatic conditions. The 1978 production of rye in Poland of 7.43 million tons was 28% of the world rye production. In spite of its many deficiencies, rye will continue to be an important crop, because it has a number of advantages over other cereals. Owing to its extensive root system, rye is the most drought resistant crop among the cereals. It is more winter hardy than wheat and it yields better on poor, sandy soils, where no other useful crops can even be grown. For these reasons, rye is grown up to the subpolar regions in Scandinavia, in the southerly latitudes of Chile and at altitudes higher than 4.000 meters above sea level, for example in the Himalayas. Rye is also able to tolerate a range of soil acidity. The role of rye has now increased with the development of the *Triticale* breeding programmes. For years, the breeders improved the rye varieties by elevating their yield potential, resistance to different diseases, drought, frost and lodging. Therefore, rye varieties can now serve as very good sources, with many features, for the breeding of more valuable *Tritivale* cultivars. In Poland, rye production is very important, since 70% of rye grains are used as a fodder for animal nutrition. The most important consideration is that the protein content in rye and its nutritive value are the main problems in the improvement of new rye varieties.

## Collections of rye genetic resources

Taking into consideration all the present advantages of rye, the Botanical Garden of the Polish Academy of Sciences started to collect the genetic resources of this crop. The main source of seed stock is an exchange with other research institutions, including crop plants research institutes, plant breeding stations, other botanical gardens and genetics and agronomy departments of universities. Totally, we have received 1.324 seed samples from 73 different institutes. The most valuable seed samples were obtained from the Vavilov Institute of Plant Industry in Leningrad, the Central Institute of Genetics and Crop Research of the Academy of Sciences of German Democratic Republic in Gatersleben, and the United States Department of Agriculture Agricultural Research Center in Beltsville, Maryland.

We began the collection of rye in 1971 in the former Department of Crop Germ Plasm Conservation with 134 taxa of rye, and in 1980 our field collection of rye had increased to 1.193 taxa, including 788 different rye cultivars and 12 species of *Secale*. This rye collection also contains specially selected materials, such as 74 plots of high protein varieties, 106 plots of materials distinguished by their genetical and morphological properties, 120 plots of dwarf forms of rye and 15 short straw forms of rye. The Botanical Garden in Poland is able to keep the leading collection of wild species and primitive forms of rye. This part of our collection, grown on 77 plots, contains 32 taxa of 12 *Secale* species.

Because of rye xenogamy, the most important problem in keeping the large collection of this genus was to prepare exact isolation. Space isolation with such a large collection is impossible, therefore, linen bags isolation is used. Preliminary experiments showed the influence of different shapes of isolators and densities of sowing on several factors of grain yield. Therefore, methodological experiments on these topics were carried out. The first purpose of the investigations was to find the optimal conditions for reproduction. A six year experiment let us prove that the density of sowing 20×3 cm gives the highest yield from a plot 1,0×1,5 m and is best for reproduction purposes. However, using the sowing density 12,5×3,0 cm produced a very similar yield of grains and harvested seeds, simultaneously originating from more single plants, represented a wider genotypic spectrum of defined taxon. Triangular isolators (called tent shaped) always gave a higher yield of grains than the rectangular shaped because of better contact within pollinating spikes.

Moreover, the protein content in rye grains was affected to a lower extent when compared with the unisolated control plants (MOLSKI et al. 1980). Thus, the sowing density 12,5×3 cm and triangular linen isolators are used in our rye collection.

Whole field collection of rye is estimated in phenological observations and analyses of some elements of the yield structure. The selected rye species and cultivars are used for the scoring of protein resources and taxonomic investigation.

## Conservation of rye genetic resources

Most of our rye seed stock is stored in the Botanic Garden Seed Bank. The seeds, dehydrated up to low levels of moisture content (5–6%), are kept in glass containers, in vacuum, neutral gases or with silica gel, at a very low air humidity; and in this state, 773 samples of rye taxa are stored.

Our efforts are based on the assumption that a long term storage of rye seeds dehydrated to minimal percentage of water content gives as good a cryptobiotic effect as a low temperature (MOLSKI 1975).

The dehydration of seeds can cause disadvantageous changes in seed viability. Therefore, the biochemical, physiological and genetical analyses of dried seeds were performed. It was found that rye seeds may be safely dried to the moisture content of about 3,0% with further dehydration causing some somatic damage (GRZESIK and MOLSKI 1975). It seems that for long term storage the rye seeds should be pre-dried at 40 °C to a moisture content of about 5,0% and then dried at a temperature of 50–60 °C to 3,0% of moisture content. These applied conditions only reduced the germination percentage to 97% and seedling emergence in EGGEBRECHT's test to 92%.

More information fot the estimation of safe and optimal seed conservation conditions were obtained due to the results obtained in biochemical investigations. For example, it was found that the content of ATP in dried rye seeds was correlated with seed viability (BARTKOWIAK and MOLSKI 1977). Temperatures above 60 °C caused a high decrease of the ATP level in seeds, and also, a similar reduction of seed germination was observed. On the other hand, it was shown that soaling the seeds in ATP solution restores their viability.

Further investigations showed that rye enzymes are very thermostable during the dehydration of seeds at temperatures below 100 °C, and no changes in ribonuclease activity were observed (PIETRZAK and MOLSKI 1977). The studies on RNA, ribosomes and aminoacyl – tRNA synthetase in rye seeds dried at different temperatures proved that temperatures not exceeding 60 °C did not cause any significant changes (CUDNY et al. 1977).

## Evaluation of rye genetic resources

The large collection of rye provides many good possibilities for evaluations and selections the most valuable materials. On the basis of evaluated rye materials, the phylogenetic studies on rye were carried out and for this purpose, the isoenzyme technique was adapted. Isoenzymes, especially esterase, proved to be very sensitive and valuable markes of genetic variation within, and between, rye populations. It was found that rye cultivars exhibited a very high level of polymorphism (Puchalski and Molski 1975). The support of esterase isoenzymes made it possible to estimate the reciprocal affinity of analysed rye cultivars. Further analyses showed that the genetic purity of rye inbred lines is easily scored by isoenzyme analysis. The isoenzyme variation was also analysed in the wild species of *Secale*, and it was found that the breeding system has a large effect on the degree of intra-specific variation (Puchalski and Molski 1981). The self-fertile species, *Secale silvestre*, was completely monomorphic for its isoenzyme forms, and different levels of polymorphism in other species of *Secale* were estimated. On the basis of obtained isoenzyme patterns, the phylogenetic relationships between rye species were concluded. Besides the biochemical techniques, the taxonomic studies on *Secale* species are performed in support of morphological traits, and numerical taxonomy methods for the description of results are used.

Isoenzymes could also serve as markers for the evaluation of breeding properties in crop plants. In our studies, we are looking for the eventual correlation between the resistance of rye materials to snow mold (*Fusarium nivale, F. culmorum, F. avenaceum*) and peroxidase patterns. Preliminary studies showed that varieties of rye, resistant to snow mold, grown in conditions favourable for pathogen, exhibited a higher activity of particular isoperoxidase electrophoretic form, in comparison to susceptible rye varieties (Puchalski 1980, unpubl.).

In spite of the opinion that between cultivars of *Secale cereale* L. a wide variation in protein content is not to be expected and should be lower than the variation caused by site and weather (Kotvics 1970), the determination of grain protein content and its amino acid composition actually proved such a large variation, not only between rye varieties (Kubiczek and Rakowska 1974), but also between single grains within the variety (Kubiczek et al. 1980). By the way, the absence of negative correlation between the grain size and weight and their protein content (%) within the rye varieties was found. There is also a great potential of high protein, rich in lysine and methionine, resources in other species than *Secale cereale*.

Therefore, theoretically, the increase by breeding rye grain protein content of higher nutritive value, without diminishing of grain size and grain yield, should be possible.

For the breeding program, there had to be a variety of quantitative and qualitative evaluation methods developed. For screening purposes of numerous samples, in the first stages of the breeding programme, the combined methods of DBC (in mg of dye binded by a constant amount of protein) with the total protein by Kjeldahl method, determinations were proven to be suitable and adequate (Kubiczek and Chojnacki 1976). The limiting nutritive values for rye grain protein were not only lysine, but also other essential amino acids mainly methionine (Kubiczek et al. 1975). Therefore, for the evaluation of nutritive value grain protein in selected materials received as a result of breeding work, the whole amino acid composition of protein should being done. However, rye grains are rich in unnutrient substances, diminishing the digestibility of storage protein and starch; for that reason, the biological test for the final determination of the nutritional value of the best-bred materials must be performed. Such a biological test using a small laboratory bird, the Japanese quail, as a testing organism was developed and is still used in Poland for evaluation purposes, therefore making it possible to assay the small samples of grains. Using this biological method, is was found that higher concentrations of resorcinols in rye grains, compared with other cereals, do not influence significantly its nutritional value, as was suggested some years ago. We theorized from the beginning that the decisive role of pentosans which appear in rye caryopses in larger amounts diminishes the digestibility and biological value of rye grains components (Kubiczek 1975). The final results received confirmed our previous theories (Marquardt 1980; Kubiczek 1980).

Simultaneously for genetic experiments there was a simple and rapid nondestructive microscopy test developed using the histochemical staining of proteins with Light green on cross-grain sectioned slices (Łuczak 1980). It is used by the Warsaw Agricultural University and our Botanical Garden for the evaluation of protein distribution, and its approximate amounts in single rye grains, with the possibility of growing plants from the embryo containing the second-half of the grain.

To establish what is the influence of environmental (like soil and climatic conditions) and genetic factors on the protein content in rye grains, and the protein's nutritive value inheritance, several experiments were carried on. Using the phenomenon of xenia, the influence of cross-pollination on rye grains protein content was found (Małuszyńska et al. 1980). Very interesting results have been obtained, in as yet uncompleted experiments, on the storage of proteins potential by rye varieties differing strongly in grain protein content with the same conditions of growth, under different nitrogen fertilisers doses, and with optimal soil humidity conditions (lysimeter experiments).

There are also some investigations dealing with changed protein proportions and their distribution within the kernel performed by means of scanning electron microscopy techniques. Higher protein rye grains have more proteinaceous matrix in deeper endosperm cells, rich in protein matrix subaleuron region, and are wider than low protein ones, alike in *S. cereale* L. cultivars and in wild *Secale* species (Kubiczek et al. 1981). This technique is also used for pentosans action observations during its swelling in water-soaked rye grains, which is probably the main reason for the low digestibility of rye proteins and starch (Kubiczek 1980).

## Zusammenfassung

Beurteilung der genetischen Ressourcen des Roggens im Botanischen Garten der Polnischen Akademie der Wissenschaften in Warschau

Der neue Botanische Garten der Polnischen Akademie der Wissenschaften wurde 1974 als eine Institution gegründet, die sich der Sammlung, Erhaltung und Beurteilung der genetischen Ressourcen von Kulturpflanzen und deren Wildformen widmen soll. Hauptgegenstand dieser Untersuchung ist der Roggen (*Secale* L. spp.) – ein Getreide, das für Polen von großer Bedeutung ist. Die genetischen Ressourcen des Roggens werden sowohl in lebenden Sammlungen auf dem Felde als auch in einer Samenbank erhalten. Die lebende Sammlung enthält 1193 Taxa (12 Roggen-Arten und 788 Roggen-Sorten). Die Samenbank umfaßt 773 Roggen-Taxa in Glasbehältern (im Vakuum, neutralen Gasen oder Silikagel) mit optimal niedrigem Samenfeuchtigkeitsgehalt. Biochemische, physiologische und genetische Untersuchungen werden durchgeführt zur Ermittlung sicherer und passender Bedingungen der Samenentwässerung und Langzeitlagerung. Diese genetischen Ressourcen des Roggens dienen als Material für taxonomische und phylogenetische Studien sowie für die Beurteilung ihrer züchterischen Eigenschaften, besonders Eiweißgehalt der Samen und sein Nährwert. Bei den taxonomischen Studien werden die Isoenzym-Technik und numerische Methoden angewendet. Die enzymatische Variation innerhalb der Roggen-Arten, -Sorten und -Inzuchtlinien werden untersucht. Für die Beurteilung der Protein-Ressourcen werden verschiedene Techniken angewendet: histochemische Färbung von Samenschnitten, Beobachtungen mit dem Raster-Elektronenmikroskop und chemische Analysen der gesamten Aminosäure-Zusammensetzung der Samen-Proteine. Diese Ergebnisse werden benutzt, um die wertvollsten Taxa aufzufinden, die dann zur Anwendung in Roggen-Zuchtprogrammen vorgeschlagen werden.

## Краткое содержание

**Оценка генетических ресурсов ржи в Ботаническом саду Польской академии наук в Варшаве**

Новый Ботанический сад Польской АН был основан в 1974 г. как учреждение, которое должно заняться сбором, сохранением и оценкой генетических ресурсов культурных растений и их дикорастущих форм. Главным объектом исследований является рожь, имеющая для Польши очень важное значение. Генетические ресурсы ржи сохраняются как в живых полевых коллекциях, так и в семенном генном банке. Живая коллекция насчитывает 1193 образца (12 видов и 788 сортов ржи). В семенном банке находятся 773 образца ржи в стеклянных сосудах (в вакууме, нейтральных газах или силикогеле) при оптимально низкой влажности семян. Проводятся биохимические, физиологические и генетические исследования, с целью выявления надёжных и подходящих условий для обезвоживания семян и их долгосрочного хранения.

Эти генетические ресурсы ржи служат материалом для таксономических и филогенетических исследований, а также для оценки их селекционных качеств, в особенности содержания белка семян и их питательности. Для таксономических целей применяют нумерические методы и исследования изоэнзимов (их изменчивость внутри видов, сортов и инбредных линий ржи). Для оценки белковых ресурсов применяют различные методы: гистохимическую окраску тонких срезов семян, наблюдения сканирующим электронным микроскопом и химические анализы общего состава аминокислот семенных протеинов. Результаты исследований используются для выявления наиболее ценных образцов, которые будут предложены для программы селекции ржи.

## Literature


Bartkowiak, S., and B. Molski, 1977: Changes of ATP content in rye grains of varied viability caused by their thermic dehydration and storage in different conditions (in Polish). – Zesz. Probl. Post. Nauk Roln. (in press).

Cudny, H., B. Wojtowicz, and B. Molski, 1977: Some biochemical problems of rye grains preparation for long-term storage by their dehydration (in Polish). – Zesz. Probl. Post. Nauk Roln. (in press).

Grzesik, M., and B. Molski, 1975: The effect of temperature and dehydration of the dried seeds on the biological properties of rye. – Hod. Rośl. Aklim. Nasien. **19**, 449–458.

Kotvics, G., 1970: Investigations on increasing the protein content of *Secale cereale* L. – In: Protein growth by plant breeding, pp. 89–98. Ed. A. Balint, Akademiai Kiado, Budapest.

Kubiczek, R., 1975: Studies on application of Japanese quails (*Coturnix coturnix japonica*) for protein estimation in small samples of cereal grains (in Polish). – PhD thesis, Food Technol. Faculty, Agriculture University, Warsaw.

–, 1980: Pentosans of rye grain as a factor obstructing the evaluation of its nutritional value for poultry. – Acta Alimen. Polon. (in press).

– , and G. Chojnacki, 1976: Characteristic of the nutritive value of the protein of rye caryopses. II. Evaluation of rye grain protein quality by Dye-Binding Capacity method. – Acta Agrobot. **29**, 267–272.

–, and M. Rakowska, 1974: Characteristic of the nutritive value of the protein from rye caryopses. I. Amino acid composition of protein and the nitrogen forms in caryopses of ten rye varieties from the breeding collection. – Acta Agrobot. **27**, 105–113.

Kubiczek, R., W. Łuczak, and B. Molski, 1981: Protein resources of wild *Secale* species. – Kulturpflanze **29**, 159–167.

–, B. Molski, and M. Rakowska, 1975: Amino acids limiting the nutritional value of rye grain protein. – Hod. Rośl. Aklim. Nasien. **19**, 617–631.

– , W. Łuczak, B. Molski, and J. Moczydłowski, 1980: Comparison of the amount and distribution of seed proteins in endosperm cells of low- and high-protein varieties of rye (*Secale cereale* L.). – Acta Alim. Polon. (in press).

–, –, –, and J. Zajaczkowski, 1980: Intravarietal variability of rye grains protein content in dependence on their size and fulfilment (in Polish). – Hod. Rośl. **6**, 15–20.

Łuczak, W., 1980: Histochemical evaluation of protein distribution within cereal grains (in Polish). – Hod. Rośl. **6**, 20–25.

Małuszynska, M., B. Molski, R. Kubiczek, and W. Łuczak, 1980: Xenia phenomenon application in rye grain protein inheritance studies (in Polish). – Hod. Rośl. **6**, 25–28.

Marquardt, R. R., 1980: Identification and effects on chick performance of the antinutritional factor in rye. – EUCARPIA Conference on Rye and *Triticale*, Poznań, 1980.

Molski, B., 1975: Rye genetic resources conservation. – Hod. Rośl. Aklim. Nasien. **19**, 447–448.

Pietrzak, M., and B. Molski, 1977: Rybonuclease activity changes during drying of rye seeds being prepared for long-term storage (in Polish). – Zesz. Probl. Post. Nauk Roln. (in press)
Puchalski, J., and B. Molski, 1975: Esterases variability within some Polish rye cultivars. – Hod. Rośl. Aklim. Nasien. **19**, 479–485.
–, –, 1981: Isoenzyme variation within the wild *Secale* L. species. – Kulturpflanze **29**, 391–399.

Dr. B. Molski
Botanical Garden of the Polish Academy of Sciences
ul. Prawdziwka 2
00-973 Warszawa p - 84, Poland

Kulturpflanze XXIX · 1981 · S. 137–145

# Primitive and modern forms of *Vicia faba*

José I. Cubero and Maria-José Suso (Córdoba, Spain)


## Summary

Two subspecies were recognized by Muratova in *V. faba*: *faba* and *paucijuga*. *Paucijuga* is hardly represented in present collections, but some of its accessions present a very primitive aspect, resembling other *Vicia* species, both because of their growing habit (very small height, many branches, no principal stem, etc.) and their genetic architecture. But some other characteristics (indehiscent pods, extreme self-fertility) suggest some degree of domestication.

Principal Component Analysis using forms belonging to the two Muratova's subspecies showed that some key characteristics in the taxonomy of *V. faba* have only a secondary discriminatory value. The best clustering was obtained using seed size, a fact suggesting that the evolution of the known cultivated forms has occurred under domestication and that main seeds were the main objective in selection. Genetic studies indicate that these taxonomic characters (number of leaflets per leaf, thickness/length of the seed) are under polygenic control.

These primitive forms cross readily with all the other botanical types of *faba* beans. In particular, crosses with the most extreme forms of modern cultivars (belonging to the *major* group) are perfectly possible.

As a conclusion, only one subspecies has to be recognized for the cultivated forms (the only known one up to date) of *faba* bean: *Vicia faba faba*. The *paucijuga* forms described here probably are those most similar to the wild types.


*Faba* beans (*Vicia faba* L.) is an old cultivated species belonging to the Near East agricultural complex (Harlan 1975) but, up to now, its wild ancestor remains unknown. The morphologically most similar *Vicia* species (*V. narbonensis*, *V. galilea*, *V. johannis*) do not cross with *V. faba*; furthermore, the protein profile of these three appears to be rather different from that of the latter (Ladizinsky 1975). Taxonometric studies (Bueno 1976) show the great distance separating *V. faba* from its nearest *Vicia* species. DNA studies (Chooi 1971) confirm this isolation: the DNA content per nucleus is twice as large in *Vicia faba* as that in the other *Vicia* species.

The only possible way to understand the changes produced by domestication and selection would be study of primitive forms. First of all, we have to define what we will consider a 'primitive form'. There are two ways of defining this concept. The first one is to consider as 'primitive' any landrace cultivated by people living in a 'primitive' way; that is, that kind of vegetal material to which very little selective effort has been put by men. Such cultigens will also be used in a 'primitive' way; it is possible to list a set of uses of *faba* beans from the most primitive to the most modern ones:

(a) Use of dry seeds to get flour for human food. This is in fact the most primitive use of both cereal and pulses, and is restricted today to the most underdeveloped regions.

(b) Use of dry seeds to be cooked and seasoned, in the same way as chickpeas, lentils and common beans are today.

(c) Use of the crop as green manure.

Both uses were known by Romans, and are restricted today to regions with a low husbandry level.

(d) Use of dry seeds to feed animals. Only possible where and when animals were abundant and both the farming systems and the human food were of better quality. One of the modern aims in the modern breeding of *faba* beans is just this one. Most of the selected cultivars belong to the *minor* and *equina* groups.

(e) Use of green seeds for human consumption either without any special preparation -just salt- or cooked- mainly boiled and seasoned. Cultivars or landraces used for this purpose belong to the *major* group, whose seeds were the latest ones appearing in the archaeological record: by the 10$^{th}$ century (HANELT 1972, and personal communication; they had to be obtained some centuries before, perhaps by the end of the Roman period). Furthermore, green seeds used in such a way have not the characteristic bitter taste of all the other seeds: the absence of this bitter principle means a further step in selection.

In some places, cultigens with long pods containing up to 8–9 large seeds, weighing up to 2 grammes each, and with thick, fleshy valves have been obtained. In some cases, these pods are all eaten. 'Aguadulce' is a very popular landrace in this sense.

(f) Use of green seeds for canning or freezing. This is the most modern use of *faba* beans.

Unfortunately, the collections of living forms have not a record of how the different accessions being used at the time of picking them up. The reason is, not only because collectors are rarely anthropologists, but also because uses are not in a pure state. So, an undoubtedly primitive people will sow *major* cultigens to use their dry seeds as the staple for a soup, in the same way as they will use lentils. Or in a rather developed farm *major* seeds will be mechanically harvested to be used in the animal food industry. As a conclusion, the primitiveness of any cultigen because of its cultivation by a primitive community and/or by its use has to be supported by other data, for example, anthropological ones.

A second way to define a cultigen as 'primitive' is the botanical one: the stronger the similarity to wild related species (*Vicia* species in this case), the more primitive it is. It is a common fact, in other species, that these forms are strikingly different from the 'normal' ones, even if the latter are primitive concerning their use.

We have localized in our collection three accessions of a striking aspect which makes them look very different from any other accession: very low height (25–30 cms), many branches (up to 12–15) with no principal stem, weak and tiny aspect, small number of leaves and leaflets per plant, very small seeds and pods, very low position of the first pod (at the soil level in some cases). Rather than the best description, it is their general aspect which makes them distinctive. If is tempting to consider them as authentic wild forms, but they are not: the pods of the three accessions are truly indehiscent.

Unfortunately, the three accessions came from exchanges, so we do not know their exact original place of cultivation. The only ecological datum we have noticed is that they grow and yield much better in hot than in cool environment, and at a high sowing density than at a low one. The three were classified as belonging to the ssp. *paucijuga* (MURATOVA 1931). MURATOVA split her ssp. *faba* in three varieties: *minor*, *equina* and *major*. HANELT (1972) – considered a different system: ssp. *minor* and *faba*. Without discussing these systems, we will accept here for convenience the four names; as we will justify later, we do not accept the existence of two subspecies in the cultivated forms (the only ones known up to now) of *V. faba*.

Worthy of mention, the putative use of these three accessions had to be (a) and/or (b) among those listed above. The scarcity of their green mass does not make them suitable as green manure and their poor yield per plant makes them unsuitable to feed animals. This is not the case of other apparent *paucijuga* accessions of our collections, showing a bigger height and a more luxurious vegetative aspect.

We have studied these accessions comparing them with modern cultivars from *equina* and *major* groups, selected either for dry seeds as fodder or for green seeds for human consumption. We did not include any *minor* cultivar (sensu MURATOVA) because there were not selected cultivars adapted to our conditions (MORENO and MARTINEZ 1980). We will give here a concise glimpse of the main aspects of our study.

## (1) Paucijuga vs Faba

Eight *paucijuga* accessions were compared to eight *faba* ones by means of an analysis of the variance according to the model

$$x_{ijk} = m + s_i + c_{j(i)} + e_{ijk}$$

where $x_{ijk}$ represents the individual observation, m the great mean, $s_i$ the effect due to the subspecies i, $c_{j(i)}$ that of the cultigen j within the i subspecies and $e_{ijk}$ the experimental error. The results are summarized in table 1; it can be seen that differences were significant for all the characters studied within subspecies, but not between them. An important Muratovian taxonomic characteristics as the number of leaflets per leaf did not show differences (to tell the complete truth, the MURATOVA's basic character was the maximum number of leaflets per leaf, but there is a strong correlation between the maximum and the average). The lack of significance concerns mainly vegetative characters.

Table 1

*Paucijuga* vs *Faba* analysis of variance

| Character | Significance* between ssp. | within ssp. | *Faba* (F) vs *Paucijuga* (P) |
|---|---|---|---|
| 1. Leaflet/leaf | NS | (2) | – |
| 2. Leaflet Density | NS | (2) | – |
| 3. Rachys length | (2) | (2) | F > P |
| 4. Leaflet length L | (1) | (1) | F > P |
| 5. Leaflet width W | (2) | (2) | F > P |
| 6. Leaflet W/L | NS | (2) | – |
| 7. Flowers/raceme | NS. | (2) | – |
| 8. Ovules/ovary | NS | (2) | – |
| 9. Pod length | (2) | (2) | F > P |
| 10. Seed length L | (2) | (2) | F > P |
| 11. Seed width W | (2) | (2) | F > P |
| 12. Seed Thickness | (2) | (2) | F > P |
| 13. Seed T/W | (2) | (2) | F > P |
| 14. Seed T/L | (2) | (2) | F < P |
| 15. Number of branches | NS | (2) | – |

* (1) Significant differences at the 5% level
(2) Significant differences at the 1% level
NS No significant differences.

## (2) Principal Component Analysis (PCA)

Data from fourteen characters and the sixteen cultigens just mentioned were studied by PCA. Table 2 shows the three main eigenvalues and eigenvectors

Table 2

Principal Component Analysis

| | Vector | v1 | v2 | v3 |
|---|---|---|---|---|
| | Eigenvalues | 7.56 | 3.23 | 1.21 |
| % | Variation explained | 54 | 23 | 9 |
| % | Variation accumulated | 54 | 77 | 86 |
| | **Character** | **Eigenvector** | | |
| 1 | Leaflets/leaf | 0.42 | 0.68 | −0.25 |
| 2 | Leaflet density | −0.43 | **0.83** | 0.28 |
| 3 | Rachis length | 0.67 | −0.50 | −0.51 |
| 4 | Leaflet length | 0.71 | −0.38 | −0.29 |
| 5 | Leaflet width | 0.78 | −0.51 | 0.30 |
| 6 | Leaflet W/L | 0.51 | −0.41 | 0.74 |
| 7 | Flowers/raceme | 0.41 | **−0.81** | 0.06 |
| 9 | Pod length | **0.84** | 0.47 | 0.08 |
| 10 | Seed length | **0.94** | 0.31 | 0.01 |
| 11 | Seed width | **0.94** | 0.30 | 0.08 |
| 12 | Seed thickness | **0.88** | 0.31 | −0.07 |
| 13 | Seed T/W | **0.94** | 0.20 | 0.19 |
| 14 | Seed T/L | **−0.90** | −0.22 | 0.07 |
| 15 | Number of branches | 0.51 | −0.04 | −0.15 |

and the proportion of the total variation explained by each one. First vector is mainly determined by pod and seed characters (roughly two thirds of the first eigenvalue), and is largely the most important one, determining by itself more than the 50% of the total variation. The second vector is much less important than the first one: leaflet density and flowers per raceme are its most important characteristics. Third vector is unimportant.

Figure 1 shows the distribution of the 16 cultigens when the mean values of the characteristics studied are transformed using the first two eigenvectors. The sixteen dots can be grouped in five clusters, and five parameters (mean values per cluster) are given as a help for the physical interpretation. It can

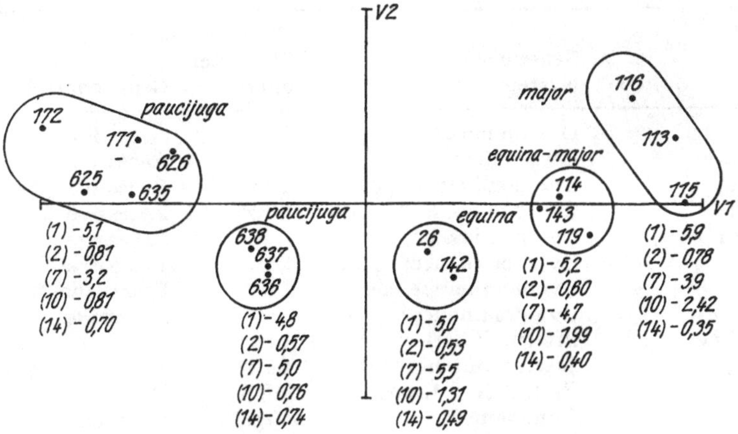


Fig. 1 Principal Component Analysis

be seen that from right to left, the characters showing the strongest variation are those of the seed (all the seed and pod characters are strongly correlated; MARTIN 1976), and from top to bottom, leaflet density and flowers per node. It can be seen how leaflets per leaf is practically unimportant. Undoubtedly, the biggest variation is shown by the first vector.

According to these results it seems that the really important discriminatory characters are those of seed and pod, in fact only one given the high correlations found among them, both at the phenotypic and genotypic level (MARTIN 1976). Other key characteristics, namely the vegetative ones, seem to be of a secondary value if any. This importance of seed characters suggests that seed was the main interest for primitive farmers, and as a consequence, that the evolution of cultivated forms has occurred under domestication.

## (3) Genetic analysis

We have performed extensive crosses between lines belonging to all the groups mentioned above; crosses are possible irrespective of the botanical group out of which the lines have been obtained (CUBERO 1973, 1974, 1976). We will refer here only to those concerning crosses including two out of the three primitive forms described above (VF 171 abd 172). The genetic systems listed in table 3 have been obtained by using diallel analysis in $F_1$ and $F_2$ as well as by the Cavalli analysis with different segregating generations (CUBERO 1976; MARTIN 1976; MARTIN and CUBERO 1978). For the sake of brevity we cannot discuss here the particularities of each case, but it is worthy of mention that, in general, *equina* and *minor* lines from cool regions have a tendency to show overdominance in

Table 3

Genetic architecture of *V. faba* primitive forms

| | Character | Genetic system (system, sense) | Primitive forms Character expression | Genic control |
|---|---|---|---|---|
| 1 (1) | Leaflets/leaf | Overdominance, pos. | Low | Recessive |
| 2 (2) | Leaflet/density | Partial dom., neg. | High | Recessive |
| 3 (3) | Rachis length | Overdominance, pos. | Low | Recessive |
| 4 (4) | Leaflet length | Overdominance*, pos. | Low | Recessive |
| 5 (5) | Leaflet width | Overdominance, pos. | Low | Recessive |
| 6 | Leaflet area | Overdominance*, pos. | Low | Recessive |
| 7 (6) | Leaflet W/L | Overdominance, neg. | Low | Dominant |
| 8 | Leaves/plant | Overdominance*, pos. | Low | Recessive |
| 9 (7) | Flowers/raceme | Interm. inheritance* | Low | – |
| 10 (8) | Ovules/ovary | Interm. inheritance* | Low | – |
| 11 | Pods/flower | Partial dom.*, pos. | High | Dominant |
| 12 | Seeds/ovule | Dominance*, pos. | High* | Dominant* |
| 13 | First pod node | Dominance, pos. | Low | Dominant |
| 14 | Pods/node | Overdominance*, pos. | High* | Dominant* |
| 15 | Seeds/pod | Overdominance, pos. | High | Dominant |
| 16 | Pod T/W | Interm. inheritance | High | – |
| 17 | Pods/plant | Overdominance*, pos. | High | Dominant |
| 18 | Seeds/Plant | Overdominance, pos. | High | Dominant |
| 19 (10) | Seed length | Partial dom.*, neg. | Low | Dominant |
| 20 (13) | Seed T/W | Partial dom., neg. | High | Recessive |
| 21 (14) | Seed T/L | Partial dom., neg. | High | Recessive |
| 22 | Seed weight | Partial dom.*, neg. | Low | Dominant |
| 23 | % seed protein | Partial dom., neg. | Low | Dominant |
| 24 | Yield/plant | Overdominance, pos. | Low | Recessive |
| 25 | Plant height | Partial dom.*, pos. | Low | Recessive |
| 26 (15) | Number of branches | interm. inheritance | High | – |

* Heterogeneity. The genetic system can show differences according to the parentals.

some characters and, on the other hand, Mediterranean *major* and *equina* lines, as well as the two *paucijuga* lines included in the crossing plans, intermediate inheritance or weak partial dominance, giving sometimes the appearance of partial dominance mainly if the study is undertaken by using diallel analysis.

Thus, table 3 is a rather rough glimpse of the genetic systems of some characters to be perfected in future works, mainly by studying individual crosses.

Be that as it may, it is striking that the two primitive forms show recessive systems for all the vegetative characters (excepting the leaflet shape, lanceolate for them) and dominant ones for those concerning seeds and pods (excepting seed shape). One possible explanation is to suppose that selection was performed by the first farmers looking for yield. International trials on course are showing that within a large range of conditions, cultivars with large seeds are, on the average, more productive than those with small seeds (*minor* types). So, recessive genes for bigger seeds were fixed in the more and more productive genotypes. (Yield/plant is a very complex character composed of all the others; it is not difficult to understand that a plant with fewer but bigger seeds than others can be more productive than these ones). So, because of a correlated response to selection for higher yield, cultigens with bigger seeds and a cluster of correlated characters were produced. Perhaps forms with more and bigger leaves were the best adapted to new conditions.

## (4) Use of the primitive forms in faba bean breeding

Our primitive accessions show a cluster of characteristics, some totally unsuitable to be agronomically used, but some, of the greatest interest for the same purpose. Among the latter, their extreme self-fertility is perhaps the most interesting one. Being given the low level of genetic variability of seed size etc. in these populations (at least in the samples we have studied), the only possibility to increase yield had to be to select for extreme fertility; now, the insect population being independent of primitive man, the main way was to obtain self-fertile varieties.

Self-fertility is being transferred by crossing to modern *major* and *equina* cultivars; lines with up to five pods per node are being derived; the indehiscency of these lines (another way of increasing yield) is a help, but not all other characters as the smallness of the seed, the low position of the first pod, their susceptibility to rust etc. These lines show that the cluster of characters present in our primitive forms can be broken up: forms with more leaves, greater seeds, lower number of branches than the parentals are on the way.

## (5) The systematics of Vicia faba

If we accept the biological concept of species, the only possible systematics for cultivated plants is that proposed by Harlan and De Wet (1971): subspecies are defined by fertility barriers, including generally in only one subspecies all the cultivated forms and grouping the wild ones in one or several distinct subspecies.

According to this idea, all the known forms of *V. faba* have to be grouped in only one subspecies, because there are not fertility barriers among them, even between modern forms and those that seem primitive by both their morphology and potential use. Not only are there no fertility barriers, but it is easy to transfer genetic characteristics to get new cultivars.

The wild ancestor of *V. faba* has to be very similar to the primitive cultigens described here, but with at least two important differences: dehiscent pods and a certain degree of allogamy. The most probable colour of seed coats was brown, even black, as most of the *Vicia* species. VF 172 has black-coloured seeds; when crossed with large seed varieties, the seeds in the descendence are brown, but the intensity is lost.


## Zusammenfassung

Primitive und moderne Formen von *Vicia faba*

Bei Vicia faba werden von Muratova zwei Subspecies unterschieden: *faba* und *paucijuga*. *Paucijuga* ist in den derzeitigen Kollektionen kaum vertreten; einige dieser Proben vermitteln aber auf Grund ihrer Wuchsform (sehr geringe Höhe, starke Verzweigung, fehlender Hauptsproß u. a.) und ihrer genetischen Beschaffenheit einen sehr primitiven Eindruck. Einige andere Merkmale (nichtplatzende Hülsen, extreme Selbstfertilität) weisen aber auf ein gewisses Maß der Domestikation hin.

Die Hauptkomponenten-Analyse bei Formen, die zu beiden von Muratova unterschiedenen Unterarten gehören, zeigte, daß einige Schlüssel-Merkmale in der Taxonomie von *Vicia faba* nur sekundäre Bedeutung für die Unterscheidung haben. Die beste Gruppierung wurde bei Verwendung der Samengröße erhalten, eine Tatsache, die einen Hinweis darauf gibt, daß sich die Evolution der bekannten kultivierten Formen unter Domestikationsbedingungen vollzog und daß vor allem die Samen den Hauptgegenstand der Selektion bildeten. Genetische Untersuchungen zeigen, daß diese taxonomischen Merkmale (Anzahl der Blättchen je Blatt, Dicke/Länge der Samen) polygen bedingt sind.

Diese Primitiv-Formen kreuzen sich leicht mit allen anderen botanischen Typen der *faba*-Bohnen. Insbesondere sind Kreuzungen mit extremen Formen der modernen Sorten (die zur *major*-Gruppe gehören) durchaus möglich.

Daraus ergibt sich die Schlußfolgerung, daß die kultivierten Formen der *faba*-Bohne nur zu einer (z. Zt. der einzigen) Unterart zu rechnen sind: *Vicia faba faba*. Die hier beschriebenen *paucijuga*-Formen haben wahrscheinlich die größte Ähnlichkeit mit den Wildtypen.



## Краткое содержание

Примитивные и современные формы *Vicia faba* L.

Муратова различает у *Vicia faba* два подвида: *faba* и *paucijuga*. Последний почти не представлен в современных коллекциях; некоторые образцы ssp. *paucijuga* производят впечатление очень примитивных форм по их габитусу (низкорослость, сильная ветвистость, отсутствуе главного стебля и пр.) и генетической конституции. Однако другие признаки (невскрывающийся боб, полная самофертильность) указывают на известную степень одомашненности.

Анализ главных компонентов у форм, принадлежащих к обоим подвидам Муратовой показал, что некоторые ключевые признаки в таксономии *Vicia faba* имеют только вторичное значение для их различаемости. Наиболее удачная группировка получилась на основании величины семян. Этот факт указывает на то, что эволюция известных культурных форм протекала в условиях одомашненности и, что отбор вёлся, в первую очередь, по признакам семян. Генетические исследования показали, что такие таксономические признаки как отношение толщины к длине семени, число листочков в сложном листе – обусловлены полигенно.

Описанные примитивные формы легко скрещиваются со всеми прочими типами конских бобов. В частности – вполне возможны скрещивания с самыми современными сортами, принадлежащими к группе *major*.

Можно сделать заключение, что культурные формы конских бобов принадлежат только к одному современному подвиду *Vicia faba faba*. Описанные здесь формы *paucijuga*, очевидно, имеют наибольшее сходство с исходным дукорастущим типом.

## Literature


Bueno, M. A., 1976: Citotaxonomía y cariología en el género *Vicia*. – Ph. D. Thesis, Facultad de Biología, Universidad Complutense, Madrid.

Chooi, W. Y., 1971: Variation in nuclear DNA content in the genus *Vicia*. – Genetics **68**, 195–211.

Cubero, J. I., 1973: Evolutionary trends in *Vicia faba*. – Theor. Appl. Genetics **43**: 59–65.

–, 1974: On the evolution of *Vicia faba*. – Theor. Appl. Genetics **45**, 47–51.

–, 1976: Herencia de caracteres cuantitativos en *Vicia faba* L. – Servicio de Publicaciones INIA, 113 pp. Madrid.

Hanelt, P., 1972: Die intraspezifische Variabilität von *Vicia faba* L. und ihre Gliederung. – Die Kulturpflanze **20**, 75–128.

Harlan, J. R., 1975: Crops and man. – American Society of Agronomy, Madison.

–, and J. M. J. De Wet, 1971: Toward a rational classification of cultivated plants. – Taxon **20**, 509–517.

Ladizinsky, G., 1975: Seed protein electrophoresis of the wild and cultivated species of section *Faba* of *Vicia*. – Euphytica **24**, 785–788.

Martin, A., 1976: Genética de los componentes del rendimiento en *Vicia faba*. – Ph. D. Thesis. Escuela Técnica Superior de Ingenieros Agrónomos. Córdoba.

–, and J. I. Cubero, 1978: Inheritance of quantitative characters in *Vicia faba*, in Some current research on *Vicia faba* in Western Europe, ed. by the European Communities.

Moreno, M. T. and A. Martinez., 1980: The divided world of *Vicia faba*. – Fabis 2: 18.

Muratova, V., 1931: Common beans (*Vicia faba* L.). – Bulletin of Applied Botany of Genetics and Plant Breeding, Supplement **50**, Leningrad.



Dr. J. Cubero
Departamento de Genética, Escuela Técnica Superior de
Ingenieros Agrónomos, Apartado 246, Córdoba, Spain.

Dr. M.-J. Suso
Centro Regional de Andaluciá, Instituto Nacional de
Investigaciones Agrarias. Apartado 240, Córdoba, Spain.

Kulturpflanze XXIX · 1981 · S. 147–150

# Origin and evaluation of cultivated plums of Georgia

R. K. BERIDZE and M. V. KVATCHADZE (Tbilissi, USSR)

## Summary

The paper presents the results of microevolutionary studies within the genus *Prunus* L. New biological data have proved the earlier assumption that some forms of the local domestic plums (*Pr. domestica*) might have developped directly from wild alychas (cherry-plum, *Pr. divaricata*) and not by hybridization between sloe and cherry-plum (*Pr. spinosa* × *Pr. divaricata*).

The earlier system of the genus *Prunus* L. based on purely morphological indications which was formed by the taxonomists, failed to explain all the various manifestations of the coacting genetic factors. This forced us to undertake microevolutionary studies to find out the phylogenetic relations in this polymorphous plant group. Therefore we decided to investigate the structure of the species belonging to this genus from the territory of Georgia, which is not very large, but is considered to be the birthplace for many cultivated plants, including plums.

Great polymorphism of the cultivated plum (*Pr. domestica* L.) is due to the original polymorphism of its wild relatives and to the ecological conditions of Georgia where the plum originally had been developped and had been cultivated since long times.

In this relation we would like to emphasize the even greater polymorphism of *Pr. divaricata* LDB., one of the wild relatives of *Pr. domestica*, pure plantations of which can be often found in Georgia. Such prevalence of alycha (*Pr. divaricata*) is undoubtedly historically conditioned and indicates the antiquity of this species (the genus *Prunus* itself is considered to be a relic of arctotertiary flora of Middle Lands) and that it was cultivated much earlier than the house plum.

Our further studies of the polymorphous forms of wild and cultivated species of this genus were made using complex methods of caryology, cytology and palynomorphology the results of which have been published in a series of papers (BERIDZE and BAJAŠVILI 1971, 1975).

The main conclusion we have drawn is that two ways of *Pr. domestica* evolution can be distinguished: 1. natural hybridization of sloe and alycha which was experimentally proved by RYBIN, 1936; 2. direct descendence from wild alycha which we supposed have been proceeded after having discovered in Georgia the 48-chromosome race of this normally diploid (2n = 16) species.

The hexaploid interspecific hybrids (sloe × alycha) show morphological characters of alycha and sloe, their meiosis is significantly changed and fertility decreased. Moreover the flower of this hybrid shows interesting morphological peculiarities, i. e. in the pollen corn we find ores instead of pores which is characteristic for the supposed parental form. Thus the hybrid pollen resembles the pollen of the cultivated plum.

When we refer to the second way of the *Pr. domestica* evolution we propose as ancestor the 48-chromosome race of alycha (*Pr. divaricata*). Typical forms of this species don't show any characters of the sloe, it is highly fertile and – according to morphological and cytological indices – is rather a polymorphous chromosome race.

Because the 48-chromosome *Pr. divaricata* morphologically doesn't significantly differ from some diploid forms of this species (it is likewise not similar to other species of the genus as *Pr. spinosa, Pr. domestica, Pr. insititia*) and because it has increased indices of several character-grades we consider it as an alloautopolyploid, developped from different polymorphous forms of alycha.

The discovery of polyploid alycha in Georgia allows us to suppose that some forms of the highly polymorphous plums might have originated from hexaploid alycha types which might be their direct ancestors.

On the second stage of our studies using biochemical methods we tried to get a new confirmation of the experimental data we mentioned above and to distinguish the regularities of the evolutionary process for different species as well as for the whole genus.

Our aim was to define the variability of the content of phenolic compounds in particular flavones, anthocyanin and catechin, in extracts of fruit peel cells from different genotypes and from material growing in different environmental conditions (different regions or vertical zones).

We used different areas of Georgia occupied by forms we were interested in as living collections for the experiments.

In our studies we used the chromatographic method described by HARBORNE (1958, 1959) the main point of which was that separated substances–peculiar for every case – migrating from the solution to the paper with different mobility are revealed on the paper as discrete spots, their successive disposition being analysed by two dimensioned chromatography.

By the analysis of the absorption spectrum, the definition of Rf-values and their comparison for different species and different forms within a species it could be shown that all examined taxa of the genus (*Pr. divaricata, Pr. spinosa, Pr. institia* and *Pr. domestica*) contain flavones and anthocyanins but no catechins. Flavones and anthocyanins in the above mentioned species are represented in the form of complex bands, within the limits of which more narrow bands can be found.

All the 100 examined samples of *Prunus* (alycha-63, house plum-30, sloe-7) contain flavones and anthocyanins except alycha types with yellow coloration, which don't contain anthocyanins.

As we can not give here all the results of our work for every species separately we shall try to give a general view of the discovered diversity of the examined characters. With the help of paper chromatography we have revealed a great number of substances of different qualities in forms of spots, which according to the Rf indices formed 13 groups (alycha-72, house plum-50, sloe-10). Thus studies of chemical differences of the biochemical process, i. e. the synthesis of phenolic compounds in different organisms of one species, enables us to discover the smallest differences of the molecular structure of these substances in different species in order to distinguish these forms with extremely low or high diversity regarding the content of these substances within one species.

Least diversity in this respect was found in wild alycha and sloe forms (1–2 types); it gradually increased in natural and artificial interspecific hybrids (3–4–5 types) and reached the maximum in the house plum and in samples of cultivated alycha (5–6 types).

The obtained results distinctly show the influence of hybridization and of natural mutation process on the structure of the studied compounds which are typical for the species of this genus (alycha, sloe) and also demonstrate a gradual enrichment of these compounds during the course of evolution. We have discovered forms which have the ability of synthesizing variants of compounds which are absolutely lacking in the ancestral taxa.

The influence of the environment can be also easily seen. The polymorphism of some taxa (even of wild plums) increases with the height above the sea level (from 200 m above s. l. in Kutaisi and Tskhaltubo to 1000–1700 m in Racha and Shovi).

Besides the above mentioned we succeeded in distinguishing by the studied biological indices the whole range of transitional forms of natural hybrids. One part of these forms had no compounds peculiar for sloe and we attributed them to hybrid forms, developped from different types of alycha. Another part of these hybrids was joint by us into groups in which wild forms of sloe could be also found.

Analogous results have been obtained in the group of the house plum. Different forms of *Pr. domestica* s. str. and *Pr. insititia* show certain similarities in some types of their components with the wild sloe and they can be often found in one group with it. According to the contents of these substances local forms of red and yellow plums approximate to alycha forms.

An interesting fact was discovered. One of these chromatographically revealed spots of yellow plums completely coincided to one of them in wild alycha. This fact shows the complete identity of one of the components of these taxa.

The obtained biochemical results have confirmed once again our previous assumption that some forms of local house plum might have developped from strains of wild alycha without any hybridizations with *Pr. spinosa.*

## Zusammenfassung

Entstehung und Evaluierung der Kulturpflaumen von Georgien

Die Arbeit stellt Ergebnisse mikroevolutionärer Untersuchungen in der Gattung *Prunus* L. vor. Neue biologische Daten bestätigten die frühere Annahme, daß manche Lokal-Formen der Hauspflaume (*Pr. domestica*) direkt aus Wildformen der Kirschpflaume (*Pr. divaricata*) und nicht aus einer Bastardierung von Schlehe und Kirschpflaume (*Pr. spinosa* × *Pr. divaricata*) hervorgingen.

**Краткое содержание**

Происхождение и оценка культурной сливы Грузии

В работе представлены результаты микроэволюционных исследований внутри рода *Prunus* L. Новые биологические данные подтвердили прежнее мнение о том, что некоторые местные формы сливы (*Pr. domestica*) произошли прямо от дикорастущей алычи (*Pr. divaricata*), а не путём гибридизации тёрна и алычи (*Pr. spinosa*× *Pr. divaricata*).

## Literature

Beridze, R. K., and E. I. Bajašvili, 1971: Sravnitelnoe citologičeskoe izučenie sliv Kartli. Bjull. naučn. tech. inf., ser. bjull. n. 700462218.

–, 1975: Novye dannye o proischoždenii kul'turnoj slivy. – Abstr. pap. XII Intern. Bot. Congr. **II**, 514. – Leningrad.

Harborne, J. B., 1958: The chromatographic identification of anthocyanin pigments. – J. chromatogr. **1**, 473–488.

–, 1959: The chromatography of the flavonoid pigments. – J. chromatography **2** (6), 581–604.

Rybin, V. A., 1936: Gibridi terna i alyči i problema proischoždenija kul'turnoj slivy. – Trudy prikl. bot., genet., selekc., ser. II, no. **10**, 5–44.

Dr. R. K. Beridze
Department of Cultivated Flora,
Institute of Botany of the
Georgian Academy of Sciences,
Kodjorskoe Road,
SU - 380007 Tbilissi, USSR

Kulturpflanze XXIX · 1981 · S. 151–157

# Preservation of gene resources of potato species of different ploidy levels for the building up of breeding material

Horst Tiemann (Groß Lüsewitz, GDR)

## Summary

In the Institute for Potato Research, Groß Lüsewitz, a collection of wild and cultivated potato species of different valence levels is preserved in living state. Chiefly the preservation of the collection is done by seeds.

By pollination of tetraploid varieties and promise clones with pollen of *Solanum phureja* a collection of dihaploids was produced. This makes it easier to utilize the gene reservoirs of the 24 chromosome wild forms for the creation of basic material. Useful gene combination can be created by intra- and interspecific hybridisation provided that the dihaploid genotypes show efficient blossom intensity and pollen fertility. To increase the yield of diplogametic tetraploids for the genetic breedings it has to be searched for genotypes in the dihaploid collection that have a higher frequency of diplogametes.

The potato *Solanum tuberosum* L. has four genomes and is by this tetraploid. Starting from the basic number with $x=12$ chromosomes, in the free nature a polyploid series of tuber-bearing *Solanum* species is to be found: diploid $2n=2x=24$, triploid $2n=3x=36$, tetraploid $2n=4x=48$, pentaploid $2n=5x=60$, hexaploid $2n=6x=72$ (Rybin 1933).

Hence a large genetic and taxonomic variability of forms follows from the variety of valence levels. To analyse them in full scale and utilize them for breeding extensive collections were gathered as gene resources in many countries.

Especially in potato breeding the small genetic width of the European potato was enlarged successfully by interspecific crossings. So genes could be inserted for the resistance against phytophthora, nematodes, cancer and others, this was done by wild and cultivated species into the *Solanum tuberosum*-genome.

Higher demands to the new potato varieties require the systematic search of further gene resources from the genepool of the natural genesis and propagation area and the analysis of the inheritance value of the potato varieties.

Now it is reported about the preservation of the collection of wild and cultivated potato species as well as the production and utilization of dihaploids in our institute.

## Extend and preservation of the collection of wild and cultivated potato species (G-LKS)

According to the level of June 1980 the G-LKS includes 2718 samples in living state. The samples are descended directly from the natural propagation area of the potato or taken over from other collections, respectively. This was only possible by a broad international co-operation. Our collection is composed in valencematic in the following way:

| | |
|---|---|
| diploid | 2n=2x=24=34,7% |
| triploid | 2n=3x=36= 0,3% |
| tetraploid | 2n=4x=48=57,4% |
| pentaploid | 2n=5x=60= 0,4% |
| hexaploid | 2n=6x=72= 7,2% |

The preservation of the collection is done vegetatively by tubers and generatively by seeds.

By production technical reasons and the common high virus susceptibility the largest part of the collection is preserved with seed in a 6 years rhythm (ROTHACKER 1966, 1968; HINZE 1976). Following possibilities are used for seed yield: According to the sample in spring about 30 seeds are sown in a greenhouse, after that 9 potted seedlings are put in a plastic basin in a frame. At least 300 seeds per sample must be available as a constant genereservoir.

Samples which form neither seeds from selfing nor from sibcrosses are crossed within the species from other localities, and if this is unsuccessful, too, they are crossed with other species. If the number of seeds is not sufficient, the tubers of the seedlings will be grown in greenhouses respectively in the fields in the next year. Provided that then the production of berries is still unsatisfactory the tuber cultivation will be repeated. If the formation of blossom is impaired by too extensive virusinfections, virusfree basic material will be gained by the method of meristem culture for the new cultivation cycles respectively they can be get as in-vitro-culture (PETT 1974). The samples are checked morphologically for purity of species at the new starting point of the growing rhythm after 6 years. For this reason a herbar is used for every sample if possible in flowering state at the first growing. The seeds are stored at 6 degrees centigrade in bags on silica gel in industrial glasses, that are sealed air-tight. With this procedure the germinating capacity had only decreased from 85 to 76 percent on average after a storage of 8 years (HINZE et al. 1978). In that way the number of multiplications can be reduced essentially and the loss of genetic diversity can be minimized.

## Production and utilization of dihaploids of Solanum tuberosum L. for the creation of basic material

About 20 years ago the methodical production and utilization of dihaploids, stimulated by papers of HOUGAS and PELOQUIN (1958), was started in many countries.

By pollination of tetraploid varieties with pollen of clones of the 24 chromo-

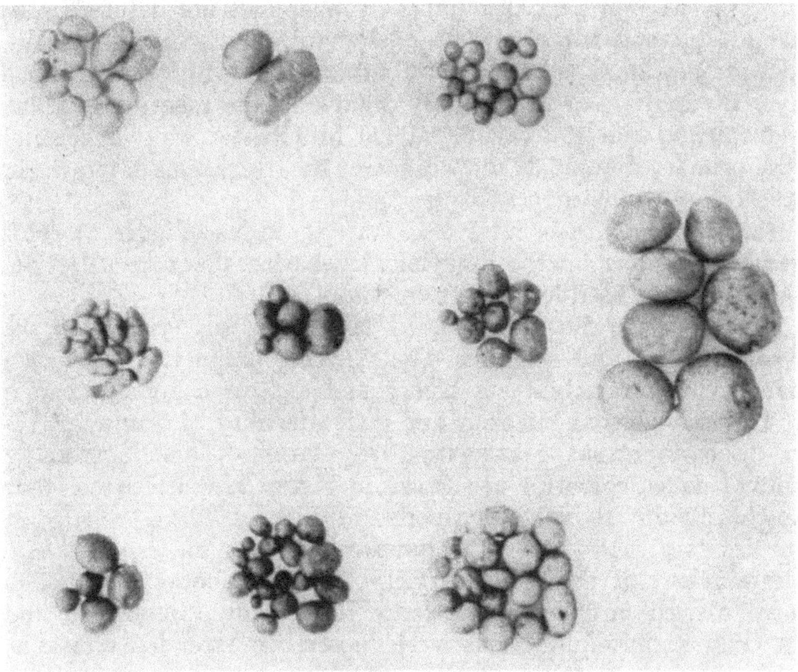

Fig. 1 Variability of the tuber properties of a primary dihaploid seedlings population. Right: the 4x parental variety 'Xenia'

Fig. 2 Growth habitus of primary genotypes in comparison with their parental form. Left: dihaploid genotype. Middle: parental form 'Mariella'. Right: dihaploid genotype

somic potato *Solanum phureja* a small number of dihaploids are induced. Also in our institute this method led to the production of a dihaploid collection, that is completed by new crossings yearly. The utilization of dihaploids is done in different ways. By analysis of the primary dihaploids the inheritance value of the tetraploid starting material can be valued in a better way. According to the descent the primary dihaploids show different high variation in largeness, form, number and colour of the tubers (Figure 1).

They show less vitality to their tetraploid starting forms (Figure 2). This can be improved by breeding on the dihaploid level with adequate selection. Only the best genotypes are used for further crossings.

The disome inheritance of dihaploids shows favourable supposition to the tetraploids for combination and selection. The splitting proportions are more simple, the phaenotypical variability is higher and extensive homozygoty is easier to reach. The less blossom intensity and pollen fertility of primary dihaploids is often disadvantageous (TIEMANN 1976; TIEMANN and SCHREITER 1976). The results of pollen colouring are shown in Figure 3. It is obvious that secondary dihaploids already show a better pollen fertility.

To create the crossing success of still unknown partners more effectively, besides the ascertainment of the pollen fertility, investigations of the pollen tube growth were carried out by luminescence microscopy (SCHREITER and TIEMANN 1977). Only such combinations were chosen for extended crossings,

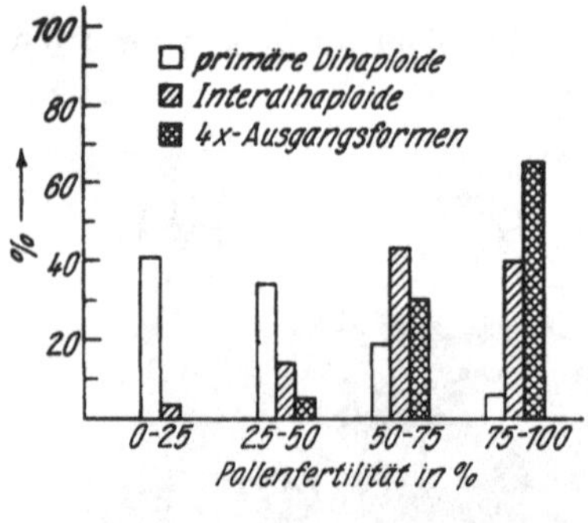


Fig. 3 Percentage of pollen fertility of primary dihaploids, secondary dihaploids and tetraploid parental forms as an average of the years 1976–1978

that showed an intensive pollen tube growth already 24 hours after the pollination. Figure 4 demonstrates a very good pollen tuber growth with the dihaploid-combination 7332×4721.

Dihaploids show further advantages using the genreservoirs of the 24 chromosomic wild species. The transmission of required genes into the *Solanum tuberosum*-genom is facilitated remarkably. Finally the influence of the chromosome number on the productivity of the potato can be found out by the dihaploid valence level. Though the yield and quality level increases with the secondary dihaploids, dihaploid genotypes did not excel their tetraploid parent forms. The yield varied according to the genotype from 10 to 80 per cent. This small tuber yield shows clearly the influence of the valence level for the productivity. Returning to the 48 chromosome level therefore is necessary on a certain level of dihaploid breeding. To be able to compare the way of origin of polyploid forms the mitotic and meiotic retetraploidisation is carried out. The

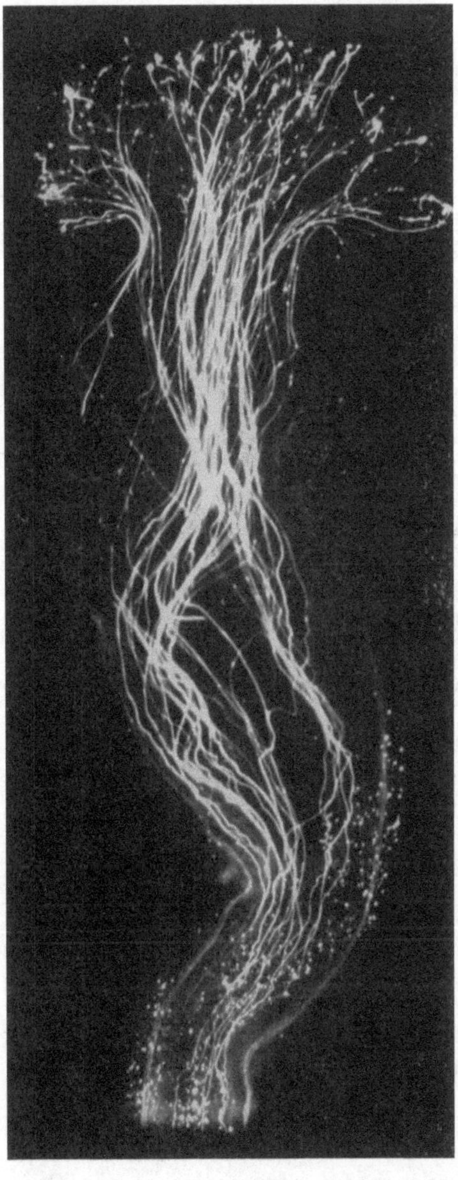

creation of mitotic polyploids is done chiefly by colchicine-treatment.

Meiotic polyploids are produced by 2x×2x, 2x×4x and 4x×2x crossings. The formation of diplogametes is necessary. Varieties serve as crossing partners. The output of diplogametic tetraploids is however very small. We reached only 0,3 seed on an average of all combinations per pollinated blossom with our 2x×4x crossings.

As not only the yield potential of potatoes but also the combination of a high genetic yield potential with good quality and resistance features stands in the foreground, retetraploidized genotypes are already preselected for their properties that are valueable for breeding. In Figure 5 results of two years of mitotic and meiotic (2x×4x) produced tetraploid genotypes of different time of ripening and origin are shown. The middle late ripening variety 'Mariella' was chosen as standard.

**Fig. 4 Intensive pollen tube growth 24 hours after pollination, microphotography, enlargement 30 times**

The results show a variation scale from 40 to 130 per cent. A tuber yield of 85 per cent was reached on an average to the standard variety 'Mariella'. 10 genotypes are superior significantly to the tetraploid Standard. From the submitted results it can be derived, that the small yield potential of dihaploid level increases again by retetraploidisation on the level of the tetraploid parental forms and more. Fundamentals are the presence of a genetic wide original that makes it possible to combine the valueable properties. In further tests the value of the high-yield-genotypes as breeding parents must be analysed in complex for their important properties for breeding.

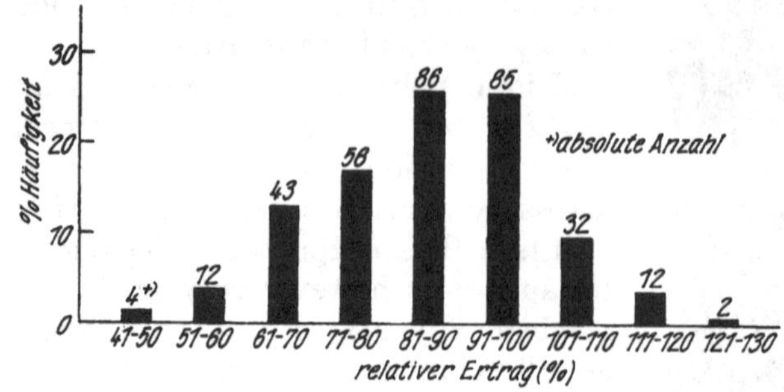


Fig. 5 Relative tuber yield of 332 retetraploidized genotypes in comparison to standard variety 'Mariella' as an average of the years 1977/78

## Zusammenfassung


Die Erhaltung der genetischen Ressourcen von Kartoffelarten verschiedener Ploidiestufen für die Entwicklung von Zuchtmaterial

Im Institut für Kartoffelforschung Groß Lüsewitz wird ein lebendes Sortiment von Wild- und Kulturkartoffeln verschiedener Valenzstufen erhalten. Die Sortimentserhaltung geschieht hauptsächlich durch Samenvermehrung. Durch Bestäubung von tetraploiden Sorten und von aussichtsreichen Klonen mit Pollen von *Solanum phureja* wurde ein Sortiment von Dihaploiden erzeugt. Dadurch wird es leichter möglich, das Genreservoir der 24chromosomigen Wildformen zur Entwicklung von Ausgangsmaterial zu nutzen. Eine brauchbare Genkombination kann durch intra- und interspezifische Hybridisierung erreicht werden, vorausgesetzt, daß die dihaploiden Genotypen ausreichend blühen und hohe Pollenfertilität besitzen. Um die Ausbeute an diplogametischen Tetraploiden für züchterische Zwecke zu erhöhen, muß im dihaploiden Sortimentsmaterial nach Genotypen gesucht werden, bei denen Diplogameten häufiger auftreten.


## Краткое содержание


Сохранение генетических ресурсов видов картофеля различных степеней плоидии для получения селекционного материала

В н.-и. институте картофеля в Грос-Люзевице сохраняется живая коллекция дикорастущих и культурных форм картофеля различных степеней плоидии. Коллекция сохраняется, главным образом путём размножения семян. Опылением тетраплоидных сортов и перспективных клонов пыльцой *Solanum phureja* был создан сортимент дигаплоидных растений. С их помощью можно будет облегчить использование генофонда 24-хромосомных дикорастущих форм для получения исходного материала. Подходящая комбинация генов может быть получена путём видовой и внутривидовой гибридизаций в том случае, если дигаплоидные генотипы будут хорошо цвести и обладать высокой фертильностью пыльцы. Чтобы для селекционных целей повысить выход диплогаметных тетраплоидов, надо в сортименте дигаплоидов найти такие генотипы, у которых будет чаще наблюдаться образование диплогамет.

## Literature

HINZE, E. 1976: Untersuchungen über die rationelle Erhaltung eines Sortiments wilder und kultivierter Kartoffelspecies. – (Report.) Groß Lüsewitz.

–, H. LUDWIG and W. JUNGES, 1978: Das Keimverhalten von Kartoffelsamen nach 8jähriger Lagerung unter verschiedenen Bedingungen. – Arch. Züchtungsforsch., Berlin, **8**, 205–209.

HOUGAS, R. W., and S. J. PELOQUIN, 1958: The potential of potato haploids in breeding and genetic research. – Am. potato J. **35**, 701–707.

PETT, B., 1974: Untersuchungen zur Meristemkultur bei Kartoffeln. – Arch. Phytopathol. und Pflanzenschutz, Berlin, **10**, 81–88.

ROTHACKER, D., 1966: Sortiment wilder und kultivierter Kartoffelspecies des Instituts für Pflanzenzüchtung Groß Lüsewitz (G-LKS), Teil 1. – Groß Lüsewitz.

–, 1968: Sortiment wilder und kultivierter Kartoffelspecies des Instituts für Pflanzenzüchtung Groß Lüsewitz (G-LKS), Teil 2. – Groß Lüsewitz.

RYBIN, V. A., 1933: Cytological investigation of the South American cultivated and wild potatoes and its significance for plant breeding (Russ.). – Trud. priklad. Bot. Genet. Selekc., ser. II, **2**, 3–100.

SCHREITER, J., and H. TIEMANN, 1977: Die Prüfung des Pollenschlauchwachstums in vivo bei Dihaploiden von *Solanum tuberosum* L. – Arch. Züchtungsforsch., Berlin, **7**, 253–258.

TIEMANN, H., 1976: Einige Aspekte zur Herstellung, Auslese und Verwendung Dihaploider in der Kartoffelzüchtung. – Biol. Rdsch. **14**, 73–77.

–, and J. SCHREITER, 1976: Zur Blühintensität und Blütenbiologie bei Dihaploiden von *Solanum tuberosum* L. – Biol. Zbl. **95**, 579–588.

Dr. H. TIEMANN
Institut für Kartoffelforschung Groß Lüsewitz
der Akademie der Landwirtschaftswissenschaften
der DDR
DDR - 2551 Groß Lüsewitz, Kreis Rostock

Kulturpflanze XXIX · 1981 · S. 159–167

# Protein resources of wild *Secale* species[1]

ROMAN KUBICZEK, WIESŁAW ŁUCZAK, BOGUSŁAW MOLSKI (Warszawa, Poland)


## Summary

The aim of this study was to find the eventual sources of increased protein content and of the changed balance between exogenic amino acids in caryopses protein of 9 wild *Secale* species (*S. chaldicum* FED., *S. Kuprijanovii* GROSSH., *S. anatolicum* BOISS., *S. montanum* GUSS., *S. silvestre* HOST, *S. ancestrale* ZHUK., *S. afghanicum* VAV., *S. dighoricum* VAV., *S. segetale* ROSHEV.). A comparison was completed between the previous cited wild species and low and high protein rye cultivars of *S. cereale* L. species, from the point of their amino acid composition and the ultrastructure of endosperm proteins. The wild species surpassed the cultivated varieties in protein content, "useful" protein content, and as much as twice the direct amount of some of the most important, from a nutritional point of view, amino acids, e.g. lysine or methionine. The first limiting amino acids were for wild species isoleucine, threonine and lysine. The wild species with the highest protein content differed significantly from the cultivated species in the amounts of protein matrix surrounding the starch granules in the deeper layers of endosperm cells.


## Introduction

Rye is, for Poland, the most important cereal crop. The deficiency of fodder protein, estimated to be 700 thousand tons a year, is stimulating the search for new sources of high protein materials, which are well balanced with amino acids. These materials could be used in the breeding program of new, improved rye varieties, richer in fodder protein.

Besides its agricultural value for Poland, rye is becoming more and more internationaly interesting as one of the parents of man-made cereal – *Triticale.*

There already have been several investigations completed in the improvement of the amount and the nutritive value of rye grain protein, especially in countries interested in rye production, due to their soil and climatic conditions, i.e. Germany, USSR. and Poland. There are several factors strongly influencing the protein content in rye grains, like the nutrient supply of the soil (LAUBE and QUADT 1959), site and weather conditions (BRUMMUND 1974; GUBANOVA and EROSZENKO 1975).

Therefore, observation of numerous cultivars of rye grown in a conservation collection of the Botanical Garden of Polish Academy of Sciences in Warsaw

1 Presented as poster

proved the strong genetic background of rye grain protein content. Several varieties were chosen from the collection with high fidelity of yearly improvement in protein content in caryopses without reducing the content of essential amino acids, in comparison to low protein varieties. (Kubiczek et al. 1980a).

In spite of the opinion that between cultivars of *Secale cereale* L. a wide variation in protein content is not to be expected and should be lower than the variation caused by site and weather (Kotvics 1970), a variation for rye cultivars grown in our collection (over one thousand taxons) was found (Kubiczek and Chojnacki 1976).

The range of variation was e.g. from 6,58% (d.w.b.) in the 'Björn' variety to 19,3% in the 'Bełta' variety, in the same year of harvest (1975) and the same place of growth (Kubiczek et al. 1980a).

Such a great variation within the collected materials created a real opportunity for improving the presently grown cultivars from the point of their protein content and nutritive value. Simultaneously, we could observe a great potential for protein synthesis in the caryopses of wild *Secale* species in our collection.

Several experiments of interspecific hybridization of rye showed possibilities for developing a new variety from the hybrid, which would be superior to *S. cereale* L. in protein percentage and nutritional value. Dimenstein and Ermakov (1958) used *S. cereale* L. and *S. segetale* Roshev. as parents. Focke (1956) and Kotvics (1970) used, for the hybridization, *S. cereale* L. and *S. montanum* Guss. as parents, receiving in the $F_2$ generation plants protein content in grains up to about 22%.

*S. segetale* being taxonomically close to cultivated ryes seems to give a greater advantage to the breeding programme for increasing the protein content of cultivated rye grains.

The review of protein resources, in regard to its amounts, nutritive value and ultrastructure in caryopses of different *Secale* wild species, from perennial species to closer to *S. cereale* L. annual ones, was the main aim to this study. This research should provide possibilities to discover still unknown ways of increasing protein content of rye through interspecific crosses.

## Material and methods

Four perennial *Secale* species: *S. chaldicum* Fed., *S. kuprijanovii* Grossh., *S. anatolicum* Boiss., *S. montanum* Guss., and five annual: *S. silvestre* Host. *S. ancestrale* Zhuk., *S. afghanicum* Vav., *S. dighoricum* Vav. and *S. segetale* Roshev. harvested in 1978 from Botanical Garden in Powsin collection, for comparison with two cultivars of *S. cereale* L., low protein cv. 'Dańkowskie Złote' (Skierniewice 1975) and high protein 'Bonel' (Skierniewice 1974), were used.

The total protein content, by the Kjeldahl method expressed on dry weight basis of grains (N×5.83), the amino acid composition of grain protein by means of the ion exchange procedure of SPACKMAN et al. (1958) after oxidazing and nonoxidazing hydrolysis of protein (Shram et al. 1954; Smith and Stockell

1954) were used, and chemical score (CS) predicting the first and next limiting the nutritive value of protein amino acids were calculated (MITCHELL and BLOCK 1946). The results are presented in % of protein and mg/100 g of grains, at 100% recovery. The amount of tryptophan in protein percent was taken as 1% of protein. For the study of the ultrastructure of caryopses previously published method with a Jeol scanning electron microscope, JSM-2, was used (KUBICZEK et al. 1980b).

## Results and discussion

The total nitrogen and protein content in percentage of caryopses on dry weight basis; the 17 estimated amino acids content, in protein and grains; the level of three first limiting the nutritive value amino acids (CS %) are presented in Table 1. Perennial rye wild species exceeded the annual species in grain protein content, having from 17.9 to 21.9% of protein (on d.w.b.) with mean average 20.4%, while the annual species had from 16.8 to 21.1%, with a mean average 19.0%. Cultivars had significantly less protein in caryopses, low protein variety 'Dańkowskie Złote' 11.4% and high protein 'Bonel' 15.2%. Growing in similar field conditions in collection, high protein cultivars of rye, e.g. 'Bonel', proved to have more protein nitrogen in caryopses than the low protein ryes (KUBICZEK and RAKOWSKA 1974). Having over 30% more protein in caryopses than 'Dańkowskie Złote' and CS at the same level, the variety 'Bonel' gives much more "useful" protein in grains, which is valuable for feed animals.

What seems to be the most interesting in the wild *Secale* species caryopses is that they contain, in some instances, as much as twice the protein than the cultivar 'Dańkowskie Złote', having simultaneously even more lysine in the protein. This means that this is the more valuable species, from the nutritional point of view, in protein fractions composition. Such grains had e.g. *S. silvestre* HOST species, of 21.1% protein on d.w.b. of caryopses and 4.1% of lysine in it. It produces a 859 mg lysine yield per 100 g of dry caryopses which is more than twice as much as the cv. 'Dańkowskie Złote' (412 mg) and almost twice the level of cv. 'Bonel' (530 mg).

Interspecific hybridization of cultivated ryes with wild *S. montanum* give some promising results, but the received highest protein progenies had the perennial character of the wild parent (KOTVICS 1970). *S. silvestre* HOST is of an annual type, but is also very far from the cultivated ryes and agronomically primitive. *S. segetale* ROSHEV. is the closest to the bred varieties of *S. cereale* L. and had, in our comparison, the lowest (16.8%) between wild species of *Secale* protein content, but still quite rich in lysine (3.82%) and limiting amino acids (CS= 56.5%). The lysine yield of 643 mg per 100 g of dry grains was still over 50% larger than the cv. 'Dańkowskie Złote', and over 70% larger in the "useful" protein content in caryopses.

Beside the *S. segetale* ROSHEV., other wild rye species, like *S. afghanicum* VAV. and *S. dighoricum* VAV., being close taxonomically to the cultivated *S. cereale* L. in their morphological and anatomical shape, are characterized by

Table 1

Amino acid composition of wild *Secale* species grain protein and chosen cultivated varieties

| Amino acid \ Species variety harvested in | *S. chaldicum* FED. (Powsin 78) | | *S. kuprijanovii* GROSSH. (Powsin 78) | | *S. anatolicum* BOISS. (Powsin 78) | | *S. montanum* GUSS. (Powsin 78) | | *S. silvestre* HOST (Powsin 78) | |
|---|---|---|---|---|---|---|---|---|---|---|
| | a | b | a | b | a | b | a | b | a | b |
| Cystine + Cysteine | 2.61 | 540 | 2.58 | 569 | 2.99 | 534 | 2.71 | 567 | 3.00 | 63 |
| Methionine | 2.47 | 511 | 2.56 | 560 | 2.66 | 475 | 2.27 | 476 | 2.70 | 56 |
| Met + Cys | 5.08 | 1.151 | 5.14 | 1.120 | 5.65 | 1.010 | 4.98 | 1.040 | 5.70 | 1.20 |
| Aspartic acid | 3.71 | 768 | 3.36 | 735 | 3.64 | 650 | 3.72 | 779 | 4.24 | 89 |
| Threonine | 3.45 | 714 | 2.40 | 525 | 2.81 | 502 | 3.11 | 652 | 3.50 | 73 |
| Serine | 4.51 | 933 | 4.36 | 954 | 4.37 | 780 | 4.87 | 1.020 | 4.54 | 9.56 |
| Glutamic acid | 27.3 | 5.640 | 28.3 | 6.190 | 27.0 | 4.820 | 27.8 | 5.830 | 25.2 | 5.30 |
| Proline | 11.7 | 2.420 | 13.0 | 2.830 | 12.3 | 2.190 | 12.3 | 2.570 | 10.2 | 2.15 |
| Glycine | 4.04 | 836 | 3.87 | 847 | 3.90 | 696 | 3.75 | 786 | 4.25 | 89 |
| Alanine | 3.58 | 741 | 3.64 | 796 | 3.90 | 696 | 3.44 | 721 | 3.60 | 75 |
| Valine | 4.15 | 859 | 4.25 | 930 | 4.41 | 787 | 4.15 | 869 | 5.69 | 1.20 |
| Isoleucine | 3.91 | 809 | 3.27 | 720 | 3.37 | 602 | 2.90 | 608 | 3.43 | 72 |
| Leucine | 5.82 | 1.200 | 5.43 | 1.190 | 5.86 | 1.050 | 5.56 | 1.160 | 6.11 | 1.29 |
| Tyrosine | 3.20 | 662 | 3.43 | 751 | 3.41 | 609 | 3.48 | 729 | 3.16 | 66 |
| Phenylalanine | 5.15 | 1.070 | 5.32 | 1.160 | 4.94 | 882 | 4.62 | 968 | 4.64 | 97 |
| Histydine | 2.87 | 594 | 2.74 | 600 | 2.85 | 509 | 2.67 | 559 | 2.89 | 60 |
| Lysine | 3.62 | 749 | 3.49 | 764 | 3.71 | 662 | 3.78 | 792 | 4.08 | 85 |
| Arginine | 4.66 | 964 | 5.26 | 1.150 | 5.01 | 894 | 6.33 | 1.330 | 5.32 | 1.12 |
| Total protein % on dry weight basis (N × 5.83) | 20.7 | | 21.9 | | 17.9 | | 21.0 | | 21.1 | |
| "Useful" protein % (total protein × CS%) | 11.7 | | 10.3 | | 9.12 | | 9.20 | | 10.9 | |
| Chemical Score (CS) % | 56.6 | | 47.1 | | 51.1 | | 43.9 | | 52.0 | |
| Limiting amino acids I | LYS | | THR | | ILE | | ILE | | ILE | |
| II | VAL | | ILE | | THR | | VAL | | LYS | |
| III | ILE | | LYS | | LYS | | LYS | | THR | |

a high protein content in grains (19.1 and 17.9%) and a good concentration of lysine (3.61 and 3.66%).

In most cases, lysine was not the first amino acid limiting the nutritional value of grain protein, but was one of the first three. Isoleucine, threonine and methionine were sometimes also the first limiting. Wild rye species are much more rich in sulphur containing amino acids, specially methionine, therefore they do not become limiting nutritive value, as it happens in rye cultivars (KUBICZEK et al. 1975).

tein % – a, and mg/100 g of dry grains – b)

| *ancestrale* IUK. 'owsin 78) | | *S. afghanicum* VAV. (Powsin 78) | | *S. dighoricum* VAV. (Powsin 78) | | *S. segetale* ROSHEV. (Powsin 78) | | *S. cereale* L. cv. Dańkowskie Złote (Skierniewice 75) | | *S. cereale* L. cv. Bonel (Skierniewice 74) | |
|---|---|---|---|---|---|---|---|---|---|---|---|
| a | b | a | b | a | b | a | b | a | b | a | b |
| 2.81 | 567 | 2.82 | 538 | 2.67 | 478 | 2.89 | 486 | 2.61 | 297 | 2.50 | 381 |
| 2.55 | 515 | 2.47 | 471 | 2.49 | 446 | 2.66 | 447 | 1.60 | 182 | 1.57 | 239 |
| 5.36 | 1.080 | 5.29 | 1.010 | 5.16 | 924 | 5.55 | 934 | 4.21 | 479 | 4.07 | 620 |
| 3.96 | 799 | 3.76 | 717 | 3.98 | 713 | 4.44 | 477 | 4.46 | 735 | 6.29 | 957 |
| 2.97 | 599 | 2.63 | 501 | 2.68 | 480 | 2.88 | 484 | 3.21 | 365 | 3.20 | 487 |
| 4.20 | 848 | 4.52 | 862 | 4.31 | 772 | 4.45 | 749 | 4.47 | 508 | 4.41 | 671 |
| 6.9 | 5.420 | 28.0 | 5.330 | 29.1 | 5.220 | 25.9 | 4.360 | 27.7 | 3.150 | 26.9 | 4.090 |
| 2.1 | 2.430 | 12.8 | 2.440 | 12.0 | 2.140 | 11.5 | 1.930 | 10.7 | 1.220 | 11.9 | 1.800 |
| 3.73 | 753 | 3.81 | 726 | 3.67 | 657 | 3.84 | 646 | 4.12 | 468 | 3.85 | 586 |
| 3.63 | 733 | 3.61 | 688 | 3.51 | 629 | 3.90 | 656 | 3.84 | 437 | 3.83 | 583 |
| 4.55 | 918 | 4.34 | 827 | 4.12 | 738 | 4.36 | 733 | 4.96 | 564 | 4.93 | 750 |
| 3.44 | 694 | 3.13 | 597 | 3.39 | 607 | 4.32 | 727 | 3.23 | 367 | 3.42 | 521 |
| 5.71 | 1.150 | 5.42 | 1.030 | 5.58 | 999 | 5.70 | 959 | 5.63 | 640 | 5.82 | 886 |
| 3.37 | 680 | 3.15 | 600 | 2.94 | 527 | 2.97 | 500 | 2.77 | 315 | 2.74 | 417 |
| 4.85 | 979 | 4.97 | 947 | 4.99 | 756 | 4.57 | 769 | 4.32 | 492 | 4.72 | 718 |
| 2.94 | 593 | 2.80 | 534 | 2.69 | 482 | 2.92 | 491 | 2.12 | 241 | 2.17 | 330 |
| 3.87 | 781 | 3.61 | 688 | 3.66 | 656 | 3.82 | 643 | 3.62 | 412 | 3.48 | 530 |
| 5.54 | 1.120 | 5.14 | 980 | 5.08 | 910 | 5.92 | 996 | 5.24 | 596 | 5.05 | 769 |
| 20.2 | | 19.1 | | 17.9 | | 16.8 | | 11.4 | | 15.2 | |
| 10.5 | | 9.03 | | 9.20 | | 9.50 | | 5.56 | | 7.70 | |
| 52.1 | | 47.4 | | 51.4 | | 56.5 | | 51.1 | | 50.6 | |
| ILE | | ILE | | ILE | | THR | | ILE | | MET | |
| THR LYS | | THR LYS | | THR VAL | | LYS VAL | | MET LYS | | ILE LYS | |

All these results may help in distinguishing the most valuable wild *Secale* species as a source of high protein caryopses with well balanced amino acid composition, with the possibility of breeding improved rye cultivars.

The scanning electron micrographs of subaleuron and inner cells of endosperm in cross sections of dry – ripe seeds let us compare and note the ultrastructural differences between high protein wild rye species and low and high protein cultivars of *S. cereale* L. (Fig. 1, 2).

The intravarietal comparison of rye cultivars differing in grains protein con-

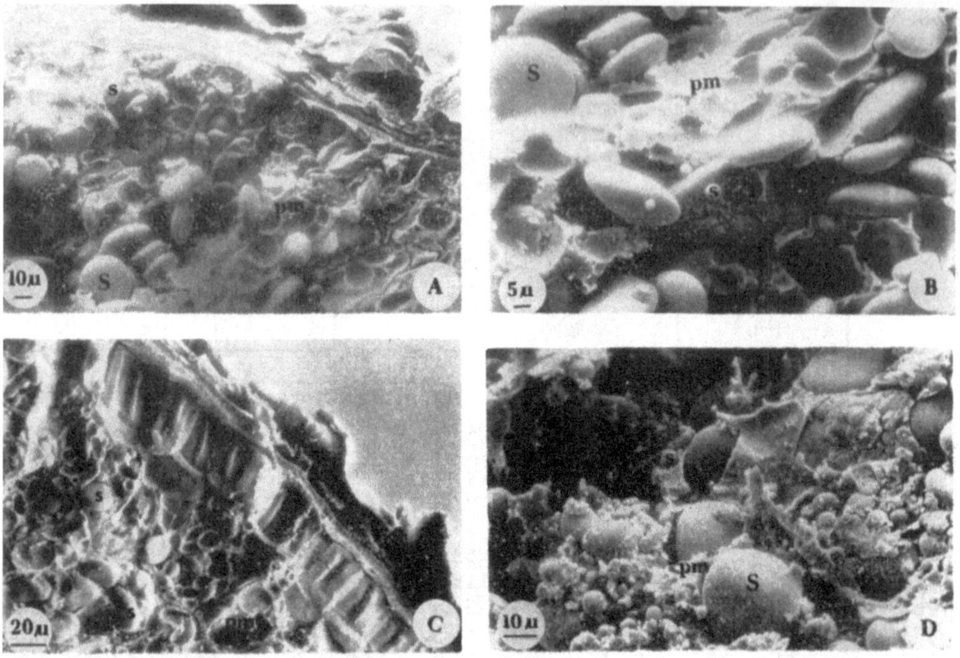


Fig. 1 Scanning electron micrographs of aleuron, subaleuron /*A*, *C*/ and deeper endosperm cells /*B*, *D*/ of caryopses of wild rye species, perennial one *S. chaldicum* FED. /1 – *A*, *B*/ and annual *S. silvestre* HOST /1*C*, *D*/; *s*-starch granules, *pm*-protein matrix

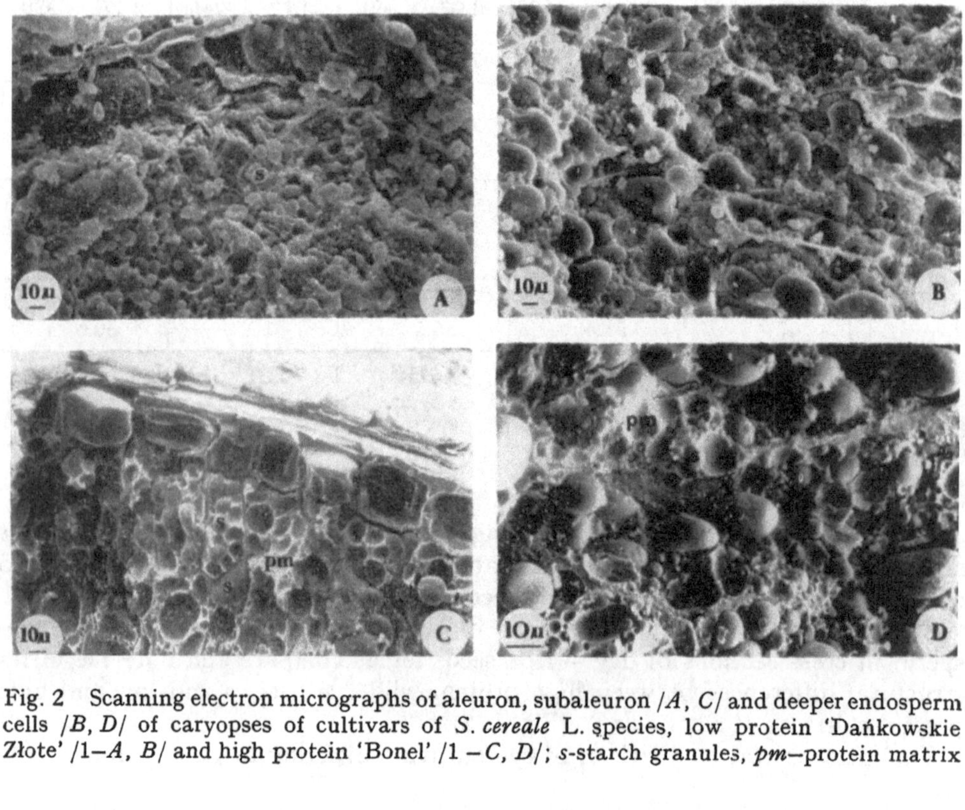


Fig. 2 Scanning electron micrographs of aleuron, subaleuron /*A*, *C*/ and deeper endosperm cells /*B*, *D*/ of caryopses of cultivars of *S. cereale* L. species, low protein 'Dańkowskie Złote' /1–*A*, *B*/ and high protein 'Bonel' /1 – *C*, *D*/; *s*-starch granules, *pm*–protein matrix

tent showed that the high protein varieties had, not only wider subaleuron cells region rich in proteins, but also higher concentrations of matrix proteins in cells of deep endosperm (KUBICZEK et al. 1980b). The same can be noticed by comparing cultivated 'Dańkowskie Złote' and 'Bonel' varieties (Fig. 2). In the caryopses of the most taxonomically distant from cultivated, perennial species, smaller amounts of large, lens shape starch granules in a dense protein matrix are noticeable (Fig. 1 – A, B).

The most interesting species in regards to high protein concentration and high lysine content, *S. silvestre* HOST (Fig. 1 – C, D) had a wide subaleuron cell region and plenty of protein matrix surrounding ovaly shaped starch granules, which are differentiated in size, but also had large amounts of cytoplasmic protein close to the endosperm cell walls. It was not possible to distinguish the ultrastructure of the proteinous substances or the presence of granular protein component, as was found in the wild – type and high – lysine barley endosperm (INGVERSEN 1975), but high lysine content of *S. silvestre* protein proves the higher proportion of glutelins to prolamines in storage proteins of this species.

All the data suggest that some of the wild *Secale* species may be of great value for the program of breeding new and improved varieties of rye, with the more protein in grains composed of high lysine fractions which are nutritionally valuable. There are still large resources of high – lysine proteins in the grains of wild *Secale* species, and their recognition becomes an important aim of our present research.

The authors thank M. MAŁUSZYŃSKA and R. IZDEBSKI for propagation of rye materials used in this research.

## Zusammenfassung

Proteinressourcen von *Secale*-Wildarten

Von 9 *Secale*-Wildarten (*S. chaldicum* FED., *S. kuprijanovii* GROSSH., *S. anatolicum* BOISS., *S. montanum* GUSS., *S. silvestre* HOST, *S. ancestrale* ZHUK., *S. afghanicum* VAV., *S. dighoricum* VAV., *S. segetale* ROSHEV.) wurde das Protein der Karyopsen daraufhin untersucht, ob es darunter Sippen mit einem verbesserten Proteingehalt und einem geänderten Verhältnis von exogenen Aminosäuren gibt. Die genannten Wildarten wurden mit proteinreichen und proteinarmen Zuchtsorten von *S. cereale* L. hinsichtlich der Aminosäure-Anteile und der Ultrastruktur des Endosperm-Proteins verglichen. Die Wildarten übertrafen die Zuchtsorten im Eiweißgehalt und im „nutzbaren" Eiweißgehalt und enthielten etwa die doppelte Menge der (unter dem Gesichtspunkt der Ernährung) wichtigsten Aminosäuren, wie Lysin und Methionin. Die ersten limitierenden Aminosäuren waren für die Wildarten Isoleucin, Threonin und Lysin. Die Wildarten mit dem höchsten Proteingehalt unterschieden sich signifikant in den Ausmaßen der Proteinmatrix, die die Stärkekörner von tiefer gelegenen Endospermzellen umgibt, von der Kulturart.

**Краткое содержание**

Белковые ресурсы дикорастущих видов *Secale* L.

У девяти дикорастущих видов ржи (*S. chaldicum* Fed., *S. kuprijanovi* Grossh., *S. anatolicum* Boiss., *S. montanum* Guss., *S. silvestre* Host, *S. ancestrale* Zhuk., *S. afghanicum* Vav., *S. dighoricum* Vav., *S. segetale* Roshev.) анализировался белок зерновок с целью найти формы с повышенным содержанием протеина и с изменённым соотношением экзогенных аминокислот. Названные дикорастущие виды сравнивались с различными селекционными сортами *S. cereale* L., обладающими как высоким, так и низким содержанием белка, в отношении их аминокислотного состава и ультраструктуры протеинов эндосперма. Дикорастущие виды оказались богаче по содержанию белка и по его питательной ценности и содержали, примерно, вдвое больше (с точки зрения питательности) необходимых аминокислот, таких как лизин и метионин. Первыми лимитирующими аминокислотами для дикорастущих видов были изолейцин, треонин и лизин. Дикорастущие виды с наибольшим содержанием протеина достоверно отличались от культурного вида размерами белковой матрицы, окружающей крахмальные зёрна глубжележащих клеток эндосперма.

## Literature

Brummund, M., 1974: Über Ergebnisse von Proteinuntersuchungen an einem Roggensortiment. – Tag.-Ber., Akad. Landwirtsch.-Wiss. DDR, **127**, 83–95.

Dimenstein, F. I. and A. I. Ermakov, 1958: Biochimija rži. Sel'skolchoz. Izd., Moskva, **1**, 165–234.

Focke, R., 1956: Über den Rohproteingehalt von *Secale cereale* und *Secale montanum* und den prozentualen Anteil einiger Aminosäuren. – Züchter, **26** (1/2), 40–41.

Gubanova, L. G., and T. T. Eroszenko, 1957: Mestnyje sorta ozimoj rži kak ischodnyj material dlja kolekcii na uveličenie soderžanija belka. – Vestnik Sel'skochoz. Nauki, **9**, 14–17.

Ingversen, J., 1975: Structure and composition of protein bodies from wild-type and high-lysine barley endosperm. – Hereditas, **81**, 69–76.

Kotvics, G., 1970: Investigations on increasing the protein content of *Secale cereale* L. – In: Protein growth by plant breeding (Ed. A. Balint), pp. 89–98. Akademiai Kiado, Budapest.

Kubiczek, R., and G. Chojnacki, 1976: Characteristic of the nutritive value of the protein from rye caryopses. II. Evaluation of rye grain protein quality by Dye-Binding Capacity method. – Acta Agrobotanica **29** (2), 267–272.

–, W. Łuczak and B. Molski, 1980a: Unpublished.

–, –, –, and J. Moczydłowski, 1980b: Comparison of the amount and distribution of seed proteins in endosperm cells of low- and high-protein varieties of rye (*Secale cereale* L.). – Acta Alimentaria Polonica, in press.

–, B. Molski, and M. Rakowska, 1975: Amino acids limiting the nutritional value of rye grain protein. – Hod. Rośl. Aklim. Nasien. **19**, 617–631.

–, and M. Rakowska, 1974: Characteristic of the nutritive value of the protein from rye caryopses. I. Amino acid composition of protein and the nitrogen forms in the caryopses of ten rye varieties from the breeding collection. – Acta Agrobotanica **27** (1), 105–113.

Laube, W., und K. Quadt, 1959: Roggen (*Secale cereale* L.). In: Handbuch der Pflanzenzüchtung (H. Kappert und W. Rudorf, Herausg.) 2. Aufl., **2**. Band, pp. 35–102, Paul Parey, Berlin und Hamburg.

Mitchell, H. H., and R. J. Block, 1946: Some relationship between the amino acid content of proteins and the nutritive values for the rat. – J. Biol. Chem. **1963**, 599–620.

Shram, E., S. Moore, and E. J. Bigwood, 1954: Chromatographic determination of cystine as cysteic acid. – Biochem. J. **57**, 33.

Smith ,R. E., and A. Stockell, 1954: Amino acid composition of crystalline carboxypeptidase. – J. Biol. Chem. **207**, (2) 501–514.

Spackman, D. H., W. H. Stein, and S. Moore, 1958: Automatic recording apparatus for use in the chromatography of amino acids. – Anal. Chem. **30**, 1190–1206.

Dr. R. Kubiczek
Botanical Garden of the Polish Academy of Sciences
ul. Prawdziwka 2
00-973 Warszawa p-84, Poland

Kulturpflanze XXIX · 1981 · S. 169–174

# Use of land-races in Czechoslovak wheat breeding[1]

Ivo Bareš and Miloslav Vlasák (Praha-Ruzyně, ČSSR)

## Summary

Four groups of Czechoslovak land-races (Czech alternate, Czech red, Moravian white, awnless, South Moravian and Slovak awned wheats) are characterized on the base of wheat collection studies in the Research Institute for Crop Production, Praha-Ruzyně, and their use in the breeding in ČSSR. A review of Czechoslovak cultivars of winter and spring wheats registered in the years 1921–1980 (table 1 and 2) and the registration period of the most important older Czechoslovak cultivars (table 3) is presented.

Of the total number of 284 registered Czechoslovak cultivars in the years 1921–1980 93 cultivars were developed by selection from land-races; 70% of 87 cultivars originated from crosses had a land-race or its derivative as one of its parents. From land-races derived cultivars 'Chlumecká 12', 'Česká Přesívka', 'Slovenská 777', 'Dobrovická 10' displayed the longest duration in the registration.

In the Czechoslovak wheat breeding land-races were used from the beginning of this century. Initially they were improved by selection of higher yielding lines; later on they were used in crosses with more productive cultivars of European breeding. In the collection of the Research Institute for Crop Production, Praha-Ruzyně, they have been preserved prevalently as selected higher yielding lines in original cultivars grown since 1910. At present, when compared with new cultivars, they show in addition to yielding high reliability conditioned by remarkable ecological adaptability following characters:

Group of Czech alternate wheats (České přesívky): Very short vernalization period (3 days), long photoperiodic response. Possibility of autumn as well as spring sowing (more limited). Taller straw; lower lodging resistance; higher tillering capacity; pyramidal, brown, awnless spikes (var. *milturum*); high winter hardiness; medium grain quality; susceptibility to rusts. Productivity by 30–40% lower.

Group of Czech red wheats (České červenky): Short vernalization period (16–18 days), short photoperiodic response. Midwinter habit. Midtall to tall straw; medium lodging resistance; medium to high tillering capacity; pyramidal to prismatic, brown, awnless spikes (var. *milturum*); medium to higher winter hardiness; medium grain quality; susceptibility to rusts. Productivity by 25–30% lower.

1 Paper not delivered during the symposium

Both groups represent a Middle European ecological group distinctly different from the European cultivars with long vernalization period (40–60 days). They tolerate later sowing in autumn without significant yield depression (Stehlík and Tymich 1920, Teltscherová 1955, Segeťa, Teltscherová and Bareš 1957).

Group of Moravian awnless white wheats (Moravské "holice", "bělky"): Midtall to tall straw; medium resistance to lodging; medium to higher tillering capacity; pyramidal, white, awnless spikes (var. *lutescens*, less often var. *aureum*); early to midseason; medium to higher winter hardiness; higher grain quality; susceptibility to rusts. Productivity by 25–35% lower. Spring forms occured as well.

Group of South Moravian and Slovak awned wheats (Jihomoravské a slovenské osinatky): Taller straw; low lodging resistance; higher tillering capacity; pyramidal, thin white (less frequently brown) spikes (var. *erythrospermum*, less frequently var. *ferrugineum*); early ripening; higher winter hardiness and grain quality. Productivity by 35–45% lower. In Moravia also spring forms occuring (Sladký 1921, Bareš 1970).

From the beginning of this century land-races were used in breeding mainly by J. Nolč and J. Dreger at Chlumec (Bohemia). In 1910 cv. Dregerova 12 and cv. Dregerova 126 originated by repeated selection from the Czech red wheats. In 1907 at Postoloprty (Bohemia) cv. Postoloprtská Přesívka was developed and in 1909 at Pyšely (Bohemia) cv. Pyšelská (Česká) Přesívka originated. In 1906 the Land Institute at Přerov (Moravia) began the selection of land-race cv. Hanácká Bělka. In Slovakia land-races from Tisza region were used in the breeding at Dioszég (now Sládkovičovo) since 1870.

Table 1

Number of registered winter wheat cultivars in Czechoslovakia (1921–1980)

| Year | Selected from land-races | Selected from crosses | Origin foreign | Origin unknown | Introduced foreign cultivars | Total |
|---|---|---|---|---|---|---|
| 1921–1940 | 80 | 41 | 27 | 14 | 20 | 182 |
| 1941–1945 | – | – | 4 | – | – | 4 |
| 1946–1980 | 2 | 29 | 7 | – | 10 | 48 |
| 1921–1980 | 82 | 70 | 38 | 14 | 30 | 234 |

Slovak cultivars originated 1941–1945 were included when again registered in 1946/1947

Table 2

Number of registered spring wheat cultivars in Czechoslovakia (1921–1980)

| Year | Selected from land-races | Selected from crosses | Foreign origin | Introduced foreign cultivars | Total |
|---|---|---|---|---|---|
| 1921–1940 | 9 | 10 | 4 | 10 | 33 |
| 1941–1945 | – | – | 1 | – | 1 |
| 1946–1980 | 2 | 7 | 1 | 6 | 16 |
| 1921–1980 | 11 | 17 | 6 | 16 | 50 |

In 1921 the concession of the originality of cultivars was regulated by law. As a consequence all original and later registered cultivars possess documentations. From these sources we present (tables 1 and 2) reviews on the number of registered cultivars in the years 1921–1980 according to the used breeding procedure. The reviews enable to evaluate the utilization of land-races in ČSSR

Table 3

Origin and period of registration (more than 25 years) of the most important wheat cultivars in Czechoslovakia

| Cultivars | Original or registered cultivars in the years | Period of registration (years) |
|---|---|---|
| Winter wheats | | |
| 'Chlumecká 12' ('Dregerova Červená 12') | from 'Česká Červenka' 1919–1970 | 52 |
| 'Česká Přesívka' (Pyšelská, later Selecty) | from 'Česká Červená Přesívka', Pyšely district (Bohemia) 1922–1960 | 39 |
| 'Slovenská 777' ('Diosecká 777') | land-race from Vrbové (Slovakia) 1923–1960 | 38 |
| 'Dobrovická 10' ('Dobrovická Česká Červená G 10') | from 'Česká Červenka', Mladá Boleslav district (Bohemia) 1927–1958 | 36 |
| 'Stupická Bastard' (Selecty) | 'Česká Červená Přesívka 23/11' ×rust resistant wheat 23/18 1927–1959 | 33 |
| 'Višňovská Hustoklasá' (Rb) | from 'Rimpaus Früher Bastard' 1925–1957 | 33 |
| 'Postoloprtská Přesívka' (58) | from 'Česká Přesívka' Postoloprty district (Slovakia) 1921–1945, 1951–1957 | 32 |
| 'Slovenská 2' ('Diosecká') | land-race from Vrbové (Slovakia) 1921–1952 | 32 |
| 'Dobrovická Červená Přesívka P2' | synthetic population of Czech alternate wheats 1922–1949 | 28 |
| 'Židlochovická (Jihomoravská) Holice' | land-races 'Jihomoravská Holice' ×'Židlochovická Jubilejná' 1932–1958 | 27 |
| 'Kaštická 53' (Bezosinná) | 'Postoloprtská 6' × 'Moravskopolní' 1935–1960 | 26 |
| Spring wheats | | |
| 'Niva' ('Novodvorská') | 'Cimbal' ×'Saumur' 1934–1968 | 34 |
| 'Ratbořská' ('Heineho Hladká Jarka') | from 'Heines Kolben' 1925–1943, 1946–1960 | 34 |
| 'Stupická (Selecty) Vouska' | from 'Huron' 1927–1960 | 34 |
| 'Dobrovická Česká Červená' (3) | from alternate wheat, Lovosice district (Bohemia) 1923–1952 | 30 |
| 'Dregerova Česká Vouska' | from Czech awned land-races 1922–1949 | 28 |

## Winter wheats

From 1921 to 1940, particularly until 1935, improved land-races prevailed in Bohemia. 46 cultivars were developed by selection mainly from Czech red and Czech alternate wheats. In Slovakia 24 cultivars were selected from Hungarian and Slovak land-races. In Moravia only 7 cultivars originated from land-races. In this period 23% of wheat cultivars were developed from crosses, particularly in Moravia (22 cultivars) where the progress of breeding was influenced by E. von TSCHERMAK-SEYSENEGG. The cultivar Marchfelder developed by him probably from a Hungarian land-race gave origin to many Moravian cultivars developed by selection and from crosses. A land-race was prevalently used as one parent in 41 cultivars selected in this period (1921–1940) from crosses.

In the post-war period 1946–1980, particularly after the nationalization of the breeding enterprises, demands on registered cultivars increased. There were 48 cultivars registered, the registration period was reduced (in the last time to only 5–6 years) and also high yielding foreign cultivars were introduced mainly from USSR but also from GDR, Poland, Yugoslavia and GFR. Two cultivars were developed in Slovakia by selection from awned land-races, however, they were registered only for a short period until 1952. From crosses with land-races were selected in Slovakia 'Slovenská Intenzívna', 'Slovenská 200' and 'Viglašská'. In the 1960s they were replaced by higher yielding Moravian awnless wheats and later by Soviet cultivars. In Bohemia and Moravia land-races were used in crosses to a larger extent.

The tradition of Czech midwinter wheats continued by the last Czechoslovak red wheat 'Draga' (registered 1965–1972). Moravian land-races or cultivars derived from them were used in the breeding of 'Kaštická Osinatka' (registered 1954–1973), 'Lada' (1964–1971), 'Pavlovická 198' (1957–1969), 'Diana I' (1960–1968), 'Iva' (1962–1969), 'Diana II' (1967–1972), 'Ostka' (1971–1976), and 'Zora' (1971–1979).

Because of low productivity pre-war cultivars developed by selection from land-races were restricted before the 1960s. The cultivars Česká Přesívka, Postoloprtská Přesívka, Dobrovická Přesívka P 2, from red wheats the cultivars Chlumecká 12, Dobrovická 10, Kaštická from Moravian cultivars Židlochovická Holice and from Slovak cultivars Slovenská 777 and Slovenská 2 had the longest registration period (table 3) (BAREŠ 1970).

## Spring wheats

Lower importance of spring wheats (6–8% of the winter wheat area) manifests also in the number of cultivars developed since 1921. In the period from 1921 to 1940 only 9 cultivars were derived from land-races; the most important were 'Dobrovická Česká Červená', 'Židlochovická Vesna', 'Staroveská Bezosinná', 'Hodonínská Bezosinná', 'Hodonínská Osinatká' and 'Dregerova Česká Vouska'. Many cultivars originated from crosses with land-races; the most important were 'Židlochovická Lada', and 'Ždánická Osinatká'. In the post-war period land-races 'Bučianská' (registered 1949–1959) and 'Vega' (1946–1959) were released in Slovakia and Moravia, respectively. The high yielding cultivars Ratbořská, Niva and Zlatka (1960–1976) were a strong competition for the cultivars derived from land-races. At present high yielding Czechoslovak cultivars Jara and Rena and cultivars introduced from GFR are grown (Bareš 1970).

## Conclusion

Czechoslovak land-races had a significant influence on wheat breeding in ČSSR. Their genetic potential was utilized to a large extent. In majority it is preserved in developed cultivars in the Czechoslovak collection. Of the total number of 284 wheat cultivars registered from 1921 to 1980 93 cultivars were selected from land-races if we disregard 46 introduced foreign cultivars and other 14 cultivars of unknown origin. Land-races were used in 70% of 87 cultivars derived from crosses.


## Zusammenfassung

Die Verwendung von Landsorten in der tschechoslowakischen Weizenzüchtung

Vier Gruppen von tschechoslowakischen Weizenlandsorten werden auf der Grundlage von Untersuchungen an der Weizenkollektion im Institut für Pflanzenproduktion, Prag-Ruzyně, und ihrer Nutzung in der Züchtung in der ČSSR beschrieben. Es sind (1) tschechische Wechselweizen, (2) tschechische rote, (3) mährische weiße, grannenlose und (4) südmährische und slowakische begrannte Weizen. Ein Überblick über die in den Jahren 1921–1980 registrierten tschechoslowakischen Winter- und Sommerweizen (Tab. 1 und 2) sowie die Zeitdauer der Registrierung der bedeutendsten älteren tschechoslowakischen Sorten (Tab. 3) schließt sich an.

Von insgesamt 284 in den Jahren 1921–1980 registrierten tschechoslowakischen Sorten entstanden 93 durch Auslese aus Landsorten. Bei den 87 Sorten, die aus Kreuzungen hervorgingen, hatten 70% entweder eine Landsorte oder den Abkommen einer Landsorte unter den Eltern. Die aus Landsorten hervorgegangenen Sorten 'Chlumecká 12', 'Česká Přesívka', 'Slovenská 777' und 'Dobrovická 10' waren am längsten registiiert.



## Краткое содержание

Использование местных сортов в чехословацкой селекции пшеницы

Описываются четыре группы местных чехословацких сортов пшеницы на коллекционном материале Института растениеводства в Праге-Рузине и их использование в селекционной работе ЧССР: (1) чешские пшеницы-двуручки, (2) чешские красные, (3) моравские белые безостые, (4) южноморавские и словацкие остистые. Затем следует обзор чехословацких озимых и яровых пшениц, зарегистрированных в 1921–1980 гг. (таб. 1 и 2), а также период времени использования (как зарегистрированных) наиболее значительных старых чехословацких сортов (табл. 3).

Из 284 сортов, зарегистрированных в 1921–1980 гг., 93 выведены путём отбора из местных сортов. Среди родительских форм 87-и сортов, выведенных путём скрещивания, 70% были или местными сортами или потомками местных сортов. Дольше всего использовались (как зарегистрированные) сорта: 'Хлумецка 12', 'Ческа пржесивка', 'Словенска 777' и 'Добровицка 10'.

## Literature

Bareš, I., 1964: Výnos světového sortimentu pšenice v ČSSR. – Rostlinná výroba **1**, 59–70.

–, 1970: Šlechtěni pšenice v ČSSR. Původ nejvýznamějších odrůd v ČSSR. – In: J. Foltýn a kol.: Pšenice. Praha. pp. 153–167.

Segeťa, V., L. Teltscherová, I. Bareš, 1957: K otázce možnosti výsevu některých ozimých odrůd pšenice na jaře. – Vědecké práce VÚRV **3**, 99–110.

Sladký, A., 1921: Krajové odrůdy a jejich význam v zušlechťování rostlin. – Moravský hospodář **23**, 11–19.

Stehlík, V., V. Tymich, 1920: Šlechtitelský význam variet a typů skládajících českou červenou přesívku. – Zemědělský archiv **11**, 335–365.

Teltscherová, L., 1955: K otázkám stadijnosti československých pšenic a ječmenů. – Vědecké práce VÚRV **1**, 9–39.

Ing. I. Bareš, CSc.
Ing. M. Vlasák
Research Institute of Plant Production
161 06 Praha-Ruzyně, Drnovká 507, ČSSR

# Section II:
# Problems of the taxonomy of cultivated plants

Chairmen:

T. B. Fursa
B. R. Baum
J. M. J. De Wet

Kulturpflanze XXIX · 1981 · S. 177–198

# Species concepts and systematics of domesticated cereals

JOHANNES M. J. DE WET (Urbana/Illinois, USA)

## Summary

Cultivated cereals are domesticated grasses. Wild progenitors of domestic cereals are aggressive natural colonizers and often become weedy in man disturbed habitats. Hybridization between cultivated cereals and their close wild relatives gives rise to mimetic cereal weeds that often accompany the domestic species well beyond its natural range. Similar weeds may also evolve from abandoned cultivars as a result of mutations that restore their ability for natural seed dispersal. Cultivated complexes of domestic species are typically variable. Man selects and isolates phenotypes to suit his fancies, with the result that different phenotypes of a cereal are often grown for different uses in the same village, and adjacent villages often grow different phenotypes for similar uses. Conventional species concepts do not hold for domestic species. It is proposed that wild, weed and cultivated complexes of individual domestic species be recognized respectively as subspecies. Major discontinuities in variation within the weed or cultivated subspecies are then identified as races without formal taxonomic recognition.

## Introduction

Systematics is a search for order in nature, without knowing the nature of this order (ADANSON 1763). Modern systematics involves three independent, though not mutually exclusive phases of study. The principal phase is taxonomy. Individuals are classified on the basis of similarities or differences into units of ever increasing complexities and given names in a prescribed hierarchy. Numbers are reduced to manageable quantities, and variation to meaningful proportions for ease in sorting and filing specimens, and population complexes are assigned names to facilitate communication. The second phase of systematic study attempts to explain origins of variation within populations, and the reasons for discontinuities or breakdown of discontinuities among populations. The third phase is a study of phylogeny, and endeavors to demonstrate how taxonomic units are related in time and space. Domestic species are ideal subjects for systematic studies. Cultivated plants are recent in origin and remain cytogenetically conspecific with their wild progenitors. It is therefore possible to study the genetics of phenotypic changes associated with changes in adaptation under domestication. Furthermore, the evolutionary history of the crop is often beautifully preserved in the archaeological records of people who domesticated the species. Phylogeny of cultivated taxa can therefore often be traced in detail.

## Evolution in the man made habitat

Organisms do not all react in the same way when man disturbs their habitats. Some flourish in the presence of man, while others have to migrate or face eventual extinction. Two kinds of organisms, weeds and domesticates prosper in ther permanently disturbed man made habitat. They are totally adapted to habitats that are continuously being disturbed by man, and can no longer compete with their wild relatives for natural habitats. Weeds are spontaneous in man-made habitats, while domesticates further require the help of man in seed dispersal or any other means of propagation.

Organisms that are adapted to habitats that are not notably disturbed by man are wild. However, among wild plants are kinds adapted to various degrees of natural disturbances of their habitats. They occupy different positions in serial succession. Those at the pioneer end of the succession spectrum can invade man disturbed as well as nature disturbed habitats, and are therefore weedy. They are not weeds, however, and are poorly adapted to withstand continuous disturbance of the habitat by man. Species at the climax end of the succession spectrum are never aggressive colonizers and have difficulties adapting to habitat disturbance.

The wild progenitors of seed crops are all aggressive colonizers. This is particularly true of domestic cereal species, and it is often argued that the domestication process involves changes in adaptation from wild to weed and eventually to cultivation (Anderson 1956). This may be true for some vegetable crops, but certainly not for most cereals. Natural colonizers are preadapted to domestication, and several grass species with excellent potential as cereals never became domesticated, probably because they lack colonizing ability. An excellent example is *Zizania aquatica* L., wild rice of the American Indian. It was harvested as a staple food for millennia, and as a commercial crop for almost three centuries before this cereal was successfully cultivated (De Wet and Oelke 1979).

Domestication is initiated when seeds from planted populations are harvested and sown by man in succeeding generations. The domestication process continues as long as man cultivates a particular crop. Food collecting, sowing or other husbandry practices of nomadic hunter-gatherers and cattle hearders do not lead to domestication (Harlan, De Wet and Price 1973; De Wet and Harlan 1975; De Wet 1975). Neither inflorescences on individual plants nor different plants in a population all mature at the same time, and sufficient seed usually escape the harvester to insure a daughter population during the next growing season. Sowing by man also will not necessarily lead to domestication. Wild cereals are often sown by nomadic food gatherers to increase the size and density of natural populations (Steward 1930; Tindale 1977). This practice may increase competative ability, but is not domestication. Plant domestication refers to changes in adaptation that insure total fitness in habitats specially prepared by man for his cultigens.

Adaptive changes associated with domestication are genetically and often also phenotypically complex. The initial ability to survive in disturbed habitats is inherent in all cereals that became domesticated. Sowing increases competative ability, and sowing in a men prepared habitat increases colonizing ability. Minor cereals such as *Brachiaria deflexa* (SCHUM.) C. E. HUBBARD (animal fonio) and *Digitaria iburua* STAPF (black fonio) of the West African savanna may actually have been agricultural weeds before they were adopted as cultivated cereals. Spontaneous weed races of animal fonio are encouraged to invade sorghum fields in parts of Angola and are harvested as cereals (DE WET 1977). In general, however, mimetic weeds of cereals are later in origin than the crop.

Sowing by man not only selects directly for competative ability, but also for uniform population maturity. Seedlings that germinated as soon as conditions became favorable have the best opportunities to develop fully, and will consequently contribute more than those that germinated later to the seed stock from which the next man sown generation will be established. Domesticated cereals therefore lack seed dormancy while weedy close relatives are characterized by various degrees of dormancy.

Harvesting wild cereals selects toward enforcing wild-type seed dispersal mechanisms. Those individuals with the most efficient means of natural seed dispersal contribute the largest proportion of seed from which a daughter population can become established. Individuals with the least efficient seed dispersal mechanisms are harvested by man, and when seeds of these are sown, selection is against seed dispersal efficency. Cereals are harvested by beating the spikelets or florets from inflorescences with a stick, by swinging a basket over mature plants to dislodge the dispersal units, by hand stripping inflorescences, by removing mature inflorescences for later threshing, or by uprooting whole plants. WILKE, BETTINGER, KING and O'CONNELL (1972) point out that complete loss of the ability of natural seed dispersal, a characteristic of all major cereals, can result only from harvesting with a knife or sickle for later threshing. Ethiopian oats (*Avena abyssinica* HOCHST.) lost the ability to naturally disperse its seeds although it is not conciously sown (LADIZINSKY 1975). The species is an accidental contaminant in cereal fields, harvested with the crop and sown with it year after year. Several cultivated cereals on the other hand never lost the ability of natural seed dispersal. Sauwi (*Panicum sonorum* BEAL) of northwestern Mexico, raishan (*Digitaria cruciata* (NEES) A. CAMUS) of the Khasi Hills in Assam, or *Brachiaria ramosa* STAPF in South India are harvested by uprooting whole plants before they are fully matured. Plants are then allowed to dry and the spikelets are shaken from the inflorescences. These cultigens are often spontaneous weeds in agricultural fields (DE WET 1979). In Malawi a cultivar of *Sorghum bicolor* (L.) MOENCH is grown that has large, sweet grains that are eaten raw before the plants are fully matured. Inflorescences are harvested, and the grains are shaken from the spikelets. This cultivar is often spontaneous in cultivated fields. It, however, disperses its seeds by floret rather than spikelet disarticulation as is characteristic in wild species of *Sorghum*. A change in harvesting technique has re-introduced natural seed dispersal.

Natural selection may also introduce new seed dispersal mechanisms in populations of abandoned cereals that are no longer harvested. Seed dispersal is absolutely essential for survival without the aid of man.

Harvesting in combination with sowing also select for uniform individual plant maturity. Two evolutionary trends are obvious (Harlan, De Wet and Price 1973). In cereals such as *Zea mays* L. (maize), *Pennisetum americanum* (L.) Leeke (pearl millet), or *Sorghum bicolor* (sorghum) apical dominance reduces tillering and lateral branching, and in extreme cases individuals with single stems and solitary inflorescences. In others such as wheat (*Triticum* species), rice (*Oryza sativa* L.) or *Eleusine coracana* (L.) Gaertn. (finger millet) sychronized tillering insures uniform individual plant maturity.

Evolution is rapid under domestication. Not only is total adaptation to man disturbed habitats achieved within a relatively few generation of sowing and harvesting (Hilu and De Wet 1980) but variation patterns within the cultivated complex are increased tremendously. Small gene pools are selected for cultivation at different times and places across the range of the wild cereal species and a multitude of phenotypes become fixed by chance as a result of genetic drift (see Wright 1966; Grant 1977). Furthermore, cultivars are transported by man beyond the natural range of the species being domesticated. This requires new adaptations that are often acquired by hybridization with close relatives from which they are naturally isolated. The most successful adaptive force in shaping evolution within the cultivated complex of domestic species, however, is concious selection by man to suit his fancies. Phenotypically different kinds of a cereal is usually grown for different uses, and people from different communities often grow different phenotypes for similar uses (De Wet and Harlan 1971). More than 300 races of maize have been described, although the total gene pool of domesticated *Zea mays* is composed of no more than six basic genetic lineages (Mangelsdorf 1974). Similarly, Stapf and Hubbard (1934) recognized 13 'species' of cultivated pearl millet, and Baum (1977) recognized phenotypic discontinuities among wild and weedy hexaploid oats as seven 'species'. Yet, these 'species' are cytogenetically conspecific and maintain their unity of phenotype primarily through man induced isolation.

## The domestic cereal species

Species are traditionally recognized as clusters of variation that are maintained as a result of more or less free gene exchange among members of a species, and the infrequent or abcence of gene exchange among different species (Mayr 1969). In other words, a species is commonly recognized as a reproductive community consisting of an array of subordinate Mendelian populations that are connected by gene flow among populations. Population systems that fit such an ideal species concept are rare in nature. Taxonomic plant species frequently "lack reality, cohesion, independence, and simple evolutionary or ecological roles" (Levin 1979). The opposite is true for clusters of variation characteristic of domestic species. Taxa recognized, be they varieties, species or even genera within wild-weed-crop complexes of domestic species are characterized by genetic cohesion, ecological unity and evolutionary independence, and they have taxonomic reality. Yet, they remain cytogenetically conspecific. It is not surprising, therefore, that traditional taxonomic hierarchies are not acceptable in classifying variation characteristic of domestic species (SINSKAYA 1948; SCHIEMANN 1949; ZHUKOVSKY 1967; JEFFREY 1968). JIRÁSEK (1966) as an example, proposes 11 subspecific categories for cultivated species, and lists over 50 names that are being used to identify phenotypic discontinuities in domestic species. HARLAN and DE WET (1971) on the other hand, indicate that such discontinuities are artifacts of mans agricultural activities, and are best recognized as races without formal taxonomic recognition. The concept race as conceived by them is a population system with a distinct cohesion of morphology, geographical distribution, ecological adaptation and often also breeding behaviour, that is maintained by disruptive selection and ethnological isolation. The six lineages of maize recognized by MANGELSDORF (1974) are basic races of domesticated *Zea mays*, as are the snowdenian species complexes of domesticated *Sorghum bicolor* recognized by SNOWDEN (1936).

**Wild progenitors of cultivated cereals.** – Evolution in the man made habitat is sympatric speciation in progress. Evolutionary divergence depends for success on isolation. BURKILL (1952) suggests that the initial isolation between wild and cultivated kinds occured when man transported his food plants beyond the natural ranges of their species. This also forced man to sow what was harvested, and to continue sowing in succeeding generations if the food crop was to be maintained in its new environment. Spacial isolation, however, is not necessary for divergent evolution of wild and cultivated races. Selection pressures associated with domestication in themselves are disruptive. MILLICENT and THODAY (1961) and THODAY (1972) conclusively demonstrate that differences in habitat preference allow for divergence to continue even where interpopulation gene flow remains possible.

Selection pressures are vastly different in the wild and cultivated habitats of domestic species. Adaptations associated with increase in fecundity under a regime of harvesting and sowing by man often have drastic effects on the phenotype of domestic cereal species. Comparative morphology therefore sometimes fails to identify the wild progenitor taxon of a particular cereal. An extreme example is *Zea mays* subsp. *mexicana* (SCHRAD.) ILTIS (Wild maize or teosinte). Teosinte is frequently treated as a species distinct from maize (BIRD 1978), or even as a different genus (RANDOLPH 1976).

Maize is unique among grasses in female inflorescence morphology. The species *Zea mays* is monoecious with male terminal inflorescences and female lateral inflorescences. Male inflorescences are composed of several noded racemes with paired spikelets at each node, arranged along an elongated primary axis and several lateral branches. The female inflorescence branch (ear) of domesticated maize is a polystichous structure with paired spikelets arranged in shallow cupules around and along a thickened central rachis, with this structure covered by modified leaves. Inflorescences are often branched with several ears developing to maturity.

The oldest known race of maize in the archaeological record comes from the

Coxcatlan culture phase of Tehuacan in Mexico and dates back some 7000 years (MANGELSDORF, MACNEISH and GALINAT 1967). Coxcatlan maize probably is the progenitor of the basic Mexican races Chapalote and Nal-Tel (WELLHAUSEN et al. 1952). It resembles these presentday races in general morphology, except that the glumes are more chartaceous and longer, and the cobs are smaller. Since no grass that resembles maize in female inflorescence morphology is known in the wild, it is often argued that Coxcatlan maize must be wild, or at least closely resembles the progenitor of domesticated maize. Arguments for and against the hypothesis that Coxcatlan maize is wild maize were extensively discussed elsewhere (BRANDOLINI 1970; GALINAT 1971; BEADLE 1972, 1979; DE WET and HARLAN 1972), and need not be repeated in this discussion. Suffice it to say that Coxcatlan maize lacks the ability of natural seed dispersal, and therefore must represent an early domesticated race of maize.

The closest living wild relative of maize is teosinte. It is widely distributed on the Central Plateau of Mexico and along the western escarpment south to Honduras (WILKES 1967, 1972). Teosinte hybridizes naturally with maize and around Chalco a typical mimetic weed race of teosinte evolved as a result of such hybridization (WILKES 1977). Evolutionary behaviour suggests that teosinte is wild maize. In contrast to the progenitors of other domesticated cereals, there is no archaeological or historical records that teosinte was ever used as a wild cereal. Furthermore, MANGELSDORF (1974) suggests that no evolutionary force is known that is capable of transforming a teosinte female inflorescence into a maize ear.

Teosinte resembles maize in vegetative and male inflorescence morphology. Teosinte shares its female inflorescence morphology only with *Zea diploperennis* ILTIS, DOEBLEY et GUZMAN and *Zea perennis* (HITCHC.) REEVES et MANGELSDORF. The female inflorescence consists of one or more racemes that are individually enclosed by a modified leaf. Racemes are distichous, with solitary fertile spikelets alternately arranged in deep cupules on the indurated rachis, each of which is closed by the strongly indurated outer glume to form a fruitcase. Seed dispersal is achieved by abscission callus that forms at maturity at the rachis nodes, with the result that the fruitcases disarticulate individually (see BEADLE 1979).

The questions that need to be answered before teosinte can be accepted as wild maize are first, are the differences that distinguish maize from teosinte genetic alternatives; second, is it likely that selection pressures associated with domestication could have resulted in mutations changing a teosinte raceme into a maize ear; and third, are these changes of the same kinds and the same order of magnitude as those distinguishing other domesticated cereals from their wild progenitors.

The genetics of differences between an ear of maize and raceme of teosinte was studied and illustrated by COLLINS (1925), LANGHAM (1940) and ROGERS (1950). From an evolutionary point of view, these differences can be attributed to adaptations associated with domestication. Wild teosinte probably was not used as a cereal, but there is archaeological evidence to suggest that this wild grass was grown as a vegetable crop (MACNEISH 1971). Young female inflorescences of teosinte are palatable when eaten raw and can be cooked as green corn. The common cereal before the appearance of Coxcatlan maize in Tehuacan was a species of *Setaria* (CALLEN 1965, 1967). The indurated fruitcase renders teosinte a very undesirable cereal. Dominant alleles of the complex tunicate allele (MANGELSDORF and GALINAT 1964), however, reduce induration and the teosinte caryopses become readily available through common methods of threshing (see BEADLE 1979). Discovery of a tunicated teosinte may have been what led to the cultivation of this grass as a cereal. It is an interesting fact that Coxcatlan maize is distinctly tunicate, and MANGELSDORF (1958) proposed that the ancestor of maize must have been tunicated.

Harvesting and sowing, as was shown earlier in this discussion, lead to loss of natural seed dispersal ability. Changing the distichous and alternate raceme of teosinte into a polystichous and opposite ear of maize can be achieved by alleles of three genes associated with loss of seed dispersal. Alternate cupules that disarticulate individually are yoked in maize. These yoked cupule pairs are further cross yoked into whorls of four, and domesticated maize lacks the ability to produce abscission callus. Selection for yield increase must have restored fertility to the rudimentary second floret at each rachis node that characterizes female racemes of teosinte. Restoring fertility to rudimentary spikelets or florets under domestication is not unique to maize. Six rowed barley (*Hordeum vulgare* L.) is characterized by three fertile spikelets at each rachis node, while in the wild ancestor the two lateral spikelets are sterile. Similarly, some sorghum cultivars are characterized by twin fertile florets in each spikelet, while in the wild progenitor of cultivated *Sorghum bicolor* the second floret is reduced.

Phenotypic differences between maize and teosinte are not more extreme than between *Pennisetum americanum* subsp. *monodii* (MAIRE) BRUNKEN and cultivars of pearl millet with racemes over one meter long (BRUNKEN 1977; BRUNKEN, DE WET and HARLAN 1977). Similarly, *Sorghum bicolor* subsp. *arundinaceum* (DESV.) DE WET et HARLAN differs significantly from races *durra, guinea, caudatum* or *kafir* of cultivated grain sorghums (DE WET and HARLAN 1971; DE WET 1978). Systematically, however, these taxa are conspecific. Archaeological evidence strongly suggests that these wild taxa are the progenitors of the phenotypically modified cultivated cereals, and in nature these cereals cross with their assumed wild progenitors to produce fertile hybrids.

**Mimetic cereal weeds.** – Annual seed crops are commonly accompanied by weeds that resemble them in habitat preference and phenotype, except hat they retain the ability of natural seed dispersal. These weeds evolve in three principal ways. First, directly from the wild progenitor of the crop through selection for adaptation to the cultivated habitat. Second, as derivatives of hybrids between the crop and its close wild relatives, Third, from abandoned cultivars through selection for new seed dispersal mechanisms. This is particularly well demonstrated in *Sorghum bicolor* (DE WET 1978).

Members of *S. bicolor* subsp. *arundinaceum* are widely distributed across the African savanna and extend into the tropical forest of West Africa and the sahel zone of North Africa. Races *verticilliflorum* (savanna), *arundinaceum* (forest) and *virgatum* (stream banks) are aggresive colonizers and often obnoxious urban or agricultural weeds. They have also been introduced to South Asia,

Australia and parts of the New World where they are strictly confined to permanently disturbed habitats. These weeds resemble their wild counterparts in phenotypic detail, and differ from them only in degree of weediness. The more common weed sorghums of Africa more or less resemble the race of grain sorghum they accompany in spikelet and sometimes overall inflorescence morphology, and extend with the cereal well beyond the natural range of the species. They are of two distinct kinds in respect to seed dispersal mechanisms. The most common mimetic weeds are characterized by wild type seed dispersal. Spikelets disarticulate at maturity because of an abscission layer that forms between the spikelet and its pedicel. These weeds represent derivatives of hybrids between the crop and members of subspecies *arundinaceum*. Hybrid swarms are common in fields where this subspecies is sympatric with sorghum cultivation. Some of these weed races breed true to type, and SNOWDEN (1936, 1955) recognizes six distinct species of African mimetic weed sorghums. SCHOLZ (1979) suggests that in pearl millet (*Pennisetum americanum*) true breeding mimetic weeds must represent cultivated kinds in which natural seed dispersal was introduced through a reverse mutation for abscission callus formation. This is unlikely. New dispersal mechanisms can evolve in cultivated cereals, but these commonly differ from those characteristic of their close wild relatives, and this has been demonstrated to occur only where the domestic species is not sympatric with close wild relatives.

Weed sorghums on the Ethiopian highlands, where subspecies *arundinaceum* is absent, disperse their seeds either by means of abscission callus or a break in the pedicel at time of maturity. Those with wild type seed dispersal probably were introduced with the crop from the African savanna. Weeds that disperse their seeds by breaking of the pedicel evidently developed without the benefit of a wild sorghum from which the dominant allele for natural seed dispersal could be introduced (DE WET, HARLAN and PRICE 1970). A similar mechanism of seed dispersal characterizes some mimetic weed sorghums of the southeastern United States.

Grain sorghums also introgress with the South Asian diploid *S. propinquum* (KUNTH) HITCHCOCK and the South Eurasian tetraploid *S. halepense* (L.) PERSOON. Mimetic weed sorghums in Thailand and the Philippines are typically characterized by small caryopses, a distinguishing feature of *S. propinquum*. Introgression between grain sorghum and *S. halepense* is particularly obvious in the warmer parts of the New World where this tetraploid has become established as a weed. Introgression is in both directions. Triploid hybrids are rhizomatous perennials and backcross with diploid as well as tetraploid parents. Tetraploid *S. halepense* benefitted from such introgression to increase its colonizing ability and adaptive range. Johnson grass of the southern and midwestern United States represents tetraploid derivatives of such introgression. Diploid derivatives were introduced with grain sorghums from Texas to the American corn belt during the last two decades and have become obnoxious weeds of maize and soybean fields from Indiana to Colorado.

Taxonomically mimetic weeds of cereals belong with the domestic species. HARLAN and DE WET (1971) propose that they be included with their closest wild relative in one subspecies distinct from the cultivated subspecies. BRUNKEN

(1977) recognizes the mimetic weeds of *Pennisetum americanum* as a third subspecies within the domestic species, and DE WET (1978) proposes that the weeds of *Sorghum bicolor* deserve subspecific rank. As a general rule it is suggested that the non-mimetic weeds be treated as part of the wild complex, and the mimetic weeds as distinct subspecies within the domestic species.

**Cultivated races.** – The cultivated complex of domestic species are extensively variable. Different ecotypes are often taken into cultivation at different times, at different places and by different people across the natural range of the species. During domestication different cultivars also introgress with different wild genotypes as they become distributed away from their centers of origin. Once the initial adaptation to cultivation is achieved, man contributes conciously to discontinuities in variation patterns. Different phenotypes are selected and isolated by man to suit his fancies.

Grain sorghums as an example, were classified by SNOWDEN (1936) into 28 species with 156 varieties and 521 distinct forms. Taxonomically these taxa are real. They are maintained through concious selection and isolation by individual farmers. Snowdenian forms are local cultivars. But, his varieties generally have geographical or ethnological unity, and his species ecological reality as well. Snowdenian species are not systematic species however. They are artifacts of mans agricultural activities. Phylogenetically the species complexes (subseries) of SNOWDEN correspond more or less with races as proposed by HARLAN and DE WET (1971).

Evolution, after domestication was initiated through harvesting and sowing, can usually be traced through comparative morphology, association with living cultures, and when available the archaeological records of the cereal. *Sorghum* was probably first taken into cultivation across a broad band of the savanna between southern Sudan and eastern Nigeria (HARLAN 1971). In this region race *verticilliflorum* is an abundant grass and is still collected as a wild cereal. This initial domestication produced race *bicolor*, now widely distributed across the savanna but never extensively cultivated. From the savanna race *bicolor* spread into the forest zone of West Africa. Selection for adaptation to a wet tropical habitat produced race *guinea* with spikelets that gape at maturity and large open inflorescences (DE WET, HARLAN and KURMAROHITA 1972). Race *kafir* is the dominant grain sorghum in the African savanna south of the equator (SHECHTER and DE WET 1975), and race *caudatum* is grown by speakers of present day Chari-Nile languages (STEMLER, HARLAN and DE WET 1975). Phenotypic differentiation probably is associated with genetic drift. *Durra* sorghums originated outside Africa from *bicolor*, or perhaps *kafir* sorghums that were introduced to the Sind-Punjab region of northwestern India at least 3000 years ago (HARLAN and STEMLER 1976). Race *durra* later entered Africa during the Islamic expansion, and is now the principal sorghum of arid West Africa. Racial evolution in sorghum is closely associated with ethnological isolation. As these barriers breakdown, races become sympatric and hybrid complexes are now as common across Africa as are basic races.

## Systematics of major cereals

Weed and cultivated races of most cereals have been studied extensively, and the wild progenitors of many cereals have been identified (DE WET 1979). Their taxonomy, however, remains uncertain (see as examples MANSFELD 1952; BOWDEN 1959; BAUM 1977). The following nomenclature is proposed for the systematically best understood cereals. Wild, weed and cultivated subspecies are listed in this order for each domestic species. Where mimetic weeds are not sufficiently known to treat them taxonomically, they are included in the wild subspecies.

1. *Avena sativa* L., Sp. Pl. 79, 1753. The synonymy of hexaploid oats is discussed by BAUM (1977).

1a. *Avena sativa* subsp. *sterilis* (L.) DE WET comb. nov. Based on *A. sterilis* L., Sp. Pl. ed. 2: 118. 1762. This subspecies includes *A. sterilis, A. athenathera* PRESL and *A. trichophylla* C. KOCH as recognized by BAUM (1977).

This subspecies includes the wild progenitors of cultivated hexaploid oats. It is widely distributed in the Near East, and the mediterranean regions of Europe and North Africa. Wild oats are characterized by dispersal units that disarticulate through abscission callus formation below the basal floret of each spikelet. Later floret separation occurs by fracturing of rachilla segments.

1b. *Avena sativa* subsp. *fatua* (L.) THELL., Vierteljahrsschr. Nat. Gesellsch. Zürich 56: 319. 1912. Based on *A. fatua* L., Sp. Pl. 80. 1753. This subspecies includes *A. fatua, A. hybrida* PETERM. and *A. occidentalis* DUR. as recognized by BAUM (1977).

This subspecies includes the common mimetic weed oats. It is characterized by florets that disarticulate individually by means of abscission callus formation. HAUSSKNECHT (1885) proposes that *fatua* oats are the progenitors of cultivated oats. This is unlikely. Lack of seed dispersal in domesticated oats is genetically dominant over fatua-type seed dispersal. In general, wild type seed dispersal is genetically dominant over the inability of natural seed dispersal that characterises cultivated cereals. A more likely explanation is that the *fatua* dispersal mechanism evolved after oats became domesticated (COFFMAN, PARKER and QUISSENBERRY 1925).

1c. *Avena sativa* subsp. *sativa*. Based on *A. sativa* L., Sp. Pl. 79, 1753. This subspecies is recognized in the sense of BAUM (1977) to include all hexaploid (2n=42) cultivated oats.

Two species of hexaploid oats are commonly recognized, *A. sativa* which is characterized by florets that separate by fracturing of the rachilla leaving a section of rachilla attached to each floret after threshing, and *A. byzantina* C. KOCH in which the basal floret leaves an abscission scar on threshing. It is widely accepted that these two races were independently domesticated (BELL 1965). A monophyletic origin of hexaploid cultivated oats, however, is equally likely (COFFMAN 1946). This becomes obvious when subspecies *sterilis* is accepted as the wild progenitor of oats. Florets of subspecies *sterilis* occur among remains of cultivated wheat and barley in agricultural settlements from Europe to China. It accompanied these crops as a weed, and was probably accidentally harvested and sown for some 7000 years before oats became an important cereal during the first millennium B. C. in northern Europe (HARLAN 1977). In the northern extremes of wheat cultivation the better adapted weed oats must have eventually been adopted as a cultivated cereal.

2. *Eleusine coracana* (L.) GAERTNER, Fruct. et Sem. 1: 8. 1788. The synonymy of this species is discussed by PHILLIPS (1972).

2a. *Eleusine coracana* subsp. *africana* (KENN.-O'BYRNE) HILU et DE WET, Econ. Bot. 30: 202. 1976. Based on *E. africana* KENNEDY-O'BYRNE, Kew Bull. 12: 65. 1957. PHILLIPS (1972) recognized wild finger millet as a subspecies of *E. indica* (L.) GAERTNER. It is, however, cytogenetically isolated from *E. indica* (2n = 18) while it produces fully fertile hybrids with tetraploid domesticated finger millet (CHENNAVEERAIAH and HIREMATH 1974). Wild as well as cultivated finger millets are commonly self pollinated and mimetic weeds or hybrid swarms are rarely encountered. The subspecies is widely distributed across the East African rift (HILU and DE WET 1976a, b).

2b. *Eleusine coracana* subsp. *coracana*. Based on *E. coracana* (L.) GAERTN., Fruct. et Sem. 1: 8. 1788. This subspecies is recognized to include all cultivated taxa of finger millet (PHILLIPS 1972).

Finger millets are widely cultivated along the rift of eastern Africa. It is a native African cereal and was probably domesticated in the East African highlands more than 5000 years ago (HILU, DE WET and HARLAN 1979). This race became widely dispersed and evolved into a coastal race with more compact inflorescences that extends from Kenya to South Africa. The lowland race was introduced to South India and from there it eventually spread across South Asia.

3. *Hordeum vulgare* L., Sp. Pl. 84. 1753. The synonymy of this species is discussed by BOWDEN (1959).

3a. *Hordeum vulgare* subsp. *spontaneum* (C. KOCH) THELL., Mem. Soc. Sci. Math. Cherbourg 38: 160. 1912. Based on *H. spontaneum* C. KOCH, Linnaea 21: 430. 1848.

Spontaneous barleys are commonly two-rowed. The two lateral spikelets of each triad at a rachis node are sterile. Two-rowed wild barleys are widespread along the Mediterranean basin and east to Afghanistan (HARLAN and ZOHARY 1966). They also occur in Tibet (SHAO et al. 1975; HSU 1975). Six-rowed spontaneous barleys, in which all spikelets of a triad at each rachis node are fertile, occur sporadically in the Near East, southern Russia and Tibet (KAMM 1954; ÅBERG 1940; BAKHTEYEV 1975). Collections from southern Russia (*H. laguncuiforme* BAKHT.) differ from Tibetan collections (*H. agriocrithon* ÅBERG) in having the lateral spikelets of each triad distinctly pedicelled.

3b. *Hordeum vulgare* subsp. *vulgare*. Based on *H. vulgare* L., Sp. Pl. 84. 1753.

This subspecies includes all cultivated barleys. Two complexes are commonly recognized (BAKHTEYEV 1962). Two-rowed barley (*H. distichon* L.) is characterized by sterile lateral spikelets and a fertile central spikelet at each rachis node. In some cultivars (*H. deficiens* STEUD.) these lateral spikelets are reduced to glumes. In six-rowed barleys (*H. hexastichon* L.) all spikelets of a triad are fertile. In some of these cultivars (*H. irregularis* ÅBERG et WIEBE) isospiculate and heterospiculate triads occur on the same inflorescence. Cultivated barleys are commonly assumed to be derived from six-rowed spontaneous barleys (SCHIEMANN 1951; TAKAHASHI 1955; BELL 1965). They argue that cultivated two-rowed barley was either derived from cultivated six-rowed kinds, or independently from spontaneous two-rowed barleys. Zohary (1959, 1960), however, suggests that six-rowed spontaneous kinds represent derivatives of hybrids between wild two-rowed barleys and cultivated six-rowed kinds. The wild ancestor of cultivated barley was probably two-rowed and the six-rowed cultivated race was probably derived under domestication from the cultivated two-rowed race (Harlan 1976).

4. *Oryza glaberrima* Steud., Syn. Pl. Glum 1: 3. 1854. The synonymy of this species is discussed by Clayton (1968, 1972).

4a. *Oryza glaberrima* subsp. *barthii* (A. Chev.) De Wet comb. nov. Based on *O. barthii* A. Chev., Bull. Mus. Hist. Nat. Paris 16: 405. 1911.

This annual wild rice of Africa is widely distributed in seasonally flooded water holes across the savanna, from Senegal to Zambia. It is often harvested as a wild cereal, and sometimes invades rice fields as a weed (*O. stapfi* Roshev.). The weed race is impossible to consistently distinguish from wild African rice on the basis of spikelet morphology.

4b. *Oryza glaberrima* subsp. *glaberrima*. Based on *O. glaberrima* Steud., Syn. Pl. Glum. 1: 3. 1854.

Cultivated African rice differs from its wild progenitor subspecies *barthii* consistently on having persistent spikelets. Further, spikelets are usually awned rather than awnless, and the lemma and palea are often glabrous rather than hispid. Portères (1956) recognizes several cultivated races.

5. *Oryza sativa* L., Sp. Pl. 333. 1753. This species has been studied extensively, but its systematics remains uncertain (Chang 1976b). It differs from African rice (*O. glaberrima*) in having a longer (15–45 mm) ligule on the lower leaves. The ligule of *O. glaberrima* is rarely over 6 mm long (Tateoka 1963).

5a. *Oryza sativa* subsp. *rufipogon* (Griff.) De Wet comb. nov. Based on *O. rufipogon* Griff., Notul. Pl. Asia 3: 5. 1851. The synonymy of this subspecies is discussed by Tateoka (1963) who indicates that the more commonly accepted *O. perennis* Moench may or may not refer to Asiatic wild rice.

This subspecies is widely distributed in Asia and Australia, and as recognized by Chang (1976b) includes both annuals and perennials. Perennials are weakly rhizomatous, extravaginal in branching, have long anthers, are adapted to continuously flooded habitats, and reproduce readily by asexual means. Annuals branch intravaginally, have shorter anthers, are adapted to shallow pools, and reproduce sexually. Sharma and Shastry (1965a, b) recognize two kinds of annuals, the common weed rice (*O. sativa* var. *fatua* Prain) and a wild taxon (*O. nivara* Sharma et Shastry). Distinguishing between these two taxa on the basis of spikelet morphology, however, is often impossible. Cytogenetic studies (Chang 1964) suggest that the perennial gave rise to annual, cultivated rice under domestication. For this reason, the spontaneous annuals of Asia are recognized as a weed subspecies of *O. sativa*. Oceanian annual races (Oka 1974) may represent truely wild *O. nivara*.

5b. *Oryza sativa* subsp. *fatua* (Prain) De Wet comb. nov. Based on *O. sativa* var. *fatua* Prain, Beng. Pl. 2: 1184. 1903.

Spontaneous weed rice occurs across the range of rice cultivation in South Asia. Cultivated rice crosses readily with members of subspecies *rufipogon*, and large hybrid swarms are often encountered in South Asia (OKA and CHANG 1961).

5c. *Oryza sativa* subsp. *sativa*. Based on *O. sativa* L., Sp. Pl. 333. 1753.

The antiquity of rice cultivation is incertain. Rice probably was collected as a wild cereal across the humid tropics of Asia where the perennial *rufipogon* is widely distributed. In parts of India and in Sri Lanka this perennial is still harvested as a wild cereal (VISHNU-MITTRE 1974). The oldest known rice remains in the archaeological record (ALLCHIN 1969; SOLHEIM 1972) are from Mohenjodaro in Pakistan (2500 B. C.), India (2300 B. C.) and Thailand (3500 B. C.).

Three races of rice are commonly recognized. Race *indica* probably originated over a broad region spreading from the Gangetic plains to Vietnam and Southern China (CHANG 1976a). It includes the floating rices of South Asia. This basic and tropical race was introduced to the Yellow river valley and lower Yangtze river basin where the cool, tolerant, temperate race *japonica* evolved. Race *japonica* was introduced to Korea and later Japan around the third century B. C. (MORINAGA 1968). Race *indica* spread south into the Malay archipelago where the large grained race *javanica* originated, and this latter race eventually became widely distributed across the islands of Southeast Asia.

6. *Pennisetum americanum* (L.) LEEKE, Zeitschr. Naturw. 79: 52. 1907. Based on *Panicum americanum* L., Sp. Pl. 56, 1753. The synonymy of this species is discussed by BRUNKEN (1977).

6a. *Pennisetum americanum* subsp. *monodii* (MAIRE) BRUNKEN, Amer. J. Bot. 64: 170. 1977. Based on *P. chudeaui* subsp. *monodii* MAIRE, Bull. Mus. Hist. Nat. Paris 2 (3): 253. 1931.

This subspecies is widely distributed from Dakar to the central Sudan across the Sahel zone of West Africa, and also occurs along the foothills of mountains in the central Sahara. This wild ancestor of pearl millet is characterized by one or rarely two spikelets per involucre, and involucres are deciduous at maturity, with bristles that are longer than the spikelets.

6b. *Pennisetum americanum* subsp. *stenostachyum* (KLOTZSCH ex A. BR. et BOUCHE) BRUNKEN, Amer. J. Bot. 64: 173 1977. Based on *P. stenostachyum* (KLOTZSCH ex A. BR. et BOUCHE) STAPF et HUBB., Kew Bull. 1933: 270. 1933.

Weed pearl millets commonly mimics the race of the crop they accompany as weeds. Inflorescences are usually compact, with involucres that are deciduous at maturity, and with usually two spikelets per involucre. Weeds are common in agricultural fields across the arid savanna of West Africa and was introduced with the crop to Angola and southwestern Africa. Some mimetic weeds can only be distinguished from the race of pearl millet they accompany by spikelets that become deciduous at maturity. SCHOLZ (1979) suggests that these weeds represent degenerate cultivars. The presence of hybrid swarms, however, suggests introgression with subspecies *monodii* (BRUNKEN, DE WET and HARLAN 1977).

6c. *Pennisetum americanum* subsp. *americanum*. Based on *Pennisetum americanum* (L.) LEEKE, Zeitschr. Naturw. 79: 52. 1907.

This subspecies includes the different cultivated species of pearl millet as recognized by STAPF and HUBBARD (1934). Inflorescences vary from cylindrical to broadly elliptic in shape, and ranges from 4 to over 200 cm in length.

Four races are recognized on the basis of grain shape and distribution (BRUNKEN, DE WET and HARLAN 1977). Race *typhoides* is characterized by ovate grains that are usually exposed between the glumes at maturity. It is cultivated across the African savanna, and was introduced to India at least 3000 years ago. Grains of race *nigritarum* are obovate and angular in cross section, and generally longer than the glumes. This race is an important cereal from Western Sudan to Nigeria, and is also grown in Senegal. In race *globosum* the grains are essentially spherical, and this race is widely cultivated from Upper Volta to the Sudan. Race *leonis* is characterized by acute, oblanceolate and terete grains, and is virtually indigenous to Sierra Leone. PORTÈRES (1976) suggests that racial evolution in pearl millet is due to several independent domestications of this cereal. Comparative morphology, however, suggests that pearl millet was domesticated across the southern margins of the Saharan highlands, and that race *typhoides* gave rise to the other three races under domestication (BRUNKEN, DE WET and HARLAN 1974).

7. *Secale cereale* L., Sp. Pl. 84. 1753. The synonymy of this species is discussed by ROSHEVITZ (1947).

7a. *Secale cereale* subsp. *ancestrale* ZHUK. in Bull. appl. bot., genet., breed. 19, 2: 54. 1928.

This subspecies is recognized to include *S. afghanicum* (VAV.) ROSHEV., *S. dighoricum* (VAV.) ROSHEV, and *S. segetale* ROSHEV. as described by ROSHEVITZ (1947). It includes two complexes. The *ancestrale* complex is characterized by small grains, and spikelets that all disarticulate at maturity. It occurs along sandy banks, edges of fields and in vineyards near the town of Aydin in the Menderes valley of Turkey (ZOHARY 1971). The *segetale* complex includes semi-brittle weed ryes, in which the lower one quarter or more of the spikelets are not deciduous at maturity. They occur across eastern Europe and extend east to Afghanistan.

7b. *Secale cereale* subsp. *cereale*. Based on *S. cereale* L., Sp. Pl. 84. 1753.

This subspecies includes all cultivated ryes. Common rye is cross fertilized, typically two-flowered, and the rachilla projects beyond the second floret. The cultivated complex sometimes recognized as *S. turkestanicum* BENSIN is self fertilized, and characterized by a well developed, but sterile third floret. It is grown in Central Asia and Transcaucasia, and probably evolved independently in Turkestan (BENSIN 1933). Cultivated rye is known from the Neolithic of Austria (WERNECK 1951), but this cereal only became widespread as a crop in Europe since the Bronze age. It is a crop that evolved out of weed rye that was introduced with wheat and barley into Europe (VAVILOV 1917).

8. *Setaria italica* (L.) P. BEAUV., Ess. Agrost. 51: 170. 1812. Based on *Panicum italicum* L., Sp. Pl. 56. 1753. The synonymy of this species is discussed by ROMINGER (1962).

8a. *Setaria italica* subsp. *viridis* (L.) THELLUNG, Mem. Soc. Sci. Nat. Cherbourg 38: 85. 1912. Based on *Panicum viride* L., Syst. Nat. ed. 10: 870. 1759.

This subspecies includes the widely distributed, cosmopolitan temperate annual commonly known as green foxtail. It is characterized by spikelets that disarticulate below the glumes to leave a cup-like scar on the rachis.

8b. *Setaria italica* subsp. *pycnocoma* (STEUD.) DE WET comb. nov. Based on

*Panicum pycnocomum* STEUD., Syn. Pl. Glum. 1: 462, 1854 (*S. viridis* (L.) BEAUV. ssp. *pycnocomum* (STEUD.) TZVEL., in Novosti sist. vysš. rast. 1968: 19. 1968; *S. viridis* var. *major* (GAUD.) POSPICH., Fl. Oesterr. Küstenl. 1: 51, 1897).

This subspecies includes the widely distributed weed foxtails of agriculture (DE WET, OESTRY-STIDD and CUBERO 1980). It differs from cultivated foxtail millets primarily in being spontaneous rather than sown. Spikelets disarticulate below the glumes and florets disarticulate above the sterile floret. These weeds represent derivatives of crosses between foxtail millets and subspecies *viridis*.

8c. *Setaria italica* subsp. *italica*. Based on *S. italica* (L.) P. BEAUV., Ess. Agrost. 51: 170. 1812.

This subspecies includes all cultivated foxtail millets. They are characterized by spikelets that are persistent on the inflorescence at maturity, but disarticulate on threshing above the sterile, lower floret. Foxtail millets are widely cultivated in Eurasia (SCHEIBE 1943). They are variously classified on the basis of inlorescence shape and fruit color (KRUSE 1972). Two well defined races can be recognized. Race *moharia* with its short, usually erect and compact inflorescences is grown in southern Europe and extends east to India. Race *maxima* is characterized by large inflorescences with distinct lateral branches. It is widely grown in southern Russia and China (DEKAPRELEVICH and KASPARIAN 1928).

9. *Sorghum bicolor* (L.) MOENCH, Meth. 207, 1794. Based on *Holcus bicolor* L., Mant. Alt. 301. 1771. The synonymy of this species is discussed by SNOWDEN (1936).

9a. *Sorghum bicolor* subsp. *arundinaceum* (DESV.) DE WET et HARLAN, in HARLAN, DE WET and STEMLER, Origins of Afr. Plt. Dom. 455, 1976. Based on *S. arundinaceum* (DESV.) STAPF, in Prain, Fl. Trop. Afr. 9: 119. 1917.

This subspecies includes the wild progenitors of grain sorghums, and their closest wild relatives. Four ecotypes are recognized. The tropical ecotype (*S. arundinaceum* (DESV.) STAPF) is widely distributed along the edges of tropical African forests from West Africa to Angola. The desert ecotype (*S. aethiopicum* (HACK.) RUPR.) occurs in the dry savanna from Mauritania to Ethiopia. The flood plain ecotype (*S. virgatum* (HACK.) STAPF) is confined to the flood plains of the lower Nile, and the savanna ecotype (*S. verticilliflorum* (STEUD.) STAPF) is a common grass across the African savanna from Senegal to South Africa.

9b. *Sorghum bicolor* subsp. *drummondii* (STEUD.) DE WET, Amer. J. Bot. 64: 481. 1978. Based on *S. drummondii* (STEUD.) MILLSP. et CHASE, Publ. Field Columb. Mus. Bot. 3: 21. 1903.

This subspecies is recognized to include all spontaneous weed sorghums of SNOWDEN (1936, 1955). It occurs in Africa wherever grain sorghums are grown. Mimetic weeds also evolved from hybridization with members of *S. halepense* (L.) PERS. in Asia (DE WET 1978).

9c. *Sorghum bicolor* subsp. *bicolor*. Based on *S. bicolor* (L.) MOENCH, Meth. Pl. 207. 1794.

This subspecies includes all cultivated species of grain sorghum recognized by SNOWDEN (1936). Origins and racial evolution were discussed by DE WET (1978). Five basic races are recognized. Race *bicolor* with its large, open inflorescences, and spikelets with long clasping glumes includes the most primitive of cultivated sorghums. It is sporadically grown across the range of grain sorghum cultivation in Africa. Race *kafir* is characterized by broadly elliptic spikelets with the glumes tightly clasping the usually longer grain. It is widely grown across the savanna of Africa south of the Equator. *Guinea* sorghums with their gaping glumes are grown primary in the wet tropics of West Africa and in Malawi. Race *caudatum* is typically characterized by turtle backed grains and is mostly confined to Chad and Uganda. *Durra* sorghums are commonly associated with Islamic people in Africa. It is characterized by somewhat flattened grains.

10. *Triticum monococcum* L., Sp. Pl. 86. 1753. The synonymy of this species is discussed by BOWDEN (1959).

10a. *Triticum monococcum* subsp. *boeoticum* (BOISS.) A. et D. LÖVE in Bot. Not. 114 (1961) 49. Based on *T. boeoticum* BOISS., Diagn. Fl. Orient. Nov. 1 (2): 69, 1854.

Wild einkorn differs from cultivated, diploid (2n=14) wheat primarily in having brittle spikes that disarticulate at maturity into individual spikelets each tipped by a sharp rachis joint. It is widely distributed across western Asia and the southern Balkans (HARLAN and ZOHARY 1966). There are two ecogeographic races, a small slender race with usually one fruited spikelets, and a robust race with usually two fertile florets per spikelet (ZOHARY, HARLAN and VARDI 1969). The slender race occurs in the northern and northwestern range of the subspecies. The robust race is more southern in distribution.

10b. *Triticum monococcum* subsp. *monococcum*. Based on *T. monococcum* L., Sp. Pl. 86. 1753.

This subspecies includes all cultivated einkorns. It is a relic crop, that is cultivated sporadically in the Balkans, Turkey and parts of southern Europe (SCHIEMANN 1948).

11. *Triticum turgidum* L., Sp. Pl. 1: 86. 1753. The synonymy of this species is discussed by BOWDEN (1959).

11a. *Triticum turgidum* subsp. *dicoccoides* (KÖRN. ex SCHWEINF.) THELL. in Naturw. Wochenschr. (Jena), N. F. 17: 470. 1918. Based on *T. dicoccoides* Körn., in SCHWEINF., Berichte Deutsch. Bot. Gesellsch. 26 (a): 309. 1908.

Wild emmer is widespread over northern Israel and Jordan, and extends east to Iran and west to Turkey (ZOHARY 1971). Weedy derivatives of hybrids between this subspecies and cultivated emmer are difficult to distinguish from wild emmer, except in habitat preference. Wild representatives are adapted to rocky habitats.

11b. *Triticum turgidum* subsp. *turgidum*. Based on *T. turgidum* L., Sp. Pl. 1: 86. 1753.

This subspecies includes all tetraploid (2n=28) cultivated wheats with a genome constitution of AABB (SEARS 1975). Except for emmer wheat, these tetraploides are free threshing. In emmer the grains remain tightly enclosed by the tough glumes. Emmer is the basic race of tetraploid wheats and probably in one of the oldest cultivated cereals of the Near East (ZOHARY 1971). Durum wheats and turgidum wheats were probably derived from cultivated emmer through the accumulation of a series of mutations that collectively allow for free threshing of the grain (FELDMAN 1976).

12. *Triticum aestivum* L., Sp. Pl. 85. 1753. This is strictly a cultivated species. It is hexaploid, and includes all commonly recognized bread wheats. These are usually divided into four races (SEARS 1975). Common bread wheat is cultivated across the temperate regions of the world. Club wheats (*T. compactum* HOST) is less common, shot wheats (*T. sphaerococcum* PERC.) are mostly grown in India, hulled wheats (*T. spelta* L.) is a relic crop in Europe and the Near East, and macha wheats (*T. vavilovi* JAKUBZ.) are local endemics in Transcaucasia and Iran.

13. *Zea mays* L., Sp. Pl. 1: 971. 1753. The synonymy of this species is discussed by HITCHCOCK (1950).

13a. *Zea mays* subsp. *mexicana* (SCHRAD.) ILTIS, Phytologia 23: 248. 1972. Based on *Euchlaena mexicana* SCHRAD., Ind. Sem. Hort. Goettingen 1832, reprinted in Linnaea 8: 25. 1833.

Teosinte is cytogenetically conspecific with maize, but differs from maize conspicuously in female inflorescence morphology. MANGELSDORF and REEVES (1939) propose that teosinte originated from maize- *Tripsacum* introgression, and MANGELSDORF (1974) suggests that it originated from maize through a series of mutations. Cytogenetic studies, however, indicate that teosinte is wild maize. It is widely distributed in Mesoamerica and crosses with maize to produce mimetic weeds, particularly around Chalco in Mexico (WILKES 1970).

13b. *Zea mays* subsp. *mays*. Based on *Z. mays* L., Sp. Pl. 971. 1753.

This species includes all the cultivated races of maize (WELLHAUSEN et al. 1952; GROBMAN et al. 1961; GOODMAN and BIRD 1977). MANGELSDORF (1974) combines all described races into six basic evolutionary lineages. These are the pointed popcorns of Mexico (Palomero Toluqueño), sweet corns of Peru (Chullpi), eight rowed corns of Peru (Confite Morocho), maize with colored aleurone from Peru (Kculli), the Chapalote-Nal Tel complex of Mexico, and the tropical flints of Colombia (Pira Naranga). Maize is native to Mesoamerica, spread south and north early in its evolutionary history, and was introduced to the Old World during historical times.

## Zusammenfassung

Artbegriff und Systematik bei domestizierten Getreiden

Kultivierte Getreide sind domestizierte Gräser. Die wilden Vorfahren der domestizierten Getreide sind aggressive natürliche Besiedler und werden oft zu Unkräutern in vom Menschen beeinflußten Standorten. Bastardierung zwischen kultivierten Getreiden und ihren wilden Verwandten läßt mimetische Getreideunkräuter entstehen, die oft die domestizierten Arten außerhalb ihres natürlichen Verbreitungsgebietes begleiten. Ähnliche Unkräuter können sich auch aus aufgegebenen Sorten entwickeln als ein Ergebnis von Mutationen, die ihre Fähigkeit zur natürlichen Samenausstreuung wiederherstellen. Kultivierte Komplexe domestizierter Arten sind in typischer Weise variabel. Der Mensch selektiert und isoliert Phänotypen nach seinen Vorstellungen mit dem Ergebnis, daß verschiedene Phänotypen eines Getreides oft zu verschiedenen Zwecken im selben Dorf angebaut werden, und benachbarte Dörfer oft verschiedene Phänotypen für ähnliche Zwecke kultivieren. Der herkömmliche Artbegriff bewährt sich nicht bei domestizierten Arten. Es wird vorgeschlagen, daß Wild-, Unkraut- und Kulturformen einer domestizierten Art als entsprechende Unterarten angesehen werden. Größere Gruppen innerhalb der Unkraut- oder Kultur-Unterart sollten als Rassen ohne formale taxonomische Bezeichnung gehandhabt werden.

**Краткое содержание**

Концепция вида и систематика одомашненных хлебных злаков

Возделываемые хлебные растения это – одомашненные злаки. Дикорастущие предки одомашненных злаков являются естественными агрессивными "колонизаторами" и часто становятся сорняками в местообитаниях, подверженных влиянию человека. Путём гибридизации культурных злаков и их дикорастущих сородичей возникают миметические формы сорняков, часто сопровождающие культурные виды и вне их естественных ареалов. Подобные же сорняки могут возникнуть и из сортов, которые больше не возделываются, в результате мутаций, восстанавливающих их способность к естественному свободному осыпанию семян. Культивируемые комплексы одомашненных видов проявляют типичную для них изменчивость. Человек селекционирует и изолирует фенотипы по своим потребностям, в результате чего различные фенотипы одного зерноврго злака часто возделываются для различных целей в одном и том же селе, а соседние сёла часто культивируют различные фенотипы для сходного использования. Общепринятое понятие вида оказывается неподходящим для одомашненных видов. Предлагается рассматривать дикорастущие, сорняковые и культурные формы сдного вида как соответствующие подвиды. Значительные группы, выделяемые внутри сорного или культурного подвида следует рассматривать как расы, не давая им формального таксономического обозначения.

## Literature

Åberg, E., 1940: The taxonomy and phylogeny of *Hordeum* L. Sect. Cerealia Auds. with special reference to Tibetan barleys. – Symb. Bot. Upsal. **4**, 1–156.

Adanson, M., 1763: Familles des Plantes. – Paris.

Anderson, E., 1956: Man as a maker of new plants and new plant communities. In: W. L. Thomas (ed.). Man's role in changing the face of the earth. – Univ. Chicago Press, Chicago.

Allchin, F. R., 1969: Early cultivated plants in India and Pakistan. In: P. J. Ucko, and G. W. Dimbleby (eds.). The domestication and exploitation of plants and animals. – Duckworth and Company, London.

Bakhteyev, F. K., 1962: A wild six rowed barley discovered for the first time. – Izvest. Akad. Nauk Turkmensk. SSR Biol. Nauk 5, 86–89.

–, 1975: Studies on barley: Present problems of the origin of barley. – Folia Quaternaria Krakow **46**, 73–81.

Baum, B. R., 1977: Oats: Wild and Cultivated. Canad. Dept. Agric. Biosyst. Res. Inst. Monogr. 14. – Thorn Press, Ottawa.

BEADLE, G. W., 1972: The mystery of maize. – Field Mus. Nat. Hist. Bull. **43**, 2–11.
–, 1979: The ancestry of corn. – Sci. Amer. **242**, 112–119.
BELL, G. D. H., 1965: The comparative phylogeny of the temperate cereals. In: J. HUTCHINSON (ed.). Crop plant evolution. – Cambridge Univ. Press, Cambridge.
BENSIN, M., 1933: Autogamous Turkestan rye (*Secale turkestanicum*). – Bull. Torrey Bot. Club **60**, 155–160.
BIRD, R. McK., 1978: A name change for Central American teosinte. – Taxon **27**, 361–363.
BOWDEN, W. M., 1959: The taxonomy and nomenclature of the wheats, barleys, and ryes and their wild relatives. – Canad. J. Bot. **37**, 657–684.
BRANDOLINI, A., 1970: Maize. In: O. H. FRANKEL and E. BENNETT (eds.). Genetic resources in plants. I. B. P. Handbook No. **11**. – F. A. O., Rome.
BRUNKEN, J. M., 1977: A systematic study of *Pennisetum* sect. *Pennisetum* (Gramineae). – Amer. J. Bot. **64**, 161–176.
–, J. M. J. DE WET, and J. R. HARLAN, 1977: The morphology and domestication of pearl millet. – Econ. Bot. **31**, 163–174.
BURKILL, I. H., 1952: Habitats of man and the origins of the cultivated plants of the Old World. – Proc. Linn. Soc. (London) **164**, 12–42.
CALLEN, E. O., 1965: Food habits of some pre-Columbian Mexican Indians. – Econ. Bot. **19**, 335–343.
–, 1967: The first New World cereal. – Amer. Antiquity **32**, 535–538.
CHANG, T. T., 1964: Present knowledge of rice genetics and cytogenetics. – Intern. Rice Res. Inst. Tech. Bull. **1**, 1–96.
–, 1976a: Rice. In: N. W. SIMMONDS, (ed.). Evolution of Crop Plants. – Longman, London and New York.
–, 1976b: The origin, evolution, cultivation, dissemination, and diversification of Asian and African rices. – Euphytica **25**, 425–441.
CHENNAVEERAIAH, M. S., and S. C. HIREMATH, 1974: Genome analysis of *Eleusine coracana* (L.) GAERTNER. – Euphytica **23**, 489–495.
CLAYTON, W. D., 1968: Studies in the Gramineae. XVII. West African wild rice. – Kew Bull. **21**, 487–488.
–, 1972: Gramineae. In: F. N. HEPPER (ed.). Flora of West Tropical Africa. Vol. **3** (2), 349–574.
COFFMAN, F. A., 1946: Origin of cultivated Oats. – J. Amer. Soc. Agron. **38**, 983–1002.
–, J. H. PARKER, and K. S. QUISSENBERRY, 1925: A study of variability in the Burr oat. – J. Agric. Res. **30**, 1–64.
COLLINS, G. N., 1925: The "metamorphosis" of *Euchlaena* into maize. – J. Hered. **16**, 378–380.
DEKAPRELEVICH, L. L., and A. S. KASPARIAN, 1928: A contribution to the study of foxtail millet (*Setaria italica* P. B. *maxima* Alf.) cultivated in Georgia (western Transcaucasia). – Bull. Appld. Bot. Plt. Breed. **14**, 533–572.
DE WET, J. M. J., 1975: Evolutionary dynamics of cereal domestication. – Bull. Torrey Bot. Club **102**, 307–312.
–, 1977: Domestication of African cereals. – Afr. Econ. Hist. **3** (Spring), 15–32.
–, 1978: Systematics and evolution of *Sorghum* sect. *Sorghum* (Gramineae). – Amer. J. Bot. **65**, 477–484.
–, 1979: Principles of evolution and cereal domestication. In: A. C. ZEVEN and A. M. VAN HARTEN (eds.). Broadening the genetic base of crops. – PUDOC, Wageningen.
DE WET, J. M. J., and J. R. HARLAN, 1971: The origin and domestication of *Sorghum bicolor*. – Econ. Bot. **25**, 128–135.
–, and –, 1972: Origin of maize: The tripartite hypothesis. – Euphytica **21**, 271–278.
–, and –, 1975: Weeds and domesticates: Evolution in the man-made habitat. – Econ. Bot. **29**, 99–107.
–, –, and B. KURMAROHITA, 1972: Origin and evolution of Guinea sorghums. – East Afr. Agric. Forest. J. **37**, 114–119.
–, –, and E. G. PRICE, 1976: Origin of variability in the *spontanea* complex of *Sorghum bicolor*. – Amer. J. Bot. **57**, 704–707.

De Wet, J. M. J., and E. H. Oelke, 1978: Domestication of American wild rice (*Zizania aquatica* L., Gramineae). – J. Agric. Tradit. Bot. Appl. **25**, 67–84.
–, L. L. Oestry-Stidd, and J. I. Cubero, 1979: Origins and evolution of foxtail millets (*Setaria italica*). – J. Agric. Tradit. Bot. Appl. **26**, 53–64.
Feldman, M., 1976: Wheats. *Triticum* spp. (Gramineae – Triticinae). In: N. W. Simmonds (ed.), Evolution of crop plants. – Longman, London and New York.
Galinat, W. C., 1971: The origin of maize. – Ann. Rev. Genet. **5**, 447–478.
Goodman, M. M., and R. McK. Bird, 1977: The races of maize. IV. Tentative grouping of 219 Latin American races. – Econ. Bot. **31**, 204–221.
Grant, V., 1977: Population structure in relation to macroevolution. – Biol. Zentralbl. **96**, 129–139.
Grobman, A., W. Salhuana, and R. Sevilla in collab. with. P. C. Mangelsdorf, 1961: Races of maize in Peru. – Nat. Acad. Sci. 3 Nat. Res. Counc. USA Publ. **915**.
Harlan, J. R., 1971: Agricultural origins: Centers and noncenters. – Science **174**, 468–474.
–, 1976: Barley. *Hordeum vulgare* (Gramineae – Triticinae). In: N. W. Simmonds (ed.), Evolution of crop plants. – Longman, London and New York.
–, 1977: Origins of cereal agriculture in the Old World. In: C. A. Reed (ed.), Origins of agriculture. – Mouton Press, The Hague, Paris.
–, and D. Zohary, 1966: Distribution of wild wheats and barley. – Science **153**, 1074–1080.
–, and J. M. J. De Wet, 1971: Towards a rational taxonomy of cultivated plants. – Taxon **20**, 509–517.
–, J. M. J. De Wet, and E. G. Price, 1973: Comparative evolution of cereals. – Evolution **27**, 311–325.
–, and A. B. L. Stemler, 1976: The races of sorghum in Africa. In: J. R. Harlan, J. M. J. De Wet, and A. B. L. Stemler (eds.). Origins of African plant domestication. – Mouton Press, The Hague, Paris.
Haussknecht, C., 1885: Ueber die Abstammung des Saathabers (Saathafers). – Mitt. Geogr. Gesellsch. Jena **3**, 231–242.
Hilu, K. W., and J. M. J. De Wet, 1976a: Racial evolution in *Eleusine coracana* subsp. *coracana* (finger millet). – Amer. J. Bot. **63**, 1311–1318.
–, and –, 1976b: Domestication of *Eleusine coracana* (L.) Gaertner. – Econ. Bot. **30**, 199–208.
–, and –, 1980: Effects of artificial selection on grain dormancy in *Eleusine*. – Syst. Bot. (in press).
–, –, and J. R. Harlan, 1979: Archaeobotanical studies of *Eleusine coracana* subsp. *coracana* (finger millet). – Amer. J. Bot. **66**, 330–333.
Hitchcock, A. S., 1950: Manual of the grasses of the United States. Revised by Agnes Chase. – USDA Miscel. Publ. **200**.
Hsu, T.-W., 1975: On the origin and phylogeny of cultivated barley with reference to the discovery of Ganze wild two-rowed barley *Hordeum spontaneum* C. Koch. – Acta Genet. Sinica **2** (2), 137.
Jeffrey, C., 1968: Systematic categories for cultivated plants. – Taxon **17**, 109–114.
Jirásek, V., 1966: The systematics of cultivated plants and their taxonomic categories. – Preslia **38**, 267–284.
Kamm, A., 1954: The discovery of wild six-rowed barley and wild *H. intermedium* in Israel. – Ann. Agric. Coll. Sweden **21**, 287–320.
Kruse, J., 1972: Beitrag zur Morphologie der Infloreszenzen von *Setaria italica* (L.) Beauv. – Kulturpflanze **19**, 53–72.
Ladizinsky, G., 1975: Oats in Ethiopia. – Econ. Bot. **29**, 238–241.
Langham, D. G., 1940: The inheritance of intergeneric differences in *Zea-Euchlaena* hybrids. – Genetics **25**, 88–107.
Levin, D. A., 1979: The nature of plant species. – Science **204**, 381–384.
MacNeish, R. S., 1971: Speculation about how and why food production and village life developed in the Tehuacan valley, Mexico. – Archaeology **24**, 307–315.
Mangelsdorf, P. C., 1958: Ancestor of corn. – Proc. Amer. Phil. Soc. **102**, 454–463.
–, 1974: Corn, its origin, evolution and improvement. – Belknap Press of Harvard, Cambridge.

MANGELSDORF, P. C., and W. C. GALINAT, 1964: The tunicate locus in maize dissected and reconstituted. – Proc. Nat. Acad. Sci. **51**, 147–150.
–, R. S. MACNEISH, and W. C. GALINAT, 1967: Prehistoric wild and cultivated maize. In: D. S. BYERS (ed.). The prehistory of the Tehuacan valley. Vol. **1**. – Univ. Texas Press, Austin.
–, and R. G. REEVES, 1939: The origin of Indian corn and its relatives. – Texas Agric. Exp. Sta. Bull. **574**, 1–315.
MANSFELD, R., 1952: Zur Systematik und Nomenklatur der Hirsen. – Züchter **22**, 304–315.
MAYR, E., 1969: Principles of systematic zoology. – McGraw-Hill, New York.
MILLICENT, E., and J. M. THODAY, 1961: Effects of disruptive selection. IV. Geneflow and divergence. – Heredity **16**, 199–217.
MORINAGA, T., 1968: Origin and geographical distribution of Japanese rice. – Jap. Agric. Res. Quat. **3** (2), 1–5.
OKA, H.-I., 1974: Experimental studies on the origin of cultivated rice. – Genetics **78**, 475–486.
–, and W.-T. CHANG, 1961: Hybrid swarms between wild and cultivated rice species, *Oryza perennis* and *O. sativa*. – Evolution **15**, 418–430.
PHILLIPS, S. M., 1972: A survey of the genus *Eleusine* in Africa. – Kew Bull. **27**, 251–270.
PORTÈRES, R., 1956: Taxonomie agrobotanique des riz cultivés *O. sativa* Linné et *O. glaberrima* Steudel. – J. Agric. Trop. Bot. Appl. **3**, 341–384, 541–580, 627–700, 821–856.
–, 1976: African cereals: *Eleusine*, Fonio, Black Fonio, Teff, *Brachiaria*, *Paspalum*, *Pennisetum* and African rice. In: J. R. HARLAN, J. M. J. DE WET, and A. B. L. STEMLER (eds.). Origins of African plant domestication. – Mouton Press, The Hague, Paris.
RANDOLPH, L. F., 1976: Contributions of wild relatives of maize ot the evolutionary history of domesticated maize: A synthesis of divergent hypotheses. – Econ. Bot. **30**, 321–346.
ROGERS, J. S., 1950: The inheritance of inflorescence characters in maize-teosinte hybrids. – Genetics **35**, 541–558.
ROMINGER, J. M., 1962: Taxonomy of *Setaria* (Gramineae) in North America. – Univ. Illinois Press, Urbana.
ROSHEVITZ, R. Y., 1947: A monograph of the wild, weedy and cultivated species of rye. – Acta Inst. Bot. Acad. Sci. URSS ser. 1, **6**, 105–163.
SCHEIBE, A., 1943: Die Hirsen im Hindukusch. – Z. Pflanzenzücht. **25**, 392–436.
SCHIEMANN, E., 1948: Weizen, Roggen, Gerste. Systematik, Geschichte und Verwendung. – Fischer Verlag, Jena.
–, 1949: Die neue Nomenklatur der Getreidearten. – Züchter **19**, 322–325.
–, 1951: Neue Gerstenformen aus Ost-Tibet und ein weiterer Fund von *Hordeum agriocrithon* Åberg. – Ber. Deutsch. Bot. Gesellsch. **64**, 56–68.
SCHOLZ, H., 1979: The phenomenon of mimetic weeds in the African *Pennisetum americanum* -a critique. In: G. KUNKEL (ed.). Taxonomic aspects of African Economic Botany. – Perez Galdos, Las Palmas.
SEARS, E. R., 1975: The wheats and their relatives. In: R. C. KING (ed.). Handbook of Genetics. Plenum Press, New York.
SHARMA, S. D., and V. S. SHASTRY, 1965a: Taxonomic studies in genus *Oryza* L. III. *O. rufipogon* GRIFF. sensu stricto and *O. nivara* SHARMA et SHASTRY nom. nov.. – Indian J. Genet. **25**, 157–167.
–, –, 1965b: Taxonomic studies in genus *Oryza* L. VI. A modified classification. – Indian J. Genet. **25**, 173–178.
SHAO, C.-C., C.-S. LI, and C. BASCHAN, 1975: Origin and evolution of the cultivated barley-wild barleys from the southwestern part of China. – Acta Genet. Sinica **2** (2), 128.
SHECHTER, Y., and J. M. J. DE WET, 1975: Comparative electrophoresis and isozyme analysis of seed proteins from cultivated races of *Sorghum*. – Amer. J. Bot. **62**, 244–261.
SINSKAYA, E. H., 1948: The principles of classification of cultivated plants with respect to the functions of infraspecific systematics. – Bot. Ž. **33**, 148–150.
SNOWDEN, J. D., 1936: A classification of the cultivated sorghums. – Kew Bull. **1935**, 221–255.
–, 1955: The wild fodder sorghums of the section *Eu-Sorghum*. – J. Linn. Soc. Bot. London **55**, 191–260.

SOLHEIM II, W. G., 1972: An earlier agricultural revolution. – Sci. Amer. **226** (4), 34–41.
STAPF, O., and C. E. HUBBARD, 1934: *Pennisetum*. In: O. PRAIN (ed.). The Flora of Tropical Africa, vol. 9.
STEMLER, A. B. L., J. R. HARLAN, and J. M. J. DE WET, 1975: Caudatum sorghums and speakers of Chari-Nile languages of Africa. – J. Afr. Hist. **16**, 161–183.
STEWARD, J. H., 1930: Irrigation without agriculture. – Mich. Acad. Sci. Arts and Letters **12**, 149–156.
TAKAHASHI, R., 1955: The origin and evolution of cultivated barley. – Adv. Genet. **7**, 227–266.
TATEOKA, T., 1963: Taxonomic studies of *Oryza*. III. Key to the species and their enumeration. – Bot. Mag. (Tokyo) **75**, 455–461.
THODAY, J. M., 1972: Disruptive selection.-Proc. Roy. Soc. London Series B **182**, 109–143.
TINDALE, N. B., 1977: Adaptive significance of the Panara or grass seed culture of Australia. In: M. R. WRIGHT (ed.). Stone tools as cultural markers.
VAVILOV, N. I., 1917: On the origin of cultivated rye. – Bull. Appl. Bot. **10**, 561–590.
VISHNU-MITTRE, 1974: Paleobotanical evidence in India. In: J. B. HUTCHINSON (ed.). Evolutionary studies on world crops. – Cambridge Univ. Press, Cambridge.
WELLHAUSEN, E. J., L. M. ROBERTS and E. HERNANDEZ-X, in collab, with P. C. MANGELSDORF, 1952: Races of maize in Mexico. – Bussey Inst. Harvard Univ. Publ. 1952, 1–223.
WERNECK, H. L., 1951: Ur- und frühgeschichtliche Roggenfunde in den Ostalpen und am Ostrande des Böhmerwaldes. – Züchter **21**, 107–108.
WILKE, P. J., R. BETTINGER, T. F. KING, and J. F. O'CONNELL, 1972: Harvest selection and domestication in seed plants. – Antiquity **46**, 203–208.
WILKES, H. G., 1967: Teosinte: The closest relative of maize. – Bussey Inst. Harvard Univ. Publ.
–, 1970: Teosinte introgression in the maize of the Nobogame valley. – Bot. Mus. Leafl., Harvard Univ. **22**, 297–311.
–, 1972: Maize and its relatives. – Science **177**, 1071–1077.
–, 1977: Hybridization of maize and teosinte in Mexico and Guatemala and the improvement of maize. – Econ. Bot. **31**, 254–293.
WRIGHT, S., 1966: Polyallelic random drift in relation to evolution. – Proc. Nat. Acad. Sci. USA **55**, 1074–1081.
ZHUKOVSKY, P. M., 1967: The nature and span of species in cultivated plants. – Bot. Ž. **52**, 1530–1539.
ZOHARY, D., 1959: Is *Hordeum agriocrithon* the ancestor of 6-rowed cultivated barley? – Evolution **13**, 279–280.
–, 1960: Studies on the origin of cultivated barley. – Bull. Res. Counc. Israel **9** D, 21–42.
–, 1971: Origin of south-west Asiatic cereals: wheat, barley, oats and rye. In: P. H. DAVIS, et al. (eds.). Plant life of southwest Asia. – Edinburgh Univ. Press, Edinburgh.
–, J. R. HARLAN and A. VARDI, 1969: The wild diploid progenitors of wheat and their breeding value. – Euphytica **18**, 58–65.

Dr. J. M. J. DE WET
Crop Evolution Laboratory,
Department of Agronomy,
University of Illinois at Urbana-Champaign
N 214 Turner Hall
Urbana, Illinois 61801, USA

Kulturpflanze XXIX · 1981 · S. 199–207

# Comments on the basic principles of crop taxonomy

JAMES MAC KEY (Uppsala, Sweden)

## Summary

Cultivated plants are subjected to an intensive evolution and a need for precise classification due to rules for seed certification. These circumstances call for flexibility in approved taxonomical schemes, i. e. complex and rigid hierarchical systems should be avoided. The biological species concept should govern the grouping at higher taxon levels, since such an approach is more likely to produce classification schemes that are able to maintain actuality and are more informative for use in practice. Grouping at lower taxon levels should concentrate on the cultivar. Use in this connection of character symbol formulas has the advantage of being both descriptive and flexible. The relevance of the recommendations is exemplified by comparing different approaches in wheat taxonomy.

Biological taxonomy aims fundamentally at identification and classification, with identification meaningful only in the presence of a worthwhile classification. The Linnean system of classification is originally based on the concept of a static and discontinuous differentiation. It did not acknowledge the evolutionary processes which discard and add, and which might break through isolating barriers. Degree of recognizability was given decisive importance. The ease of recombination between taxonomic markers was not taken into account. Neither was the selective advantage of a specific trait, i. e. its chance to be maintained and steadily actual, decisive for its introduction as a taxonomic criterion. These drawbacks hardly appear in the slow and restricted evolutionary process that normally proceeds in natural ecosystems. But they may be annoyingly clear in connection with cultivated plants, where man accelerates and drastically redirects the evolutionary processes. A considerable number of characteristics offering adaptation to the original, natural habitat are being replaced by entirely new features fitting demands under domestication.

It is not only that taxonomy has to adapt to the faster and more divergent evolutionary processes of the domesticated as compared to the wild plants. The demand on taxonomy has become more and more different especially in recent time. Uniform cultivars have allowed and seed certification rules and plant breeder' rights have made it necessary to be able to identify and thus to classify with a much higher precision than sufficient for the wild flora.

The different trends for wild and cultivated plants do not necessitate different basic principles in taxonomy. The frame set allows sufficient steps for proper adaptation, but these steps must also be taken and accepted by taxonomic authorities and especially those interconnected with gene bank operations.

In order to be able to meet the demand for flexibility, crop taxonomy must primarily be simple. A complex hierarchical grouping so often used just in connection with cultivated plants must be avoided owing to its inherent rigidity. Well established man-made intergeneric and interspecific hybrids must be given the same consideration as natural ones. Decisions as to where to set borders and how to subdivide must be made with greatest possible foresight. The biological species concept (cf. LÖVE 1964) should govern the grouping at higher taxon levels, since such an approach is more likely to produce classification schemes that are able to maintain actuality and are more informative for use in practice. It should be remembered that evolution in crop plants tends to enhance the impression of a morphological discontinuous variation due to the fact that macromutants have played such an important role. Partly due to their recent induction, partly due to the way man handles his crops, drastic organic reconstructions have seldom as in nature been followed by an isolating, true speciation process.

Different treatments of the genus *Triticum* may be taken to illustrate the

Table 1

Subdivision in Species of *Triticum* L. According to DOROFEEV and KOROVINA (1979)

| | | | |
|---|---|---|---|
| Subgenus *Triticum* | | Subgenus *Boeoticum* | |
| Sect. *Urartu* | | Sect. *Monococcum* | |
| *T. urartu* ($A^u$) | 6 | *T. boeticum* ($A^b$) | 61 |
| | | *T. monococcum* ($A^b$) | 14 |
| Sect. *Dicoccoides* | | *T. sinskajae* ($A^b$) | 0 |
| *T. dicoccoides* ($A^uB$) | 25 | Sect. *Timopheevii* | |
| *T. dicoccum* ($A^uB$) | 64 | | |
| *T. karamyschevii* ($A^uB$) | 3 | *T. araraticum* ($A^bG$) | 13 |
| *T. ispahanicum* ($A^uB$) | 2 | *T. timopheevii* ($A^bG$) | 4 |
| *T. turgidum* ($A^uB$) | 69 | *T. zhukovskyi* ($A^bA^bG$) | 0 |
| *T. jakubzineri* ($A^uB$) | 0 | *T. militinae* ($A^bG$) | 2 |
| *T. durum* ($A^uB$) | 120 | | |
| *T. turanicum* ($A^uB$) | 20 | Sect. *Kiharae* | |
| *T. polonicum* ($A^uB$) | 21 | *T. kiharae* ($A^bGD$) | 0 |
| *T. aethiopicum* ($A^uB$) | 203 | | |
| *T. persicum* ($A^uB$) | 18 | | |
| Sect. *Triticum* | | | |
| *T. macha* ($A^uBD$) | 14 | | |
| *T. spelta* ($A^uBD$) | 54 | | |
| *T. vavilovii* ($A^uBD$) | 7 | | |
| *T. compactum* ($A^uBD$) | 96 | | |
| *T. aestivum* ($A^uBD$) | 194 | | |
| *T. sphaerococcum* ($A^uBD$) | 17 | | |
| *T. petropavlovskyi* ($A^uBD$) | 4 | | |

Species often subdivided in ssp., convar., subconvar., and var. Figures above give number of latter category in each species.

Table 2

Gene Formulas for Some Tetraploid Wheat "Species"

| Common name | | Gene formula |
|---|---|---|
| Macaroni wheat | *T. durum* DESF. | *qqpp* |
| Polish wheat | *T. polonicum* L. | *qqPP* |
| Dika wheat | *T. carthlicum* NEVSKI | *QQpp* |

*Q* gives squareheadedness and inhibits spelting and rachis brittleness.
*P* gives large, papyraceous glumes.

Table 3

Gene Formulas for Some Hexaploid Wheat "Species"

| Common name | Latin name | Gene formula |
|---|---|---|
| Spelt wheat | *T. spelta* L. | $qqccS_1S_1$ |
| Bread wheat | *T. aestivum* L. em. FI. et PA. | $QQccS_1S_1$ |
| Club wheat | *T. compactum* HOST | $QQCCS_1S_1$ |
| Indian dwarf wheat | *T. sphaerococcum* PERC. | $QQccs_1s_1$ |

*Q* gives squareheadedness and inhibits spelting and rachis brittleness, located to chromosome 5A.
*C* gives compact ears, located to chromosome 2D.
$s_1$ gives round glumes and seeds, located to chromosome 3D.

implication of the above specific aspects on crop taxonomy. DOROFEEV and KOROVINA (1979) give one of the very latest examples of a detailed hierarchical subdivision of the genus comprising no less than 27 different species. They recognize the existence of genetic isolating barriers and modern cyto- and chemotaxonomic evidences on phylogenetic genomic relationships at subgenus and section level. Within each section, crossability between species is perfect. Borders are often indistinct and of clear dependent only on individual, spectacular major genes (cf. Tables 2 and 3). In a population, such species can be made anew over and over again owing to simple recombination. The species level within a section is not chosen arbitrarily, but it is symptomatic for this approach in taxonomy that DOROFEEV and KOROVINA (l. c.) use ear compactness caused by gene C to separate species, while MANSFELD (1951) has chosen to use this distinction at the lowest level separating varieties pairwise.

Already in 1918 THELLUNG reacted against such an almost purely morphological species concept and recognized only three wheat species, according to nomenclatural rules named *T. monococcum*, *T. turgidum* and *T. aestivum*. The very same year, SAX (1918) and SAKAMURA (1918) showed this grouping to reflect a ploidy series, i. e. it had a definite phylogenetic significance. This drastic lumping is to be preferred to a very detailed splitting into morphologically separated units, since it is an approach that allows to reflect interrelationships as well as barriers giving genetic complications at plant breeding, outcrossing in connection with seed production, etc. In addition, as said before, simplicity implies flexibility and thus automatically a preadaptation for future events.

Table 4

Subdivision in Species of *Triticum* L. em. BOWDEN made in 1959

| | |
|---|---|
| I. Diploid species | |
| *T. monococcum* ($A^b$) | *T. bicorne* ($S^b$) |
| | *T. speltoides* (S) |
| | *T. comosum* (M) |
| | *T. uniaristatum* ($M^u$) |
| | *T. longissimum* ($S^1$) |
| | *T. umbellulatum* ($C^u$) |
| | *T. tripsacoides* ($M^t$) |
| | *T. dichasians* (C) |
| | *T. aegilops* (D) |
| II (a). Allotetraploid wheats | |
| *T. turgidum* ($A^uB/A^bG/A^bA^bG$) | |
| II (b). Allohexaploid wheats | |
| *T. x aestivum* ($A^uBD$) | |
| II (c). Allopolyploid species of interspecific origin | |
| | *T. ovatum* ($C^uM^o$) |
| | *T. triaristatum* ($C^uM^t$) |
| | *T. kotschyi* ($C^uS^1$) |
| | *T. triunciala* ($C^uM^t$) |
| | *T. cylindricum* (CD) |
| | *T. macrochaetum* ($C^uM^b$) |
| | *T. crassum* ($DM^{cr}/DD^2C^{cr}$) |
| | *T. turcomanicum* |
| | *T. juvenale* ($DC^uM^j$) |
| | *T. ventricosum* ($DM^v$) |
| III. Other artificial and natural interspecific hybrids | |

Wheat species in left, goat grasses in right column.

Table 5

Subdivision in Species of *Triticum* L. em. MAC KEY in 1968, 1975

Sect. *Monococca*
*T. monococcum* ($A^b$)
*T. urartu* ($A^u$)
Sect. *Dicoccoidea*
*T. timopheevi* ($A^bG$)
*T. turgidum* ($A^uB$)
Sect. *Speltoidea*
*T. zhukovskyi* ($A^bA^bG$)
*T. aestivum* ($A^uBD$)
Sect. *Triticale*
*T. turgidocereale* ($A^uBR$)
*T. rimpaui* ($A^uBDR$)
Sect. *Trititrigia*
*T. turgidomedium* ($A^uBX$)
*T. aestivomedium* ($A^uBDX$)

THELLUNG (l. c.) did not have the complete information, and BOWDEN (1959) accepting his grouping ignored new discoveries. He bulked the more recently discovered *timopheevi*(4×)-*araraticum*(4×)-*zhukovskyi*(6×)-complex into *T.*

*turgidum* in spite of it being less crossable with emmer than emmer is with *T. monococcum* and *T. aestivum*. There is in fact two speciation processes superimposed upon each other giving a biphyletic origin of wheats. Prior to the allopolyploidization and surely also after its first step, a genomic differentiation occurred. To be more accurate there are thus six good species in biological sense among true wheats, two at each ploidy level. A taxonomy accepting this situation will best reflect genetic barriers and phylogenetic relations (cf. Table 5; MAC KEY 1968, 1975).

BOWDEN's arrangement of the genus *Triticum* is here mentioned more because it illustrates the problem of setting borders at generic level for plants where man continues an evolutionary trend started by nature. LINNAEUS (1753) originally had a wider concept than merely the true wheats, but already in 1823 DUMORTIER narrowed the genus to its present, mostly accepted size. Since the progenitors of the polyploid wheats are found in both *Triticum* (L.) DUM. and *Aegilops* L., BOWDEN (l. c.) felt obliged to follow the nomenclatural rule for intergeneric hybrids and group the two genera into one. MORRIS and SEARS (1967) accepted this emendation and rejected another proposal from CHENNAVEERAIAH (1960) as being inconsistent. He wanted, for sound phylogenetic reasons, to include only the section *Sitopsis* JAUB. et SPACH of *Aegilops* into *Triticum*. Such a rearrangement will of course do injustice to the hybrid relationships between *Sitopsis* and the rest of *Aegilops*. BOWDEN's lumping is, however, not free from such objections either. The polyploids of the whole tribe *Triticeae* DUM. form a network of introgressions and other interconnections. If the nomenclatural rule for intergeneric hybrids should be followed without violation, the whole tribe must be lumped into one single genus (HYLANDER 1945; STEBBINS 1956; MAC KEY 1966, 1968, 1975; RUNEMARK and HENEEN 1968). No realistic taxonomist is willing to be that obedient.

The mistake with BOWDEN's approach is, however, more that it is basically retrospective. Had he only dealt with an evolutionarily stagnant complex, the weakness would not immediately be apparent. In trying to be foresighted (cf. his group III in Table 4), he immediately broke the rule he felt he must obey.

It is not difficult to observe a discontinuity developing in the Aegilops-*Triticum*-complex. *Aegilops* is evolving towards weediness and will further elaborate on the original principle of fruit dissemination. Under the control of man, *Triticum* will further develop its greater ability to utilize fertile land and will follow a completely different trend in ear construction. It thus appears more appropriate to maintain a generic separation between *Aegilops* and *Triticum* and use the nomenclatural code to declare *Triticum* a hybrid genus (cf. Table 5; MAC KEY 1968, 1975). Such a treatment will allow further genomic manipulations to be considered such as the development of triticales, wheat-*Agropyron*-syntheses, etc. The present work with building up an intergraded third genome in the hexaploid triticale (GUSTAFSON and ZILLINSKY 1979) by part genome R, part genome D favours such a taxonomic treatment of the ryewheats.

The demand for flexibility in crop taxonomy does not agree with ambitions to build up complex and detailed hierarchical classification schemes. The example taken from the arrangement made by DOROFEEV and KOROVINA (l. c.) may again be used as a basis for discussion. They subdivide the genus *Triticum* into subgenera, sections, species, subspecies, convarieties, subconvarieties, varieties, and sometimes even formae. The infraspecific system is artifical and based on a number of pre-established taxonomic markers limited to ears, glumes, awns, and seeds. The 27 species are subdivided into 1031 varieties. From phylogenetic, geographic, ecologic, and agronomic points of view, this systematically admirable scheme gives more disorder than order. It allows the user to differentiate his wheat material only from a specific and fixed aspect. It serves to differentiate primitive material represented by typical samples preserved in botanical gardens or world collections. It is highly inefficient in differentiating between modern European wheat cultivars, i. e. to serve practical interests. Selective advantage for intermediate ear density, glabrous and white glumes, red seeds in bread wheat and white (amber) seeds in durum make them fall in just a very few groups at the lowest taxon level.

According to classical taxonomic traditions, the 1031 varieties are given often long, complicated, frequently revised and not seldom meaningless Latin

Table 6

Formula Symbols for Varietal Descriptions in Wheat Suggested by GANDILIAN (1980)

| Taxonomic criterion | | Symbol |
|---|---|---|
| Completely awnless | *eumuticus* | *eum* |
| Awnless | *muticus* | *mu* |
| Short-awned | *subaristatus* | *sar* |
| Awned | *aristatus* | *ar* |
| Glabrous | *nudus* | *nu* |
| Rough | *tuberculatus* | *tu* |
| Hairy | *pubescens* | *pu* |
| White | *albus* | *al* |
| Yellow | *luteus* | *lu* |
| Spotted | *maculatus* | *m* |
| Ashy to smoky grey | *cinereus* | *ci* |
| Red | *ruber* | *ru* |
| Marginally dark coloured | *tristis* | *tri* |
| Brownish grey | *fumigatus* | *fu* |
| Black | *niger* | *ni/n* |
| Purple | *violaceus* | *vi* |
| Green | *viridis* | *vir* |

Characterization based on awn and glume characteristics and seed colour. E. g. *T. aestivum* var. *erythroleucon* KÖRN. is characterized as *T. aestivum arnuru(al)* describing type as awned, with glabrous and red glumes and white seeds.

names. GANDILIAN (1980) has recently suggested a considerable improvement for varietal denomination. Each taxonomic marker is characterized by a symbol (cf. Table 6), and each variety is described by a formula indicating the relevant traits involved. This approach reveals the system of classification, describes the group directly, and makes the principle of hierarchical grouping unnecessary.

GANDILIAN is, however, a traditional taxonomixt and does not deviate from the classical taxonomic markers that at least in more developed agricultural countries have lost much of their actuality. But his principle is applicable for more relevant traits, e. g. if a wheat has spring or winter habit or if it is conventionally tall or semi-dwarf.

Following the above recommendation in connection with taxonomy of cultivated plants to use an extremely simple classification system as to number of taxon levels, it appears more appropriate to discard the variety concept and go directly to the cultivars. This is what the plant breeder does (Åberg 1943; Mac Key 1954; Blixt 1979). He arranges his basic material for each species country-wise as to its primary origin. It gives him a handy system which takes ecological adaptation and breeding values fairly well into consideration. It is a grouping which, in contrast to a morphological principle of classification, will improve by continued plant breeding. Whenever the name of the cultivar is unknown or the entry does not have cultivar status, he uses the name of the collection site. The FAO world catalogues of genetics stocks are examples of this kind of classification. A key based on morphological, or other distinct and easily determinable characters, for the restricted variation and considerations of every single country, will then fulfil even the most exacting demands. Such an arrangement offers the possibility to introduce any kind of appropriately separating character in the system. This is important whenever seed certification rules necessitate that every cultivar in authorized seed trade must be distinguishable. In certain situations, chemotaxonomical markes or precise patterns of race-specific disease resistance may even be needed to be introduced. The universality is reached as long as the taxonomical marker symbols are internationally authorized and used. For most crops, such a symbol system has already been developed by geneticists for characters governed by known genes.

## Zusammenfassung

Bemerkungen über die Grundprinzipien der Kulturpflanzentaxonomie

Kultivierte Pflanzen unterliegen einer intensiven Evolution und bedürfen einer präzisen Klassifizierung nach den Festlegungen für Saatgutzertifizierung. Diese Umstände erfordern Flexibilität bei den bisher benutzten taxonomischen Schemata, d. h. komplexe und starre hierarchische Systeme sollten vermieden werden. Der Gruppenbildung der höheren Rangstufen sollte der biologische Artbegriff zugrunde liegen, da solch ein Verfahren besser imstande ist, Klassifikationsschemata aufzustellen, die geeignet sind, die Aktualität zu wahren, und die informativer sind zum Gebrauch in der Praxis. Die Gruppenbildung der niederen Rangstufen sollte sich auf die Sorte konzentrieren. In diesem Zusammenhang hat die Anwendung von Symbolformeln den Vorteil, sowohl beschreibend als auch flexibel zu sein. Die Bedeutung der Empfehlungen wird durch den Vergleich verschiedener taxonomischer Gliederungen von *Triticum* erläutert.

**Краткое содержание**

Замечания об основных принципах таксономии культурных растений

Культурные растения проделывают интенсивную эволюцию и нуждаются в точной классификации для соблюдения правил сертификации посевного материала. Эти обстоятельства требуют от таксономических схем, которые применялись до сих пор, больше гибкости; иными словами – следует избегать построения сложных и жёстких систем. Для образования групп более высокого ранга надо основываться на концепции биологического вида; такой подход благоприятствует созданию классификационных схем, лучше отражающих действительное положение и более информативных, в применении их на практике. Для группообразования на более низком уровне следует руководствоваться концепцией сорта. Применение символических формул для характеристики признаков может оказаться полезным и для описания форм и для гибкости системы. Значение приведённых рекомендаций поясняется сравнением различных траксономических систем пшеницы.

## Literature

Åberg, E., 1943: Problems in the classification of cultivated plants. – Chron. Bot. **7**, 375–378.

Blixt, S., 1979: Systematic botany and plant breeding. – In: Systematic botany, plant utilization and biosphere conservation (Edit. I. Hedberg) pp. 13–16. Almqvist & Wiksell Int., Stockholm.

Bowden, W. M., 1959: The taxonomy and nomenclature of the wheats, barleys, and ryes and their wild relatives. – Can. J. Bot. **37**, 656–684.

Chennaveeraiah, M. S., 1960: Karyomorphological and cytotaxonomic studies in *Aegilops*. – Acta Hort. Gotoburgensis **23**, 85–178.

Dorofeev, V. F., and O. N. Korovina, 1979: (Wheat). – Flora of cultivated plants **1**. – Kolos, Leningrad.

Dumortier, B. C., 1823: Observations sur les Graminées de la Flora Belgique. – Agrost. Belg. Tentamen, Tournay/Casterman.

Gandilian, P. A., 1980: (Determiner to the wheats, aegilopses, ryes and barleys). – Isdatel' stvo AN Armjanskoj SSP, Erevan.

Gustafson, J. P., and F. J. Zillinsky, 1979: Influences of natural selection in the chromosome complement of hexaploid triticale. – Proc. 5th Int. Wheat Genet. Symp., New Dehli 1978, Indian Soc. Genet. Pl. Breed., Indian Agric. Res. Inst., New Delhi, **2**, 1201–1207.

Hylander, N., 1945: Nomenklatorische und systematische Studien über nordische Gefässpflanzen. – Uppsala Univ. Årsskr. 1945, No. **7**, 337 pp.

Linnaeus, C., 1753: Species plantarum. – Holmiae.

Löve, A., 1964: The biological species concept and its evolutionary structure. – Taxon **13**, 33–45.

Mac Key, J., 1954: The taxonomy of hexaploid wheat. – Svensk Bot. Tidskr. **48**, 579–590.

–, 1966: Species relationship in *Triticum*. – Proc. 2nd Int. Wheat Genet. Symp., Lund 1963, Hereditas Suppl. **2**:237–276.

–, 1968: Relationships in the *Triticinae*. – Proc. 3rd Int. Wheat Genet. Symp., Canberra 1968, pp. 39–50 Aust. Acad. Sci., Griffin Press. Netley.

–, 1975: The boundaries and subdivision of the genus *Triticum*. – Proc. 12th Int. Bot. Congr., Leningrad 1975

MANSFELD, R., 1951: Das morphologische System des Saatweizens, *Triticum aestivum* L. s. 1. – Der Züchter **21**, 41–60.

MORRIS, R., and E. R. SEARS, 1967: The cytogenetics of wheat and its relatives. – In: Wheat and wheat improvement (Edit. K. S. QUISENBERRY and L. P. REITZ) pp. 19–87. Amer. Soc. Agron. Inc., Madison, Agron. Ser. 13.

RUNEMARK, H., and W. K. HENEEN, 1968: *Elymus* and *Agropyron*, a problem of generic delimitation. – Bot. Not. **121**, 51–79.

SAKAMURA, T., 1918: Kurze Mitteilung über die Chromosomenzahlen und die Verwandtschaftsverhältnisse der *Triticum*-Arten. – Bot. Mag., Tokyo **32**, 151–154.

SAX, K., 1918: The behaviour of chromosomes in fertilization. – Genetics **3**, 309–327.

STEBBINS, G. L., 1956: Taxonomy and the evolution of genera, with special reference to the family Gramineae. – Evolution **10**, 235–245.

THELLUNG, A., 1918: Neuere Wege und Ziele der botanischen Systematik, erläutert am Beispiele unserer Getreidearten. – Naturw. Wochenschr., N. F. **17**, 449–458, 465–474.

Prof. Dr. J. MAC KEY
Department of Plant Breeding
Swedish University of Agricultural Sciences
S - 750 07 Uppsala, Sweden

Kulturpflanze XXIX · 1981 · S. 209–239

# Taxonomy of the infraspecific variability of cultivated plants

BERNARD R. BAUM (Ottawa, Canada)

## Summary

The needs and aims of taxonomy of cultivated plants at the infraspecific level are dealt with briefly in the context of taxonomy in general. The components of infraspecific taxonomy of cultivated plants are formulated and defined. These components are: identification, classification, genealogy, nomenclature – that are complementary to each other. The desired requirements of each component are discussed in the light of a review, conducted for this paper, on the subject covering the period (1975–) 1976–1979.

Developing automatic identification schemes for cultivars is becoming increasingly important. Identification needs to be accurate, rapid and devoid of human error. A classification system must be based on as many characters as possible and use numerical taxonomic techniques to generate and evaluate groupings. A suitable system would be composed of polythetic, non-hierarchical, informal groupings. The most useful genealogies are those that trace individual pedigrees as far back as possible. They can most profitably be done with the aid of computer systems, such as those used for oat and barley cultivars, by the author and his associates. Because there is no one to one correspondence between genealogies and classifications, pedigrees or genealogies enable one to draw inferences about cultivars different from those drawn from classifications. A nomenclature system for cultivars must be as complete as possible. This is achieved by listing all the cultivars names, commercial synonyms, translations and transliterations, and cross referencing. Examples of confusions due to faulty nomenclature are provided. In the author's opinion, a nomenclature system for cultivar groups is not needed because it is too cumbersome and because groups need not be formalized.

It is desirable to work on the four components of infraspecific taxonomy at a world-wide basis to increase their usefulness for growers, merchants, breeders and scientists concerned with cultivar development and improvement. An exception might be in identification within a country whereby a local system might satisfy most current needs.

## Introduction

Taxonomy is the study of diversity and of relationships between objects. Taxonomy includes also the study of the principles, methods and rules that one uses to study diversity. Some authors, e. g. SIMPSON (1961), DAVIS and HEY-

WOOD (1963), call the latter taxonomy and the former systematics. Both terms are used interchangeably in the literature and regarded by many as synonyms.

Biological taxonomy is a particular case of taxonomy.

The study of diversity and relationships between objects necessarily involves classification. Classification can be categorized into three classes: diagnosis, categorization and characterization (GOOD, 1965). These three classes are, however, interdependent with time.

In biological taxonomy organisms (the objects) are classified into taxa. Those taxa are then characterized by descriptions often called diagnoses. Subsequently the descriptions are named. The resulting taxa form a system. The latter is then used for identification purposes, or diagnosis in the sense of GOOD (1965).

In taxonomic activities one classifies knowledge about objects rather than the objects themselves (HILL, 1973); one characterizes classes based on that knowledge; and one develops identification devices as aids to recognize previously unknown items into those classes from the knowledge acquired about these items. Why then are we engaged in taxonomic studies? Briefly, for mental clarification and communication (GOOD, 1965). Consequently, the taxonomic process helps us to acquire and organize knowledge for utilitarian purposes. In biological taxonomy we aquire knowledge about organisms for the aforesaid purposes.

The need for taxonomy at the infraspecific level arose probably first with domestication of plants and animals. Man in his quest for satisfying his needs selected genetic stocks from the pool of variability in the plants and animals he had domesticated.

Recognition of differentiation below the species level was very real in domestication out of necessity; while it was greatly ignored in non-domestic biota until the 19th century. At this time botanists, such as JORDAN, KERNER, BONNIER and DARWIN, made attempts to relate variation to natural selection, while MENDEL pioneered the study of hereditary laws underlying variation. TURRESSON in the 1920's pioneered a new approach to study variation by means of experimental cultures. This, and related approaches, climaxed with the multidisciplinary studies by CLAUSEN, KECK and HIESEY (1940, 1945, 1948). At about the same period, PEARSON (1926) developed the coefficient of racial likeness and MAHALANOBIS (1936) developed the generalized distance statistic. Many investigators who will not be cited here, and from varied disciplines, such as botanical and zoological systematics, anthropology, statistics, all contributed to the development of biometrics (biometry) and numerical taxonomy. Nowadays, biometrics and numerical taxonomy are extremely useful in studies of variation analysis.

I stated beforehand that the need for infraspecific taxonomy became very real with the incipience of domestication. It has increasingly gained importance especially in view of the very large numbers of cultivars that have been produced.

## The Components of Infraspecific Taxonomy of Cultivated Plants

What does infraspecific taxonomy of cultivated plants entail? What needs does it answer?

First and foremost, there is a very real need to recognize one cultivar from another. The price that a farmer gets for his crop is determined by the kind of cultivar. Merchants and buyers must be very careful in many instances in dealing with correct cultivars, e. g. barley cultivars designed for breweries, flowers for specific purposes, tomato cultivars for canned soup. Distinction of one cultivar from another has recently gained more attention in connection with breeders' rights legislation in various countries.

Second, there is a need to classify cultivars. Article 7 of the Code for cultivated plants (GILMOUR et al., 1969) lists three main categories, namely: genus, species, and cultivar. Furthermore, according to Article 25 and 26 supplementary categories, e. g. subgenus, subspecies and cultivar groups are mentioned. The Code is only a set of rules governing names given to cultivars or to classes of cultivars. The Code itself does not have anything to do with classification. It is clearly understood by most that there is a need to classify cultivars by the species to which they belong. But, do we really need to classify them into complex hierarchical subdivisions, such as subspecies, variety, subvariety, form, etc. ... as was initiated by VAVILOV? HAWKES (1970) was right when he stated that we „must try to steer a middle course". Let us examine briefly why we need classes of cultivars. Firstly, classes might assist us in constructing and formulating identification schemes. These classes have an important potential use for classifying genetic material which does not belong to named cultivars. Secondly, cultivar classes may be useful, provided they are meaningful in some ways in the sense that other facts can be integrated into these as they become available. Cultivar classes serve the users for communication purposes. Cultivar classes may reflect genetic content, coherence, traits possessed in common, and indicate to which group desirable genotypes belong, thus playing an important part in introduction and utilization of genetic resources.

Third, a phylogenetic system is extremely desirable. By this I mean genealogical relationships among cultivars. Breeders, geneticists, pathologists, and other investigators need information on genealogical relationships for sensible planning of their breeding and related research programmes. All the cultivars on a global basis, past and present, consist of a great part of the total gene pool of a species. Genealogies can have significant use in relation to gene pool utilization and conservation. When comprehensive reference to literature is provided for each parent on the retrieved pedigrees, considerable assistance in gene pool investigation is given.

Fourth, a nomenclature system is absolutely necessary for cultivars. The system should include (1) the correct name and commercial synonyms in various countries, (2) translations and transliterations, and (3) cross referencing. Nomenclature is intimately connected with the three components above. Correct identification is useless without accurate naming. Inaccurate naming leads to breakdown in communication. Classifications and geneologies are meaningless when names are misapplied.

**Factors in Studying Infraspecific Taxonomy of Cultivated Plants – Some Problems and Examples.**

*Identification*

The requirements for achieving an identification scheme are of two different but complementary aspects. These are:

(1) techniques of observation and data acquisition; and (2) methods of data analysis.

(1) Techniques. From incipient domestication until recently, various morphological features have been used as characters. Identification of cultivars by morphological characters became increasingly disappointing in different species because of inherent variability found in many characters. This has been recently coupled with the unfamiliarity with, or unwillingness to use, modern methods of data analysis. This further raised the need to seek for other techniques. In the last decade various chemical approaches became fashionable. Among these techniques, polyacrylamide gel electrophoresis of the gliadin in wheat was found to be most promising.

For the purpose of this paper I conducted a literature search covering the past four years in the Commonwealth Agricultural Bureau data base. The pertinent articles are listed in Appendix 1 by taxa. A survey of these publications revealed that besides gross morphology the following approaches were undertaken to obtain characters for identification: Electron microscopy of cuticles and trichomes; dry matter of various plant parts; electrophoresis of various enzyme systems; thin layer chromatography of various compounds; spectrophotometry; phenologic traits; phenol testing; anatomical traits; crystallization of extracts from various plant parts; UV radiation, and morphometric measurements.

Furthermore, the survey of these publications (Appendix 1) has disappointedly shown that no identification means was developed by the various investigators in Appendix 1 except for very few, such as AUTRAN and BOURDET (1975), ELLIS and BEMINSTER (1977). Of course many authors, not in Appendix 1, and who used morphological methods provided keys. Keys have been elaborated for many cultivars in the last 200 years, but those were usually restricted to a country or a district only.

(2) Data Analysis. According to SVIRIDOV (1978) the principles for constructing a key from character data were laid down by LAMARCK as early as 1778. Since then and until recently keys were constructed essentially in the same manner, i. e. by inspection of the data, and by neural assessment[1] of the most important diagnostic characters from the data.

In the last decade many developments have taken place in the methods and principles for generating keys from data. These are reviewed in PANKHURST

1 see DAVIS and HEYWOOD 1963: 113, 119, 139 re: neural assessment

(1975, 1978). Although much of this recent development was due to the availability of computers, the data at the cultivar level are in many cases unsuitable for application of the principles and methods of generating keys, whether these are traditional or recent. The reason for this is that cultivars, although bred for "uniformity and consistency", in fact vary, their boundaries are fuzzy, and consequently it is difficult to find suitable characters.

The chief *desideratum* for cultivar identification is automated identification. Producers and merchants are increasingly subjected to national and international legislations and regulations. The role of seed trade for example is becoming increasingly complex. Correct varietal description can only be checked by trained crop inspectors examining the growing crop before harvest. Great reliance must be placed on contract growers and seed processors to ensure that seed parcels remain true to cultivar name and free from contamination from other seed lots. Identification is therefore of prime concern. Thus, means of identification must be accurate, reliable, objective and speedy. Because of these requirements, most promising are automatic data gathering methods combined with methods of identification based on matching and/or probabilistic approaches, or various statistical multivariate analytic techniques. Interface between the kind of data gathering and this latter kind and combination of methods can easily and most profitably be achieved by means of electronic computers.

BAUM and BRACH (1975) used fluorescence spectrography to identify oat (*Avena* L.) cultivars by the seed and continue their efforts with the aim of achieving full automation. BAUM and THOMPSON (1976) used morphology of the seed, and, although their observations were performed by visual means, they were tailored to measure size-shape by means of an automatic scanner. GREENHALGH and BAUM (1980) used pyrolysis gas liquid chromatography to identify five Canadian cultivars by their seeds. Recently BAUM et al. (1980) successfully used the technique of image analysis to identify barley (*Hordeum* L.) species and a few cultivars on the basis of endospermic starch granules. In all these four above mentioned attempts, discriminant analysis (SEAL, 1964) and the computations of related classification function coefficients (COOLEY and LOHNES, 1962) proved most useful. RUBY and WRIGHT (1977) (Appendix 1) in their studies of *Pinus sylvestris* L. varieties, stated that differences between varieties must be based on biometric data. MAIER (1977) successfully used a classification procedure based on discriminant analysis to identify hops cultivars employing 30 different chemical extracts as variables. MOSS and BROWN (1976) (Appendix 1) used classification function coefficients for 154 potato cultivars. To conclude on identification: the need for automatic identification requires that implementation of modern technology, such as scanning devices and automatic chemical analytic devices, be explored, and that the data obtained be used for identification using advanced numerical and statistical meth ds. The two may be interfaced easily with the aid of electronic computers. Automatic data recording and numerical techniques of analysis are extremely efficient when carried out with the aid of computers, but if the observations and analyses were done by hand instead they would be extremely time consuming, prone to human error and bias, and less reliable.

*Classification*

As already mentioned, classification and identification are interrelated. Although classification contributes to identification by providing groupings, the two are conceptually different (see GOOD, 1965). The requirements for achieving a classification depend on (1) techniques for data acquisition, and (2) data analysis.

(1) Techniques. These can be morphological, chemical, physiological etc. . . . as in identification, except that there is no time limit for data acquisition. The more and diverse the characters that are used, the greater the likelihood that the classifications generated from the data will reflect the "true relationships" among the cultivars and any other non cultivar material which is a member of the same gene pool being classified.

Many classification schemes of cultivars were based on morphology and done of course in conjunction with identification. Because of difficulties encountered in morphology due to variation, recently attempts were made to explore other techniques of observations. Appendix 1 lists references to the different approaches in the last four years and includes morphology. So, for instance SCHILDRICK (1976) used a combination of chromosome number, rhizome formation and heading dates to classify *Festuca rubra* L. cultivars into groupings; ELEKES (1975) used UV radiation. SCRIBAN and STROBBEL (1978) used electrophoresis of the prolanine fraction to group *Hordeum vulgare* L. s. l. cultivars. Chemical techniques were used mainly, but not only, for identification; not for classification as can be deduced from references in Appendix 1.

Although the reasons are not known, it appears that, at least at the cultivar level, classification cannot reflect phylogeny (in the sense of genealogy). BAUM (1970) found and documented this in *Avena* cultivars. BAUM and LEFKOVITCH (1973) described the lack of a one to one correspondence between genealogical relationships of *Avena* cultivars and their classifications based on morphological and agronomical characters, and documented the complexities associated with this. Subsequently, AUTRAN and BOURDET (1975) and AUFIAU et al. (1976) found a similar lack of one to one correspondence between pedigrees and enzymatic composition.

(2) Data Analysis. Most cultivars groupings so far published were done by neural assessment. With the introduction of numerical taxonomy and electronic computers, it became increasingly fashionable to use multivariate statistical and numerical techniques for classification. Appendix 2 provides a list, by species, of investigators who used one or a combination of several of these techniques for cultivar classification or for infraspecific classification of cultivated material which does not belong to named cultivars.

The classifications based on numerical and related multivariate techniques appear to yield more insight into relationships of cultivars within groups and among groups, and of course the same applies to unnamed genetic resources. The resulting groupings by many methods are polythetic (SOKAL and SNEATH 1963: 13) in contrast to the monothetic groupings that were hitherto created by neural assessment.

The concept of polythetic classification, or polytypic aggregations (BECK-

NER, 1959) became a recognized principle in a range of natural sciences (NEEDHAM, 1975). A polytypic aggregation or polythetic class is defined by BECKNER as follows:

A class is ordinarily defined by reference to a set of properties which are both necessary and sufficient (by stipulation) for membership in the class. It is possible, however, to define a group **K** in terms of a set **G** of properties $\mathbf{f}_1, \mathbf{f}_2, \ldots, \mathbf{f}_n$ in a different manner. Suppose we have an aggregation of individuals (we shall not as yet call them a class) such that:
1) Each one possesses a large (but unspecified) number of the properties in **G**.
2) Each **f** in **G** is possessed by large numbers of these individuals and
3) No **f** in **G** is possessed by every individual in the aggregate.
By the terms of (3), no **f** is necessary for membership in this aggregate; and nothing has been said to warrant or rule out the possibility that some **f** in **G** is sufficient for membership in the aggregate.

A direct consequence of this to classification is overlapping extensions, i. e. recognition of borderline cases. This is the price for endorsing this concept. I consider this advantageous over the traditional view of forcing taxa into mutually exclusive groups. The relationships revealed and thus portrayed are more true to the information (in terms of characters) used for doing this. That does not mean, of course, that among the possible classifications some carry more information than others, based on the same data. For an illuminating discussion on the topic of polythetic classification see NEEDHAM (1975) and references cited in NEEDHAM's paper.

Monothetic classifications can be viewed as particular cases of polythetic classifications. A diagnostic key of the conventional and traditional type is also a monothetic classification. Monothetic classes may be useful for one's particular purpose. But, in the same manner, a particular polythetic classification may be useful for a specific purpose even if it possesses less information content than some other possible classifications that can be obtained from the data.

The allowance of overlapping extensions causes problems in identification to classes by traditional means. However, identification can be effected by probabilistic methods (e. g. BAUM and LEFKOVITCH, 1972, for *Avena* cultivars), discriminant functions (examples are legion), similarity matching (e. g. SNEATH, 1979) and related methods (see PANKHURST, 1975) which are becoming increasingly fashionable in medical diagnosis, geology, bacteriology and physical anthropology.

Another aspect related to data analysis is whether classification ought to be hierarchic or just a non-hierarchic partition. I believe that the latter alternative is more suitable for infraspecific taxonomy of cultivated plants. See further details on this below in the 'Nomenclature (b) Cultivar Groups' section.

To conclude on classification: numerical taxonomic methods have a definite advantage over traditional methods of cultivar classification. This applies regardless of the kind of data used. Numerical methods enable one to create classifications based on all the available data, that is to take data of different kinds together. The classes created by many methods are polythetic.

*Genealogical Relationships*

Many attempts done in this regard were restricted to individual countries alone. Very few investigators made efforts to trace back the most early parentages. MILATZ (1936), for example, made a special effort to construct a genealogy, complete as far as possible, of all the German varieties of oats.

The preparation of pedigrees is an elaborate procedure. Electronic data processing can be used to facilitate efficient retrieval of the required information input. Furthermore, cultivars have been introduced and reintroduced from country to country, and this necessitates that pedigrees be done on a world-wide basis.

Since it is the names of the cultivars that appear on the pedigrees, nomenclature is of crucial importance. For this reason the two must be done together, and both on a world-wide basis. BAUM and THOMPSON (1970) have developed

```
CODE          FULL NAME

GOLF        GOLDEN DROP
              REFERENC:  214
GOLG        GOLD MEDAL
              REFERENC:  214
GOLH    *   GOLD(GERMANY 1973)
              SEE ALSO:  GOL3
              TRNSLTNS:  ORG1(FRE)
              REFERENC:  238
GOL1    *   GOLIAD
              REFERENC:    4 214
GOL2    *   GOLDFOIL
              REFERENC:    4  35  68  96 213 214
GOL3    *   GOLD(SWEDEN 1936)
              SEE ALSO:  GUL
              SYNONYMS:  GOLH(238)
              TRNSLTNS:  ORG1(FRE)
              REFERENC:   12 101 117 150 176 177 178 186 193 198 243
GOL4    *   GOLDEN ARCHER
              REFERENC:   12 101 102 198
GOL5        GOLDEN PROMISE
              REFERENC:   20  93 191 214
GOL6        GOLDEN GRAIN
              REFERENC:   44 174 214
GOL7    *   GOLDTHORPE
              REFERENC:   57 101 102 104 121 174 204 213 214
GOL8    *   GOLDEN QUEEN
              REFERENC:   89 214
GOL9    *   GOLDEN MELON
              REFERENC:  101 104 214 239
GOO         GOODWILL
              REFERENC:  214
GOP         GOPAL
              REFERENC:   80 206
GOR     *   GORDON
              REFERENC:    4 146 214
GOR1        GORSDORFER
              REFERENC:  244
GOS         GOSPECK
              REFERENC:   19 214
GOY         GOYANGJAERAE
              REFERENC:  239
GRA         GRANDE
              REFERENC:  214
GRA1        GRANDPA
              REFERENC:  214
GRE         GREAT BEARDLESS
              SEE ALSO:  HOR
              REFERENC:    4
GRE1        GREENOUGH
              REFERENC:  172
GRI         GRIGNON
              REFERENC:  214
GRO         GRODKOWICKI
              REFERENC:  214
```

1

Fig. 1 Portion of output of list of names from Barley register. Names are arranged according to alphabetical order of their codes (left column). An asterisk indicates that a pedigree is known and will appear in the output. The appropriate references are found in the numerical reference list (Fig. 3). Translations, synonyms and cross-references are referred to their codes and may be looked for in the same list elsewhere

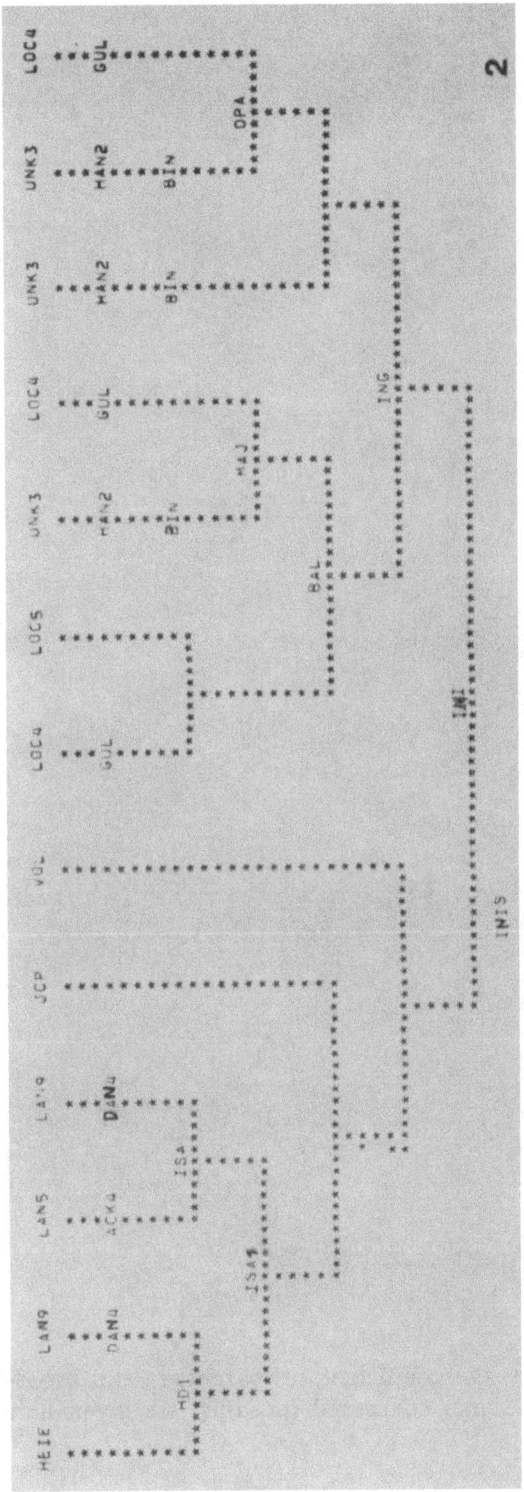


Fig. 2 Pedigree of 'Inis' Barley, as generated by the computer system. See text

a computer system general enough to be applicable to most cultivars. They have refined it recently (THOMPSON and BAUM, 1978) and added to its capabilities. This system was implemented for oats (*Avena*) cultivars (BAUM, 1972a), and is currently used for barley (*Hordeum*) cultivars. The two papers also comment on other attempts by several investigators and point to serious limitations in these. The system of BAUM and THOMPSON, besides generating pedigrees, consists of nomenclature data (see below), computes the coefficients of parentage and inbreeding (in some special sense), and lists the pertinent literature references for each name chronologically. Examples of output follow. In Fig. 1 appears a portion of the listing of names with their cross references, synonyms, translations and pertinent references to literature. So for instance, 'Gold' from Sweden (1936), the name itself may be a synonym for 'Gull' but 'Gold' itself has synonyms and translations. The pedigree of 'Inis' is shown in Fig. 2. The names appear in code form listed in the name list (Fig. 1). The references to literature are listed according to number sequence to facilitate finding them (Fig. 3) with respect to names, however the references also appear

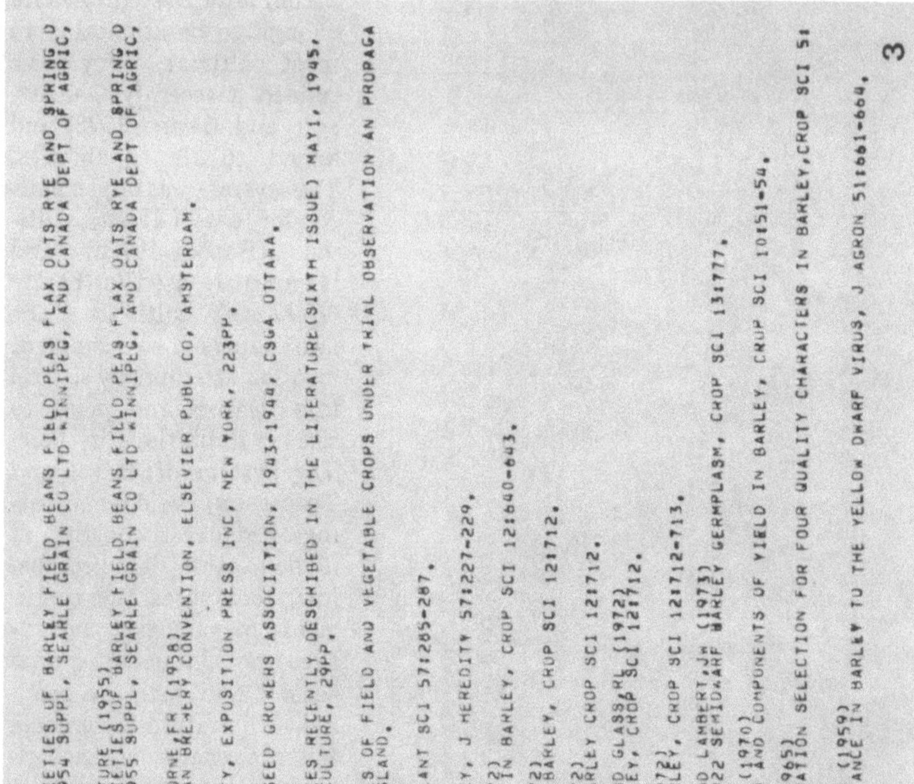

```
         HANDBOOK OF CANADIAN VARIETIES OF BARLEY FIELD BEANS FIELD PEAS FLAX OATS RYE AND SPRING D
         URUM AND WINTER WHEAT, 1954 SUPPL, SEARLE GRAIN CO LTD WINNIPEG, AND CANADA DEPT OF AGRIC,
          OTTAWA.
149.  46 CANADA DEPARTMENT OF AGRICULTURE (1955)
         HANDBOOK OF CANADIAN VARIETIES OF BARLEY FIELD BEANS FIELD PEAS FLAX OATS RYE AND SPRING D
         URUM AND WINTER WHEAT, 1955 SUPPL, SEARLE GRAIN CO LTD WINNIPEG, AND CANADA DEPT OF AGRIC,
          OTTAWA.
150.   6 AUFHAMMER,G, BERGAL,P, AND HORNE,FR (1958)
         BARLEY VARIETIES, EUROPEAN BREWERY CONVENTION, ELSEVIER PUBL CO, AMSTERDAM.
151. 105 HARLAN,HV (1957)
         ONE MAN'S LIFE WITH BARLEY, EXPOSITION PRESS INC, NEW YORK, 223PP.
152. 190 OTTAWA (1944)
         ANNUAL REPORT, CANADIAN SEED GROWERS ASSOCIATION, 1943-1944, CSGA OTTAWA.
153.  80 ENGLAND (1945)
         NEW AND PROMISING VARIETIES RECENTLY DESCRIBED IN THE LITERATURE(SIXTH ISSUE) MAY1, 1945,
         CAMBRIDGE SCHOOL OF AGRICULTURE, 29PP.
154.  23 CAMBRIDGE (1960)
         N I A B GUIDE TO VARIETIES OF FIELD AND VEGETABLE CROPS UNDER TRIAL OBSERVATION AN PROPAGA
         TION, NIAB(CAMBRIDGE) ENGLAND.
155. 246 WOLFE,RI (1977)
         KLONDIKE BARLEY, CAN J PLANT SCI 57:285-287.
156. 197 RANA,RS (1966)
         INDUCED TWINNING IN BARLEY, J HEREDITY 57:227-229.
157. 196 RASMUSSON,DC, AND BYRNE,I (1972)
         FITNESS OF HETEROZYGOTES IN BARLEY, CROP SCI 12:640-643.
158. 110 HOFFMAN,EN, AND FITCH,LA (1972)
         REGISTRATION OF VALUE 70 BARLEY, CROP SCI 12:712.
159.  87 FOOTE,H, AND KRONSTAD,WE (1972)
         REGISTRATION OF CASBON BARLEY CROP SCI 12:712.
160. 198 RASMUSSON,DC, BANTTARI,EE, AND GLASS,RL (1972)
         REGISTRATION OF CREE BARLEY, CROP SCI 12:712.
161. 109 HOCKETT,EA, AND ESLICK,RF (1972)
         REGISTRATION OF ERBET BARLEY, CROP SCI 12:712-713.
162. 199 RASMUSSON,DC, BANTTARI,EE, AND LAMBERT,JW (1973)
         REGISTRATION OF M21 AND M22 SEMIDWARF BARLEY GERMPLASM, CROP SCI 13:777.
163. 200 RASMUSSON,DC, AND CANNELL,RO (1970)
         SELECTION FOR GRAIN YIELD AND COMPONENTS OF YIELD IN BARLEY, CROP SCI 10:51-54.
164. 201 RASMUSSON,DC, AND GLASS,RL (1965)
         EFFECTIVE OF EARLEY GENERATION SELECTION FOR FOUR QUALITY CHARACTERS IN BARLEY,CROP SCI 5:
         389-391.
165. 202 RASMUSSON,DC, AND SCHALLER,CW (1959)
         THE INHERITANCE OF RESISTANCE IN BARLEY TO THE YELLOW DWARF VIRUS, J AGRON 51:661-664.
                                                                                                 3
```

Fig. 3 Portion of output of the numerical list of references. See text

in alphabetic order (Fig. 4). Finally, the coefficients of parentage and breeding are given for each pedigree; the ones computed for 'Inis' are given here (Fig. 5).

```
                BARLEY AND MALT STUDIES:1, DEVELOPING NEW VARIETIES OF BARLEY FOR MALTING AND THEIR PROPER
                TIES, CEREAL CHEM 12:596-609.
78.     31   DICKSON,P (1965)
                CEREAL VARIETIES FOR 1966, SEED TRADE REV 17:284-265.
79.     35   DIETZ,SM (1930)
                THE VARIETAL RESPONSE AND INHERITANCE OF RESISTANCE IN BARLEY TO ERYSIPHE GRAMINIS HORDEI
                PF4, IOWA J SCI 5:25-33.
80.    153   ENGLAND (1945)
                NEW AND PROMISING VARIETIES RECENTLY DESCRIBED IN THE LITERATURE(SIXTH ISSUE) MAY1, 1945,
                CAMBRIDGE SCHOOL OF AGRICULTURE, 29PP.
81.     36   FEDAK,G (1974)
                FINGERPRINTING BARLEY CULTIVARS, CANADA AGRIC 19(1):36-37.
82.    226   FEDERAL REPUBLIC OF GERMANY (1967)
                WHAT'S NEW IN THE FIELD, SAATGUT-WIRTSCHAFT 19(4):115-116.
83.    227   FEDERAL REPUBLIC OF GERMANY (1967)
                WHAT'S NEW IN THE FIELD, SAATGUT-WIRTSCHAFT 19(10):322-333.
84.     37   FIDDIAN,WEH (1966)
                BARLEY VARIETIES PRESENT AND FUTURE, J NAT INST AGRIC BOT, SUPPL 10:8-11.
85.     38   FIDDIAN,WEH (1967)
                WHEAT BARLEY AND OAT VARIETIES COMPLETING PRIMARY TRIALS IN 1966, J NAT INST AGRIC BOT 11:
                1-27.
86.     39   FIDDIAN,WEH (1973)
                WHEAT BARLEY AND OAT VARIETIES COMPLETING PRIMARY TRIALS IN 1972, J NAT INST AGRIC BOT 13:
                95-122.
87.    159   FOOTE,H, AND KRONSTAD,WE (1972)
                REGISTRATION OF CASRON BARLEY CROP SCI 12:712.
88.    241   FREISLEBEN,R (1940)
                THE PHYLOGENETIC IMPORTANCE OF ASIAN BARLEYS, DER ZUCHTER 12:257-272.
89.     40   GANGWAR,LC, AND BHATNAGAR,PS (1967)
                K-70 OUTSHINES OTHER BARLEYS IN EASTERN UP, INDIAN FARMING 17(3):42.
90.     41   GARDENHIRE,JH, AND ATKINS,IM (1969)
                REGISTRATION OF TAMU-ERA BARLEY, CROP SCI 9:678.
91.     42   GAUL,H, AND GRUNEWALDT,J (1971)
                INDEPENDANT VARIATION OF CULM LENGTH AND SPIKE-INTERNODE LENGTH OF A BARLEY-ERECTOIDES MUT
                ANT IN A CHANGED GENETIC BACKGROUND, BARLEY GENETICS 2, PROCEEDINGS OF THE SECOND INTERNAT
                IONAL BARLEY GENETICS SYMPOSIUM, WASHINGTON STATE UNIV, PULLMAN(1970).
92.     43   GILBERT,WW (1948)
                ARIZONA DEVELOPS NEW ARIVAT BARLEY STRAIN, CROPS AND SOILS 1(3):29.
93.     49   GORE,UR, REID,DA, AND MOSEMAN,JG (1968)
                REGISTRATION OF MILLER BARLEY, CROP SCI 8:513.
94.     44   GOTHARD,PG (1974)
                SCREENING OF BARLEY VARIETIES FOR ALPHA-AMYLASE CONTENT, J INST BREW 80(4):387-390.
```

4

Fig. 4 Portion of output of the alphabetical list of references. See text

271. CODE NAME= INI FULL NAME= INIS

| | | UNK3 | LAN9 | LAN5 | LOC4 | MAN2 | DAN4 | ACK4 | HEIE | GUL | BIN | |
|---|---|---|---|---|---|---|---|---|---|---|---|---|
| 225 | UNK3 | 0.5000 | | | | | | | | | | UNKNOWN CVAR=AUSTRIAN |
| 689 | LAN9 | 0.0 | 0.5000 | | | | | | | | | LANDRACE=NIEDERBAYER |
| 682 | LAN5 | 0.0 | 0.0 | 0.5000 | | | | | | | | LANDRACE=LOWER BAVARIA |
| 457 | LOC4 | 0.0 | 0.0 | 0.0 | 0.5000 | | | | | | | LOCAL CVAR (GOTLAND) SWEDEN |
| 101 | MAN2 | 0.5000 | 0.0 | 0.0 | 0.0 | 0.7500 | | | | | | HANNA |
| 646 | DAN4 | 0.0 | 0.5000 | 0.0 | 0.0 | 0.0 | 0.7500 | | | | | DANUBIA |
| 645 | ACK4 | 0.0 | 0.0 | 0.5000 | 0.0 | 0.0 | 0.5 | 0.7500 | | | | ACKERMANN'S BAVARIA |
| 693 | HEIE | 0.0 | 0.0 | 0.0 | 0.0 | 0.0 | 0.0 | 0.0 | 0.5000 | | | HEILS |
| 299 | GUL | 0.0 | 0.0 | 0.0 | 0.5000 | 0.0 | 0.0 | 0.0 | 0.0 | 0.7500 | | GULL |
| 300 | BIN | 0.5000 | 0.0 | 0.0 | 0.0 | 0.7500 | 0.0 | 0.0 | 0.0 | 0.0 | 0.8750 | BINDER |
| 648 | LOC5 | 0.0 | 0.0 | 0.0 | 0.0 | 0.0 | 0.0 | 0.0 | 0.0 | 0.0 | 0.0 | LOCAL CVAR SCANDANAVIA |
| 227 | ISA | 0.0 | 0.2500 | 0.2500 | 0.0 | 0.0 | 0.3750 | 0.3750 | 0.0 | 0.0 | 0.0 | ISARIA |
| 666 | HD1 | 0.0 | 0.2500 | 0.0 | 0.0 | 0.0 | 0.3750 | 0.0 | 0.2500 | 0.0 | 0.0 | HD 16 MEHLTAURESISTENTE |
| 223 | OPA | 0.2500 | 0.0 | 0.0 | 0.2500 | 0.3750 | 0.0 | 0.0 | 0.0 | 0.3750 | 0.4375 | OPAL |
| 297 | MAJ | 0.2500 | 0.0 | 0.0 | 0.2500 | 0.3750 | 0.0 | 0.0 | 0.0 | 0.3750 | 0.4375 | MAJA |
| 5144 | | 0.0 | 0.0 | 0.0 | 0.2500 | 0.0 | 0.0 | 0.0 | 0.0 | 0.3750 | 0.0 | |
| 268 | JCP | 0.0 | 0.0 | 0.0 | 0.0 | 0.0 | 0.0 | 0.0 | 0.0 | 0.0 | 0.0 | JCP |
| 267 | ISA1 | 0.0 | 0.2500 | 0.1250 | 0.0 | 0.0 | 0.3750 | 0.1875 | 0.1250 | 0.0 | 0.0 | ISARIA NOVA |
| 5084 | | 0.3750 | 0.0 | 0.0 | 0.1250 | 0.5625 | 0.0 | 0.0 | 0.0 | 0.1875 | 0.6563 | |
| 333 | BAL | 0.1250 | 0.0 | 0.0 | 0.2500 | 0.1875 | 0.0 | 0.0 | 0.0 | 0.3750 | 0.2188 | BALDER |
| 269 | VOL | 0.0 | 0.0 | 0.0 | 0.0 | 0.0 | 0.0 | 0.0 | 0.0 | 0.0 | 0.0 | VOLDAGSEN |
| 5052 | | 0.0 | 0.1250 | 0.0625 | 0.0 | 0.0 | 0.1875 | 0.0938 | 0.0625 | 0.0 | 0.0 | |
| 270 | ING | 0.2500 | 0.0 | 0.0 | 0.1875 | 0.3750 | 0.0 | 0.0 | 0.0 | 0.2813 | 0.4375 | INGRID |
| 5053 | | 0.0 | 0.0625 | 0.0313 | 0.0 | 0.0 | 0.0938 | 0.0469 | 0.0313 | 0.0 | 0.0 | |
| 271 | INI | 0.1250 | 0.0313 | 0.0156 | 0.0938 | 0.1875 | 0.0469 | 0.0234 | 0.0156 | 0.1406 | 0.2188 | INIS |

TABLE CONTINUED CODE NAME= INI

| | | LOC5 | ISA | HD1 | OPA | MAJ | | JCP | ISA1 | | BAL | |
|---|---|---|---|---|---|---|---|---|---|---|---|---|
| 648 | LOC5 | 0.5000 | | | | | | | | | | LOCAL CVAR SCANDANAVIA |
| 227 | ISA | 0.0 | 0.5000 | | | | | | | | | ISARIA |
| 666 | HD1 | 0.0 | 0.1875 | 0.5000 | | | | | | | | HD 16 MEHLTAURESISTENTE |
| 223 | OPA | 0.0 | 0.0 | 0.0 | 0.5000 | | | | | | | OPAL |
| 297 | MAJ | 0.0 | 0.0 | 0.0 | 0.4063 | 0.5000 | | | | | | MAJA |
| 5144 | | 0.2500 | 0.0 | 0.0 | 0.1875 | 0.1875 | 0.5000 | | | | | |
| 268 | JCP | 0.0 | 0.0 | 0.0 | 0.0 | 0.0 | 0.0 | 0.5000 | | | | JCP |
| 267 | ISA1 | 0.0 | 0.3438 | 0.3438 | 0.0 | 0.0 | 0.0 | 0.0 | 0.5938 | | | ISARIA NOVA |
| 5084 | | 0.0 | 0.0 | 0.0 | 0.4688 | 0.4219 | 0.0938 | 0.0 | 0.0 | 0.7188 | | |
| 333 | BAL | 0.1250 | 0.0 | 0.0 | 0.2969 | 0.3438 | 0.3438 | 0.0 | 0.0 | 0.2578 | 0.5938 | BALDER |
| 269 | VOL | 0.0 | 0.0 | 0.0 | 0.0 | 0.0 | 0.0 | 0.0 | 0.0 | 0.0 | 0.0 | VOLDAGSEN |
| 5052 | | 0.0 | 0.1719 | 0.1719 | 0.0 | 0.0 | 0.0 | 0.2500 | 0.2969 | 0.0 | 0.0 | |
| 270 | ING | 0.0625 | 0.0 | 0.0 | 0.3828 | 0.3828 | 0.2188 | 0.0 | 0.0 | 0.4883 | 0.4258 | INGRID |
| 5053 | | 0.0 | 0.0859 | 0.0859 | 0.0 | 0.0 | 0.0 | 0.1250 | 0.1484 | 0.0 | 0.0 | |
| 271 | INI | 0.0313 | 0.0430 | 0.0430 | 0.1914 | 0.1914 | 0.1094 | 0.0625 | 0.0742 | 0.2401 | 0.2129 | INIS |

TABLE CONTINUED CODE NAME= INI

| | | VOL | | ING | | INI | |
|---|---|---|---|---|---|---|---|
| 269 | VOL | 0.5000 | | | | | VOLDAGSEN |
| 5052 | | 0.0 | 0.5000 | | | | |
| 270 | ING | 0.0 | 0.0 | 0.6289 | | | INGRID |
| 5053 | | 0.2500 | 0.2500 | 0.0 | 0.5000 | | |
| 271 | INI | 0.1250 | 0.1250 | 0.3145 | 0.2500 | 0.5000 | INIS |

INBREEDING COEFFICIENT= 0.0

5

*Nomenclature*

(a) Cultivars. As required for species, at the cultivar level all the nomenclatural aspects are necessary, and ought to be done therefore on a global basis. The necessary aspects are: the correct name, synonyms including all the commercial synonyms, translations and transliterations, and of course all the cross references among them for each cultivar. It is desirable also to provide the date of registration, or at least the earliest known date of the use of a particular name. The latter of course might not be possible for landraces and is therefore an unrealistic requirement for them. The system of BAUM and THOMPSON (1970) and THOMPSON and BAUM (1978) accomodates all the requirements mentioned. Fig. 1 illustrates this and is taken from part of an output of barley cultivars.

There is a grey area between synonyms and translations because, in some sense, a translation might also be a synonym in another language. A few cases from oats (*Avena*), taken from BAUM (1972b) might illustrate the nomenclature problem. 'Dawn' oats (Fig. 6) were reported from Australia in about 1959; yet a completely different cultivar, but with the same name 'Dawn' (Fig. 7) was registered in North America in 1967. Similarly, the Canadian oat 'Kent' (Fig. 8) is entirely different from the Australian 'Kent' (Fig. 9). 'Abondance' oats (Avoine Abondance) was introduced in 1887–1888 from Germany, where it was called 'Ueberfluss' Hafer (Fig. 10), to France. In the 1890's an 'Abundance' oat (Fig. 11), different from the former 'Abondance', was bred in England. Subsequently 'Abundance' was introduced in France and also the name

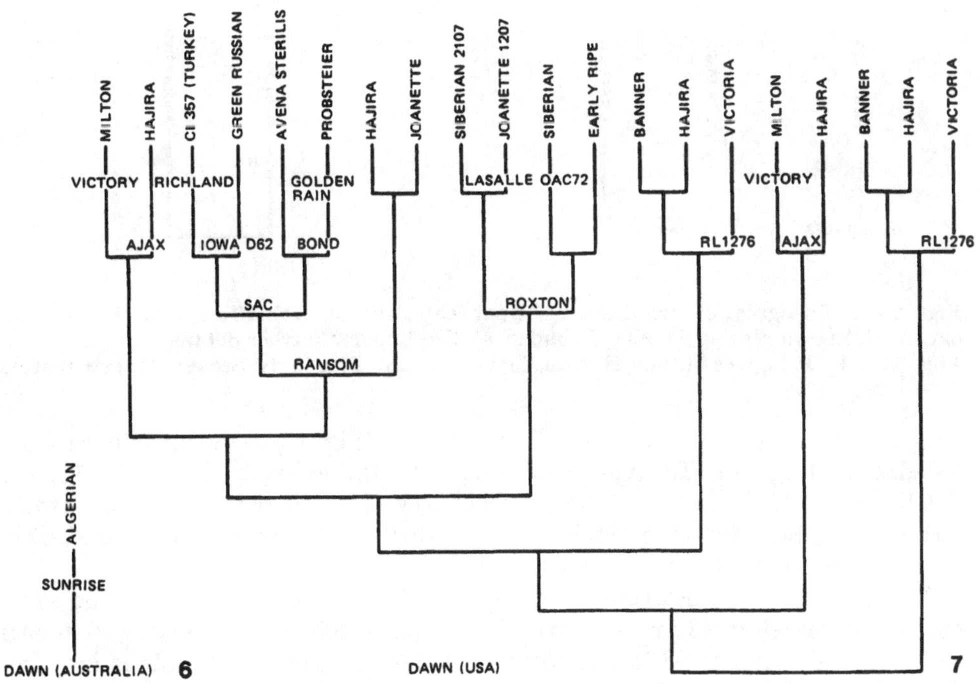


Figs. 6, 7 Pedigrees of two different 'Dawn' oats. See text for details

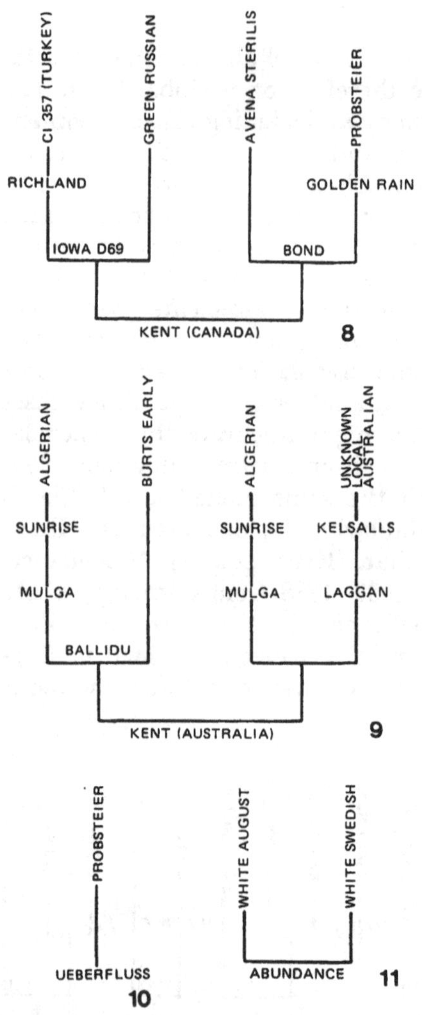


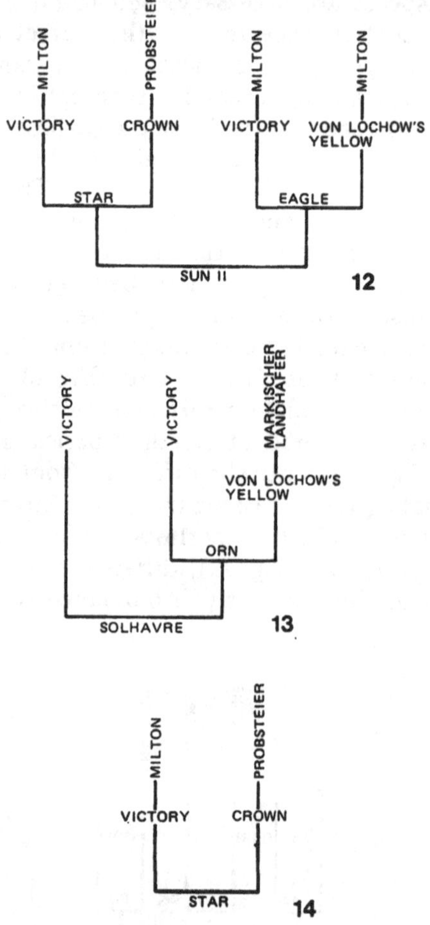


Figs. 8–11 Pedigrees of two different 'Kent' oats, and of 'Ueberfluss' and 'Abundance' oats both known also as Avoine 'Abondance'. See text for further details
Figs. 12–14 Pedigrees of 'Sun II' oats, 'Solhavre' and 'Star' oats. See text for explanation

was translated to 'Abondance'. 'Sund 2' oats (Fig. 12) were introduced from Sweden to England. To avoid confusion with the existing 'Sun' (Fig. 13) the English gave a new name to 'Sun 2', viz. 'Steel'. But the name 'Steel' is a translation of the Scandinavian 'Stål' (Fig. 14) which is different from 'Sun 2'. Still more confusion might have been created.

The Code of Nomenclature of Cultivated Plants permits reuse of the same cultivar name after 10 years "have elapsed since the first cultivar was known to be grown" (Article 48 in Gilmour et al. 1969). Clearly, this rule is inappropriate in view of the global nomenclature picture because it leads to confusion in retrieval and induces misinformation. I made a proposal (Baum, 1972b) to delete Article 48 and to amend Article 44; but, to my knowledge no serious step has been taken until recently. Relevant information about a particular cultivar is attached to it by means of its names. The nomenclature system is thus important not only for communication purposes but as a carrier of information.

(b) Cultivar Groups. Nomenclature of cultivar groups is intimately related to the problem of recognition of cultivar groups, i. e. to classification of cultivars. A number of investigators developed an urge to recognize infraspecific categories, and obviously some differed in their opinions as to the number of ranks that ought to be recognized, while others differed in naming the various categories, while still others ignored what had already been done before them and reinvented categories and names. In all 57 categories have been coined according to the excellent summary by JIRASEK (1961). Of the 57 some designate more or less the same. JIRASEK (1966), however, felt an urge to propose a system of 12 ranks as follows: Specioid; Subspecioid; Cultiplex; Subcultiplex; Convarietas; Subconvarietas; Provarietas; Subprovarietas; Concultа; Subconculta; Cultivar; Subcultivar. Furthermore, this system has actually been implemented in a study of *Callistephus chinensis* (L.) NEES cultivars (NECHANSKY and JIRASEK, 1967) in which Latin names were assigned to the different taxa of the different infraspecific ranks.

The practice of recognizing infraspecific categories above the cultivar level (and below sometimes also, as in JIRASEK), and the practice of giving the resulting taxa formal Latin names still thrives. It is my opinion that the efforts required to develop such systems, together with the problems associated with their practical use, do not warrant the undertaking of erecting them. Besides the category "Cultivar" it may be practical to use the "Subspecies" category in some cases.

It is to be expected that comparisons of genealogies with the respective classifications, the latter consisting of systems of hierarchic categories, will reveal discordance, as in BAUM (1970) and AUTRAN and BOURDET (1975). The lack of concordance stems from the following facts. At the infraspecific level biological isolation does not exist as a rule. Furthermore, gene exchange is to a great extent promoted by man by bringing geographically distant populations into contact through introductions or by actually breeding them for the production of new cultivars. Thus a complex web of anastomosing genealogical reticula is formed. Clearly, hierarchical infraspecific classification is not appropriate because it cannot reflect the true genealogy and its purpose becomes meaningless.

For most cases instead, recognition of informal non-hierarchical groupings of cultivars, including unnamed genetic material, is perfectly satisfactory especially when these groupings are clearly defined and are provided with means for their identification. A few examples, taken from Table 2, of such informal cultivar groupings are: RHODES et al. (1966, 1969), BAUM and LEFKOVITCH (1972a, 1972b), MARTIN et al. (1977), McK. BIRD and GOODMAN (1977), GOODMAN and McK. BIRD (1977), OCHI and NAKAJIMA (1978), MARTIN and RHODES (1979).

For most purposes China aster (*Callistephus chinensis*) cultiplex *planus* convarietas *concolor* provar *sulcatus* cultivar 'Feuerteufel Riesen Bukett', is still *C. chinensis* cvar. 'Feuerteufel Riesen Bukett', irrespective of its classification. For specific purposes, such as selection of genetic material with specific traits for breeding or introduction, a special purpose classification of informal groupings, which in some cases are polythetic and overlap, might be satisfactory. Multipurpose classifications, if at all attainable, might better fulfill a number of specific purposes. I believe that non-hierarchical partitions of whole gene pools of cultivated plants, i. e. cultivars together with unnamed material, are sufficient for many purposes.

To conclude on nomenclature: a system of nomenclature that includes a listing of all names given to all the cultivars for each cultivated species, together with commercial synonyms, translations and transliterations, all cross-referenced, is a prime requirement for fulfiling the needs of those who are concerned with cultivars. In contrast, systems of cultivar groups or infraspecific groupings should preferably be informal and nomenclature does not fulfill any useful purpose in that case.

## General conclusions

1. Infraspecific taxonomy of cultivated plants is useful and important to growers, buyers, merchants, breeders, scientists, and various others working with or concerned with cultivated plants.

2. In order to be useful an infraspecific taxonomic treatment should provide the following desirable systems, which complement each other:

(a) an identification system; (b) a classification system; (c) a genealogical system; (d) a nomenclatural system.

3. Ideally an identification system should enable automatic identification of cultivars. Its aim is to provide accuracy, speed and reliability in identifying samples to cultivars.

4. A classification system consisting of single non-hierarchic partitions is probably the most suitable. Such a system might be based on different premises from those of an identification system. Its aim is to provide a summary of information of relationships based on characters (or traits).

5. A genealogical system is based on individual pedigrees, and provides a summary of information of phylogenetic relationships among cultivars. At the cultivar level it is expected that a genealogical system might not have a one to one correspondance with a classification system.

6. A nomenclatural system of cultivars should provide complete information on names, synonyms, translations and transliterations concerning all the cultivars, past and present, of each particular crop. Such a system facilitates communication among users of these cultivars and provides a framework on which all the pertinent information can be tagged.

7. There is no apparent need for a nomenclatural system of cultivar groups. The groups names may remain informal since reference to them is considerably less than to cultivar names.

8. Taxonomic investigations of cultivated plants at the infraspecific level

should preferably be done on a world-wide basis, or at least serious attempts must be carried out with this in mind. An identification system might fulfill all the local needs when done on a local basis. A classification system, a genealogical system and a nomenclatural system of cultivar names instead, must be done on a global basis to fulfill their usefulness. Moreover, these three systems can be comprehensively achieved only when done on a world-wide basis.

## Zusammenfassung


Taxonomie der infraspezifischen Variabilität von Kulturpflanzen

In der vorliegenden Arbeit wird kurz auf die Notwendigkeit und die Ziele einer Taxonomie von Kulturpflanzen auf dem infraspezifischen Niveau unter allgemein-taxonomischen Gesichtspunkten eingegangen. Die Komponenten einer infraspezifischen Taxonomie von Kulturpflanzen – Identifikation, Klassifikation, Genealogie und Nomenklatur – werden aufgeführt und definiert. Anhand einer speziell für die vorliegende Arbeit zusammengestellten Literaturübersicht, die den Zeitraum von (1975) 1976 bis 1979 umfaßt, werden anzustrebende Richtlinien für die einzelnen Komponenten diskutiert.

Die Entwicklung automatisierter Systeme zur Identifikation von Sorten gewinnt immer mehr an Bedeutung. Eine Identifikation muß exakt, schnell und unbeeinflußt von menschlichen Irrtumsmöglichkeiten erfolgen. Ein Klassifikationssystem muß auf der größtmöglichen Anzahl von Merkmalen beruhen und numerisch-taxonomische Methoden zur Erzeugung und Bewertung von Gruppierungen verwenden. Ein entsprechendes System würde sich aus polythetischen, nichthierarchischen, informalen Gruppen zusammensetzen. Unter den möglichen Genealogien sind solche, die den Stammbaum einer einzelnen Sorte so weit wie möglich zurückverfolgen, am brauchbarsten. Am besten lassen sie sich mit Hilfe von solchen Computersystemen erzeugen, wie sie vom Autor und seinen Mitarbeitern für Hafer- und Gerstensorten angewendet wurden. Stammbäume oder Genealogien lassen andere Schlußfolgerungen über Sorten zu als Klassifikationen, da keine eindeutige wechselseitige Zuordnung zwischen beiden besteht. Ein Nomenklatursystem für Sorten muß so vollständig wie möglich sein. Dies kann durch Aufzählung aller Sortennamen, kommerziellen Synonyme, Übersetzungen und Transliterationen, einschließlich der entsprechenden Verweise erreicht werden. Es werden Beispiele für Fälle angegeben, wo eine fehlerhafte Nomenklatur zu Verwirrungen führte. Der Autor ist der Meinung, daß ein Nomenklatursystem für Gruppen von Sorten nicht benötigt wird, da es zu schwerfällig ist, und weil solche Gruppen nicht unbedingt formalisiert sein müssen.

Eine internationale Fortführung der Arbeiten an den vier Komponenten der infraspezifischen Taxonomie ist anzustreben, um ihre Nützlichkeit für den Pflanzenbau, den Handel, die Züchtung und die Wissenschaft, die sich mit der Entwicklung und Verbesserung von Sorten beschäftigt, zu erhöhen. Eine Ausnahme könnte die Sortenidentifikation innerhalb eines Landes darstellen, wo ein lokales System den meisten aktuellen Bedürfnissen genügen würde.

**Краткое содержание**

Таксономия внутривидового многообразия культурных растений

С точки зрения таксономии в целом, в работе кратко излагаются цели и необходимость таксономии культурных растений на внутривидовом уровне. Перечисляются и определяются составные части внутривидовой таксономии культурных растений: опознавание (идентификация), классификация, генеалогия и номенклатура. Используя обзор литературы, составленный специально для этой работы, охватывающий период времени от 1975–76 по 1979 гг., автор обсуждает желательные направления развития перечисленных составных частей таксономии.

Разработка автоматизированных систем для опознавания сортов приобретает всё большее значение. Эта идентификация должна быть точной, быстрой и свободной от возможных ошибок, проводящего её человека. Система классификации должна основываться на возможно большем числе признаков и применении методов нумерической таксономии для образования группировок и их оценки. Такую систему можно было бы составить из политетических, неиерархических, неформальных групп. Наиболее пригодной генеалогией окажется такая, по которой удастся возможно глубже проследить родословную отдельного сорта. Их составление лучше всего проводить с помощью систем на базе ЭВМ, напр. таких, которые применялись автором и его сотрудниками для сортов овса и ячменя. На основании родословных или генеалогий получаются иные данные о сортах, чем на основании классификаций, так как методы не однозначно соотносимы. Номенклатурная система сортов должна быть возможно более полной. Этого можно достичь перечислением всех названий сортов, торговых синонимов, переводных названий и транслитераций, включая соответствующие ссылки. Приводятся примеры случаев, в которых ошибочная номенклатура приводила к неразберихе. Автор считает, что номенклатурная система для групп сортов не нужна, так как она слишком громоздка и так как такие группы не должны быть обязательно формализированы.

Надо стремиться расширять работы по упорядочению четырёх составных частей внутривидовой таксономии в международном масштабе, чтобы повысить её значение для растениеводства, торговли, селекции и науки, занимающейся улучшением сортов. Как исключение можно допустить идентификацию сортов внутри какой-нибудь одной страны, в которой местная система достаточна для большинства актуальных требований.

## Literature

Appendices 1 and 2 contain all their bibliographic references. Some articles from those appendices are cited in the text and they are listed here also.

Autran, J. C., and A. Bourdet, 1975: L'identification des variétés de blé: établissement d'un tableau général de détermination fondé sur le diagramme électrophorétique des gliadines du grain. – Annales de l'amélioration des Plantes **25**, 277–301.

BAUM, B. R., 1970: The problem of classifying cultivars with special emphasis on oat (*Avena*) cultivars. – Canadian Journal of Botany **50**, 1373–1381.
–, 1972a: Material for an international oat register. – Agriculture Canada. Ottawa, 266 p.
–, 1972b: Comments on rules of publication and use of cultivar names. – Taxon **21**, 299–301.
–, and E. J. BRACH, 1975: Identification of oat cultivars by means of fluorescence spectrography – A pilot study aimed at automatic identification of cultivars. – Canadian Journal of Botany **53**, 305–309.
–, and L. P. LEFKOVITCH, 1972a: A model for cultivar classification and identification with reference to oats (*Avena*). I. Establishment of the groupings by taximetric methods. – Canadian Journal of Botany **50**, 121–130.
–, and –, 1972b: A model for cultivar classification and identification with reference to oats (*Avena*). II. A probabilistic definition of cultivar groupings and their Bayesian identification. – Canadian Journal of Botany **50**, 131–138.
–, and –, 1973: A numerical taxonomic study of phylogenetic and phenetic relationships in some cultivated oats, using known pedigrees. – Systematic Zoology **22**, 118–131.
–, W. PETRUK, and L. G. BAILEY, 1980: Assessment of the value of endospermic starch granules for the taxonomy of barley (*Hordeum*) species and cultivars with special emphasis on their identification, using the technique of image analysis. – Z. Pflanzenzücht. **85**, 212–225.
–, and B. K. THOMPSON, 1970: Registers with predigree charts for cultivars: their importance, their contents, and their preparation by computer. – Taxon **19**, 762–768.
–, and –, 1976: Classification of Canadian oat cultivars by quantifying the size-shape of their 'seeds': a step towards automatic identification. – Canadian Journal of Botany **54**, 1472–1480.
BECKNER, M., 1959: The biological way of thought. – Columbia University Press, New York.
CLAUSEN, J., D. D. KECK, and W. M. HIESEY, 1940: Experimental studies on the nature of species I. Effect of varied environments on Western North American plants. – Carnegie Institute Washington, Publ. No. 520.
–, –, and –, 1945: Experimental studies on the nature of species II. Plant evolution through amphiploidy and autoploidy, with examples from Madiinae. – Carnegie Institute Washington, Publ. No. 564.
–, –, and –, 1948: Experimental studies on the nature of species III. Environmental responses of climatic races of *Achillea*. – Carnegie Institute Washington, Publ. No. 581.
COOLEY, W. W., and P. R. LOHNES, 1962: Multivariate procedures for the behavioral sciences. – John Wiley and Sons, New York.
DAVIS, P. H., and V. H. HEYWOOD, 1963: Principles of Angiosperm Taxonomy. – D. Van Nostrand Company Inc., Princeton, New Jersey, New York.
GILMOUR, J. S. L., F. R. HORNE, E. L. LITTLE, F. A. STAFLEU, and R. H. RICHENS. 1969: International code of nomenclature of cultivated plants – 1969. – International Bureau for Plant Taxonomy and Nomenclature, Utrecht.
GOOD, I. J., 1965: Categorization of classification. – In: Mathematics and computer science in biology and medicine, pp. 115–128. – Medical Research Council.
GREENHALGH, R., and B. R. BAUM, 1980: Feasibility study of the identification of cultivars by pyrolysis gas chromatography using oat (*Avena*) seed kernels. – Seed Science and Technology (In Press)
HAWKES, J. G., 1970: The taxonomy of cultivated plants. – In: Genetic resources in plants – their exploration and conservation (Eds. O. H. FRANKEL and E. BENNETT), pp. 69–85). – Blackwell Scientific Publ., Oxford.
HILL, L. R., 1973: A fault in Pratt's critique of numerical taxonomy. – Journal of Theoretical Biology **40**, 397–398.
JIRASEK, V., 1961: Evolution of the proposals of taxonomical categories for the classification of cultivated plants. – Taxon **10**, 34–45.
–, 1966: Systematika kulturnich rostlin a jeji tridici kategorie. – Preslia **38**, 267–284.
MAHALANOBIS, P. C., 1936: On the generalized distance in statistics. – Proc. Nat. Inst. Sci. India **2**, 49–55.

MAIER, J., 1977: Versuche zur Bestimmung von Hopfensorten nach chemischen Kennzeichen bei Naturhopfen, Hopfenextrakten und in Gemischen. – Hopfen-Rundschau **28**, 140–144.
MILATZ, R. VON, 1936: Der Hafer im Sortenregister. – Landwirt. Jahrb. **83**, 1–152.
NECHANSKY, F., and V. JIRASEK, 1967: Systematische Studie über kultivierte Sommerastern (Gartenastern) – *Callistephus chinensis* (L.) NEES (Asteraceae). – Preslia **39**, 122–150.
NEEDHAM, R., 1975: Polythetic classification: convergence and consequences. – Man. The Journal of the Royal Anthropological Institute **10**, 349–369.
PANKHURST, R. J. (Ed.), 1975: Biological identification with computers. – Academic Press, London. 339 pp.
–, 1978: Biological identification. The principles and practice of identification methods in biology. – University Park Press, Baltimore. 104 pp.
PEARSON, K., 1926: On the coefficient of racial likeness. – Biometrika **18**, 105–177.
RUBY, J. L., and J. W. WRIGHT, 1977: A revised classification of geographic varieties in Scots Pine. – Silvae Genetica **25**, 169–175.
SEAL, H. L., 1964: Multivariate Statistical Analysis for Biologists. – Methuen and Co. Ltd., London. 207 pp.
SIMPSON, G. G., 1961: Principles of Animal Taxonomy. – Columbia University Press, New York.
SNEATH, P. H. A., 1979: Basic program for identification of an unknown with presence-absence data against an identification matrix of percent positive characters. – Computers and Geosciences **5**, 195–213.
SOKAL, R. R., and P. H. A. SNEATH, 1963: Principles of Numerical Taxonomy. – W. H. Freeman and Co., San Francisco and London. 359 pp.
SVIRIDOV, A. V., 1978: O nekotorych aktual'nych voprosach teorii identifikacii biologičeskich ob''ektov s pomošč'ju ključej. – Biologičeskie nauki. Naučnye doklady vysšej školy. Moscow, USSR, **10**, 15–28.
THOMPSON, B. K., and B. R. BAUM, 1978: Preparation of a barley register: an information retrieval system for cultivars. – Taxon **27**, 471–477.

Dr. B. R. BAUM,
Principal Research Scientist
Biosystematics Research Institute
Agriculture Canada, Research Branch
Saunders Building
Central Experimental Farm
Ottawa, Ontario, Canada
K1A OC6

1980-05-13 PAGE 1

| Taxon / Title | Author(s) | Source | Year, Vol. | Pages |
|---|---|---|---|---|
| ACER RUBRUM CULTIVARS — IDENTIFICATION OF FOUR RED MAPLE CULTIVARS WITH SCANNING ELECTRON MICROSCOPY | KRAUSE, C.R., | HORTSCIENCE | 1978, 13 | 586-588 |
| ALLIUM CEPA CULTIVARS — A VOROSHAGYMA FAJTAK SZARAZANYAGTARTALMANAK TAXONOMIAI JELENTOSEGE | SARADA, K.S., TOMOSKOZI, M., | KERTGAZDASAG | 1977, 9 | 53-60 |
| ALLIUM CULTIVARS — STUDIES ON THE DETERMINATION OF SPECIES AND CULTIVARS ON THE BASIS OF ELECTROPHORETIC PATTERNS OF SEED PROTEIN AND SEED ENZYMES II. ALLIUM SPP., CUCUMBER AND MELON, PEA AND GARDEN BEAN | NAKAMURA, S., | JOURNAL OF THE JAPANESE SOCIETY FOR HORTICULTURAL SCIENCE | 1977, 46 | 233-244 |
| ALLIUM SPECIES — STUDIES ON THE DETERMINATION OF SPECIES AND CULTIVARS ON THE BASIS OF ELECTROPHORETIC PATTERNS OF SEED PROTEIN AND SEED ENZYMES II. ALLIUM SPP., CUCUMBER AND MELON, PEA AND GARDEN BEAN | NAKAMURA, S., | JOURNAL OF THE JAPANESE SOCIETY FOR HORTICULTURAL SCIENCE | 1977, 46 | 233-244 |
| AVENA SATIVA CULTIVARS — DEVELOPMENT OF A VARIETAL PURITY TESTING SYSTEM FOR OATS (AVENA SATIVA L.).IN DISSERTATION ABSTRACTS INTERNATIONAL 35(6)NO 74-28,936 | ALLEN, J.R., | THE PENNSYLVANIA STATE UNIVERSITY | 1974 | 94 |
| DIE BESCHREIBUNG DER LANDWIRTSCHAFTLICHEN WERTEIGENSCHAFTEN VON GETREIDESORTEN | HRON, R., | JAHRBUCH DER BUNDESANSTALT FUR PFLANZENBAU UND SAMENPRUFUNG IN WIEN | 1975 | 153-159 |
| BRASSICA CULTIVARS — STUDIES ON THE DETERMINATION OF SPECIES AND CULTIVARS ON THE BASIS OF ELECTROPHORETIC PATTERNS OF SEED PROTEIN AND SEED ENZYMES I.BRASSICA SPP., RADISH AND CVCURBITA SPP. | NAKAMURA, S., | JOURNAL OF THE JAPANESE SOCIETY FOR HORTICULTURAL SCIENCE | 1977, 46 | 32-47 |
| BRASSICA SPECIES — STUDIES ON THE DETERMINATION OF SPECIES AND CULTIVARS ON THE BASIS OF ELECTROPHORETIC PATTERNS OF SEED PROTEIN AND SEED ENZYMES I.BRASSICA SPP., RADISH AND CVCURBITA SPP. | NAKAMURA, S., | JOURNAL OF THE JAPANESE SOCIETY FOR HORTICULTURAL SCIENCE | 1977, 46 | 32-47 |
| CAPSICUM CULTIVARS — STUDIES ON SOME HORTICULTURAL CHARACTERS INFLUENCING QUALITY AND YIELD IN THE PEPPER (CAPSICUM ANNUUM L.) | JAE-MOON, H., BYOUNG-YIL, L., | JOURNAL OF THE KOREAN SOCIETY FOR HORTICULTURAL SCIENCE | 1977, 18 | 173-180 |
| CARTHAMUS OXYACANTHA ECOTYPES — IDENTIFICATION AND POLYMORPHISM OF CULTIVARS AND WILD ECOTYPES OF SAFFLOWER BASED ON ISOZYME PATTERNS | BASSIRI, A., | EUPHYTICA | 1977, 26 | 709-719 |
| CARTHAMUS TINCTORIUS CULTIVARS — IDENTIFICATION AND POLYMORPHISM OF CULTIVARS AND WILD ECOTYPES OF SAFFLOWER BASED ON ISOZYME PATTERNS | BASSIRI, A., | EUPHYTICA | 1977, 26 | 709-719 |
| CERASUS CULTIVARS — MORPHOLOGICAL CHARACTERS OF CHERRY LEAVES AND THEIR UTILIZATION IN CULTIVAR CLASSIFICATION | YUSHEV, A.A., | VSESOIUZNYI INSTITUT RASTENOVODSTVA, BIULLETEN. | 1975, 54 | 34-40 |
| A CLASSIFICATION OF CHERRY VARIETIES AND CHEMICAL COMPOSITION OF THE FRUITS | YUSHEV, A.A., SHAROVA, N.I., | BULLETIN OF APPLIED BOTANY, GENETICS AND PLANT BREEDING | 1976, 57 | 86-92 |

Appendix 1 Taxonomic studies of variability below the species level, of cultivated plants an d mostly of cultivars. These are arranged according to taxa. Sources extracted from Commonwealth Agricultural Data Base of the last four years. Because it is the data base of the last four years, occasionally earlier articles also appear.

1980-05-13 PAGE 2

| Subject / Title | Author | Journal | Year / Vol. | Pages |
|---|---|---|---|---|
| CICER ARIETENUM CULTIVARS<br>ESTUDIO MORFOLOGICO DE CULTIVARES DE GARBANZO | BAZAN, L.C., | ANALES CIENTIFICOS (PERU) | 1974<br>12 | 65-67 |
| CITRULLUS VULGARIS CULTIVARS<br>THE CLASSIFICATION OF WATERMELONS | FURSA, T.B., | BULLETIN OF APPLIED BOTANY, GENETICS AND PLANT BREEDING | 1977<br>61 | 3-10 |
| CITRUS CULTIVARS<br>CHARACTERIZATION OF CITRUS CULTIVARS AND SEPARATION OF NUCELLAR AND ZYGOTIC SEEDLINGS BY THIN LAYER CHROMATOGRAPHY | TATUM, J.H., BERRY, R.E., HEARN, C.J., | PROCEEDINGS OF THE FLORIDA STATE HORTICULTURAL SOCIETY | 1974<br>87 | 75-81 |
| CHARACTERIZATION OF CITRUS CULTIVARS BY CHEMICAL DIFFERENTIATION | TATUM, J.H., HEARN, C.J., BERRY, R.E., | JOURNAL OF THE AMERICAN SOCIETY FOR HORTICULTURAL SCIENCE | 1978<br>103 | 492-496 |
| CITRUS SPECIES<br>CHARACTERIZATION OF CITRUS CULTIVARS AND SEPARATION OF NUCELLAR AND ZYGOTIC SEEDLINGS BY THIN LAYER CHROMATOGRAPHY | TATUM, J.H., BERRY, R.E., HEARN, C.J., | PROCEEDINGS OF THE FLORIDA STATE HORTICULTURAL SOCIETY | 1974<br>87 | 75-81 |
| CUCUMIS SPECIES<br>STUDIES ON THE DETERMINATION OF SPECIES AND CULTIVARS ON THE BASIS OF ELECTROPHORETIC PATTERNS OF SEED PROTEIN AND SEED ENZYMES II. ALLIUM SPP., CUCUMBER AND MELON, PEA AND GARDEN BEAN | NAKAMURA, S., | JOURNAL OF THE JAPANESE SOCIETY FOR HORTICULTURAL SCIENCE | 1977<br>46 | 233-244 |
| CUCURBITA CULTIVARS<br>ESTUDO POMOLOGICO COMPARATIVO DAS CULTIVARES DE CUCURBITA | ROCHELLE, L.A., | ANAIS DA ESCOLA SUPERIOR DE AGRICULTURA "LUIZ DE QUEIROZ" | 1976 | |
| CUCURBITA MAXIMA CULTIVARS<br>DESCRICAO TAXONOMICA DE CULTIVARES DE CURCUBITA MAXIMA DUCHESNE E CUCURBITA PEPO LINNEU | ROCHELLE, L.A., | ANAIS DA ESCOLA SUPERIOR DA AGRICULTURA "LUIZ DE QUEIROZ" | 1974<br>31 | 667-700 |
| CUCURBITA PEPO CULTIVARS<br>DESCRICAO TAXONOMICA DE CULTIVARES DE CURCUBITA MAXIMA DUCHESNE E CUCURBITA PEPO LINNEU | ROCHELLE, L.A., | ANAIS DA ESCOLA SUPERIOR DA AGRICULTURA "LUIZ DE QUEIROZ" | 1974<br>31 | 667-700 |
| CUCURBITA SPECIES<br>STUDIES ON THE DETERMINATION OF SPECIES AND CULTIVARS ON THE BASIS OF ELECTROPHORETIC PATTERNS OF SEED PROTEIN AND SEED ENZYMES I.BRASSICA SPP., RADISH AND CVCURBITA SPP. | NAKAMURA, S., | JOURNAL OF THE JAPANESE SOCIETY FOR HORTICULTURAL SCIENCE | 1977<br>46 | 32-47 |
| CULTIVARS<br>PREPARATION OF A BARLEY REGISTER: AN INFORMATION RETRIEVAL SYSTEM FOR CULTIVARS | THOMPSON, B.K., BAUM, B.R., | TAXON | 1978<br>27 | 471-477 |
| DAUCUS CAROTA CULTIVARS<br>ALLOMETRIC GROWTH STUDIES OF THE CARROT CROP I.EFFECTS OF PLANT DEVELOPMENT AND CULTIVAR | STANHILL, G., | ANNALS OF BOTANY | 1977<br>41 | 533-540 |
| DAUCUS CULTIVARS<br>STUDIES ON THE DETERMINATION OF SPECIES AND CULTIVARS ON THE BASIS OF ELECTROPHORETIC PATTERNS OF SEED PROTEIN AND SEED ENZYMES II. ALLIUM SPP., CUCUMBER AND MELON, PEA AND GARDEN BEAN | NAKAMURA, S., | JOURNAL OF THE JAPANESE SOCIETY FOR HORTICULTURAL SCIENCE | 1977<br>46 | 233-244 |

Appendix 1 (continued)

1980-05-13 PAGE 3

| | | | | |
|---|---|---|---|---|
| DIANTHUS CULTIVARS<br>CARACTERISATION DES CULTIVARS A L'AIDE DE DIFFERENCES QUANTITATIVES DE TENEUR EN COMPOSES POLYPHENOLIQUES: APPLICATION AU MAIS ET A L'OEILLET | MONTIES, B., CHESNEAUX, M.-T., MUTIN, C., | ANNALES DE L'AMELIORATION DES PLANTES | 1976<br>26 | 495-516 |
| FESTUCA RUBRA CULTIVARS<br>A PROVISIONAL GROUPING OF CULTIVARS OF FESTUCA RUBRA L. | SHILDRICK, J.P., | THE JOURNAL OF THE SPORTS TURF RESEARCH INSTITUTE | 1976<br>52 | 9-13 |
| FICUS CARICA CULTIVARS<br>L'IDENTIFICATION VARIETALE DU FIGUIER, FICUS CARICA L. PAR L'ETUDE DU POLYMORPHISME ENZYMATIQUE PAR ELECTROPHORESE | VALDEYRON, G., VALIZADEH, M., | ACADEMIE D'AGRICULTURE DE FRANCE, COMPTES RENDUS, | 1976<br>62 | 170-175 |
| UTILISATION DU POLYMORPHISM PROTEIQUE POUR L'ETUDE DES VARIETES DE FIGUIER (FICUS CARICA L.) | VALIZADEH, M., RIVALS, P., VALDEYRON, G., | SEANCES DE L'ACADEMIE D'AGRICULTURE DE FRANCE | 1977<br>63 | 647-655 |
| GLYCINE MAX CULTIVARS<br>ADATOK NEHANY KOZTERMESZTETT SZOJAFAJTA ALAKTANI JELLEMZESEHEZ | KURNIK, E., GAL, M., SZABO, L., | TAKARMANYTERMESZTESI KUTATO INTEZET KOZLEMENYEI | 1974<br>14 | 5-15 |
| IDENTIFICATION OF SOYBEAN CULTIVARS BY THE SURFACE RELIEF OF THE SEED COAT | LINSKENS, H.F., PFAHLER, P.L., KNUIMAN-STEVENS, E.L., | THEORETICAL AND APPLIED GENETICS | 1977<br>50 | 147-149 |
| GOSSYPIUM BARBADENSE CULTIVARS<br>THE USE OF TECHNOLOGICAL PROPERTIES OF COTTON VARIETIES IN THE CONFIRMATION OF THEIR IDENTITY | SALLOUMA, B.M., EL-GAZZAR, A., ABDELLAH, M.E., | PHYTOLOGIA | 1975<br>31 | 264-266 |
| HEVEA BENTHAMIANA<br>PROTEIN AND ENZYME VARIATION IN SOME HEVEA CULTIVARS | YEET, Y.M., GHANDIMATHI, H., PARANJOTHY, | JOURNAL OF THE RUBBER RESEARCH INSTITUTE OF MALAYSIA | 1977<br>25 | 9-18 |
| HEVEA BRASILIENSIS<br>PROTEIN AND ENZYME VARIATION IN SOME HEVEA CULTIVARS | YEET, Y.M., GHANDIMATHI, H., PARANJOTHY, | JOURNAL OF THE RUBBER RESEARCH INSTITUTE OF MALAYSIA | 1977<br>25 | 9-18 |
| HORDEUM CULTIVARS<br>PREPARATION OF A BARLEY REGISTER: AN INFORMATION RETRIEVAL SYSTEM FOR CULTIVARS | THOMPSON, B.K., BAUM, B.R., | TAXON | 1978<br>27 | 471-477 |
| HORDEUM VULGARE CULTIVARS<br>PRIJEDLOG NOVE KLASIFIKACIJE LODICULA JECMA | AGANOVIC, Z., | RADOVI POLJOPRIVREDNOG FAKULTETA UNIVERZITETA U SARAJEVU | 25 | 47-54 |
| BARLEY CULTIVAR IDENTIFICATION BY USE OF ISOZYME ELECTROPHORETIC PATTERNS | BASSINI, A., | CANADIAN JOURNAL OF PLANT SCIENCE | 1976<br>56 | 1-6 |
| DIE BESCHREIBUNG DER LANDWIRTSCHAFTLICHEN WERTEIGENSCHAFTEN VON GETREIDESORTEN | HRON, R., | JAHRBUCH DER BUNDESANSTALT FUR PFLANZENBAU UND SAMENPRUFUNG IN WIEN | 1975 | 153-159 |
| CHEMOTAXONOMICKA STUDIE VYBRANYCH JECMENE (HORDEUM SATIVUM L.) | HUBACEK, J., LACHMAN, J., | ROSTLINNA VYROBA | 1977<br>23 | 151-158 |
| ROZLISOVANI CESKOSLOVENSKYCH SORT JARNIHO JECMENE PODLE KLASU | KUHN, F., | ACTA UNIVERSISTATIS AGRICULTURAE | 1976<br>24 | 419-433 |
| IDENTIFICATION OF AUSTRALIAN BARLEY CULTIVARS BY LABORATORY METHODS: GEL ELECTROPHORESIS AND GEL ISOELECTRIC FOCUSING OF THE ENDOSPERM PROTEINS | MCCAUSLAND, J., WRIGLEY, C.W., | AUSTRALIAN JOURNAL OF EXPERIMENTAL AGRICULTURE AND ANIMAL HUSBANDRY | 1977<br>17 | 1020-1027 |
| SUR LA CHIMIOTAXONOMIE DE L'ORGE ET DU MALT ETUDIEE PARELECTROFOCALISATION SUR GEL | SCRIBAN, R., STROBBEL, S., | COMPTES RENDUS DE LA SOCIETE DE BIOLOGIE | 1978<br>172 | 647-650 |

Appendix 1 (continued)

| | | | | |
|---|---|---|---|---|
| 1980-05-13 | | | PAGE 4 | |
| HORDEUM VULGARE CULTIVARS<br>VARIETAL IDENTIFICATION OF SINGLE SEEDS OF BARLEY BY ANALYSIS OF HORDEIN POLYPEPTIDES | SHEWRY, P.R., PRATT, H.M., MIFLIN, B.J., | JOURNAL OF THE SCIENCE OF FOOD AND AGRICULTURE | 1978<br>29 | 587-596 |
| MODIFIED GRAPHICAL ANALYSIS: A TECHNIQUE FOR GERMPLASM CLASSIFICATION | SINGH, V.P., CHAUDHARY, B.D., | GUJARAT AGRICULTURAL UNIVERSITY RESEARCH JOURNAL | 1978<br>3 | 59-63 |
| BARLEY IDENTIFICATION BY GRAIN CHARACTERS IN NEW ZEALAND | SPARKS, G.A., MALCOLM, J.P., | NEW ZEALAND JOURNAL OF EXPERIMENTAL AGRICULTURE | 1978<br>6 | 1-10 |
| HUMULUS LUPULUS CULTIVARS<br>VERSUCHE ZUR BESTIMMUNG VON HOPFENSORTEN NACH CHEMISCHEN KENNZEICHEN BEI NATURHOPFEN, HOPFENEXTRAKTEN UND IN GEMISCHEN | MAIER, J., | HOPFEN-RUNDSCHAU | 1977<br>28 | 140-144 |
| JUGLANS ILLINOENSIS CULTIVARS<br>THE ORIGIN AND DEVELOPMENT OF PECAN VARIETIES | MILLER, G., | FRUIT VARIETIES JOURNAL | 1977<br>31 | 16-21 |
| JUNIPERUS CULTIVARS<br>CHEMICAL IDENTIFICATION OF WOODY ORNAMENTAL PLANTS | FRETZ, T.A., | OHIO REPORT ON RESEARCH AND DEVELOPMENT | 1978<br>63 | 23-25 |
| LINUM USITATISSIMUM STRAINS<br>THE USE OF ANATOMICAL PROPERTIES OF FLAX VARIETIES ON THE CONFIRMATION OF THEIR IDENTITY | MOMTAZ, A., EL-GAZZAR, A., GAAFAR, S., | PHYTOLOGIA | 1976<br>33 | 474-479 |
| LOLIUM CULTIVARS<br>EVALUATION AND CLASSIFICATION OF RYE GRASS VARIETIES IN THE COLLECTION | SHUTOVA, Z.P., | BULLETIN OF APPLIED BOTANY, GENETICS AND PLANT BREEDING | 1977<br>59 | 70-87 |
| MALUS DOMESTICA CULTIVARS<br>MORPHOLOGY OF THE FLOWER IN THE APPLE MALUS DOMESTICA BORKH. AND THE CLASSIFICATION OF ITS VARIETIES | PONOMARENKO, V.V., | BULLETIN OF APPLIED BOTANY, GENETICS AND PLANT BREEDING | 1976<br>57 | 25-33 |
| MORUS CULTIVARS<br>VARIETAL AFFINITIES IN MULBERRY (MORUS SPP.) ASSESSED BY PEROXYDASE ISOZYMES | HIRANO, H., | JAPANESE JOURNAL OF BREEDING | 1977<br>27 | 350-358 |
| MUSA CULTIVARS<br>SOME GUIDELINES IN THE CLASSIFICATION OF GHANIAN PLANTAIN (MUSA AAB GROUP)(ABSTRACT ONLY). | KARIKARI, S.K., ABAKAH-GYENIN, A.K., | PARADISIACA | 1976<br>1 | 15 |
| NARCISSUS CULTIVARS<br>A NACISZFAJTAK RENDSZERTANA | JAMBOR, E., | KERTGAZDASAG | 1977<br>9 | 67-79 |
| ORYZA GLABERRIMA CULTIVARS<br>CHARACTERISTICS OF RICE CULTIVARS | CHANG, T.T., PARKER, M.B., | RISO | 1976<br>25 | 195-201 |
| ORYZA SATIVA CULTIVARS<br>CHARACTERISTICS OF RICE CULTIVARS | CHANG, T.T., PARKER, M.B., | RISO | 1976<br>25 | 195-201 |
| PANICUM SPECIES<br>CHARACTERISTICS AND CLASSIFICATION OF SPECIES AND STRAINS OF PANICUM I.AGRONOMIC CHARACTERISTICS AND CLASSIFICATION OF PANICUM ACCESSIONS INTRODUCED TO JAPAN | OCHI, M., NAKAJIMA, K., | BULLETIN OF THE NATIONAL GRASSLAND INSTITUE | 1977<br>11 | 78-97 |

Appendix 1 (continued)

1980-05-13 PAGE 5

| Subject | Author(s) | Title | Journal | Year | Vol. | Pages |
|---|---|---|---|---|---|---|
| PANICUM SPECIES AND STRAINS | OCHI, M., NAKAJIMA, K., | CHARACTERISTICS AND CLASSIFICATION OF SPECIES AND STRAINS OF PANICUM II. CLASSIFICATION OF SPECIES AND STRAINS OF PANICUM BY PRINCIPAL COMPONENT ANALYSIS AND THEIR EVALUATION | BULLETIN OF THE NATIONAL GRASSLAND INSTITUTE | 1978 | 13 | 41-62 |
| PANICUM STRAINS | OCHI, M., NAKAJIMA, K., | CHARACTERISTICS AND CLASSIFICATION OF SPECIES AND STRAINS OF PANICUM I.AGRONOMIC CHARACTERISTICS AND CLASSIFICATION OF PANICUM ACCESSIONS INTRODUCED TO JAPAN | BULLETIN OF THE NATIONAL GRASSLAND INSTITUE | 1977 | 11 | 78-97 |
| PHASEOLUS CULTIVARS | NAKAMURA, S., | STUDIES ON THE DETERMINATION OF SPECIES AND CULTIVARS ON THE BASIS OF ELECTROPHORETIC PATTERNS OF SEED PROTEIN AND SEED ENZYMES II. ALLIUM SPP., CUCUMBER AND MELON, PEA AND GARDEN BEAN | JOURNAL OF THE JAPANESE SOCIETY FOR HORTICULTURAL SCIENCE | 1977 | 46 | 233-244 |
| PHASEOLUS LUNATUS CULTIVARS | BAUDET, J.C., | THE TAXONOMIC STATUS OF THE CULTIVATED TYPES OF LIMA BEAN (PHASEOLUS LUNATUS L.) | TROPICAL GRAIN LEGUME BULLETIN | 1977 | 7 | 29-30 |
| PHASEOLUS VULGARIS CULTIVARS | ABRAHAO, I.O., | DISTINCAO DE CULTIVARES DE FEIJOEIRO (PHASEOLUS VULGARIS L.) PELO METODO DA CRISTALIZACAO SENSITIVA. | ANAIS DA ESCOLA SUPERIOR DE AGRICULTURA "LUIZ DE QUEIROZ" | 1976 | 33 | 329-335 |
| PINUS SYLVESTRIS VARIETIES | RUBY, J.L., WRIGHT, J.W., | A REVISED CLASSIFICATION OF GEOGRAPHIC VARIETIES IN SCOTS PINE | SILVAE GENETICA | 1977 | 25 | 169-175 |
| PISUM CULTIVARS | NAKAMURA, S., | STUDIES ON THE DETERMINATION OF SPECIES AND CULTIVARS ON THE BASIS OF ELECTROPHORETIC PATTERNS OF SEED PROTEIN AND SEED ENZYMES II. ALLIUM SPP., CUCUMBER AND MELON, PEA AND GARDEN BEAN | JOURNAL OF THE JAPANESE SOCIETY FOR HORTICULTURAL SCIENCE | 1977 | 46 | 233-244 |
| PISUM SATIVUM CULTIVARS | HORVATH, Z., | BORSOFAJTAK FAJTARENDSZERTANI VIZSGALATA I. | AGROBOTANIKA | 1974 | 15 | 199-208 |
| POA PRATENSIS CULTIVARS | ELEKES, P., | DIFFERENTIATION OF CULTIVARS AND CULTIVAR GROUPS OF POA PRATENSIS BY THE FLUORESCENCE OF THEIR ALKALINE EXTRACTS | SEED SCIENCE AND TECHNOLOGY | 1975 | 3 | 643-645 |
| | NIEMYSKI, K., GRZELAK, K., | IDENTIFYING KENTUCKY BLUEGRASS (POA PRATENSIS) CULTIVARS BY SEED MORPHOLOGICAL CHARACTERS | SEED SCIENCE AND TECHNOLOGY | 1975 | 3 | 625-641 |
| | WEHNER, D.J., DUICH, J.M., WATSCHKE, T.L., | SEPARATION OF KENTUCKY BLUEGRASS CULTIVARS USING PEROXIDASE ISOENZYME BANDING PATTERNS | CROP SCIENCE | 1976 | 16 | 475-480 |
| PRUNUS AMYGDALUS CULTIVARS | FANIZZA, G., BOGYO, T.P., | A CLUSTER ANALYSIS OF ALMOND VARIETIES IN APULIA | RIVISTA DELLA ORTOFLOROFRUTICOLTURA ITALIANA | 1976 | 60 | 277-281 |
| PRUNUS PERSICA CULTIVARS | BELLINI, E., GIANELLI, G., | SUL VALORE TASSONOMICO DI ALCUNI CARATTERI DEL RAMO NEL PESCO | RIVISTA DELLA ORTOFLOROFRUTICULTURA ITALIANA | 1975 | 59 | 440-458 |
| | FILIPOVICH, S.D., | ACRYLAMIDE GEL-ELECTROPHORETIC TECHNIQUE AS A TOOL IN THE TAXONOMY OF PEACH CULTIVARS (ABSTRACT ONLY) | PROCEEDINGS OF THE XIX INTERNATIONAL HORTICULTURAL CONGRESS, WARSZAWA | 1974 | 1 | 13 |

Appendix 1 (continued)

```
1980-05-13                                                                                                                    PAGE 6

PRUNUS PERSICA CULTIVARS        NITRANSKY, S.,                                                                                1977
  NIEKTORE RASTOVE A MORFOLOGICKE ZNAKY VYBRANYCH KULTIVAROV  PIESTANY VYSKUMNY USTAV RASTLINNEJ VYROBY VEDECKE PRACE         14   149-154
  BROSKYN

RAPHANUS SATIVUS CULTIVARS      NAKAMURA, S.,                                                                                 1977
  STUDIES ON THE DETERMINATION OF SPECIES AND CULTIVARS ON    JOURNAL OF THE JAPANESE SOCIETY FOR HORTICULTURAL              46    32-47
  THE BASIS OF ELECTROPHORETIC PATTERNS OF SEED PROTEIN AND   SCIENCE
  SEED ENZYMES I.BRASSICA SPP., RADISH AND CVCURBITA SPP.

ROSA CULTIVARS                  KUHNS, L.J., FRETZ, T.A.,                                                                     1978
  DISTINGUISHING ROSE CULTIVARS BY POLYACRILAMIDE GEL         JOURNAL OF THE AMERICAN SOCIETY FOR HORTICULTURAL             103   503-508
  ELECTROPHORESIS, I.EXTRACTION AND STORAGE OF PROTEIN AND    SCIENCE
  ACTIVE ENZYMES FROM ROSE LEAVES
                                KUHNS, L.J., FRETZ, T.A.,                                                                     1978
  DISTINGUISHING ROSE CULTIVARS BY POLYACRILAMIDE GEL         JOURNAL OF THE AMERICAN SOCIETY FOR HORTICULTURAL             103   509-516
  ELECTROPHORESIS. II.ISOENZYME VARIATION AMONG CULTIVARS     SCIENCE
                                KUHNS, L.J., FRETZ, T.A.,                                                                     1978
  A STUDY OF THE POTENTIAL USE OF ELECTROPHORESIS IN          OHIO AGRIC RES DEVELOP CENTRE, RESEARCH BULLETIN 1094.               1-33
  DISTINGUISHING ROSE CULTIVARS

SECALE CEREALE CULTIVARS        STEINBERGER, J.,                                                                              1975
  BEITRAG ZUR IDENTIFIZIERUNG UND BESCHREIBUNG VON            AUSTRIA, BUNDESANSTALT FUR PFLANZENBAU UND SAMENPRUFUNG             113-150.
  ROGGENSORTEN (SECALE CEREALE L.)                            IN WIEN

SOLANUM TUBEROSUM CULTIVARS     LOESCHCKE, V., STEGEMANN, H.,                                                                 1978
  IDENTIFICATION OF POTATO CULTIVARS BY ELECTROPHORESIS       3RD INTERNATIONAL CONGRESS OF PLANT PATHOLOGY, MUNCHEN              302
  (ABSTRACT ONLY)                                             16-23 AUG 1978
                                HOSS, J.P., BROWN, E.,                                                                        1977
  A MULTIPLE-ENTRY KEY SYSTEM USING EDGE-PUNCHED CARDS TO     ECONOMIC BOTANY                                                31   294-297
  IDENTIFY CULTIVARS OF POTATO
                                STEGEMANN, H., LOESCHCKE, V.,                                                                 1977
  DAS EUROPAISCHE KARTOFFELSORTIMENT UND SEINE INDEXIERUNG    POTATO RESEARCH                                                20   101-110

TRIFOLIUM ALEXANDRINUM CVARS    GABER, S.D., HAMMAH, A.M.,                                                                    1979
  LABORATORY METHODS FOR THE RECOGNITION OF SEEDS OF THE      SEED SCIENCE AND TECHNOLOGY                                    7    125-130
  'MISKAWY' AND 'FAHL' CULTIVARS OF EGYPTIAN CLOVER
  (TRIFOLIUM ALEXANDRINUM)

TRIFOLIUM SUBTERRANEUM CULTIVR  QUINLIVAN, B.J., COLLINS, A.J., BROWN, M.L.,                                                  1976
  VARIETAL IDENTIFICATION OF SEEDLING SUBTERRANEAN CLOVER     SEED SCIENCE AND TECHNOLOGY                                    4    185-190
  USING THIN LAYER CHROMATOGRAPHY

TRITICUM AESTIVUM               KONAREV, V.G., GAVRILJUK, I.P., GUBAREVA, N.K.,                                               1976
  CULTIVAR IDENTIFICATION AND DOCUMENTATION OF WHEAT GENETIC  VESTNIK SELSKOKHOZIAS TVENNOI NAUKI                            9    113-125
  RESOURCES BY GLIADIN ELECTROPHORETIC SPECTRA

TRITICUM AESTIVUM CULTIVARS     AUFIAU, P., AUTRAN, J.C., CHARBONNIER, L., DOUSSINAULT, G., FEILLET, P., GODON, B.,           1976
                                GRIGNAC, P., JOUDRIER, P., ET AL.
  LA VARIABILITE GENETIQUE DE LA COMPOSITION DES GLIADINES,   ANNALES DE L'AMELIORATION DES PLANTES                          26    51-66
  GLUTENINES, B-AMYLASES, A-ESTERASES, PEROXYDASES ET
  PHOSPHATASES ACIDES DU BLE (T. AESTIVUM)
                                AUTRAN, J.-C., BOURDET, A.,                                                                   1975
  L'IDENTIFICATION DES VARIETES DE BLE: ETABLISSEMENT D'UN    ANNALES DE L'AMELIORATION DES PLANTES                          25   277-301
  TABLEAU GENERAL DE DETERMINATION FONDE SUR LE DIAGRAMME
  ELECTROPHORETIQUE DES GLIADINES DU GRAIN
                                AUTRAN, J.-C.,                                                                                1975
  IDENTIFICATION DES PRINCIPALES VARIETES COMMUNAUTAIRES DE   BULLETIN DES ANCIENS ELEVES DE L'ECOLE DE MEUNERIE            270   316-324
  BLE TENDRE PAR ELECTROPHORESE DES GLIADINES DU GRAIN
```

Appendix 1 (continued)

1980-05-13 PAGE 7

| | | | | |
|---|---|---|---|---|
| TRITICUM AESTIVUM CULTIVARS<br>VARIETES DE BLES COMMENT LES RECONNAITRE ? | BEAUX, Y. | BULLETIN DES ANCIENS ELEVES DE L'ECOLE DE MEUNERIE | 1975<br>268 | 185-188 |
| PRAKTISCHE ERFAHRUNGEN MIT ERKENNUNGSMETHODEN AUF GRUND AUSSERERKORNMERKMALE | BOLLING, H., MEYER, D. | GETREIDE MEHL UND BROT | 1976<br>30 | 201-203 |
| WHEAT CULTIVAR IDENTIFICATION BY GLIADIN ELECTROPHOREGRAMS. 1. APPARATUS, METHOD AND NOMENCLATURE | BUSHUK, W., ZILLMAN, R.R. | CANADIAN JOURNAL OF PLANT SCIENCE | 1978<br>58 | 505-515 |
| LABORATORY METHODS FOR IDENTIFYING NEW ZEALAND WHEAT CULTIVARS | COLES, G.D., WRIGLEY, C.W. | NEW ZEALAND JOURNAL OF AGRICULTURAL RESEARCH | 1976<br>19 | 499-503 |
| THE IDENTIFICATION OF UK WHEAT VARIETIES BY STARCH GEL ELECTROPHORESIS OF GLIADIN PROTEINS | ELLIS, J.R.S., BEMINSTER, C.H. | JOURNAL OF THE NATIONAL INSTITUTE OF AGRICULTURAL BOTANY | 1977<br>14 | 221-231 |
| ZUR PROBLEMATIK DER ELEKTROPHORETISCHEN SORTENUNTERSCHEIDUNG AM GLIADINMUSTER DES WEIZENS | GUNZEL, G. | GETREIDE MEHL UND BROT | 1976<br>30 | 210-212 |
| DIE BESCHREIBUNG DER LANDWIRTSCHAFTLICHEN WERTEIGENSCHAFTEN VON GETREIDESORTEN | HRON, R. | JAHRBUCH DER BUNDESANSTALT FUR PFLANZENBAU UND SAMENPRUFUNG IN WIEN | 1975 | 153-159 |
| TRITICUM AESTIVUM HYBR, STRAIN<br>ELECTROPHORESIS OF GLIADIN AS A METHOD FOR IDENTIFICATION OF WHEATS IN WHICH B-CHROMOSOME IS COMPLETELY OR PARTIALLY REPLACED BY RYE R-CHROMOSOME 1. | POPERELYA, F.A., SOZINOV, A.A. | SOVIET AGRICULTURE SCIENCES | 1977<br>2 | 1-4 |
| TRITICUM ARARATICUM<br>CULTIVAR IDENTIFICATION AND DOCUMENTATION OF WHEAT GENETIC RESOURCES BY GLIADIN ELECTROPHORETIC SPECTRA | KONAREV, V.G., GAVRILJUK, I.P., GUBAREVA, N.K. | VESTNIK SELSKOKHOZIAS TVENNOI NAUKI | 1976<br>9 | 113-125 |
| TRITICUM POETICUM<br>CULTIVAR IDENTIFICATION AND DOCUMENTATION OF WHEAT GENETIC RESOURCES BY GLIADIN ELECTROPHORETIC SPECTRA | KONAREV, V.G., GAVRILJUK, I.P., GUBAREVA, N.K. | VESTNIK SELSKOKHOZIAS TVENNOI NAUKI | 1976<br>9 | 113-125 |
| TRITICUM CULTIVARS<br>MODIFIED PHENOL TEST FOR THE VARIETAL IDENTIFICATION OF WHEAT SEED | BANERJEE, S.K., CHANDRA, S. | SEED SCIENCE AND TECHNOLOGY | 1977<br>5 | 53-60 |
| WHEAT CULTIVARS QUICKLY IDENTIFIED BY MODIFIED PHENOL TEST | WRIGLEY, C.W. | JOURNAL OF THE AUSTRALIAN INSTITUTE OF AGRICULTURAL SCIENCE | 1977<br>43 | 148 |
| TRITICUM DURUM CULTIVARS<br>THE IDENTIFICATION OF CULTIVATED PLANTS.II. THE GENUS TRITICUM L. | BADAWI, A.A., EL-GAZZAR, A. | ALLAM, M.A.<br>PHYTOLOGIA | 1978<br>38 | 267-279 |
| THE IDENTIFICATION OF CULTIVATED PLANTS.III. CONFIRMATORY KEYS TO SOME WHEAT VARIETIES. | BADAWI, A.A., EL-GAZZAR, A. | ALLAM, M.A.<br>PHYTOLOGIA | 1978<br>38 | 280-286 |
| TRITICUM PYRAMIDALE CULTIVARS<br>THE IDENTIFICATION OF CULTIVATED PLANTS.II. THE GENUS TRITICUM L. | BADAWI, A.A., EL-GAZZAR, A. | ALLAM, M.A.<br>PHYTOLOGIA | 1978<br>38 | 267-279 |
| THE IDENTIFICATION OF CULTIVATED PLANTS.III. CONFIRMATORY KEYS TO SOME WHEAT VARIETIES. | BADAWI, A.A., EL-GAZZAR, A. | ALLAM, M.A.<br>PHYTOLOGIA | 1978<br>38 | 280-286 |
| TRITICUM VULGARE CULTIVARS<br>THE IDENTIFICATION OF CULTIVATED PLANTS.II. THE GENUS TRITICUM L. | BADAWI, A.A., EL-GAZZAR, A. | ALLAM, M.A.<br>PHYTOLOGIA | 1978<br>38 | 267-279 |

Appendix 1 (continued)

```
1980-05-13                                                                                                    PAGE 8

TRITICUM VULGARE CULTIVARS      BADAWI, A.A., EL-GAZZAR, A., ALLAM, M.A.,                                     1978
  THE IDENTIFICATION OF CULTIVATED PLANTS.III. CONFIRMATORY  PHYTOLOGIA                                       38   280-286
  KEYS TO SOME WHEAT VARIETIES.

VACCINIUM CORYMBOSUM CULTIVARS  GOUGH, R.E., HINDLE, R.J., SHUTAK, V.G.,                                      1976
  IDENTIFICATION OF TEN HIGHBUSH BLUEBERRY CULTIVARS USING    HORTSCIENCE                                     11   512-514
  MORPHOLOGICAL CHARACTERICS

VITIS VINIFERA CULTIVARS        DRAWERT, F., GORG, A.,                                                        1974
  UBER DIE ELEKTROPHORETISCHE DIFFERENZIERUNG UND           ZEITSCHRIFT FUR LEBENSMITTELUNTERSUCHUNG UND FORSCHUNG   154  328-338
  KLASSIFIZIERUNG VON PROTEINEN III.DISK-ELEKTROPHORESE UND
  ISOELEKTRISCHE FOKUSSIERUNG IN POLYACRYLAMID-GELEN VON
  PROTEINEN UND ENZYMEN AUS TRAUBEN

ZEA MAYS CULTIVARS              MONTIES, B., CHESNEAUX, M.-T., MUTIN, C.,                                     1976
  CARACTERISATION DES CULTIVARS A L'AIDE DE DIFFERENCES     ANNALES DE L'AMELIORATION DES PLANTES             26   495-516
  QUANTITATIVES DE TENEUR EN COMPOSES POLYPHENOLIQUES:
  APPLICATION AU MAIS ET  A L'OEILLET
```

Appendix 1 (continued)

Appendix 2
Numerical taxonomic studies carried out on infraspecific variability of cultivated plants.

*Arachis hypogaea* L. cultivars

BROWN, D. F., C. M. CATER, K. F. MATTIL, and J. G. DARROCH, 1975: Effect of variety, growing location and their interaction on the fatty acid composition of peanuts. – Journal of Food Science **40**, 1055–1060.

*Armoracia rusticana* GAERTN., MEY., SCHERB. cultivars

RHODES, A. M., S. G. CARMER, and J. W. COURTER, 1969: Measurement and classification of genetic variability in horseradish. – Journal of the American Society of Horticultural Science **94**, 98–102.

*Avena sativa* L. cultivars

BAUM, B. R., and L. P. LEFKOVITCH, 1972a: A model for cultivar classification and identification with reference to oats (*Avena*). I. Establishment of the groupings by taximetric methods. – Canadian Journal of Botany **50**, 121–130.

–, and –, 1972b: A model for cultivar classification and identification with reference to oats (*Avena*). II. A probabilistic definition of cultivar groupings and their Bayesian identification. – Canadian Journal of Botany **50**, 131–138.

–, and –, 1973: A numerical taxonomic study of phylogenetic and phenetic relationships in some cultivated oats, using known predigrees. – Systematic Zoology **22**, 118–131.

–, and E. J. BRACH, 1975: Identification of oat cultivars by means of fluorescence spectography – A pilot study aimed at automatic identification of cultivars. – Canadian Journal of Botany **53**, 305–309.

–, and G. F. ESTABROOK, 1978: Application of compatibility analysis in numerical cladistics at the infraspecific level. – Canadian Journal of Botany **56**, 1130–1135.

*Brassica* L. cultivars

COLE, R. A., and K. PHELPS, 1979: Use of canonical variate analysis in the differentiation of swede cultivars by gas-liquid chromatography of volatile hydrolysis products. – Journal of Science of Food and Agriculture **30**, 669–676.

*Dioscorea alata* L. cultivars

MARTIN, F. W., and A. M. RHODES, 1977: Intraspecific classification of *Dioscorea alata*. – Tropical Agriculture **54**, 1–13.

*Hordeum vulgare* L. cultivars

MOLINA-CANO, J. L., 1976: A numerical classification of some European barley cultivars (*Hordeum vulgare* L. s. l.). – Zeitschrift für Pflanzenzüchtung **76**, 320–333.

–, and J. M. ELENA ROSSELLO, 1978: A further contribution to the classification of barley cultivars: Use of numerical taxonomy and biochemical methods. – Seed Science and Technology **6**, 593–615.

SNEATH, P. H. A., 1976: Some applications of numerical taxonomy to plant breeding. – Zeitschrift für Pflanzenzüchtung **76**, 19–46.

*Lolium* L. varieties

THOMSON, A. J., 1974: The examination of some winter hardiness components in grasses using canonical variate analysis. – Journal of Agricultural Science, Cambridge, **83**, 545–550.

*Medicago sativa* L. cultivars

YAMADA, T., and S. SUZUKI, 1974: Classification of alfalfa cultivars by clustering method based on quantitative characters and its significance in introduction and conservation of genetic resources. – Proceedings of the XII International Grassland Congress (1974), 374–379.

*Mimulus* L. hybrids

WELLS, H., 1980. A distance coefficient as a hybridization index: An example using *Mimulus longiflorus* and *M. flemingii* (Scrophulariaceae) from Santa Cruz Island, California. – Taxon **29**, 53–65.

*Oryza* L. cultivars

JANORIA, M. P., A. M. RHODES, and M. N. SHRIVASTAVA, 1976: Phenetic similarity among dwarf rice cultivars, grown in two environments, as estimated by numerical taxonomic methods. – Annals of Botany **40**, 845–849.

*Oryza* L. cultivars and species

MORISHIMA, H., 1969. Phenetic similarity and phylogenetic relationships among strains of *Oryza perennis*, estimated by methods of numerical taxonomy. – Evolution **23**, 227–247.

*Petunia*×*hybrida* VILM. cultivars

NATARELLA, N. J., and K. C. SINK, 1974: A chromatographic study of phenolics of species ancestral to *Petunia hybrida*. – The Journal of Heredity **65**, 85–90.

*Phaseolus vulgaris* L. cultivars

ADAMS, M. W., 1977: An estimation of homogeneity in crop plants, with special reference to genetic vulnerability in the dry bean, *Phaseolus vulgaris* L. – Euphytica **26**, 665–679.

*Pinus contorta* DOUGL. ex LOUD.

JEFFERS, J. N. R., and T. M. BLACK, 1963: An analysis of variability in *Pinus contorta*. – Forestry **36**, 199–218

*Poa ampla* MERR.

WILLIAMSON, C. J., and R. J. KILLICK, 1978: Multivariate methods as an aid in identifying *Poa ampla* × *P. pratensis* hybrids from maternal-type offspring. – Heredity **41**, 215–225.

*Portulaca oleracea* L.

GORSKE, S. F., A. M. RHODES, and J. H. HOPEN, 1979: A numerical taxonomic study on *Portulaca oleracea*. – Weed Science **27**, 96–102.

*Quercus* L. hybrids

RUSHTON, B. S., 1978: *Quercus robur* L. and *Quercus petraea* (MATT.) LIEBL.: A multivariate approach to the hybrid problem, 1. Data acquisition, analysis and interpretation. – Watsonia **12**, 81–101.

–, 1979: *Quercus robus* L. and *Quercus petraea* (MATT.) LIEBL.: A multivariate approach to the hybrid problem, 2. The geographical distribution of population types. – Watsonia **12**, 209–224.

*Rosa* L. cultivars

KUHNS, L. J., and T. A. FRETZ, 1978: Distinguishing rose cultivars by polyacrilamide gel electrophoresis. II. Isoenzyme variation among cultivars. – Journal of the American Society for Horticultural Science **103**, 509–516.

*Saxifraga* L. hybrids

JAWORSKA, H., and N. NYBOM, 1967: A thin-layer chromatographic study of *Saxifraga caesia, S. aizoides*, and their putative hybrids. – Hereditas **57**, 159–177.

*Solanum melongena* L. cultivars

MARTIN, F. W., and A. M. RHODES, 1979: Subspecific grouping of eggplant cultivars. – Euphytica **28**, 376–383.

## *Solanum tuberosum* L. cultivars

Brown, E., and J. P. Moss, 1976: The identification of potato varieties from tuber characters. – Journal of the National Institute of Agricultural Botany **14**, 49–69

Raeuber, A., H. Tiemann, and I. Griess, 1978: Die Anwendung der Clusteranalyse in der Pflanzenzüchtung, dargestellt an dihaploiden Genotypen von *Solanum tuberosum*. – Archiv für Züchtungsforschung **8**, 181–192

## *Sorghum* L. cultivars

Shechter, Y., 1975: Biochemical-systematic studies in *Sorghum bicolor*. – Bulletin of the Torrey Botanical Club **102**, 334–339.

## *Triticum* L. cultivars

Sneath, P. H. A., 1976: Some applications of numerical taxonomy to plant breeding. – Zeitschrift für Pflanzenzüchtung **76**, 19–46.

## *Tsuga canadensis* (L.) Carrière morphs

Kessell, S. R., 1979: Adaptation and dimorphism in eastern Hemlock, *Tsuga canadensis* (L.) Carr. – The American Naturalist **113**, 333–350.

## *Ulmus minor* Mill. em. Richens× *U. glabra* Huds.

Jeffers, J. N. R., and R. H. Richens, 1970: Multivariate analysis of the English Elm population. – Silvae Genetica **19**, 31–38.

## *Vigna* Savi cultivars

Mascherpa, J.-M., 1975: Taxonomie numérique: Chimie taxonomique et ordinateur au service de la systématique. – Saussurea **6**, 171–185.

## *Zea mays* L. cultivars

Camussi, A., 1979: Numerical taxonomy of Italian populations of maize based on quantitative traits. – Maydica **24**, 161–174.

Goodman, M. M., 1967: The races of maize: I. The use of Mahalanobis' generalized distances to measure morphological similarity. – Fitotecnia Latinoamericana **4**, 1–22.

–, 1968: The races of maize: II. Use of multivariate analysis of variance to measure morphological similarity. – Crop Science **8**, 693–698.

–, and R. McK. Bird, 1977: The races of maize: IV. Tentative grouping of 219 Latin American races. – Economic Botany **31**, 204–221.

McK. Bird, R., and M. M. Goodman, 1977: The races of maize V: Grouping maize races on the basis of ear morphology. – Economic Botany **31**, 471–481.

Rhodes, A. M., and S. G. Carmer, 1966: Classification of sweet corn inbreds by methods of numerical taxonomy. – Proceedings of the American Society for Horticultural Science **88**, 507–515.

Tarcicio, C. S., M. M. Goodman, C. D. Eduardo, and J. O. Rawlings, 1978: Use of genetic effects and genotype by environmental interactions for the classification of Mexican races of maize. – Genetics **90**, 339–348.

Kulturpflanze XXIX · 1981 · S. 241–250

# Numerical taxonomic studies in the genera *Triticum* L. and *Pisum* L.

JÜRGEN SCHULTZE-MOTEL and DORIS MEYER


## Summary

Up till now numerical taxonomy has been applied to cultivated plants only to a limited extent though this method seems to be especially suitable for handling the great diversity of these plants. Moreover the theory of numerical taxonomy has not been studied sufficiently.

Therefore studies with cultivated plants were begun with the aim to investigate the fitness of different numerical taxonomic techniques for their classification. Materials used: 1) 24 species of *Triticum* and additionally 24 species of *Aegilops* 2) 1225 *Triticum* taxa from Iran (mostly *T. aestivum*) 3) 312 taxa of *Pisum* (mostly *P. sativum*).

Using 10 coefficients and 7 clustering methods in *Triticum* and *Aegilops* in most cases a clear separation of both genera resulted as well as a species arrangement within *Triticum* according to the genomes.

The main aim when dealing with the wheat from Iran was the investigation of the influence of different character sets. At this the coefficient of GOWER and the unweighted pair-group method using arithmetic averages (UPGMA) were applied. The separation of the different species is satisfying clear while the representatives of the different varieties of *T. aestivum* could not always be found in one cluster. This effect increased when adding further characters sets.

In *Pisum* the influence of 11 similarity coefficients to the dendrogram formation has been investigated. With two-state characters the simple matching coefficient and with mainly multistate characters the Canberra metric gave best results (using UPGMA and information measures SUMRAT and SAMRAT).

The hitherto existing results show the complicatedness of the evaluation of numerical taxonomic methods in their application to cultivated plants. Therefore generalizations are not yet possible.


It is a well known fact that cultivated plants are much more difficult to classify than wild plants. The cause for this phenomenon is the manipulation by man as already expressed in the name. Man has adapted his needs to some degree to the plants, but has altered to a much wider degree the plants according to his needs and is continuing to alter them nowadays by selection, hybridisation and mutation. As consequence of this human influence the variability of cultivated plants is very great and the taxa of cultivated plants are often very similar. In many cases the taxa can be discerned by one character only. This high degree of similarity between taxa of cultivated plants makes their classification very difficult.

Furthermore the choice of characters for the botanical classification of a group of cultivated plants may be difficult and biased by subjectivity.

In the late fifties and early sixties a new method was created which was enthusiastically believed to solve the problems in taxonomy. This method, the numerical taxonomy, is handling many characters in form of numbers by using a computer. Furthermore this method is said to be characterized by objectivity (SNEATH and SOKAL 1973). It seems logical to expect this method especially apt for managing the problems of classifying cultivated plants. In a numerical taxonomic study one can use besides the traditional morphological and anatomical characters also physiological, agronomical and other ones. The investigator yet has to select the characters critically and at the end of the computing to evaluate the grouping, which only by him becomes a classification of the plant group studied. Despite of the obvious advantages described above numerical taxonomy has been applied to cultivated plants up till now only to a limited extent though this method seems to be especially suitable for handling the great diversity of these plants.

This limited application of numerical taxonomy to cultivated plants is, we feel, a shortcoming. But there is another shortcoming. The methodology of numerical taxonomy is in our opinion not so perfect, experienced and fully elaborated as to use it without any restrictions. Any material cannot be handled by any mathematical procedure. By using numerical taxonomic methods new problems arise for example when coding the characters, when electing a similarity coefficient for quantifying the similarity between the taxonomic units as well as when electing a clustering method.

Scientists of several disciplines dealt with procedures for the classification of different materials and for the computing of similarities. Thus a great number of similarity coefficients and of clustering methods were created. This variability is very entangling and often it is not clear if a mathematical procedure, being exact in itself, shows a biological relevance. In our opinion the theory of numerical taxonomy has not been studied sufficiently. Therefore studies with cultivated plants were begun with the aim to investigate the fitness of different numerical taxonomic techniques for their classification. The materials used are the following:

1. 24 species of *Triticum* and additionally 24 species of *Aegilops*
2. 1225 *Triticum* taxa from Iran (mostly *Triticum aestivum*)
3. 312 taxa of *Pisum* (mostly *Pisum sativum*).

Let us begin with the *Triticum*/*Aegilops* material. This investigation goes back to a suggestion made by Prof. HAWKES during his last visit to Gatersleben. The material comprises almost all known species of *Triticum* and of *Aegilops*. The genus *Triticum* is represented by 27 operational taxonomic units (OTU's) comprising 24 species. Only the species *T. sinskajae*, *T. petropavlovskyi* and *T. militinae* are lacking. The genus *Aegilops* is represented by 25 OTU's comprising 24 species. Here the newly described species *Ae. searsii* is also included. The number of characters used in our investigation is 72. From these 58 are qualitative characters, the other 14 are quantitative characters.

The characters are as follows: characters which are used traditionally in determination, that are characters of the spike, besides this characters of the leaves, of the culm, characters of the genome, number of chromosomes, anatomical characters of kernel, culm and root, some chemical characters as lysin and protein content and so on.

For computing it was necessary to transform in some cases the quantitative in qualitative characters.

The mathematical methods used were the following: coefficients of similarity: JACCARD (reading A), JACCARD (reading D), simple matching coefficient, coefficient of ROGERS and TANIMOTO, phi coefficient, mean square distance, coefficient of CLARK, Canberra metric, mean character difference and GOWER's similarity coefficient. The clustering methods used were the following: nearst neighbour, furthest neighbour, centroid method, median method, WARD's method, LANCE and WILLIAMS method, and unweighted pair-group method using arithmetic averages (UPGMA).

Between the similarity coefficients on the one side and the clustering methods on the other side all combinations have been computed. Thus a total of 70 dendrograms or phenograms were constructed based on the same data matrix.

Looking at these 70 dendrograms one can notice that in almost all cases a clear separation between *Triticum* and *Aegilops* is evident (Fig. 1). There have been some discussions in the past if it is feasible, favourable or necessary to combine the species of *Aegilops* and *Triticum* in one genus. If one takes into consideration the characters used traditionally then the species must be put into two genera. When using many characters–as in our case–the result is similar. Only when considering the character of crossability between the species one could be forced to unite the species in one genus. But we feel it not feasible to overestimate this one character which–if used consequently–would force us to include also species of other genera into the genus *Triticum* (e.g. *Secale* and *Agropyron*). In some cases there is an exception concerning the species *Aegilops mutica* which is included either in the dendrogram part of *Triticum* or is standing apart from *Triticum* and *Aegilops*. This would support the opinion that *Aegilops mutica* is to be considered as a member of a separate genus *Amblyopyrum*.

Within the *Aegilops* species the section *Sitopsis* is the group of which the species in most cases can be found together in one dendogram part. The tendency that other sections or genome types are reflected in the structure of the dendograms can be seen only to a limited extent.

In *Triticum* there is a tendency that in the groups represented in the dendrograms the ploidy level and the genomes are reflected to a varying degree.

In summing up the results of this investigation we can state that the methods used are able to separate the species of the genera *Triticum* and *Aegilops* sufficiently. Within the genera *Triticum* and *Aegilops* the separation of the several subgroups, that is the ploidy level in *Triticum* or the sections in *Aegilops* is realized to a varying degree of exactness.

Generally spoken, the best results are reached if using the mean character difference and the GOWER coefficient. The clustering method combined with these coefficients seems to be without any influence.

In the second pait of our investigations we dealt with *Triticum* too but at the infraspecific level. The main aim was to investigate the influence of different character sets. At this the coefficient of GOWER and the unweighted pair-group method were applied. The material is a collection of wheats from Iran which we

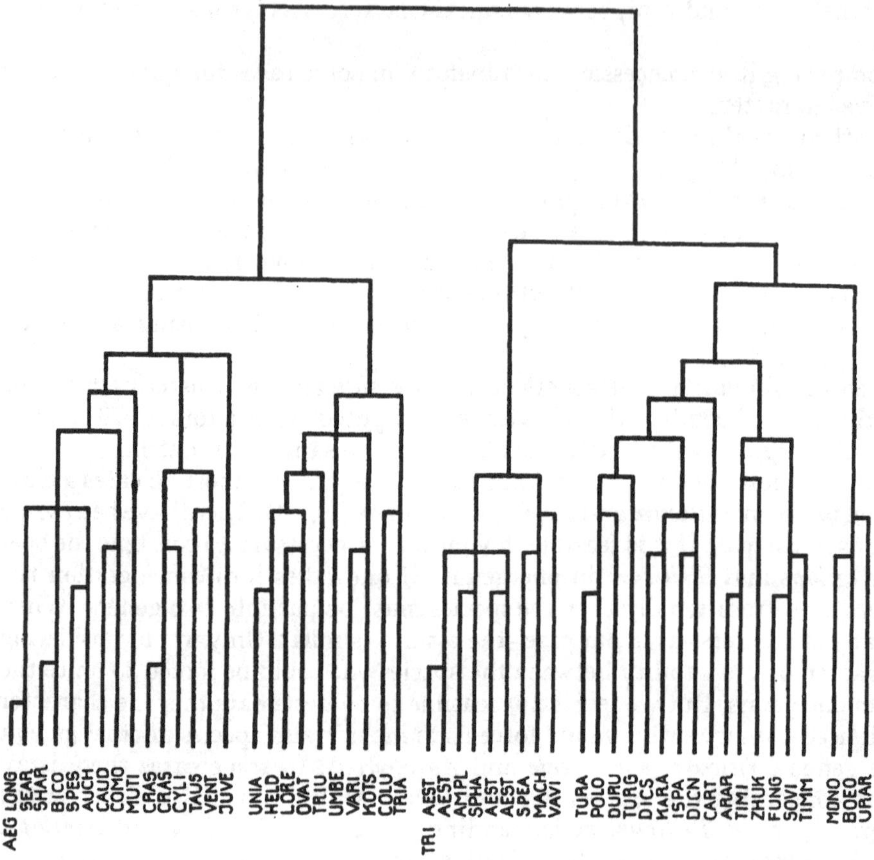


Fig. 1 Dendrogram of 24 *Triticum* species and 24 *Aegilops* species using GOWER's coefficient and UPGMA clustering

handled by means of methods of numerical taxonomy in order to get an insight into the diversity of wheat and to find ways to arrange the parts of the diversity.

The Iran is a country with a very long history and also a territory of great diversity of wheat. The material consists of samples which come from 500 different localities of the Iran. It was collected by Prof. KUCKUCK whilst acting as an FAO expert on cereal breeding in Iran. In the fields there is often a mixture of several species and the material of one species is likewise inhomogenous. Therefore at first some separations must be made resulting in 1212 different forms which are looked at as Operational Taxonomic Units. The material was cultivated in small beds of 2 square meters and during several years many evaluations and measurements were made. In the systematics of cultivated plants mostly such characters are chosen for building systems which are accessible and come from those parts of the plant which are used as food in cereals consequently the ears. In general only a small number of characters is treated.

In contrary to this we tried in our investigations to fulfil the basic demand of systematics according which as many characters as possible should be taken into consideration. Therefore we investigated more than 100 characters. These are partly qualitative, partly quantitative characters; besides the normally used morphological characters of the ear also characters of the vegetative parts were used for example length and breadth of the flag leaf and of the second leaf, leaf area, attitude of the leaf, characters which may be important for breeding purposes. Besides these characters some phenological characters like time of earing, flowering, yellow ripeness were used.

Other characters used: anatomical characters of the leaf, resistance against mildew, yellow rust, stem rust and such characters which are important in judging the yield potentialities that is: weight of ears, weight of grains, number of grains per ear, 1000 kernel weight, protein content, lysin content. As a whole besides many morphological characters also anatomical, chemical and agricultural characters are used. The basic data material consists of more than half a million data and measurements and is thus a good basis for the investigations. It is selfevident that such a vast data material can only be handled by means of electronic data processing. Using electronic data processing is here clearly no modernism but the only way to tackle such problems.

Before discussing the own material and the computings we must shortly speak about the methods by which it was tried up to now to handle the great diversity of the common wheat.

Up till now mostly artificial systems were created which are exclusively based on morphological characters. In 1951 MANSFELD made a survey of the then existing botanical varieties of the polymorphic common wheat which comprises 404 forms. These were discriminated by the colour of the ears and of the grains, the hairiness of the glumes, the length of awns and some other characters of the ears.

Let us now discuss the Iranian wheat material. The composition of this material is as follows: *Triticum aestivum* (1088 OTU's), *Tr. spelta* (2), *Tr. durum* (106), *Tr. turgidum* (4), *Tr. turanicum* (3), *Tr. ispahanicum* (2), *Tr. orientale* (1), *Tr. polonicum* (1) and *Tr. dicoccon* (5).

In order to get an insight into the manner in which a dendrogram reflects the basic data and thus the diversity of a given material at first several small computings were made operating with 150 OTU's only.

The OTU's were chosen in the following way:

From common wheat 10 members from 8 of the most important botanical varieties were chosen which simultaneously represent the 4 basic groups of the MANSFELD system: 1) ears awned and glumes hairy 2) ears awned and glumes glabrous 3) ears without awns and glumes with hairs 4) ears without awns and glumes glabrous. Besides this some single members of 10 other botanical varieties were chosen. The species *Triticum durum* is represented with 30 numbers and also the resting species with several members each. For comparison also 13 high yielding varieties were used.

The OTU's were computed in 3 variants:

1) Those characters were computed which were used in the system of MANSFELD, that are characters of the ear. The resulting arrangement of the varieties fits indeed well with the MANSFELD system.
2) In this case besides the characters used in case 1 further characters of the ear were used which are partly marking the yield. These are ear weight, grain

weight, number of grains per ear and so on. In this case there is also a good conformity with the botanical species and varieties.

3) In this variant 80 qualitative and quantitative characters from all parts of the plants were used. In this case there is no or only little conformity with the botanical varieties surely because of the many quantitative characters used.

Summarizing this we can see that all three variants are different from one another, but this is not astonishing because they were based on different character complexes.

In another computing 740 OTU's and 80 characters were used. The resulting 7 groups show no conformity with the then existing system. This could not be expected because again many more characters were used that normally.

Generally it is to be seen that with the methods and the characters now used such groups result which are not in conformity with the botanical varieties and which also do not harmonize with the groups in the system of Mansfeld and others. The resulting groups are above the level of the botanical varieties. For practical use it would be advantageous if the groups created were not too numerous as more as the traditional systematist working with cultivated plants challenged by their great diversity mostly tends to overclassify. These and other classifications which are not in conformity with the botanical varieties would cause nomenclatural consequences if one does not prefer to abandon the naming of groups at the infraspecific level at all.

In the third part of our investigations we dealt with *Pisum* (Meyer 1980). The aim of this investigation was to examine a number of similarity coefficients to test their taxonomical usefulness. The genus *Pisum* had been chosen because the respective material in the Gatersleben collection is very comprehensive. Furthermore a great experience exists in evaluating the relevant diagnostic characters. Moreover the *Pisum* strains are genetically homogenous lines because of their strict autogamy. Last not least the existence of a modern classification of the genus supported the choice of this material.

The collection used comprised 312 taxa and consisted of the following groups. As the *Pisum* material of the Gatersleben assortment mainly belongs to *Pisum sativum* ssp. *sativum*, above all taxa of this taxonomical group were investigated namely 307. 18 strains belonged to convar. *axiphium*, 91 to convar. *medullare*, 2 strains to convar. *medullosaccharatum*, 139 to convar. *sativum* and 57 to convar. *speciosum*. Of the remaining taxa 2 belonged to ssp. *transcaucasicum* and 3 were members of the species *abyssinicum*.

Within this material 71 characters were scored. There were 43 qualitative and 28 quantitative characters which consisted of morphological and phenological ones. Because not all characters could be calculated with all chosen similarity coefficients we had to form different character sets. Besides this the characters were used in their original form as well as in a standardized form.

The similarity coefficients are those which were already used in the *Triticum*/*Aegilops* investigation. The first 6 coefficients–namely the coefficient of Jaccard (reading A and reading D), the simple matching coefficient, the coefficient of Rogers and Tanimoto, the phi coefficient and the coefficient of Gower–belong of the group of association measures. The measure the number of matching characters of two taxonomic units compared with the number of all characters. The four next coefficients–namely the mean character difference, the mean square distance, the Canberra metric and the coefficient of divergence of CLARK–are measures which assess differences between character values. They are called distance measures. The last coefficient is from the group of correlation coefficients, it is the product-moment correlation coefficient of PEARSON.

These similarity coefficients were used with the different character sets in the following way.

To the first character set comprising all 71 characters the coefficient of GOWER, the distance measures and the correlation coefficient were applied. With the exception of the coefficient of GOWER these coefficients were also tested on the base of the second character set, which were the 28 quantitative characters. The qualitative characters were used on the one hand as a whole, on the other hand as a set of 21 characters. The 21 characters are those which dominate in the determination key of LEHMANN (1954). Furthermore we discerned between a multistate and a two-state character coding. While the multistate qualitative characters were mainly used for the evaluation of the distance measures and of the correlation coefficients, the two-state qualitative characters were especially applied to test the association measures.

Altogether the application of 11 similarity coefficients to the several character sets resulted in 50 different dendrograms. In all these cases the unweighted pair-group method using arithmetic averages was used.

The interpretation of the dendrograms was practised in two steps. In the first step the dendrograms were evaluated according to their outer appearance. A dendrogram was accepted if it contained distinct groups.

In the second step the accepted dendrograms were investigated according to their internal structure. As the dendrograms were very similar in the constitution of their groups the so-called best dendrogram could not found out without using mathematical methods. In connection with the numerical taxonomic investigation and evaluation of the *Avena* species by BAUM (1974) we used the two related information-theoretic measures SUMRAT and SAMRAT. They calculate the information values of the dendrograms on the base of the groups contained in the dendrograms. The name SUMRAT is derived from "sum of ratios". SAMRAT is a measure related to SUMRAT. The dendrogram which in our investigation had the highest SUMRAT and SAMRAT values was considered as the "best" dendrogram.

According to this interpretation 32 dendrograms were rejected and only 18 dendrograms were considered as taxonomically acceptable.

Among the rejected dendrograms were those which were based on all 71 characters as well as those which were based on the quantitative characters alone. Only the multistate and two-state qualitative characters resulted in taxonomically acceptable dendrograms.

Consequently the choice of characters has a greater influence to the dendrogram formation than the similarity coefficient. The same is true for the character coding. Among the 18 taxonomically acceptable dendrograms there were all types of similarity measures with the exception of GOWER's coefficient. The evaluation of these 18 dendrograms by means of the information-theoretic measures SUMRAT and SAMRAT gave the following results.

As the formation of the dendrograms is heavily influenced by the characters, the evaluation of the dendrograms was performed on the one hand with regard to all dendrograms and on the other hand with regard to the character sets.

In respect to the 21 multistate qualitative characters the dendrogram of the Canberra metric had the highest information content. This dendrogram was not only the best one within this character set but the best at all. Furthermore the product-moment correlation coefficient and the mean square distance gave good results. Concerning the 43 and 21 two-state qualitative characters respectively the coefficient of JACCARD (reading D) and the simple matching coefficient produced dendrograms with highest information values. The Canberra metric gave moderate information values.

The figuring of multidimensional units in two-dimensional spaces–as it is done by formation of dendrograms–always results in more or less biased relations of these units. Therefore these dendrograms were supplemented by a principal component analysis. The principal component analysis was performed by using simultaneously the 18 taxonomically acceptable dendrograms. As a result of applying this method the taxonomic units were separated in 2 distinctly main groups.

One group consisted of convar. ***sativum*** and convar. ***medullare*** as well as of convar. ***medullosaccharatum*** and the white-flowered strains of convar. ***medullosaccharatum*** and the white-flowered strains of convar. ***axiphium***. The other group comprised the convar. ***speciosum***, the two representatives of ssp. ***transcaucasicum***, the reddish-flowered strains of convar. ***axiphium***, and the species ***abyssinicum***. While the taxa of the first group all have white flowers, all of the members of the second group have reddish ones. Within both groups there were subgroups based on the formation of parchment-layers.

A further diagnostic character within the first group was the seed shape and within the second group the marking of the seed coat.

The sequence of the diagnostic characters–colour of flower, parchment-layer, seed shape, and marking of the seed coat respectively–was not fixed by subjective statements but by means of mathematical objective consideration of many morphological and physiological characters. This should be regarded when forming a new classification of the genus *Pisum*.

Looking at the material which we investigated one can state the following. In the cases of *Triticum* from Iran and of *Pisum*, which comprise taxa mostly at the infraspecific level, the dendrograms became the more unclear the more quantitative characters were added. Theoretically then it can be expected that–if using some more characters–no acceptable dendrogram or system will result at all despite of the fulfilment of a basic demand of numerical taxonomy namely the use of many characters. Evidently the use of as many characters as possible is neither theoretically nor practically feasible. Furthermore the characters have no equal weight in themselves. Using mathematical–that means objective–methods neither automatically produces dendrograms nor are the resulting dendrograms in every case better than a hand-made system. Up till now there are no cooking recipes for all materials and all opportunities. Much work has to be done before generalizations can be made.

*Acknowledgements*

To the mathematicians Dr. E. TABBERT and M. CERULLA (VEB Datenverarbeitungszentrum Schwerin) we are very grateful for their programming and computing work.

## Zusammenfassung

Numerisch-taxonomische Studien in den Gattungen *Triticum* L. und *Pisum* L.

Die numerische Taxonomie ist bisher bei Kulturpflanzen nur wenig angewendet worden, obwohl sie besonders geeignet erscheint, die große Vielfalt dieser Pflanzen zu handhaben. Außerdem ist die Theorie der numerischen Taxonomie noch nicht genügend studiert worden.

Deshalb wurden Studien an Kulturpflanzen begonnen mit dem Ziel, die Eignung verschiedener numerisch-taxonomischer Techniken für ihre Klassifikation zu untersuchen. Benutzte Materialien: 1) 24 Arten von *Triticum* und zusätzlich 24 Arten von *Aegilops*, 2) 1225 *Triticum*-Taxa aus dem Iran (meist *T. aestivum*), 3) 312 *Pisum*-Taxa (meist *P. sativum*).

Unter Benutzung von 10 Koeffizienten und 7 Cluster-Methoden ergab sich bei *Triticum* und *Aegilops* in den meisten Fällen ein klare Trennung beider Gattungen sowie innerhalb von *Triticum* eine Anordnung der Arten nach den Genomen.

Hauptziel beim Studium des Iran-Weizens war die Untersuchung des Einflusses verschiedener Merkmals-Sätze. Hierbei wurden der Koeffizient von GOWER und die unbewichtete Paar-Gruppen-Methode mit arithmetischen Mittelwerten (UPGMA) angewendet. Die Trennung der verschiedenen Arten war hinreichend klar, während die Vertreter der verschiedenen Varietäten von *T. aestivum* nicht immer in einer Gruppe zu finden waren. Diese Wirkung verstärkte sich bei Hinzufügen weiterer Merkmalssätze.

Bei *Pisum* wurde der Einfluß von 11 Ähnlichkeitskoeffizienten auf die Dendrogrammbildung untersucht. Mit zweistufigen Merkmalen gab der simple matching-Koeffizient und mit hauptsächlich vielstufigen Merkmalen die Canberra-Metrik die besten Ergebnisse (unter Benutzung der UPGMA und der Informationsmaße SUMRAT und SAMRAT).

Die bisherigen Ergebnisse zeigen die Kompliziertheit der Einschätzung numerisch-taxonomischer Methoden bei ihrer Anwendung auf Kulturpflanzen. Deshalb sind Verallgemeinerungen noch nicht möglich.

### Краткое содержание

Нумерически-таксономические исследования внутри родов *Triticum* L. и *Pisum* L.

До сих пор нумерическая таксономия редко применялась для культурных растений, хотя она и кажется особенно подходящей для овладения многообразием этих растений. Теория нумерической таксономии также ещё мало изучена.

Это дало повод начать исследования на культурных растениях, с целью, выяснить пригодность различных приёмов нумерической таксономии для их классификации. Использовался следующий материал: 1) 24 вида *Triticum* и, добавочно, 24 вида *Aegilops*; 2) 1225 образцов пшениц из Ирана, главным образом *T. aestivum*; 3) 312 образцов гороха, гл. обр. *P. sativum*.

Применяя 10 коэффициентов и 7 различных методов группировки можно было, в большинстве случаев, чётко разграничить роды *Triticum* и *Aegilops*, а внутри рода *Triticum* – распределить виды по их геномам.

Главной целью обработки иранских пшениц было исследование влияния различных наборов признаков. Здесь применялись коэффициент Гоуера и метод невзвешенных парных группировок с арифметическим средним (UPGMA). Разграничение отдельных видов было достаточно чётким, но представители различных разновидностей мягкой пшеницы не всегда оказывались в одной и той же группе. Это обстоятельство было ещё более заметным, когда число наборов признаков увеличивалось.

У гороха изучалось влияние 11 коэффициентов сходства на образование дендрограмм. Для двухступенчатых признаков лучшие результаты дал коэффициент "simple matching", а для, главным образом, многоступенчатых – "Canberra metric" (применялись UPGMA а как меры информации – SUMRAT и SAMRAT).

Результаты работы, полученные до сих пор, показывают сложность оценки методов нумерической таксономии, когда они применяются для культурных растений. Таким образом, делать обобщения – ещё рано.

## Literature

Baum, B. R., 1974: Classification of the oat species (*Avena*, Poaceae) using various taximetric methods and an information-theoretic model. – Can. J. Bot. **52**, 2241–2262.

Lehmann, C. O., 1954: Das morphologische System der Saaterbsen (*Pisum sativum* L. sens. lat. Gov. ssp. *sativum*). – Züchter **24**, 316–337.

Mansfeld, R., 1951: Das morphologische System des Saatweizens, *Triticum aestivum* L. s. l. – Züchter **21**, 41–60.

Meyer, D., 1980: Numerisch-taxonomische Untersuchungen an *Pisum sativum* L. – Kulturpflanze **28**, 285–340.

Sneath, P. H. A., and R. R. Sokal, 1973: Numerical Taxonomy. The principles and practice of numerical classification. – San Francisco, XV + 573 pp.

Dr. J. Schultze-Motel
Zentralinstitut für Genetik und Kulturpflanzenforschung
DDR 4325 Gatersleben
Corrensstraße 3

Kulturpflanze XXIX · 1981 · S. 251–266

# Character analysis and numerical classification in a wheat collection from Sicily

ENRICO PORCEDDU, STEFANO VANNELLA and PIETRO PERRINO (Bari, Italy)

## Summary

A study was undertaken using 48 characters observed in plants, spikes and seeds in order to determine the amount of information of different characters and their taxonomic importance in tetraploid wheat samples from Sicily. A classification of the material was also attempted using similarity indices.

The most important characters appear: leaf pubescence, glume colour and length, and seed colour. Using all characters, similarity indices among samples were computed and a classification was worked out. This classification, which is very similar to those proposed by various scientists, produces hierarchical groups, showing different levels of merging, and offers very interesting hypotheses on the evolution of the Sicilian *durum* wheats.

## Introduction

The need to classify the available material in order to get a satisfactory analysis and presentation has always been apparent to all those who have to study broad collections of individuals or objects.

During these past years, work on preserving plant genetic resources has resulted in the creation of large germplasm collections. On the one hand, these must be classified and catalogued and, on the other hand, they must provide an opportunity to study existing relations between different taxonomic groups within one species, as well as the variations of a given character at varying distances from the centres of origin, etc.

Both these aspect, however, require the most appropriate characters to be selected, namely those which contain the largest possible amount of information. Also, methods of classification and comparison have to be selected with the utmost care so that groups, each containing as much similarity as possible are formed.

Regarding the first aspect, i.e. the choice of the most appropriate characters, the following items described by ROGERS and APPAN (1969) should be considered:

1) What is the taxonomically significant information content of each character?
2) What is the amount of correlation between characters, as present in the ways of the same genetic cause?
3) Is there any redundancy as a consequence of the description in different ways of the same genetic cause?
4) Should the character states be redefined?

5) Which characters are of diagnostic value?
6) Which characters can be eliminated because they are of marginal interest?

Regarding the second aspect, the problem is one of choosing the proper criteria for classification. Should the selected method use just one character at a time, or should it use several characters simultaneously. Whereas the states of each character will devide the plants into non-overlapping groups, classification utilizing one character at a time produces groups of individuals showing heterogeneous relationships, with the exception of a single character. The overall similarity principle does not only produce a better representation of total gene content in a given material, but also reduces to a bare minimum the fact that a character may not reflect the natural affinity of individuals in an appropriate manner.

It is the objective of this paper to determine the importance of characters observed in plants, spikes and caryopses from a collection of *durum* wheats from Sicily and to produce a classification emphasizing existing relationships both between and within the groups into which the collection can be divided.

## A brief history of Sicilian wheats

During their westward migration, many Mediterranean wheats have–for some reason or other–landed in Sicily. Naturally, not all of them have been brought under cultivation over large areas, but many have certainly stayed there or have undergone spontaneous crossings. Therefore, the patrimony of characters in local wheat populations used to be a very rich one.

Populations were made up essentially of tetraploid types and occasionally of hexaploids. The former were represented by a vast array of types which used to be cultivated under the most different environmental conditions, whereas the hexaploids were restricted to mountain areas and mainly used for making cakes. Bread would be made exclusively of *durum* wheat.

Unfortunately, too little is known about old land races from Sicily. Due to lack of exhaustive material, the authors who have previously discussed the subject gave some rather scanty or contradictory information. Among the first to deal with this subject were PATANÈ and STRINGHER (1884) who described 4 cultivars ('Realforte', 'Gigante', 'Giustalisa' for *durum* wheats and 'Maiorca' for bread wheats) which were practically distributed all over Sicily, 7 cultivars ('Sammartinara', 'Castigghiuni', 'Ciciredda', 'Scavuzza' for *durums* and 'Piluseddu', 'Cuccitta' and 'Tumminia' for bread wheats) growing in the territories of Palermo, Girgenti and 'Trapani, and 5 cultivars ('Farro' and 'Trentino' for *durums* and 'Urria', 'Cicirella' and 'Giustalisa' for bread wheats) that were common in the provinces of Catania and Siracusa. Unfortunately, no other information is provided by the two authors. On the contrary, it would seem that they were not seriously interested in names, possible synonyms and systematic presentation of the material since certain cultivars, such as 'Tumminia', are listed among bread wheats. Nor do they mention 'Biancuccia' which LORENZONI (1910) describes as one of the commonest wheats at the turn of the century along with 'Realforte', 'Gigante' or 'Sammartinara' and 'Timilia'.

The first true description, however, was made by PRESTIANNI (1926) who listed 52 cultivars and their synonyms and briefly described the characteristics of about half of them.

The following year, E. DE CILLIS (1927) made a maiden attempt to describe the material organically and reported 23 cultivars belonging to three species, *T. vul-*

Table 1
Classification of Sicilian wheats proposed by *de Cillis* (1942)

| Species | Botanical variety | Cultivar | Code No. |
|---|---|---|---|
| *T. durum* DESF. | *leucurum* KOERN. | Cannizzara | 1 |
| | | Francesa | 2 |
| | | Francesone | 3 |
| | | Gigante | 4 |
| | | Regina | 5 |
| | | Sammartinara | 6 |
| | | Tripolino | 7 |
| | *affine* KOERN. | Timilia a reste bianche | 8 |
| | *leucomelan* KOERN. | Biancuccia | 9 |
| | | Bidi | 10 |
| | | Chiattulidda | 11 |
| | | Girgentana | 12 |
| | | Inglesa | 13 |
| | | Trentino | 14 |
| | *reichenbachii* KOERN. | Scavuzza | 15 |
| | | Timilia a reste nere | 16 |
| | | Tunisina | 17 |
| | | Urria | 18 |
| | *melanopus* KOERN. | Farro lungo | 19 |
| | | Giustalisa | 20 |
| | | Martinella | 21 |
| | | Realforte | 22 |
| | | Semenzella | 23 |
| | *hordeiforme* KOERN. | Russello | 24 |
| | *erythromelan* KOERN. | Castiglione glabro | |
| | | Cotrone | 25 |
| | | Gioia | 26 |
| | | Lina | 27 |
| | | Pavone | 28 |
| | | Ruscia | 29 |
| | | Vallelunga glabra | |
| | *apulicum* KOERN. | Vallelunga pubescente | 30 |
| | *niloticum* KOERN. | Castiglione pubescente | 31 |
| | *coerulescens* KOERN. | Scorzonera | 32 |
| *T. turgidum* L. | *siculum* KOERN. | Ciciredda | 33 |
| | *speciosissimum* KOERN. | Bivona | 34 |
| | | Bufala rossa corta | 35 |
| | | Paola | 36 |
| | *mertensii* KOERN. | Bufala rossa lunga | 37 |
| | *iodurum* KOERN. | „ nera corta | 38 |
| | | „ „ lunga | 39 |
| *T. vulgare* HOST | *erythrospermum* KOERN. | Maiorcone | |
| | *albidum* KOERN. | Maiorca | |
| | *lutescens* KOERN. | Romano | |
| *T. compactum* HOST | *humboldtii* KOERN. | Cuccitta | |

*gare* ('Maiorca', 'Cuccitta', 'Maiorcone'), *T. turgidum* ('Bufala') and *T. durum* ('Farro Lungo', 'Farro', 'Biancuccia', 'Sambucara', 'Realforte', 'Sammartinara', 'Russello', Martinella', 'Scorzonera', 'Bidi', 'Vincitutti', 'Francesa', 'Chiattulidda', 'Tripolino', 'Triminia', 'Farro Corto', 'Ruscia', 'Gigante', 'Trentino') and grouped them according to spike characteristics, cold requirement, earliness, tillering, lodging resistance, shrivelling, rusts, and productivity.

DE CILLIS (1927) supplements his list with a very detailed description of synonyms, zones of diffusion, etc. Three years later, BUFALINO (1930) confirms that the cultivars are 23 in number and reports biometric data for the spike and caryopses. However, he described some additional cultivars ('Cannavacante', 'Giustalisa', 'Regina', 'Scavuzza', 'Semenzella', 'Cicirello' and 'Grano del Miracolo') other than those described by DE CILLIS (1927) and, quite obviously, left out as many.

In the same year, AZZI (1930) reports the following cultivars: 'Sammartinara' or 'Gigante' and 'Biancuccia', 'Scorzonera', 'Francesa' and 'Ruscia "Bianca" and "Coccitta"'. Other commonly growing cultivars were 'Sambucara', or 'Scavarella' or 'Scavolidda', and 'Cannizzara', whereas 'Maiorca' and 'Castiglione' were predominant in the mountain areas.

Worth noting is 'Trimenia' (or 'Tumminia' or 'Timilia'), which used to be cultivated all over the area though in comparatively small amounts.

SCHREIBER (1934) describes 160 specimens (collected by VAVILOV in 1927) and lists 4 species: *T. vulgare*, separated into 7 botanical varieties, *T. durum* grouped into 6 sections and 10 botanical varieties, *T. turgidum* (3 sections and 7 botanical varieties). This description already lists some of the new cultivars, obtained by selection, such as 'Sen. Cappelli', 'Mentana', 'Ardito', etc.

In 1942, U. DE CILLIS published a monographic study on Sicilian wheats, starting with specimens collected in several areas of Sicily which had been multiplied and evaluated for several years. He also attempted a botanic and varietal classification, based upon the characters of the plant, spike, and caryopses, as suggested by PERCIVAL (1921) and reported names and characteristics of as many as 45 cultivars.

The classification proposed by DE CILLIS (1942) is reported in table 1.

## Material and methods

The material used in this study is a collection of tetraploid wheats from Sicily. In practice, this collection is made up of all the wheats reported by DE CILLIS (1942) except for 'Castiglione glabro', 'Vallelunga glabra' and hexaploid forms.

The characters considered are those suggested by PERCIVAL (1921), already used by DE CILLIS (1942). The characters reflect different aspects of the plant in general, and of the spike and caryopsis as shown in table 2.

The method of character analysis is the one elaborated by EASTABROOK and ROGERS (1966). It is assumed that the overall information possessed by a group of objects (by cultivars, in this particular instance) can be represented by the area of a rectangle and that the information held by one character corresponds to a subset that is identificable in the area of a circle, as represented in a Venne diagram:

Table 2

Characters observed in 39 varieties of Sicilian tetraploid wheats included in the analysis

| | |
|---|---|
| 1. Growth habit of young plants | 25. Glume pubescence |
| 2. Leaf colour | 26. Mucron width |
| 3. Leaf pubescence (upper face) | 27. „ length |
| 4. „ „ (lower face) | 28. „ form |
| 5. Leaf edge hairiness | 29. „ coursing |
| 6. Leaf sheath hairiness | 30. Shoulder width |
| 7. „ „ colour | 31. „ form |
| 8. „ „ veins | 32. Tooth form |
| 9. Auricles colour | 33. Rachis toughness |
| 10. „ hairiness | 34. „ fragility |
| 11. Culm solidity | 35. Awn colour base |
| 12. „ colour | 36. „ „ apex |
| 13. „ consistency | 37. „ length |
| 14. „ brittleness | 38. „ persistence |
| 15. Spike form-profile | 39. Caryopsis colour |
| 16. „ form-face | 40. „ length |
| 17. „ density | 41. „ profile |
| 18. „ section | 42. „ furrow width |
| 19. „ angle | 43. „ „ depth |
| 20. Glume colour-apex | 44. „ „ edges |
| 21. „ „ base | 45. „ structure |
| 22. „ veins | 46. Brush length |
| 23. „ length | 47. „ width |
| 24. „ width | 48. Embryo size |

The amount of information possessed by a character, irrespective of the other characters, is called unconditional entropy. According to Shannon (1948), it is the result of

$$H(I) = \sum_{i=1}^{n} pi \lg pi$$

and is linked by two factors:

1. Number of subsets, or of the groups that a character is able to make up, that is of its states, meaning that as the number of states increases so does the amount of information possessed by the character. The more states there are, wider is the possibility of choice and the more difficult is to make a guess; the greater the degree of uncertainty, the larger is the amount of information gained when the state of the character is known.
2. Distribution of the objects in the states, i.e. in the sub-sets. The more evenly the objects are distributed in the states, the more difficult it is to make a guess, and therefore the larger is the amount of information gained when the state of a character is known. Conversely, if most of the objects are contained in one state, the smaller is the degree of uncertainty, the smaller is the information gained when the group to which the object belongs, is known.

When two characters I and J have one part of the information in common, it is possible to distinguish between a part of information pertaining solely to the individual characters (area A and B), called conditional entropy and a common part (C) called redundancy, as indicated below:

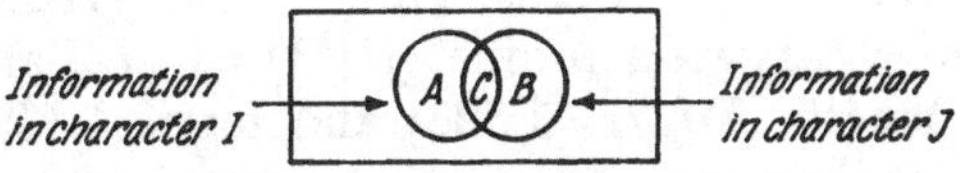


The value of conditional entropy in character J in the presence of character I, i.e. of the amount of information supplied by character J after character I has been observed on the same material is given by:

$$H(J/I) = \sum_{g=1}^{i} H(J/I)/p(I)$$

where i is the number of classes in character I.

The ratio, total sum of conditional entropies in two characters $H(I/J)+H(J/I)$ to the amount $(A+B+C)$ of information held by them $H(I, J)$, is the distance between the two characters:

$$D(I/J) = \frac{H(J/I)+H(I/J)}{H(I, J)}$$

When several characters are considered, one can estimate the average distance between one character and all the other characters in the analysis.

Therefore, the analysis of characters has been directed at providing a general indicator of the "goodness" of a character, i.e. it has been intended to decide which characters represent the others better. For this reason, for each character I, we have calculated the sums of relatios between the common amount of information that character I shares with any other character J and the total amount of information held by character J:

$$\text{SAMRAT (I)} = \sum_{j=1}^{m} \frac{H(I)-H(I/J)}{H(I)}$$

and the sum of ratios between the amount of information that character I has in common with any other character J and the amount of information possessed by I itself:

$$\text{SUMRAT (I)} = \sum_{J=1}^{n} \frac{H(I)-H(I/J)}{H(J)}$$

The clustering method used here is a technique based on a model in graph theory developed at the Taximetrics Laboratory, Boulder (WIRTH et al., 1966; EASTABROOK, 1966; EASTABROOK and ROGERS, 1966) which, according to PRANCE et al. (1966), yields the most useful and informative results.

The method consists of two parts:

1) The similarity between all the pairs of objects in the study is calculated;
2) the objects are connected in various clusters, starting with the objects that are more similar and then proceeding in a decreasing order until all the objects form a single cluster.

The method of similarity measure is the one by which, in its simplest form, the similarity, S(a,b) attributed to the pair of objects *a* and *b* is equal to the ratio between the number of characters common to *a* and *b* and the total number of characters.

## Results

### *Character analysis*

The total amount of information contained in each of the 48 analyzed characters is reported in table 3. The characters providing the largest amount of information are glume colour, both at the base [H(21) = 2.304] and apex [H(20) = 2.234] and the caryopsis colour [H(39) = 2.222], followed by shoulder form [H(31 = 1.916] and width (H(30) = 1.420], glume veins [H(22 = 1.821], culm [H(12) = 1.763] and auricles colour [H(9) = 1.348], caryopsis profile [H(41) = 1.772] and furrow depth [H(43) = 1.465], mucron length [H(27) = 1.526] and form [H(28) = 1.301] and tooth form [H(32) = 1.314]. Those giving the least amount of information are growth habit of the young plants [H(1) = 0.172] and colour of the leaf sheath [H(7) = 0.172].

Table 3

Information – unconditional entropy – in the analyzed characters

| Char. | H(I) | Char. | H(I) | Char. | H(I) | Char. | H(I) |
|---|---|---|---|---|---|---|---|
| 1 | .172 | 13 | .918 | 25 | .858 | 37 | 1.262 |
| 2 | .996 | 14 | 1.049 | 26 | 1.228 | 38 | .619 |
| 3 | 1.118 | 15 | .858 | 27 | 1.526 | 39 | 2.222 |
| 4 | .956 | 16 | .918 | 28 | 1.301 | 40 | 1.146 |
| 5 | .619 | 17 | .988 | 29 | 1.215 | 41 | 1.772 |
| 6 | .677 | 18 | .779 | 30 | 1.420 | 42 | 1.139 |
| 7 | .172 | 19 | .391 | 31 | 1.916 | 43 | 1.465 |
| 8 | .999 | 20 | 2.234 | 32 | 1.314 | 44 | .918 |
| 9 | 1.348 | 21 | 2.304 | 33 | .999 | 45 | .018 |
| 10 | .719 | 22 | 1.821 | 34 | .961 | 46 | 1.215 |
| 11 | 1.216 | 23 | 1.251 | 35 | .896 | 47 | .477 |
| 12 | 1.763 | 24 | 1.075 | 36 | .462 | 48 | .732 |

Table 4

Distance and conditional entropy for correlated and independent characters

| Characters I–J | Distance D(I/J) | Conditional entropy H(J/I) | H(I/J) |
|---|---|---|---|
| 3– 4 | .145 | .000 | .162 |
| 3–10 | .612 | .206 | .605 |
| 3–38 | .581 | .106 | .605 |
| 4–10 | .558 | .206 | .442 |
| 4–38 | .517 | .106 | .442 |
| 6–10 | .707 | .403 | .361 |
| 10–19 | .684 | .124 | .453 |
| 10–38 | .631 | .258 | .358 |
| 13–14 | .124 | .130 | .000 |
| 19–38 | .693 | .382 | .154 |
| 20–21 | .129 | .192 | .122 |
| 5–13 | 1.000 | .918 | .619 |
| 13–30 | 1.000 | 1.420 | .918 |
| 38–44 | 1.000 | .918 | .619 |

The less distant, or more closely associated characters are (table 4) stem consistency and brittleness [D(13–14) = 0.124], colour at the apex and base of glumes [D(20–21) = 0.129], pubescence of the upper and lower face of the leaf [D(3–4) = = 0.145].

There is also high correlation between auricles hairiness on the one hand, and pubescence of the upper [D(3–10) = 0.612] and lower face of the leaf [D(4–10) = = 0.558], sheath hairiness [D(6–10) = 0.707], spike angle tilt [D(10–19) = 0.684] and awn persistance [D(10–38) = 0.631] on the other hand. To continue, awn persistence is closely correlated with pubescence of the upper [D(4–38) = 0.517] and lower [D(3–38) = 0.581] face of the leaf and with spike angle [D(19–38) = 0.693].

Conversely, there seems to be complete independence (D = 1.0)–at least in the analyzed material–between culm consistency (13) and hairiness of the leaf edge

Table 5
SUMRAT and SAMRAT values for different characters

| CHAR. | SUMRAT | SAMRAT | CHAR. | SUMRAT | SAMRAT | CHAR. | SUMRAT | SAMRAT | CHAR. | SUMRAT | SAMRAT |
|---|---|---|---|---|---|---|---|---|---|---|---|
| 1 | 1.0 | 6.4 | 13 | 3.8 | 4.3 | 25 | 2.5 | 2.9 | 37 | 5.8 | 4.5 |
| 2 | 3.1 | 3.1 | 14 | 4.9 | 4.8 | 26 | 3.5 | 3.1 | 38 | 5.6 | 9.5 |
| 3 | 8.9 | 8.4 | 15 | 3.1 | 3.4 | 27 | 4.0 | 2.6 | 39 | 13.0 | 6.4 |
| 4 | 8.1 | 8.9 | 16 | 2.4 | 2.9 | 28 | 3.2 | 2.7 | 40 | 6.0 | 5.3 |
| 5 | 2.1 | 3.9 | 17 | 3.7 | 3.8 | 29 | 3.7 | 3.3 | 41 | 7.7 | 4.3 |
| 6 | 5.4 | 8.9 | 18 | 2.0 | 2.6 | 30 | 3.7 | 2.9 | 42 | 3.4 | 3.2 |
| 7 | 1.3 | 8.5 | 19 | 3.7 | 10.0 | 31 | 8.2 | 4.3 | 43 | 5.4 | 3.8 |
| 8 | 2.7 | 3.1 | 20 | 10.9 | 5.8 | 32 | 3.7 | 2.8 | 44 | 2.2 | 2.7 |
| 9 | 4.5 | 3.6 | 21 | 12.4 | 6.1 | 33 | 2.5 | 2.5 | 45 | 3.9 | 4.1 |
| 10 | 6.7 | 9.8 | 22 | 7.9 | 4.7 | 34 | 2.1 | 2.4 | 46 | 2.8 | 2.6 |
| 11 | 6.6 | 5.3 | 23 | 7.7 | 6.2 | 35 | 3.8 | 5.0 | 47 | 2.3 | 4.9 |
| 12 | 8.0 | 4.8 | 24 | 4.7 | 4.5 | 36 | 2.3 | 6.0 | 48 | 2.5 | 3.8 |

(5) and shoulder width (30) and also between awn persistence (38) and the edges of the caryopsis furrow (44).

A closer insight into the strictly associated characters shows the extreme interest of the characters, glume colour at the apex (20) and base (21). As seen earlier, these are the characters which contain the largest amount of information yet they are so closely associated that any one of them is sufficient to express over 90% of the information. The amount of information supplied by the colour of the glume apex, after observing the base colour, H(20/21), is 0.122, or 5% of the original amount, whereas the amont of information supplied by the base colour, after observing the apex H(21/20) is 0.192, or 8% of the original information, whereas the common information is equal to 2.112.

A characteristic example is also provided by the characters culm consistency (13) and brittleness (14): the total amount of information supplied by consistency [H(13) = 0.918] is lower than brittleness [H(14) = 1.049] (Table 3) and all the information relevant to the consistency is contained in the brittleness, so that the amount of information left in the former character, after analyzing the latter, is equal to 0. Thus it would seem altogether useless to measure the material for the consistency when brittleness is already known, unless we want to lay special emphasis upon both characters.

The situation with the two characters, pubescence of the upper (3) and lower (4) face of the leaf is similar: H(3) = 1.118; H(4) = 0.956; H(3/4) = 0.152; H(4/3) = = 0.000.

In all of the independent characters (D = 1), the situation is reversed, i.e. conditional entropy is equal to unconditional entropy and the amount shared in common is, therefore, nil.

Finally, the analysis was directed at

determining which character is (or which characters are) best suited to represent all the other characters by means of SUMRAT and SAMRAT determinations.

Tabulated results (Table 5) show that the characters which best represent the others are as follows: pubescence of the upper (3=8.9–8.4) and lower face (4=8.1–8.9) of the leaf, hairiness of the leaf sheath (6=5.4–8.9), apex (20=10.9–5.8) and base colour (21=12.4–6.1) of glume, caryopsis colour (39=13.0–6.4), auricles hairiness (10=6.7–9.8), solidity (11=6.6–5.3) and colour (12=8.0–4.8) of stem, length (40=6.0–5.3) and base profile (41=7.7–4.3) of caryopsis, awn length (37=5.8–4.5) and persistence (38=5.6–9.5), glume veins (22=7.9–4.7) and length (23=7.7–6.2).

Instead, it would seem that the following are less representative: growth habit of young plants (1=1.0–6.4), sheath colour (7=1.3–8.5), tilt (19=3.7–10.0), awn colour, especially at the apex (18=2.3–6.0). Their SAMRAT values are essentially due to the small amount of unconditional entropy of the characters appearing at the denominator.

*Cluster analysis*

All the data, irrespective of their amount of information and representativeness, were processed by cluster analysis in order to find out whether the cultivars belonging to one botanical variety (b.v.) tend to join in one and the same group. The results are given in figures 1 and 2.

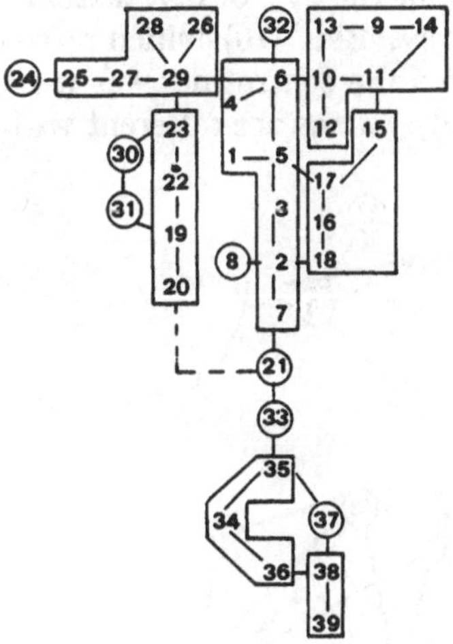


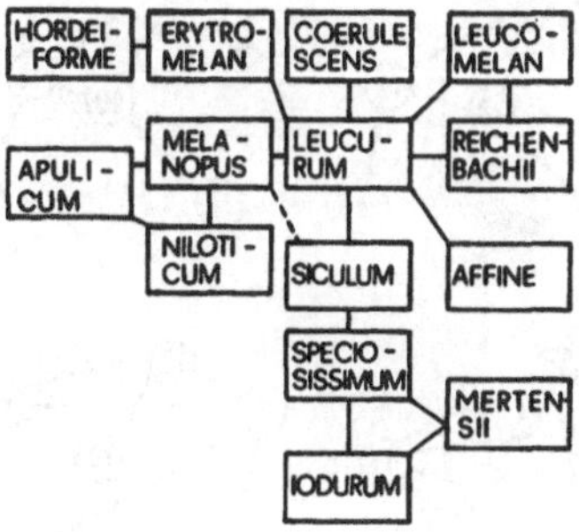


Fig. 1 Results of the cluster analysis (for the numbers cf. table 1)
Fig. 2 Results of the cluster analysis

By observing the figure(s) it becomes immediately obvious that there is a clear-cut separation between *durum* and *turgidum* wheats and that there are equally clear subdivisions inside each group. As a matter of fact, the analysis reflects the traditional classification rather faithfully, with most cultivars being grouped according to the different botanical varieties (b.vs.) to which they belong. An exception to this rule are the cultivars belonging to b.v. *melanopus* which are separated into two groups: one comprising a single variety, 'Martinella', at the meeting point between *durum* and *turgidum* wheats, the other comprising the remaining cultivars (of the b.v. *melanopus*), close to b.v. *erythromelan*.

Thanks to this similarity of results between the traditional classification and cluster analysis, it is possible to investigate the relationships between the cultivars and the b.vs. on the assumption that there is a logical sequence between them, their possible points of contact, the cultivars nuclei from which other cultivars may have originated, etc. Thus, for example, b.v. *speciosissimum* appears to be the central nucleus in *turgidum* wheats, connected with the *durums* through b.v. *siculum*, and with b.v. *mertensii* and *jodurum* as some sort of further diversification.

In the *durum* wheat cluster, b.v. *leucurum* (1–7) seems to be the central nucleus from which as many as 6 groups may have been derived. The transition is through cultivar 'Sammartinara' (6) to b.v. *leucomelan* and through cultivar 'Francesa' (2) and 'Regina' (5) to b.v. *reichenbachii*. In turn, the latter two b.vs. are in contact with each other through cvs. 'Chiattulidda' (11) and 'Scavuzza' (15). Through cv. 'Francesa' (2), b.v. *leucurum* is also in contact with the only cv. of the *affine* b.v. the white-awned 'Timilia' (8), whereas cv. 'Sammartinara' (6) provides a point of contact with 'Scorsonera' (32), the only cv. of b.v. *coerulescens*, and with b.v. *erythromelan*. The latter is linked to b.v. *melanopus* and to 'Russello' (24) the only cv. in b.v. *hordeiforme*.

Also the relationship among cvs. within each b.v. seems to be particularly interesting. For example if the clusterforming process for the cvs. of b.v. *leucomelan* (Figure 3), is analyzed, it is possible to find out that cv. 'Bidì' (10), which provides the point of contact with b.v. *leucurum*, also forms the central nucleus from which the other cvs. may, possibly, have differentiated. Things are different with

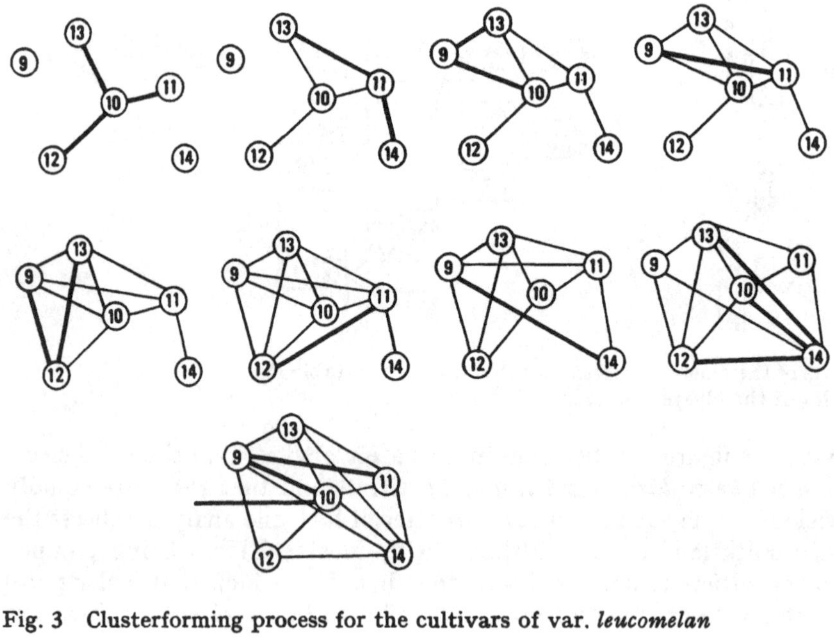

Fig. 3 Clusterforming process for the cultivars of var. *leucomelan*

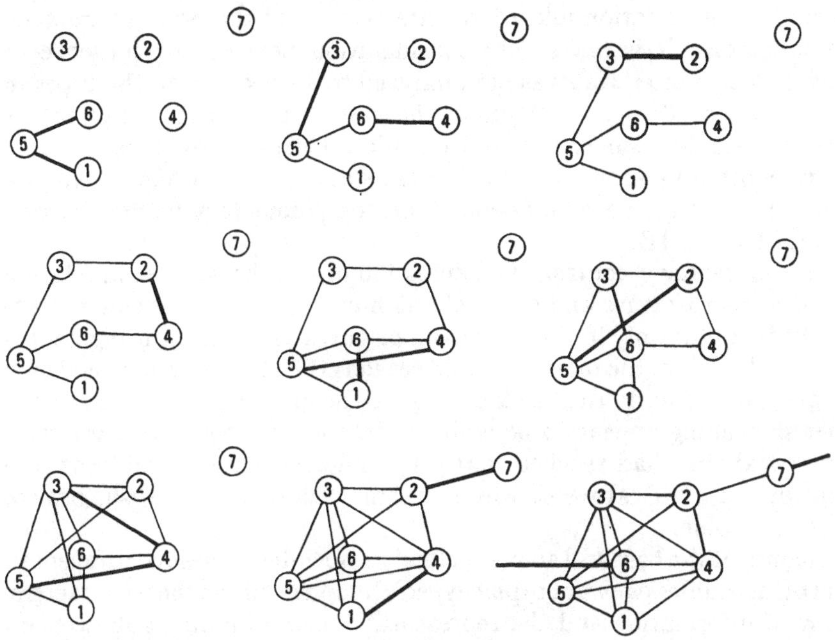


Fig. 4 Clusterforming process for the cultivars of var. *leucurum*

b.v. *leucurum* (Figure 4), in which the cluster is soon obtained from an initial group consisting of 'Cannizzara' (1), 'Regina' (5) and 'Sammartinara' (6). The only exception here is cv. 'Tripolino' (7) which joins the cluster only when the links between the other cvs. have become quite strong. It should also be noted that whereas the cvs. of b.v. *leucomelan* are in contact with other clusters through 'Bidì' (10), the nodal cv., those belonging to the b.v. *leucurum* establish their first contact with another cluster through the last joining cv. Rather, it establishes a link with another b.v. even before establishing one link with another cv. of the b.v. to which it belongs.

## Discussion and conclusions

The analysis allows to quantify the information contained in the various characters and possibly to speculate on the results, although there is little doubt that they are related to the specific and numerically rather limited material considered in the analysis.

Generally speaking, present results confirm that the characters of the spike used by most workers as a basis for their botanical classifications are of great systematic importance. In fact, half the characters of the spike have unconditional entropy values between 2.3 and 1.2, i.e. between the highest calculated value and 50% of the same value. Only four out of fourteen characters of the plant in general and four out of ten characters of the caryopsis fall within this range of values.

Glume colour is no doubt the most important character, both in terms of absolute amount of information and representativeness of the other characters. In most classifications, it is taken into consideration next to the presence or absence of awns–a character that was not analyzed by us because all the types in our study were awned. The base colours of the glumes are white, red or black, with intermediate shades. Monogenic and digenic inheritance for these characters have been reported by some authors (AUSEMUS et al., 1946), while others (ASLAM, 1958) indicate that glume colour depends on complementary factors, located in chromosomes 1A and 1B.

Whatever the hereditary control, the expression of the character is in relation to light intensity, temperature and atmospheric humidity during ripening (PERCIVAL, 1921). In fact, it is only in the warm and bright seasons that the pigment is fully developed, whereas in the moist and cold seasons the glumes have a tendency towards ash-grey, particularly the black ones. Also the glumes of spikes that have ripened under shrivelling appear to be whitish. In our analysis, which was performed on material that had reached maturity under optimum conditions, the colours found in the material were: white, cream, yellow, red, reddish, brown, black, blackish and blue.

Since the colours of the base and apex are closely associated, it seems sufficient to observe either of them unless we want to place special emphasis upon these characters.

The amount of information and the representativeness of glume pubescence–listed as the third character in botanical classifications–is rather small. This character is extremely evident and does not leave any doubts during observations. Still, in the material considered in this study, its frequency and association with other characters are such that it can be regarded as quite negligible. Material from very different regions need to be analysed for a careful evaluation of information in this character.

Much the same can be said about the colour of awns, which some authors believe to be of remarkable systematic importance. As a matter of fact, as pointed out by PERCIVAL (1921), this character behaves rather strangely: while in certain wheats, among which the various forms of *melanopus*, colour–particularly black–is very steady, elsewhere it is strongly conditioned by light, temperature and humidity after ear formation. Precisely because it is dependent on external conditions, the study of the character's inheritance is rendered more complicated and its taxonomic value is greatly diminished as pointed out by METZGER (1824), KOERNICKE (1873), HOWARD (1912) and PERCIVAL (1921), among others.

The analysis confirms the importance of another character, glume shoulder shape, both in terms of total amount of information and representativeness. The shape of the glume shoulder, i.e. the apical margin between the mucro and the external margin, however, shows some variation even within the same spike so that observations must always be carried out with great precision and always in the central part of the inflorescence.

Yet another important character is the presence of veins in the glumes (22). However, what has been said about environmental influence upon glume colour applies all the more to this latter character which is mainly present in tetraploid wheats and can therefore be profitably used for variety classification. Here again, the analysis of a larger and more diversified number of specimens will help to produce a more definite answer.

A certain amount of information is also contained in the characters of the mucro, especially length, which is also its more stable character. On the other hand, it is scarcely representative because it is differently associated with other characters.

Of extremely controversial interest is spike density, one of the easiest characters to be surveyed and also one of remarkable agronomic importance. Some workers–SÉRINGE (1819), ERIKSSON (1894) and VAVILOV (1931)–attach a great importance to this character and include it among the items to be considered for a botanical classification whereas others, like CLARK and BAYLES (1935), DE CILLIS (1942) regard it as rather unimportant. The results of the present analysis seem to agree with this latter group. Also agronomically important is the size of glumes which, from a systematic point of view, presents a certain amount of intermediate information and is fairly representative. In fact, this character is rather stable within the species and within the botanical variety.

Concerning the stem, colour alone shows a satisfactory amount of information: the characters pubescence of the leaf base, hairiness of the sheath and earlets and stem consistency are adequately representative. It should be noted, however that this representativeness is essentially due to specific associations of characters as shown in Table 2. Unfortunately, stem colour is also strongly affected by the environment, hence its systematic value is rather limited.

Things are different for the characters of the caryopsis whose colour and profile are among those which contain the greatest amount of information and which are most representative. It is a known fact that caryopsis colour is due to a resinous oil which develops in the ripening stage. However, grain colour does not only depend on the amount of material that has been accumulating, but also on the thickness, transparency and, possibly, colour of the pericarp in which it is enveloped as well as on the vitreous or mealy characteristics of the endosperm. White wheats exhibit a range of shades from light cream to amber and yellow, whereas in red wheats the colour ranges from discoloration to brick red and brown. Shrivelling and attacks by fungi will often distort true colours. In *durum* wheats in which the endosperm is translucid, certain classes can only be observed by means of very careful examination and by placing the caryopses against a dark background. During this analysis, the scale of evaluation was changed several times in order to work out those states that would best fit the material being examined. Also the considerations that can be drawn from cluster analysis are different.

In contrast with conventional classifications, where the various botanical varieties form different entities without any mutual relationship except for the characters they have in common, the analysis enables the various b.vs. to be regarded as being part of a single entity, closely related to one another so that their relationships can be studied also in the light of their geographic distribution. Thus, for example, figures 1 and 2 clearly describe the central role of b.v. *leucurum* around which practically all of the other b.vs. of *durum* wheat are clustered. The latter may or may not be directly related, like *leucomelan* and *reichenbachii*, or *erythromelan* and *melanopus* etc. In this connection, it seems interesting to note that 'Sammartinara' (6), 'Regina' (5) and 'Tripolino' (7)–i.e. cultivars through which b.v. *leucurum* has certain points of contact with other b.vs.–at one time used to be scattered all over the island. Could this mean that they were among the first to reach Sicily or were most suited to the environmental conditions of the island? And that other b.vs. could have differentiated from them later on, possibly after crossing with other materials that were introduced into the island but were unable to find the right environment for their development? An answer could be found by elaborating the analysis further–also with the aid of crossings–both on our material and on material obtained from the whole Mediterranean region. Today in Sicily cultivars belonging to b.v. *leucurum* are rare, cultivar 'Capeiti', as a modern race, being an exception (PERRINO et al., 1973).

Also worth pointing out is the central core formed by cultivars 'Sammartinara' (6), 'Bidì' (10) and 'Russia' (29) through which contact is established among b.vs. *leucurum*, *leucomelan* and *erythromelan*. Also 'Russia', like 'Sammartinara', is already reported in the earliest works as being present all over Sicily and is often mistaken for 'Russello'. A curious example is 'Bidì' which, according to DE CILLIS (1942), seems to have been introduced from North Africa as late as 1874 where upon it spread immediately to the whole island. Now, while there are no doubts its being introduced in 1874, it is very likely that it was the last of a long series of transfers from and into Sicily, at a time when both Sicily and North Africa were Greek colonies. Of course, it may also be that these transfers date back to older migrations. The central role of this variety within b.v. *leucomelan*, as shown in Figure 3, seems to be certain rather than a casual coincidence derived from the analysis, all the more so, since the other cultivars in the b.v. used to be grown in smaller environments.

Turgid wheats too seem to respond to the environment. With turgid wheats, the analysis indicates proximity not only of the two black 'Bufale' (both belonging to the same b.v.) but also of the two red 'Bufale'. These are attributed to different b.v. in the conventional classification although in practice both of them were propagated mainly because of the length of their spike–a character that the first analysis had shown to possess a small amount of information. It would certainly be more interesting, in this connection, to consider in the analysis, the physiological characters and those pertaining to resistance, in connection with adaptation, since these would be of practical value for breeding work.

Despite the limitations and peculiar characteristics of the analyzed material, the results obtained provide some useful indications concerning the information contained in the characters that have been analyzed and the existing relationships among Sicilian cultivars. No doubt, a more comprehensive understanding of the evaluation of these materials could be obtained by analysing larger collections that would include materials from various regions of South West Asia and from the Mediterranean Basin where wheats used to have their centres of diversification.

## Zusammenfassung

Merkmalsanalyse und numerische Klassifikation an einer Weizen-Kollektion aus Sizilien

Untersucht wurden 48 Pflanzen-, Ähren- und Kornmerkmale an Proben tetraploider Weizen aus Sizilien, um den Informationsgehalt unterschiedlicher Merkmale und ihren taxonomischen Wert zu erfassen. Anhand von Ähnlichkeitsindices wurde auch eine Klassifikation des Materials vorgenommen.

Als bedeutsamste Merkmale erscheinen: Blattpubeszenz, Spelzenfarbe und -länge und Kornfarbe. Auf der Grundlage aller Merkmale wurden Ähnlichkeitsindices zwischen den Proben errechnet und eine Klassifikation ausgearbeitet. Diese ist den Klassifikationen anderer Wissenschaftler sehr ähnlich, ergibt hierarchische Gruppen mit unterschiedlichem Ähnlichkeitsniveau und ermöglicht die Aufstellung einer interessanten Hypothese zur Evolution der sizilianischen *durum*-Weizen.

## Краткое содержание

Анализ признаков и нумерическая классификация пшениц Сицилии

Исследовались 48 признаков растения, колоса и зерна на образцах тетраплоидных пшениц Сицилии с целью выяснения их информационной и таксономической ценности. При помощи коэффициентов сходства была предпринята и классификация материала.

Наиболее существенными признаками оказались: опущённость листа, цвет и длина плёнок, цвет зерна. При учёте всех признаков оценивались коэффициенты сходства между образцами и строилась их классификация. Эта классификация оказалась весьма сходной с таковыми других авторов, выявила иерархические группировки на различных уровнях сходства и позволила выдвинуть интересную гипотезу эволюции сицилианских твёрдых пшениц.

## Literature

Aslam, M. C., 1958: Genetic studies in interspecific crosses between *durum*, *sphaerococcum* and *vulgare* types of wheat. – Agriculture (Pakistan) **9**, 109–119.

Ausemus, E. R., J. B. Harrington, L. P. Reitz and W. W. Worzella, 1946: A summary of genetic studies in hexaploid and tetraploid wheats. – J. Amer. Soc. Agron. **38**, 1082–1099.

Azzi, G., 1930: Le climat du blé dans le monde. Les bases ècologiques de la culture mondiale du blé. – Institut International d'Agriculture. – Rome.

Bufalino, R., 1930: I frumenti siciliani. – Enna.

Clark, J. A. and B. B. Bayles, 1935: Classification of wheat varieties grown in the United States. – U. S. Dept. of Agriculture, Techn. Bulletin **459**. – Washington.

de Cillis, E., 1927: I grani d'Italia. – Ministro dell'Economia Nazionale. – Roma.

de Cillis, U., 1942: I frumenti siciliani. – Stazione Sperimentale Granicoltura Siciliana, pubbl. **9**. – Catania.

Estabrook, G. F., 1966: A mathematical model in graph theory for biological classification. – J. Theoret. Biol. **12**, 297–310.

–, and D. J. Rogers, 1966: A general method of taxonomic description for a computed similarity measure. – BioScience **16** (11), 789–793.

Eriksson, J., 1894: Über die Spezialisierung des Parasitismus bei den Getreiderostpilzen. – Ber. Deut. Bot. Gesell. **12**, 292–331.

Howard, A. and G. L. C., Howard, 1912: On the inheritance of some characters in wheat. I. – India Dept. Agr. Mem. Bot. Ser. **5**, 1–47.

Körnicke, F., 1873: Systematische Übersicht der Cerealien und monocarpischen Leguminosen aus dem ökonomisch-botanischen Garten zu Poppelsdorf bei Bonn, ausgestellt in Wien im Jahre 1873. – Bonn.

LORENZONI, 1910: Relazioni del Delegato Tecnico Sicilia. – In: Inchiesta parlamentare sulle condizioni dei contadini nelle province meridionali e nella Sicilia. Tomo **1**, Parte 1 e 2. – Roma.

METZGER, J., 1824: Europäische Cerealien. – Landwirtschaftliche Pflanzenkunde. 2 vols – Heidelberg.

PERCIVAL, J., 1921: The wheat plant, a monograph. – London.

PATANÈ e STRINGHER, 1884: La Coltivazione dei cereali in Sicilia. – Atti della Giunta per l'inchiesta agraria e sulle condizioni della classe agricola, Vol. **III**, tomo **I**, fasc. 3. – Roma.

PERRINO, P. and F. MARTIGNANO, 1973: The presence and distribution of old wheat varieties in Sicily. – Ecologia Agraria, **9** (1), 1–11.

PRANCE, G. T., D. J. ROGERS and F. WHITE, 1969: A taximetric study of an angiosperm family: generic delimitation in the Chrysobalanaceae. – New Phytol. **68**, 1203–1234.

PRESTIANNI, 1926: I frumenti coltivati in Sicilia. – Commissione provinciale per la Propaganda Granaria di Caltanisetta. – Caltanisetta.

ROGERS, D. J. and S. G. APPAN, 1969: Taximetric methods for delimiting biological species. – Taxon **18** (6), 608–624.

SCHREIBER, D., 1934: (Wheats of the Islands of the Mediterranean.) – Bull. Appl. Bot. Genet. Pl. Breed., Serie **V**, No. **2**, 41–240.

SÉRINGE, N. C., 1819: Mélanges botaniques. Monographie. Céréales de la Suisse. – Berne et Leipzig. Descriptions et figures des céréales européennes. Second article, Gen. iii. Froment. Annal. des sc. phys. et natur. d'agriculture et d'industrie, Lyon, v. **103**.

SHANNON, C. E., 1948: A mathematical theory of communications. – Bell System Technical Journal **27**, 379–423, 623–656.

VAVILOV, N. I., 1931: (The wheats of Abyssinia and their place in the general systems of wheats. A contribution to the knowledge of the 28 chromosomes group of cultivated wheats.) – Bull. Applied Botany and Plant Breeding, Supp. **51**.

WIRTH, M., G. F. ESTABROOK and D. J. ROGERS, 1966: A graph theory model for systematic biology, with an example for the *Oncidiinae* (*Orchidaceae*). – Systematic Zoology **15** (**1**), 59–69.

Prof. E. PORCEDDU and Dr. P. PERRINO,
Istituto del Germoplasma, C.N.R.,
Via G. Amendola 165/A, I–70 126 Bari, Italy
Dr. S. VANNELLA, Plant Breeding Institute, Faculty of Agriculture,
University of Bari, Via G. Amendola 165,
I - 70 126 Bari, Italy

Kulturpflanze XXIX · 1981 · S. 267–272

# Problems of the infraspecific taxonomy of oat

FRANTIŠEK KÜHN (Brno, ČSSR)


## Summary

A classification of types of *Avena sativa* in Czechoslovac land-races, based on features used in the system of MALCEV (1930), is presented.


Oat is a young crop, whose evolution passed for a long time in a weed stage. The oldest carbonised grain sample with prevailing of oat is from Červeník near Hlohovec in Slovakia from early iron age.

Oat as a crop cultivated previously for fodder has been less selected as other crops. In breeding are now prefered self-pollinating thick-kernel types, which replace now other types of oat. The taxonomy of oat is difficult, and until now only the varieties, based on colour of paleas and on presence of awns, which are noted in KÖRNICKE and WERNER (1885) are commonly used. The system of MALCEV (1930) is but more progressive, but for classification it is necessary to have whole ripe plants.

In 1977, KARL HAMMER and I have gathered in Slovakia 370 samples of oat from local populations, which contain nearly the whole known variability of this species. I show there a classification of them. I describe also interesting findings, collected by me in 1972–1980 in Moravia and Slovakia, which illustrate the borderlines between *Avena sativa, A. fatua* and *A. sterilis*.

Oat was in the crop collection from East Slovakia 1977 the most frequently collected and the morphologically most differenciated crop. Oat was in all villages one of the main crops; in Hrabová Roztoka oat was the only spring cereal. The variability of the types is great, and they contain many transitional types. The features vary in a great degree in dependence on environmental conditions. Most of the characters allow to estimate only the degree or level of evolution from weed to crop. In several types, mainly in types with narrow grain, cross pollination is possible. Many populations are a complex of for example 8–10 very different types. In the collection were some combinations of features not yet mentioned in existing oat systematics.

The collection has been divided into different types mostly according to an analyse of unripe plants, grown in Gatersleben 1978 and 1979. The base for the division of the collection was the combination of features mentioned in the system of MALCEV (1930) and in MORDVINKINA (1936). The hairyness of lemma and the colour of lemma have not been included in typisation, the system would be much complicated. The fracture or semiabscission of the basal grain of the spikelet is probably an important feature for systematics, but its importance must yet be examined.

I give first a commentation of the used morphological terms:
– as subsp. *macrantha* have been designated samples with the longer glumes longer than 30 mm.
– as subsp. *nodipilosa* have been designated samples with hairs on all or on a part of the nodes (upper or lower nodes, on main or lateral stalks). The number of hairs shall not be lower than 10 on a node. The length of hairs must be more than 0,2 mm; stalks which are rough with 0,1 mm long dents or bristles are not included here.
– as subsp. *praegravis* have been subsumed samples with ripe corticated basal grains at least 2,5 mm broad.
– the magnitude of the panicle has been aspected as middle great, if the panicle was in the original material in Slovakia up to 140 mm long, in culture in Gatersleben 1979 up to 180 mm (sometimes 200 mm) and in 1980 up to 200 mm long. This feature varies in different years, and the size can be estimated only in comparison with samples grown in the same place and in the same year. The same can be stated for the plant heigh.
– as conv. *contracta* have been designated plants, in which at least some panicles of the plant (on main or lateral stalks) have adpressed branches.
– dense panicles with main branches standing off less than 45° were called "Steifrispe". Sometimes they have single branches adpressed.
– panicles with main branches standing far more than 45° from panicle axis have been designated as broad or effuse.
– as awned have been designated types with at least a half of the spikelets with an awn.
– lemmas have been designated as short with a length under 14 mm, and as long with more than 18 mm.
– the colour of unripe plants was waxy-malachite green, on vaginas sometimes plumbum-grey, or in other cases non-waxy, dark green to light green or pale green. The colour of ripe lemmas was white, yellow, sometimes ochre, brown to black, or grey (in one panicle have been often partly grey, partly white lemmas).
– the hairyness of the lemma differed in the collection from lemmas with hairy back to types with only the base of the lemma with long (more than 3 mm) or short (less than 3 mm) hairs, or with glabrous lemma. Some times the rhachilla was hairy.

Between the awned and awnless types was no greater difference; in one population were often awned, partly awned and awnless plants of the same type, and in the progency this character is not very constant. The colours of the lemma white and yellow are often not distinguishable. In the populations mostly are together plants with white and with yellow lemma, and the plants often do not differ in other features. Very frequently there were in the same very homogeneous populations plants with white and yellow, awned and awnless lemmas.

The types with broad grains in the collection are often earlier ripening and cleistogameous, and have dense panicles, but the types with narrow grains and effuse panicle ripe late and are chasmogameous; in some samples the stigmas are curving out over the bords of the lemma. The cleistogameous group is more uniform (5 types); the chasmogameous group is more differenciated (20 types). The anthers are in subsp. *macrantha* 4 mm long, in the other types mostly 2–2,4 mm long. Grey or brown lemma or bristle base of lemma are mostly combined with thin stalk, laxe panicle and slender grain. A bristle base of lemma is often combined with hairy nodes. The stalks have had in Gatersleben 1979 in most samples 4 leaves (in 1980 5 leaves); in some samples the stalks have had more leaves.

The sequence of features used in typisation of the collection was:–length of glumes-thickness of grains-thickness of grains-hairyness of nodes-character of panicle-erectness of stalks-waxyness of vagina and of panicle.

The types in the collection from East Slovakia 1977 are the following: panicle diffuse, if not mentioned as contract; the types include populations with and without semiabscission:

– subsp. *macrantha* (HACK.) MORDV.: the longer glumes are longer than 30 mm

– – nodes without hairs: Type I: stalks semierect, high, panicle effuse, laxe, grain narrow

– – types with hairy nodes: Type II: plants dark green, waxy, ±erect, high, panicle effuse, grain narrow

– Type III: plants pale green, little waxy, panicle effuse

– Type IV: panicle contract, plant high, glumes up to 34 mm long

– subsp. *praegravis* (KRAUSE) MORDV.: corticated basal grain at least 2,5 mm broad. Mostly cleistogamic.

– – types with glabrous nodes: Type V: stalk erect, waxy, 0,6–0,9 m high. Panicle dense. Often semiabscission.

– Type VI: panicle long, broad, stalks ascendent, glumes up to 28 mm long.

– – types with hairy nodes: Type VII: panicle dense, up to 180 mm long, stalks erect, thick, 0,9–1,1 m high, waxy, lemmas often short. Is the commonst type in the collection.

– Type VIII: panicle dense. Stalk erect, 1–1,2 m high, waxy, panicle rich, up to 230 mm long, spikelets often 3-grained.

– Type IX: panicle size middle, rich, but aperte. Stalk intensively waxy, spikelets often 3-grained.

– Type X: panicle longer than 180 mm, florets chasmogamic, lemma base bristle

– Type XI: panicle contract, stalks 1,4 m high, panicle up to 280 mm long, with spikelets, glumes 26 mm long

– subsp. *sativa:* nodes glabrous, grain narrow, panicle not dense

– Type XII: stalks up to 1,7 m high, panicle effuse, plants ripening late

– Type XIII: stalks erect, up to 1,7 m high, panicle 270–320 mm long, plants pale green, glumes up to 28–29 mm long

– Type XIV: panicle 180–270 mm long, effuse, plant 0,9–1,1 m high, glumes up to 30 mm long, base of lemma often hairy

– Type XV: plant pale green, 1–1,2 m high, panicle effuse, up to 280 mm long, glumes up to 27–28 mm long, lemma grey, on the base often hairy

– Type XVa: glumes 20 mm long, lemma 14 mm long

– Type XVb: panicle dense

– Type XVI: panicle contract, plant ripening late

– subsp. *nodipilosa* (MALZ.) MORDV.: nodes hairy, grain narrow

– Type XVII: stalks erect, 1 m high, waxy, panicle dense

– Type XVIII: plants with many stalks, stalks erect, 0,9–1 m high, panicle dense, but aperte, up to 220 mm long, plants often with semiabscission, lemma often grey, often with hairy base

– Type XIX: panicle effuse, broad, glumes up to 30 mm long, lemma long
– Type XX: stalks erect, 0,9 m high, panicle effuse, broad, 210 mm long, glumes 25–29 mm long
– Type XXI: stalks semierect, 0,8–1,3 m high, panicle effuse, broad, 200–250 mm long, glumes 26–30 mm long, base of lemma often hairy
– Type XXII: as XX, but not waxy
– Type XXIII: as XXI, but not waxy
– Type XXIV: panicle contract, leaves big, panicle big (330 mm), plants lightly waxy, glumes 28–29 mm long

The collection can be of interest for selection work; there are many thick-kernel types with strong stalks, with a great number of spikelets in the panicle (type VIII). From the view of evolution of oat, in the collection are important types. The oat populations from Hrabová Roztoka, Jalová, Zboj and Nová Sedlica are very polymorph with many ancient types; for example the population from Zboj from a burned wald parcel was containing very ancient types; 8 of the 9 types have a bristle base of the lemma, three of them have semiabscission.

In the populations of oat in Slovakia and Moravia, I found also intermediate types between *Avena sativa* L., *A. fatua* L. and *A. sterilis* L., which can show the narrow relationship between these species:

1. In Rohozná near Brezno nad Hronom, I found 1972 a plant of *Avena sterilis* L. subsp. *pseudosativa* THELL. v. *subsolida* MALZ. The plant has the aspect of *A. fatua*, but has spikelets with semiabscission on the lower grain, it has a strong, shining rhachilla, which can be broken on its base; the awns have a conspicuous long upper part of the awn (over the knee). The progeny of the plant is constant. This type has a position between *Avena sativa*, *A. fatua* and *A. sterilis* subsp. *ludoviciana*; it can be an ancestral type for the mentioned three species (see SINSKAJA 1955).

2. Fatuoid mutant of *Avena sativa:* I found 1972 a plant in a field of the selected oat variety "Diadem" near Moravská Třebová. The habitus of the plant was similar to the selected variety, but each grain had a horseshoe-shaped base and an awn. The lemma was white. The progeny of the plant is constant, very uniform, with the uniformity of a selected variety. The progeny propagated itself on an experiment field for several years as a weed.

3. Intermediate types between *Avena fatua* and *A. sativa;* for example *Avena fatua* L. subsp. *fatua* v. *vilis* (WALLR.) HAUSSKN. subv. *basifixa* THELL. and *A. fatua* L. subsp. *fatua* v. *glabrata* PETERM. subv. *pseudobasifixa* THELL. I found these types in three populations of *Triticum dicoccon* DESF. in the White Carpats near Sobotište 1969, 1972, 1974, and in other summer cereals in north-eastern Moravia and in Rohozná near Brezno n. Hr. in Middle Slovakia (1972). These types have a semiabscission under the basal grain. Only a part of the spikelets have an awn. The spikelets are partly deciduous spontaneously. In some plants, the back of the lemma is slightly hairy. The second and third grain of the spikelet can be broken off only by fracture; the fracture takes place on the upper end of the rhachilla. The lemmas are brown; only in one sample from *Fagopyrum esculentum* MOENCH from Huslenky in north-eastern Moravia, the lemma was white. The progeny of these types is constant. These types can be regarded as

initial types of *Avena sativa*; they are only gradually different from types of *Avena sativa* with semiabscission; the determination as *Avena fatua* dependents on its occurence as weed in other crops. From analoguous types from Tartaria (*A. fatua* L. v. *kasanensis* VAV. etc.) the czechoslovac types differ by glabrous nodes and longer rhachilla.

4. Hybrids between *A. sativa* and *A. fatua:* I found a few plants on different places of Moravia and West Slovakia between oat with admixture of *Avena fatua*, mostly only single plants. The plants have often an intermediate character between the two species. The plants were mostly high, with laxe panicle. The basal grain of the spikelet has a semiabscission, rarely a fracture; the second and third grain remains connected with the basal grain, and can be only broken off. Only the basal grain can be awned. The base of the lemma is hairy, sometimes also the back of the lemma. The lemma is brown. The plants are fertil. Beneeds, plants can be found in fields of breed varieties of *A. sativa*, which have the size and the character of the panicle of *A. sativa*, but they have more awns than *A. sativa*, and the base of the lemma often hairy. The lemmas are brown, but sometimes white. These plants are fertile, partly fertile or sterile. The hybrid plants have a progeny cleaving in *A. sativa*, *A. fatua* and intermediate plants; the intermediate plants are more frequent as plants resembling *A. sativa* and *A. fatua*. The morphology of hybrid plants is variable. The hybrid character of these plants can be probably recognised only by sowing.

## Zusammenfassung

Probleme der infraspezifischen Taxonomie bei Hafer

Eine Klassifizierung der Typen von *Avena sativa* in tschechoslowakischen Landsorten auf der Grundlage des von MALCEV (1930) verwendeten Systems wird dargestellt.

## Краткое содержание

Проблемы внутривидовой таксономии овса

Предлагается классификация типов *Avena sativa*, встречающихся среди чехословацких местных сортов, на основе системы, разработанной МАЛЬЦЕВЫМ (1930).

## Literature

KÖRNICKE, W., 1885: Die Arten und Varietäten des Getreides. Bonn.

KROLL, H., 1975: Anmerkungen zu unseren *Avena*-Arten. – Kieler Not. Pflanzenk. Schleswig-Holstein **7**, 2–6.

MALCEV, A., 1930: Ovsjugi i ovsy. – Bull. Applied Bot., Leningrad, Supplem. **38**, 522 pp.

Mordvinkina, A. I., 1936: Oves – *Avena* (Tourn.) L. – Kul't. flora SSSR **2**, 333–436, Moskva-Leningrad.
Sinskaja, E. N., 1955: Proischoždenie pšenicy. – Problemy botaniki **2**, Moskva-Leningrad.

Dr. František Kühn
Vysoká škola zemědělská
Zemědělská ul. č. 1
66265 Brno, ČSSR

Kulturpflanze XXIX • 1981 • S. 273–286

# Specific and intraspecific classification of *Sorghum* species of the ser. *Bicoloria* (SNOWD.) IVANJUK. et DORON. and of the ser. *Caffra* (SNOWD.) IVANJUK. et DORON.

L. K. IVANJUKOVICH (Leningrad, USSR)

## Summary

The intraspecific classification of cultivated *Sorghum* species of the ser. *Bicoloria* and ser. *Caffra* is discussed. The classification is based on studies or large living and herbarium collections as well as on the typical and authentic material; published monographs on the genus being made use of. In 8 cultivated species 70 varieties are considered, 8 of them were discovered for the first time. The data received contributed to enlarging the intraspecific systematics of a number of species presented in SNOWDEN's monograph (1936).

Parallelism in the evolution for a number of characters, i.e. homologous series, was discovered when intraspecific variability of *Sorghum* was investigated.

The detailed study of the species and of the intraspecific taxons reveals new features of their affinity and differences and contributes to developing the phylogenetic system of the genus.

As a result of the studies varieties were found out which would be of practical value for breeding work.

## Introduction

At present two ways of approach to the classification of *Sorghum* are considered: one of them is based on systematic characters (SNOWDEN 1935, 1936) whereas at the base of the other one (DE WET 1967, 1978; CHANDRASEKHARIAH et al. 1969) there is the principle of absent genetic barriers for crosses of cultivated (2n = 20) and of wild *Sorghum* species (2n = 20).

With the object of systematizing cultivated *Sorghum* species some methods of numerical taxonomy had been used by us: The correlative pleiad method, the estimation method of species correlative structure divergence (IVANJUKOVICH and ARISTARCHOVA 1978). These methods being applied some quantitative morphological characters were treated, such characters which elicit the features of the growing period, appreciably effecting the area of the species, and quantitative morphologic-anatomical leaf characters (along with studies of the epidermis which is in cereal crops of taxonomic value). As a result of these studies (ARISTARCHOVA and IVANJUKOVICH 1976; IVANJUKOVICH and HAMMER 1976; IVANJUKOVICH and JAKUŠEVSKIJ 1973; IVANJUKOVICH and ARISTRACHOVA 1978) more delicate differences were discovered and taxonomic independency was confirmed in a number of *Sorghum* species investigated in the volume of species accepted by SNOWDEN (1936).

According to N. I. Vavilov's conception the Linnean species is a complex system, and when developing the intraspecific classification, we proceeded from this definition.

## Materials and objective of the studies

The genus *Sorghum* Moench is subdivided by us into 7 sections (Ivanjukovich and Doronina 1979). In the present paper we shall deal only with the sect. *Sorghum* (comprising all cultivated species). We divide it into 5 series: ser. *Bicoloria* (Snowd.) Ivanjuk. et Doron., ser. *Caffra* (Snowd.) Ivanjuk. et Doron., ser. *Guineensia* (Snowd.) Ivanjuk. et Doron., ser. *Nervosa* (Snowd.) Ivanjuk. et Doron., ser. *Durra* (Snowd.) Ivanjuk. et Doron. Of the above series only two have been thoroughly studied up to the present: ser. *Bicoloria* and ser. *Caffra*. The series *Bicoloria* and *Caffra* have been studied by us, in the first instance, as the representatives of the series *Bicoloria* have more characters of wild species, whereas the species of the series *Caffra* possess, on the contrary, features of a greater evolutionary progress.

Distinguishing features of the series *Bicoloria* are rather spreading panicles and a characteristic structure of bisexual spikelets: the glumes are leather-like/coriaceous (more seldom bark-like/crustaceous). As a rule they compactly embrace the caryopses which are completely included in the glumes, more seldom they are protruding them, but, if so, than no more than the half.

Distinguishing features of the series *Caffra* are rather compressed and dense panicles, but if loose, the branchlets are abundantly pubescent; the glumes are leather-like (coriaceous), incompactly embrace the caryopses, which are protruding them for a third, or more than the half; more seldom the caryopses are of the same length as the glumes, but then the caryopses are nearly spherical at the tops.

The objective of the studies was to develop the specific and intraspecific classification of the species comprised in the sect. *Sorghum* ser. *Bicoloria* and ser. *Caffra*. The herbarium collections of the N. I. Vavilov All-Union Research Institute of Plant Industry (VIR), the V. L. Komarov Botanical Institute of the USSR Academy of Sciences (BIN), the Institute of Botany in Dushanbe, the type specimens received by the courtesy of our colleagues from Kew (England), as well as the VIR Sorghum living collection have been used as materials for the investigations.

Literature informations including a number of monographs e.g. Hackel (1889), Stapf (1934), Snowden (1935, 1936, 1955) were also used.

Following his predecessors, Snowden (1935, 1936) based the characteristics of species on spikelet and caryopsis morphological characters; as determinative characters he considered the form, size and consistence of the glumes when fruiting, presence or absence of pubescence and its nature (soft, firm), characteristic features of lemma structure: acumintae, awnless or awned; deciduous or retaining when fruiting etc.

The intraspecific classification is elaborated by Snowden on such characters as form and size of the panicle, degree of its compactness, form of bisexual and staminate spikelets and others. Whereas Snowden (1936) not always referred to such characters as presence or absence of awns on lemmae, consistence of inferior glumes in bisexual spikelets when flowering and fruiting and anthocyan coloration of the glumes when flowering. However they have to be considered in the complex of characters. Using these characters and not neglecting those considered by SNOWDEN we made additions to the intraspecific systematics of a number of species comprised in the series *Bicoloria* and *Caffra* and suggested more convenient dichotomous keys to determine specific and intraspecific taxons instead of SNOWDEN's polytomous keys (1936).

## Improving the specific and intraspecific classification in the light of new data

As a result of our studies we are in a position to share SNOWDEN's point of view–he includes 6 species into the series *Bicoloria: Sorghum bicolor* (L.) MOENCH, *S. saccharatum* (L.) MOENCH, *S. miliiforme* (HACK.) SNOWD., *S. simulans* SNOWD., *S. elegans* (KOERN.) SNOWD., *S. notabile* SNOWD. (IVANJUKOVICH and DORONINA 1979a und b) and 5 species into the series *Caffra: S. caffrorum* (RETZ.) BEAUV., *S. caudatum* STAPF., *S. coriaceum* SNOWD., *S. dulcicaule* SNOWD., *S. nigricans* (RUIZ et PAVON) SNOWD.

In connection with our examination of the intraspecific classification of the species comprised in the series *Bicoloria*, a widely cultivated species *S. saccharatum* (L.) MOENCH–sweet Sorghum–including the history how this species was named, should be briefly considered LINNAEUS (1753 and 1763) described the species *Holcus saccharatus* characterized by a spreading panicle with a naked axis, pubescent coriaceous glumes and by the presence of the awns on lemmae. When the species *Holcus saccharatus* was described for the first time specimens of three different species were taken as its base. The fact was established by SNOWDEN (1936) and confirmed by CVELEV (1969), but later on LINNAEUS (1771) made the choice of the lectotype correctly and hence there is no reason to renounce the name *S. saccharatum* and assume it to be doubtful according to SNOWDEN's idea (1936).

FORSKÅL (1775) described a new *Sorghum* species from India *Holcus dochna* FORSK. As seen from the diagnosis: "Paniculae verticillatae ramis erectis . . . calycinae glumae villosae, flavicanthes, apice viridis ovales, muticae . . . corolla parva, . . . alba, pellucida, arista. Semen . . . ovale, compressum, ferrugineum, in calyce persistente, apice dehiscente retentum . . .", these features are identical to those of *H. saccharatus*.

The photo of a typical specimen *H. dochna* sent from Kew, and the comparison of the diagnoses of *H. dochna* and *H. saccharatus* give ground to agree with CVELEV (1969) *H. dochna* and *H. saccharatus* to be the same species.

For the first time the name *Sorgum* for LINNAEUS species of *Holcus* was used by ADANSON (1763). However this name was not legal because it was based on the generic name *Holcus* L. comprising the species *Sorghum* and the species *Holcus* known to him. Later on MOENCH (1794) gave for the first time the diagnosis of the genus *Sorghum* MOENCH and of its two species: *S. saccharatum* and *S. bicolor*. In the same paper MOENCH made also new combinations: *S. saccharatum* (L.) MOENCH and *S. bicolor* (L.) MOENCH. Though SNOWDEN (1936) considers *S. saccharatum* a nomen dubium, avoids to use it, and takes up the name *S. dochna*, to our mind it is not worth while to give up the name *S. saccharatum* widely used in literature, moreover the priority of the name, given by LINNAEUS (1753) should not be neglected.

Koernicke (1873) who considered all the diversity of cultivated *Sorghum* as one species, *Andropogon sorghum* (L.) Brot., in his classification among 8 varieties of the species indicated the variety *Andropogon sorghum* var. *technicus*–broom corn. Later on this variety was transfered by Battandier et Trabut (1895) to the rank of the species–*S. technicum* (Koern.) Batt. et Trab. whereas Snowden (1935, 1936) considered it again as a variety, but now of *S. dochna*–*S. dochna* var. *technicum* (Koern.) Snowd. As the names *S. dochna* and *S. saccharatum* had been attributed to the same species the given variety had to be considered under the priority name *S. saccharatum*.

Snowden (1935, 1936) included both sweet sorghum and broom corn into the same species and Cvelev (1969) thinks it to be correct taking into account insignificant differences in spikelet structure, their common areals, as well as the existence of transitional varieties between both groups. Numerous varieties of *S. saccharatum* have been included by Cvelev (1969) into 2 groups (convarieties) differing in the habit of panicle branching. Of these convar. *saccharatum* comprises varieties with an expressed panicle axis and lateral branchlets equal to them in length, or somewhat shorter. As to the varieties pertaining to the convar. *technicum* (Koern.) Tzvel. they have panicles with markedly shortened axis and lateral branchlets which are longer than the latter. Along with the differences in morphological structure these groups of varieties are differently used economically: the varieties pertaining to convar. *saccharatum* are grown to receive sugary substances and for forage, whereas the varieties belonging to convar. *technicum* are for the production of besoms and brushes.

The detailed study of the intravarietal composition of *Sorghum* species–pertaining to these two series–contributed to our adding new varieties to Snowden's number of varieties and also to suggesting new combinations (Doronina and Ivanjukovich, in print).

Not to enlarge the length of the report no latin diagnoses of new varieties are adduced here. Before long this material will appear in print in Bulletin of Applied Botany, Genetics and Plant Breeding.

Key for Determination of varieties of *Sorghum saccharatum* convar. saccharatum

1. Glumes 7–8 cm long; caryopses fully hidden therein . . . . . . . .
   *S. saccharatum* var. *giganteum* Doron. et Ivanjuk.
\+ Glumes up to 6 mm long . . . . . . . . . . . . . . . . . . . . 2.
2. When flowering glumes thin-crustaceous, when fruiting getting slightly more rough . . . . . . . . . . . . . . . . . . . . . . . . 3.
\+ When flowering glumes crustaceous, when fruiting getting more rough . 5.
3. Lemmae of fruit-bearing bisexual spikelets awnless; when flowering the glumes anthocyan coloured . . . . . . . . . . . . . . . . . . . .
   *S. saccharatum* var. *papyraceum* Doron. et Ivanjuk.
\+ Lemmae of fruit-bearing bisexual spikelets with awn . . . . . . . . 4.
4. Caryopses hidden in the glumes . . . . . . . . . . . . . . . . .
   *S. saccharatum* var. *pulchrum* (Burk. ex Benson et Subba Rao) Doron. et Ivanjuk.
\+ Caryopses protrude out of the glumes . . . . . . . . . . . . . .
   *S. saccharatum* var. *atrum* (Snowd.) Doron. et Ivanjuk.

5. Lemmae with awn . . . . . . . . . . . . . . . . . . . . . . . . 6.
\+ Lemmae awnless . . . . . . . . . . . . . . . . . . . . . . . . . 8.
6. Panicles very compact with a curved axis; glumes straw-coloured . . .
*S. saccharatum* var. *burmanicum* (BURK. ex BENSON et SUBBA RAO) DORON. et IVANJUK.
\+ Panicles loose with a straight axis; glumes from straw-coloured up to black . . . . . . . . . . . . . . . . . . . . . . . . . . . . 7.
7. Panicles loose spreading; inferior branchlets reach $^2/_3$ of the panicle length; glumes reddish, more seldom spotted . . . . . . . . . . .
*S. saccharatum* var. *saccharatum*
\+ Panicles though loose, but not spreading; inferior branchlets not reaching $^2/_3$ of the panicle length; glumes brown-coloured or black . . . . .
*S. saccharatum* var. *formosum* (SNOWD.) DORON. et IVANJUK.
8. Caryopses hidden in the glumes; when fruiting glumes light brown-coloured . . . . . . . . . . . . . . . . . . . . . . . . . . . .
*S. saccharatum* var. *irungu* (BURK. ex BENSON et SUBBA RAO) DORON. et IVANJUK.
\+ Caryopses protruding out of the glumes at the tops and from lateral sides
9. Caryopses whitish; when fruiting glumes black . . . . . . . . . .
*S. saccharatum* var. *melliferum* (SNOWD.) DORON. et IVANJUK.
\+ Caryopses brown-coloured; when fruiting glumes from straw-coloured up to dark brown or vine-red (more seldom spotted) . . . . . . . . . . 10.
10. Panicles obovate; inferior branchelets reach a half of the panicle length; when fruiting glumes vine-red-coloured or spotted . . . . . . . .
*S. saccharatum* var. *obovatum* (HACK.) DORON. et IVANJUK.
\+ Panicles ellipsoidal; inferior branchlets not reaching $^1/_3$ of the panicle length; when fruiting glumes from straw-coloured up to dark brown .
*S. saccharatum* var. *wightii* (HACK.) DORON. et IVANJUK.

*S. saccharatum* convar. *technicum* (KOERN.) TZVEL. comprises two varieties: var. *technicum* (KOERN.) DORON. et IVANJUK., the panicle branchlets are more or less smooth; the panicles 30–80 cm long and var. *corymbosum* (HACK.) DORON. et IVANJUK. with scabrous branchlets and panicles 20–40 cm long.

Key for determination of varieties of *S. bicolor* (L.) MOENCH

1. Panicles loose . . . . . . . . . . . . . . . . . . . . . . . . . 2.
\+ Panicles very compact; caryopses slightly protruding out of the glumes
*S. bicolor* var. *arduini* (J. F. GMEL.) SNOWD.
2. Lemmae of bisexual fruit-bearing spikelets awned . . . . . . . . . 3.
\+ Lemmae of bisexual fruit-bearing spikelets awnless . . . . . . . . .
*S. bicolor* var. *exaristatum* DORON. et IVANJUK.
3. Glumes 5–6 mm long, when fruiting widely obovate; mature caryopses very compact therein and not protruding out of the glumes . . . . .
*S. bicolor* var. *bicolor*
\+ Glumes nearly ovate; about 3,5 mm long and wide but if obovate, then the caryopses protruding out of the glumes . . . . . . . . . . . . . 4.
4. Glumes up to 3,5 mm long, panicles 10–20 cm long . . . . . . . . .
*S. bicolor* var. *subglobosum* (HACK.) SNOWD.

\+ Glumes more than 3,5 mm long, panicles more than 20 cm long . . . . 5.
5. Glumes on all the surface naved, shining, staminate or sexless spikelets, decidious when fruiting . . . . . . . . . . . . . . . . . . . . . . .
*S. bicolor* var. *picigutta* Snowd.
\+ Glumes naked, shining on the dorsal side, at the top and from lateral sides hairy; staminate or sexless spikelets non decidious when fruiting . .
*S. bicolor* var. *charisianum* (Busse et Pilg.) Snowd.

In *S. simulans* Snowd. no new varieties were discovered.

Key for determination of varieties of *S. caffrorum* (Retz.) Beauv.

1. Caryopses large (more than 5,5 mm long) . . . . . . . . . . . . . . 2.
\+ Caryopses less large (less than 5,5 mm long) . . . . . . . . . . . . 4.
2. Caryopses with rostellum . . . . . . . . . . . . . . . . . . . . . .
*S. caffrorum* var. *albidum* (Koern.) Snowd.
\+ Caryopses without rostellum . . . . . . . . . . . . . . . . . . . . 3.
3. Caryopses yellowish (of ivory colour) . . . . . . . . . . . . . . . .
*S. caffrorum* var. *ovoideum* Snowd.
\+ Caryopses reddish, more dark . . . . . . . . . . . . . . . . . . . .
*S. caffrorum* var. *ondongae* (Koern.) Snowd.
4. Caryopses coloured dark-brown . . . . . . . . . . . . . . . . . . .
*S. caffrorum* var. *melanospermum* (Hack.) Snowd.
\+ Caryopses more light-coloured . . . . . . . . . . . . . . . . . . . 5.
5. Lemmae with short awn (less than 5 mm) . . . . . . . . . . . . . 6.
\+ Lemmae awnless . . . . . . . . . . . . . . . . . . . . . . . . . . 7.
6. Caryopses nearly rounded (4,5 mm long, 4 mm wide), rostellum absent .
*S. caffrorum* var. *bicarinatum* (Hack.) Snowd.
\+ Caryopses elliptic (4,2 mm long, 2,8 mm wide) with a distinctly expressed rostellum . . . . . . . . . . . . . . . . . . . . . . . . . .
*S. caffrorum* var. *breviaristatum* Snowd.
7. Panicle branchlets and glumes with a very dense and evenly distributed pubescence . . . . . . . . . . . . . . . . . . . . . . . . . . . 8.
\+ Panicle branchlets and glumes with a less dense and evenly distributed pubescence . . . . . . . . . . . . . . . . . . . . . . . . . . . 10.
8. Caryopses light-brown with a red shade, at the top with a rostellum . . .
*S. caffrorum* var. *brunneolum* Snowd.
\+ Caryopses from whitish and rose-yellow-coloured to brown, without rostellum . . . . . . . . . . . . . . . . . . . . . . . . . . . . . 9.
9. Caryopses dorsiventrally compressed . . . . . . . . . . . . . . . .
*S. caffrorum* var. *caffrorum*
\+ Caryopses biconvex . . . . . . . . . . . . . . . . . . . . . . . . .
*S. caffrorum* var. *lasiorhachis* (Hack.) Snowd.
10. Width of the caryopses surpasses their length, caryopses yellowish, sometimes slightly oblique turned in the glumes . . . . . . . . . .
*S. caffrorum* var. *neesii* (Koern.) Snowd.
\+ Width of caryopses does not surpass their length . . . . . . . . . . 11.
11. Length of caryopses surpasses their width, caryopses elliptical . . . . 13.

\+ Length of caryopses equal or nearly equal to their width, due to this caryopses rounded . . . . . . . . . . . . . . . . . . . . . . . . . 12.

12\. Panicles very compact, glumes brown-coloured, caryopses red-coloured .
*S. caffrorum* var. *densissimum* (BUSSE et PILG.) SNOWD.

\+ Panicles less compact, glumes from light-brown to straw-coloured, caryopses whitish . . . . . . . . . . . . . . . . . . . . . . . . .
*S. caffrorum* var. *albofuscum* (KOERN.) SNOWD.

13\. Caryopses about 5 mm long and 4,2 mm wide, at the top grey-coloured, in their inferior part yellow, sometimes slightly oblique turned in the glumes . . . . . . . . . . . . . . . . . . . . . . . . . .
*S. caffrorum* var. *cinereum* SNOWD.

\+ Caryopses 3,9–4,5 mm long, 2,8–3,5 mm wide, yellowish-brown, in the glumes not turned oblique . . . . . . . . . . . . . . . . . . . .
*S. caffrorum* var. *sapidum* SNOWD.

Key for determination of varieties of *S. caudatum* STAPF

1\. Panicles with very shortened awns, caryopses in bisexual spikelets . . .
*S. caudatum* var. *scoparium* SNOWD.

\+ Panicles with non shortened or slightly shortened awns, with 1–2 caryopses in bisexual spikelets . . . . . . . . . . . . . . . . . . 2.

2\. Panicles loose . . . . . . . . . . . . . . . . . . . . . . . . . 3.

\+ Panicles compact . . . . . . . . . . . . . . . . . . . . . . . . 5.

3\. Panicles longer than 28 cm with one caryopsis in bisexual spikelets . .
*S. caudatum* var. *sudanicum* SNOWD.

\+ Panicles up to 26 cm long, with one or two caryopses in bisexual spikelets . 4.

4\. Panicles one-sided with 2 caryopses in bisexual spikelets with black-coloured glumes . . . . . . . . . . . . . . . . . . . . . . . . .
*S. caudatum* var. *dicarpum* IVANJUK.

\+ Panicles not one-sided with 1 caryopsis in bisexual spikelets, glumes coffee-coloured . . . . . . . . . . . . . . . . . . . . . . . . . .
*S. caudatum* var. *coffeatum* IVANJUK.

5\. Panicle branchlets with spikelets only in the very upper parts, caryopses purple-coloured . . . . . . . . . . . . . . . . . . . . . . . .
*S. caudatum* var. *purpureum* IVANJUK.

\+ Panicle branchlets with spikelets also in the very inferior parts, caryopses not purple-coloured . . . . . . . . . . . . . . . . . . . . 6.

6\. Glumes large, 4,5 mm and more in length . . . . . . . . . . . . 7.

\+ Glumes less large up to 4,5 mm in length . . . . . . . . . . . . 11.

7\. Glumes thin-skinned, usually fragile (more often straw-coloured, sometimes with a reddish or a chestnut tint) . . . . . . . . . . . . .
*S. caudatum* var. *fragile* STAPF

\+ Glumes hard-skinned, not fragile (usually darkish up to black) . . . . 8.

8\. Caryopses white with lilac tint . . . . . . . . . . . . . . . . .
*S. caudatum* var. *feterita* STAPF

\+ Caryopses differently coloured . . . . . . . . . . . . . . . . . 9.

9\. Caryopses with rostellum from red up to brown . . . . . . . . . .
*S. caudatum* var. *gibbum* STAPF

+ Caryopses without rostellum . . . . . . . . . . . . . . . . . . . . . . . 10.
10. Caryopses dark-rosy, yellowish-reddish . . . . . . . . . . . . . .
*S. caudatum* var. *hackelii* (CHIOV.) SNOWD.
+ Caryopses beige-coloured, lemmae awned . . . . . . . . . . . .
*S. caudatum* var. *aristatum* IVANJUK.
11. Glumes of various colour but one of them straw-coloured, sometimes with a reddish tint or spotted . . . . . . . . . . . . . . . . . . . . . . . 12.
+ Glumes of the same coloration, either brown or black (sometimes spotted) 14.
12. Caryopses assymetric, from dorsal side very convex, 3,6 mm in length, 3,4 mm in width . . . . . . . . . . . . . . . . . . . . . . .
*S. caudatum* var. *caudatum*
+ Caryopses of different structure . . . . . . . . . . . . . . . . . . . . 13.
13. Caryopses nearly rounded, compressed dorsiventrally . . . . . . . .
*S. caudatum* var. *durum* SNOWD.
+ Caryopses broad-elliptical (sometimes up to rounded), nearly biconvex . .
*S. caudatum* var. *procerum* SNOWD.
14. Caryopses broad-elliptical, glumes nearly naked or with sparse hairs at the base and on the margin . . . . . . . . . . . . . . . . . . . .
*S. caudatum* var. *colorans* (PILG.) SNOWD.
+ Caryopses of different size . . . . . . . . . . . . . . . . . . . . . . . 15.
15. Caryopses obovate . . . . . . . . . . . . . . . . . . . . . . . . . 16.
+ Caryopses flattened-rounded, if rounded then yellowish- or brownish-red. 17.
16. Caryopses light yellow-coloured, sometimes with a lilac tint . . . . . . .
*S. caudatum* var. *lens* STAPF
+ Caryopses from light–to dark-brown in coloration
*S. caudatum* var. *bantuorum* SNOWD.
17. Caryopses flattened-rounded (in breadth more than in length) . . . . . 18.
+ Caryopses rounded yellowish- or brownish-red coloured . . . . . . .
*S. caudatum* var. *natae* (KOERN.) SNOWD.
18. Caryopses reddish, 4,4 mm long, 5,4 mm wide, biconvex . . . . . . .
*S. caudatum* var. *talodianum* SNOWD.
+ Caryopses dingy-yellow, 4,7 mm long, 5,1 mm wide, biconvex . . . .
*S. caudatum* var. *kerstingianum* (BUSSE et PILG.) SNOWD.

Key for determination of varieties of *S. coriaceum* SNOWD.

1. Rachis very shortened; panicle umbraculiform . . . . . . . . . . . .
*S. coriaceum* var. *umbelliforme* SNOWD.
+ Rachis not shortened, if shortened (seldom met with) then not very strong, panicle of a different form . . . . . . . . . . . . . . . . . 2.
2. Spikelets on pedicels fall off when fruiting for the most part; caryopses coloured from dirty-white to brown (in the type) . . . . . . . . . .
*S. coriaceum* var. *tanganyikae* SNOWD.
+ Spikelets on pedicels do not fall off when fruiting . . . . . . . . . . 3.
3. When fruiting inferior glumes winged at the tops, sometimes with teeth; caryopses not turned oblique . . . . . . . . . . . . . . . . . . . 4.
+ When fruiting inferior glumes wingless at the tops and without teeth; caryopses often turned oblique . . . . . . . . . . . . . . . . . . 5.

4. Inferior glumes up to 4 mm long, broad-elliptical up to rounded; awns short (up to 8 mm) . . . . . . . . . . . . . . . . . . . . .
*S. coriaceum* var. *breviglume* SNOWD.
\+ Inferior glumes longer than 4 mm, elliptical up to long-elliptical; awns long (12 mm and more) . . . . . . . . . . . . . . . . . . . . .
*S. coriaceum* var. *coriaceum*
5. Caryopses broad-elliptical up to rounded, biconvex or slightly dorsiventrally compressed; glumes may be turned inside in the upper part
*S. coriaceum* var. *subinvolutum* SNOWD.
\+ Caryopses ovate, flattened at the facial side, convex at the dorsal side; with narrow, nearly acuminate tops; glumes not turned inside in the upper part . . . . . . . . . . . . . . . . . . . . . . . .
*S. coriaceum* var. *lualabicum* SNOWD.

Key for determination of varieties of *S. nigricans* (RUIZ et PAVON) SNOWD.

1. Panicle umbraculiform . . . . . . . . . . . . . . . . . . . . 2.
\+ Panicle not umbraculiform . . . . . . . . . . . . . . . . . . . 3.
2. Panicle 20 cm long, 15,3 cm wide, with thick branchlets; caryopses either red or purple-black, 4,1 mm long, 3,5 mm wide . . . . . . . . . .
*S. nigricans* var. *nyanzae* SNOWD.
\+ Panicle 11,5–15 cm long, 8,5 cm wide, with thin branchlets, caryopses purple-black, 3,7 mm long, 3,5 mm wide . . . . . . . . . . . .
*S. nigricans* var. *cerevisiae* (STAPF) SNOWD.
3. Panicle compact, dense, ovate, broad-fusiform, elliptical or cylindrical 11–21 cm long, 3–7,5 cm wide, caryopses 3–4 mm long, 2,8–4 mm wide, obovate . . . . . . . . . . . . . . . . . . . . . . . . 4.
\+ Panicle loose or somewhat compact, moderately dense, seldom lobed . . 5.
4. Caryopses honey-coloured . . . . . . . . . . . . . . . . . .
*S. nigricans* var. *concolor* (SCHUM.) SNOWD.
\+ Caryopses red-coloured or purple-black, seldom greyish-brownish . . .
*S. nigricans* var. *peruvianum* (HACK.) SNOWD.
5. Panicle lobed 15–20 cm long, 5–8 cm wide, caryopses 3,9 mm long, 3,9 mm wide, purple-coloured . . . . . . . . . . . . . . . . .
*S. nigricans* var. *lobatum* SNOWD.
\+ Characters of a different type . . . . . . . . . . . . . . . . 6.
6. Caryopses from red up to purple-black . . . . . . . . . . . . . 7.
\+ Caryopses dirty-white or yellowish-brown coloured . . . . . . . . 9.
7. Panicle loose, awn is seen very well along the whole of its length
*S. nigricans* var. *ussiense* (KOERN.) SNOWD.
\+ Panicle moderately dense, awn unseen or sometimes hardly to be seen . . 8.
8. Panicle obovate-oblong, caryopses 3–3,5 mm long, 2,5–3 (seldom 3,5) mm wide, at the base of the caryopsis yellow . . . . . . . . . . . .
*S. nigricans* var. *angolense* (RENDLE) SNOWD.
\+ Panicles elliptically-oblong, caryopses 3,5–4 (seldom 4,5) mm long, 3,5–4 (seldom 3) mm wide, dark, purple also at the base of the caryopsis .
*S. nigricans* var. *stuhlmannii* (KOERN.) SNOWD.

9. Caryopses dirty-white or yellow, glumes are straw-or cherry-coloured when flowering . . . . . . . . . . . . . . . . . . . . . . . . . . .
   *S. nigricans* var. *calcareum* Busse et Pilg.
+ Caryopses yellowish-brown, glumes are from brown- to black-coloured when fruiting . . . . . . . . . . . . . . . . . . . . . . . . .
   *S. nigricans* var. *suffuscum* Snowd.

The species *S. dulcicaule* Snowd. is divided by us into 2 varieties: var. *dulcicaule* with whitish caryopses and straw-coloured glumes and var. *griseo-lilacinum* Ivanjuk., with grey-violet caryopses and black-coloured glumes.

As a result of our studies on intraspecific variability of *Sorghum* affinity in the development of forms in cultivated species from different series was discovered. i.e. homologous series.

For instance, the varieties of *S. bicolor* and *S. saccharatum* of the series *Bicoloria* are as a rule characterized by loose, often spreading panicles or panicles of the umbellate type. However varieties with very compact panicles can be met with: var. *arduini* of the species *S. bicolor* and var. *burmanicum* of the species *S. saccharatum.*

Varieties of the series *Caffra* are usually characterized by compressed, comparatively dense panicles with well expressed axes, but along with the panicles of the type certain varieties (e.g. var. *sudanicum* and var. *dicarpum*–the varieties of *S. caudatum*; var. *ussiense*–*S. nigricans*) have sparse, sometimes one-sided panicles, whereas other varieties are noted for umbellate panicles (e.g. var. *umbelliforme* in the species *S. coriaceum*; var. *nyanzae* and var. *cerevisiae* in the species *S. nigricans*).

In the varieties comprised in *S. bicolor* and in *S. saccharatum* the caryopses are either completely included into the glumes (var. *bicolor* pertaining to *S. bicolor*; var. *giganteum* and var. *pulchrum* pertaining to *S. saccharatum*) or their tops are seen a little (var. *exaristatum*, var. *picigutta*, var. *subglobosum*, var. *arduini*, var. *charisianum*–varieties of *S. bicolor*; var. *atrum*, var. *melliferum*, var. *obovatum*, var. *wightii*–varieties of *S. saccharatum*), sometimes the same variety has forms possessing both characters (var. *formosum*, var. *irungu* of *S. saccharatum*) and still more seldom the caryopses are naked to the half of their length (as in *S. simulans* and in varieties of the species *S. elegans*).

In varieties comprised in the species of the series *Caffra* the caryopses are more often protruding out of the glumes for more than the half of their length, however, the varieties of the species *S. dulcicaule* have caryopses equal to the length of the glumes and not protruding out of them.

Awned and awnless varieties occur in the species of both series: ser. *Bicoloria* and ser. *Caffra*. In the series *Bicoloria* the typical species *S. bicolor* comprises five awned varieties and only one awnless variety var. *exaristatum*. Of the five awned varieties var. *bicolor* and var. *subglobosum* are characterized by long awns (more than 10–12 mm, seldom 9 mm); var. *arduini* has short awns (less than 6 mm); var. *picigutta* and var. *charisianum* sometimes have reduced awns.

The species *S. saccharatum* convar. *saccharatum* comprises: five awned varieties of which var. *saccharatum*, var. *giganteum*, var. *pulchrum*, var. *formosum* have long awns whereas var. *burmanicum* has short ones; five awnless varieties (var. *papyraceum*, var. *irungu*, var. *melliferum*, var. *obovatum*, var. *wightii*) and one variety–var. *atrum*–which may have either developed long or reduced awns.

The varieties var. *technicum* and var. *corymbosum* pertaining to convar. *saccharatum* are usually long-awned.

The varieties comprised in the species of the series *Caffra* are also noted to be either or not. Awnless varieties characteristic of the species comprised in *S. caffrorum* (var. *breviaristatum* and var. *bicarinatum* with a short awn 5 mm excepting), of *S. caudatum* (var. *aristatum* excepting) and others predominant. All awned varieties are observed in the species *S. coriaceum*, long awns being met with in the majority of them (var. *umbelliforme* with the awns 6 mm long excepting).

Sweet-culmed nature is characteristic of the varieties of the species comprised in the series *Bicoloria* and of those in the series *Caffra*.

This character arose as a parallel mutation in the varieties of cultivated species belonging to different series.

Parallelism and evolution are observed also in respect to the characters of caryopsis. Coloration, form, size, absence or presence of the rostellum etc. happen to repeat in the varieties of different species. For example the caryopses with the rostellum (adhered bases of stigmatic branches) are present in the species *S. simulans* (ser. *Bicoloria*) in which varieties are not separated; in var. *gibbum* of the species *S. caudatum*, in var. *albidum* and in var. *brunneolum* of the species *S. caffrorum* (ser. *Caffra*); assymetric caryopses characteristic of the species *S. caudatum* (var. *caudatum*, var. *dicarpum*, var. *gibbum* and others) were noted also in the species *S. coriaceum* (var. *lualabicum*).

Thus, parallel variability of very different characters is discovered in the varieties of the series compared.

The species were found to be not equivalent as to their variability. With species cultivated on account of sweet culms long continued breeding has been carried out and they are presented by a great number of intraspecific taxons.

The widely cultivated species *S. saccharatum* includes, as said above, two convarieties: convar. *saccharatum* comprising 13 varieties and convar. *technicum* with 2 varieties, i.e. 15 varieties on the whole.

The area of this species extends from Europa and Africa to Asia, India, Burma, China and America.

The closely related species *S. bicolor* has only 5 varieties, the fact being explained by a smaller scale of breeding work.

The lower fodder value of the species *S. bicolor* because of its hard-skinned glumes resulted in making it less deserving for breeding work as compared with the species *S. saccharatum* possessing crustaceous glumes. To SNOWDEN's mind some varieties of the former originated from Arabia, India or Burma and later on were introduced from there to many other countries and thus were spread all over the globe.

The species *S. caffrorum* and *S. caudatum* commonly grown for grain, comprise correspondingly 14 and 19 varieties; the species *S. nigricans* cultivated for its sweet culm comprises 10 varieties; *S. coriaceum*–6; the narrowly–localized species *S. dulcicaule* (Nigeria, Belgian Congo) has only two varieties.

The detailed study on intraspecific systematics of *Sorghum* contributed to defining the unequal value of the species in respect to their variability. Species of narrow localization are presented by a minimum number of varieties, whereas species of wide cultivation have a complex taxonomic structure.

Hence, comprehensive studies on the diversity of cultivated *Sorghum* species of the ser. *Bicoloria* and of the ser. *Caffra* contributed to define the specific composition of the series and to note the intraspecific polymorphism on the one hand, and on the other hand to discover varieties within the species which may have practical value: 1) The presence of two fruit-bearing flowers in bisexual spikelets (*S. caudatum* var. *dicarpum*); 2) the presence of a colourless seed coat in the caryopsis, and, consequently, the absence of bitter taste therein (*S. caudatum* var. *kerstingianum*, *S. caudatum* var. *durum*, *S. caudatum* var. *procerum*); 3) the presence of black-coloured glumes in many varieties of *S. saccharatum* convar. *saccharatum* noted for P-vitamin activity and availability of non-toxic pigments—cyanidine, quercetin, pelargenidine (Olifson et al., 1971), due to this fact these varieties may be used in medical and food industry. The detailed study of the species and their intraspecific diversity discovers the traits of affinity and of differences and gives ground to develop the natural phylogenetic system of the genus, as well as to discover material for breeding work.


## Zusammenfassung

Spezies- und intraspezifische Klassifikation von Arten aus der ser. *Bicoloria* (Snowd.) Ivanjuk. et Doron. und der ser. *Caffra* (Snowd.) Ivanjuk. et Doron. der Gattung *Sorghum* Moench

Die intraspezifische Klassifikation kultivierter *Sorghum*-Arten aus der ser. *Bicoloria* und der ser. *Caffra* wird diskutiert. Die Klassifikation basiert sowohl auf dem Studium umfangreicher lebender und Herbarium-Kollektionen als auch auf Typus- und Originalmaterial; publizierte Gattungsmonographien wurden genutzt. Bei 8 Kulturarten werden 70 Varietäten akzeptiert, von denen 8 zum ersten Mal gefunden wurden. Die erzielten Ergebnisse führten zu einer Erweiterung der intraspezifischen Systematik einer Anzahl der Arten, die in Snowden's Monographie (1936) aufgeführt sind.

Parallelevolution für eine Anzahl von Merkmalen, d. h. homologe Reihen, wurde bei der Untersuchung der intraspezifischen Variabilität von *Sorghum* nachgewiesen.

Die ausführliche Untersuchung der Arten und der intraspezifischen Taxa läßt neue Grundzüge ihrer Ähnlichkeiten bzw. Unterschiede erkennen und trägt zur Entwicklung des phylogenetischen Systems der Gattung bei.

Im Ergebnis der Untersuchungen wurden Varietäten gefunden, die von praktischem Wert für die Züchtung sein können.



## Краткое содержание

Видовая и внутривидовая классификация видов ser. *Bicoloria* (Snowd.) Ivanjuk. et Doron. и ser. *Caffra* (Snowd.) Ivanjuk. et Doron. рода *Sorghum* Moench

Обсуждается внутривидовая классификация культурных видов сорго ser. *Bicoloria* и ser. *Caffra*. Классификация основана как на изучении богатых живых и гербарных коллекций, так и на аутентичном материале и типах; использованы также опубликованные монографии рода. Принимаются 8 культурных видов с 70 разновидностями, из которых 8 были найдены впервые. Результаты исследований привели к расширению внутривидовой систематики ряда видов, приводимых в Снауденовской монографии (1936).

При исследовании внутривидовой изменчивости сорго для ряда признаков была показана параллельная эволюция, т. е. обнаружены гомологические ряды.

Детальное изучение видов и внутривидовых таксонов дает возможность выявить новые данные их сходства и различий и способствует построению филогенетической системы рода.

В результате исследований были найдены разновидности, которые могут получить практическое значение в селекционной работе.

## Literatur

Adanson, M., 1763: Families des plantes. – Paris.

Aristarchova, M. L. i L. K. Ivanjukovich, 1976: Korreljacionnaja struktura nekotorych količestvennych priznakov kul'turnych vidov Sorghum. – Bot. Žur. **61**, 219–226.

Battandier, J. A. et L. Trabut, 1895: Flora d'Alger et catalogue des plantes d'Algerie, Monocotyledones.

Cvelev, N. N., 1969: Zametki o zlakach flory SSSR. – Novosti sistematiki vysšich rastenij **5**, 15–30.

Chandrasekhariah, S. R. et al., 1969: Multivariate analysis of genetic divergence in Eu-Sorghum. – Proc. Nat. Inst. Sci., India, part B, **35**, 2.

de Wet, J. M. J., 1978: Systematics and evolution of *Sorghum* sect. *Sorghum* (Gramineae). – Amer. J. Bot. **65**, 477–484.

–, and J. P. Huckabay, 1967: The origin of *Sorghum bicolor*. II. Distribution and domestication. – Evolution **21**, 287–802.

Doronina, Ju. A. i L. K. Ivanjukovich: Vnutrividovaja klassifikacia *S. bicolor* (L.) Moench i *S. saccharatum* (L.) Moench, in print.

Forskål, P., 1775: Flora aegyptiaco-arabica. – Hauniae (Copenhagen).

Hackel, E., 1889: *Andropogoneae*. – In: De Candolle, Monographiae phanerogamarum **6**, 499–520. – Parisiis.

Ivanjukovich, L. K., 1977: Anatomičeskoje stroenie lista nekotorych vidov sorgo (*Sorghum* Moench subgen. *Sorghum*). – Bot. Žur. **62**, 578–588.

– i M. L. Aristarchova, 1978: Korreljacionnaja struktura količestvennych anatomičeskich priznakov lista nekotorych vozdelyvaemych vidov *Sorghum* (*Poaceae*). – Bot. Žur. **63**, 1751–1758.

– i Ju. A. Doronina, 1979a: Obzor vidov podserii *Bicoloria* Snowd. sekcii *Sorghum* roda *Sorghum* Moench (*Poaceae*). – Bot. Žur. **64**, 751–756.

– i –, 1979b: Sekcionnoje delenie roda *Sorghum* Moench (*Poaceae*). – Bot. Žur. **64**, 1672–1673.

– und K. Hammer, 1976: Bemerkungen zu einigen kult vierten *Sorghum*-Sippen auf der Grundlage ihrer Korrelationsstruktur. – Kulturpflanze **24**, 191–204.

– i E. S. Jakuševskij, 1973: Anatomičeskoje stroenie lista nekotorych vidov sorgo (*Sorghum* Moench subgen. *Sorghum*). – Bot. Žur. **58**, 1028–1037.

Koernicke, F., 1873: Systemat. Übersicht der Cerealien. – Bonn.

Linnaeus, C., 1753: Species Plantarum. – Holmiae (Stockholm), v. 1.

–, 1763: Species Plantarum. – Holmiae (Stockholm), v. 2.

–, 1771: Mantissa Plantarum altera. – Holmiae (Stockholm).

Moench, C., 1794: Methodus plantas horti botanici et agri. – Marburg.

Olifson, L. E. et al., 1971: O chimičeskoj prirode krasitelja, vydelennogo iz plenki zerna sacharnogo sorgo. – Žur. piščevaja technologija, 39–41.

Snowden, J. D., 1935: A classification of the cultivated sorghums. – Kew Bull. **5**, 221–255.

–, 1936: The Cultivated Races of *Sorghum*. – London.

–, 1955: The wild fodder sorghums of the sect. *Eu-Sorghum*. – J. Linn. Soc. **55**, 191–260.

Stapf, O., 1934: *Sorghum*. – In: Flora of Tropical Afrika **9**, 104–154.

Dr. L. K. Ivanjukovich
N. I. Vavilov All-Union Institute of Plant Industry
190000, Leningrad, USSR

Kulturpflanze XXIX · 1981 · S. 287–296

# Problems of *Papaver somniferum*-classification and some remarks on recently collected European poppy land-races

KARL HAMMER (Gatersleben, GDR)

## Summary

The principles of formal classification of cultivated plants are discussed with reference to *Papaver somniferum*. This species can be divided into 3 subspecies–the ancestral wild race ssp. *setigerum* (DC.) CORB. and the two cultivated races ssp. *somniferum* and ssp. *songaricum* BASIL. differing in geographical distribution and characters of the stigmatic lobes. In both of the cultivated races two convars are accepted on the basis of their indehiscent vs. shattering capsules. Varieties are defined according to DANERT (1958) using seed- and flower-colours.

Poppies, recently collected in parts of central Europe, are classified according to the proposed system. Unlike other crops, very variable land-races of poppies can still be found in this region. About 20 percent of the totally accepted varieties could be found, containing also races from ssp. *songaricum* and types with shattering capsules. The handling and maintaining of this material in the Gatersleben gene-bank is briefly discussed.

For the classification of poppies in the Gatersleben collection the system of DANERT (1958) is used. This is a formal taxonomic treatment of the infraspecific variability of this cultivated species.–There are tendencies in studies of the taxonomy of wild and also cultivated plants to more or less abandon infraspecific formal classifications. The reasons for this trend may be firstly some doubts and uncertainties in the use of infraspecific categories and the boundaries between them, secondly, difficulties in tracing infraspecific names and therefore nomenclatural problems, and thirdly insufficient suitability of the infraspecific systems for taxonomic investigations; consequently the International Code of Botanical Nomenclature is sometimes argued to be unsatisfactory in the information it provides for infraspecific variation (BURTT 1970, RAVEN 1974). On the other hand the importance of infraspecific classification for promoting the study of infraspecific variation has been stressed (VALENTINE 1975, STACE 1976). Such classifications have turned out to be very useful as a framework for the study of infraspecific variability in living collections of cultivated plants and their relatives (gene-banks). In these collections the populations are propagated under conditions which are different from those in their native sites and therefore some of the population constituents will be propagated whereas others may show poor development and will eventually disappear from the population. Thus populations of crop plants and their relatives in Gatersleben are separated into different morphological types, to prevent losses. Any later outcrossing and occasional mixtures between the different types can also be easily detected by this method. The principles of maintaining collections based on the study of morphological variation have been described by LEHMANN and MANSFELD (1957). In our opinion there is a need for a more detailed infraspecific classification at least as a framework for further taxonomical investigations. For this purpose informal classifications, e.g. classifying poppies according to their use for baking and cooking or as a source of oil, for medicinal and narcotic uses, and as ornamentals, are not sufficient.

A problem concerning formal infraspecific classifications in cultivated plants is the use of the categories. Cultivated plants in general show a greater amount of genetic variation then wild plants and they differ in many respects from wild species. Therefore some authors (MANSFELD 1954, JIRÁSEK 1966, JEFFREY 1968, PARKER 1978) have advocated the use of special formal categories for every or nearly every rank in infraspecitic classifications of cultivated plants in contrary to wild species. The highlight of these efforts is the classification proposed by JIRÁSEK (1966) which contains 12 categories from specioid (for species) to subcultivar. JEFFREY (1968) considers the specioid category to be unjustified because, in his opinion, the species links cultivated plant systematics with wild plant systematics. There are closer connections between the two systematics because the evolutionary mechanisms in wild and cultivated species are the same. There is also considerable variability in some wild plants, e.g. in autogamous (Aegilops–HAMMER 1980) or apomictic species. Weeds and plants in heavily grazed areas also show man's influence in promoting evolution in certain directions. Therefore, two systematics documented by completely different infraspecific or even specific categories cannot be accepted.

Some of the Gatersleben activities in cultivated plant taxonomy are shown in table 1. The number of different taxa at the lowest rank of categories depends on the variability of the crop which in turn is largely a function of the breeding system. Strongly allogamous crops very often show continuous variability and therefore the number of distinct races, e.g. on the variety-level, is low in *Brassica oleracea* and *Beta vulgaris*, whereas autogamous species have a discontinuous variability giving the possibility of distinguishing many infraspecific classes (e.g. *Hordeum vulgare* and *Triticum aestivum*). Self-compatible allogamous crops are in some respects intermediate. To this last group belongs *Papaver somniferum* which is characterized by its relatively high and classifyable variability.

To some extend this comparison (cf. table 1) is based on the subjective judgement of the authors. But most of the cited authors feel the need to fill the gap between ssp. and var. by the category convar. according to the ideas of FRIEDRICH ALEFELD who used the concept of "Varietäten-Gruppe" already in 1866 (MANSFELD 1953, HELM 1964). The convar. has turned aut to be a very useful category for polymorphous cultivated plants. A good example is in its use in the barley classification by MANSFELD (1950). Earlier authors beginning from LINNÉ considered some of the barley-races as good species. However MANSFELD put them into one species because of the lack of genetic distance between them, e.g. convar. *distichon* and convar. *intermedium*. To give the rank of subspecies to these races is not advisable because the geographical or ecological differences are not sufficient. But for the rank var. there are striking morphological differences and also some

Table 1

Some of the Gatersleben activities in cultivated plant taxonomy

| Species | Authority of classification | number of different taxa at the lowest taxonomical rank | categories used | | | | |
|---|---|---|---|---|---|---|---|
| *Brassica oleracea* L. | HELM (1963) | 31 | | convar. | var. | | f. |
| *Beta vulgaris* L. | HELM (1957) | 17 | ssp. | convar. | provar. var. | | f. |
| *Vicia faba* L. | HANELT (1972) | 6 | ssp. | | var. | subvar. | cv.-group |
| *Nicotiana rustica* L. | DANERT (1963) | 6 | | | var. | | |
| *Papaver somniferum* L. | DANERT (1958) | 52 | | convar. | var. | | |
| *Triticum aestivum* L. s. l. | MANSFELD (1951) | 404 | | | var. | | |
| *Hordeum vulgare* L. s. l. | MANSFELD (1950) | 192 | | convar. | var. | | |
| *Pisum sativum* L. s. l | LEHMANN (1954) | 101 | ssp. | convar. | var. provar. | | |
| *Lycopersicon esculentum* MILL. | LEHMANN (1955) | 24 | | convar. | var. | | |
| *Glycine max* (L.) MERR. | LEHMANN (1962) | 97 | | convar. | var. | | |
| *Linum usitatissimum* L. | KULPA and DANERT (1962) | 28 | | convar. | var. | | |

geographical patterns of diversity. For this reason the use of the category of convar. seems to be a good decision.

Based on the classification of DANERT (1958) the following ideas concerning Papaver somniferum have been developed. It was originally thought that *P. somniferum* and its possible ancestor *P. setigerum* differed in chromosome numbers, the former having $2n = 22$ and the latter $2n = 44$. Recently diploid races of *P. setigerum* have been found (HAMMER and FRITSCH 1977). This result suggests that *P. setigerum* is the ancestor of *P. somniferum*. Because of the absence of crossing-barriers as well as no gross morphological differences between them they can be accepted as subspecies and the combination *P. somniferum* ssp. *setigerum* (DC.) CORB. can be used. To bring cultivated and ancestral wild races into one species has been advocated by many authors, e.g. MANSFELD (1953), HARLAN and DE WET (1971).

In this way the autonym ssp. *somniferum* is accepted (ICBN-Article 26) for at least a part of the cultivated poppies. This part has to contain Linnaeus' type species of *P. somniferum*. It will be shown here that there is a need to accept yet another subspecies. The cultivated poppies of central Europe have sulcate lobes of the stigmatic disc with dentate margin whereas the surface of the lobes in other races preferably from southern regions is flat with entire margin. This morphological difference has been considered as taxonomically very important. ROTHMALER (1949) thought this difference to be even sufficient to enclose flat-lobed races of *P. somniferum* into *P. glaucum* BOISS. et HAUSSKN. (*P. glaucum* ssp. *indicum* (VESS.) ROTHMALER) considering that they are descended from *P. glaucum* which shows the same character.

This treatment cannot however be accepted because *P. glaucum* has a different chromosome number (2n=14) and *P. somniferum* apparently arose in the West Mediterranean region (SCHULTZE-MOTEL 1979) far away from the asiatic *P. glaucum*. But there is good reason to accept the flat-lobed races as another subspecies of *P. somniferum*.

In Europe races with sulcate stigmatic lobes predominate whereas in Asia cultivars frequently have flat ones. This tendency was originally very clearly expressed but nowadays it is less striking as will be shown in the discussion on the recently collected European land-races. However, a geographical difference can be seen even nowadays for the morphologically distinct races. Therefore the rank of subspecies is recommended in its traditional use (DU RIETZ 1930). The correct name for this subspecies has been chosen from the 7 subspecies described by BASILEVSKAJA (1928) by MOWAT and WALTERS (1964) in the "Flora Europaea" (ssp. *songaricum* BASIL.). Thus two cultivated and one wild subspecies are distinguished within *P. somniferum*.

As has been already mentioned, DANERT (1958) also used the category convar. For the classification of *Papaver somniferum* he has chosen the characters of the stigmatic disc combined with seed colours as criteria for this category. In this paper the disc characters are proposed as differential characters of the subspecies and indehiscent vs. shattering capsules are proposed as differential characters for the convar. Indehiscent capsules are typical of the progress in domestication and all modern cultivars show this character. Here the problem of reticulate variation is seen which is typical of the infraspecific level especially for cultivated plants. Seed colours are, of course, also influenced by domestication, and light seeds are mostly found in modern cultivars whereas the ancestral ssp. *setigerum* has only dark seeds. In landraces there are forms with shattering capsules and dark seeds, and forms with indehiscent capsules and light seeds as well as the reciprocal combinations. In the case of reticulate variation it is difficult to decide which is the character of highest taxonomical status. So a decision had to be made and the dehiscent vs. indehiscent capsules were judged as more important characters for delimiting the category convar. in *P. somniferum*. In this connection it has to be stated that classifications on the infraspecific level are often artificial. MANSFELD (1954) has recommended that phylogenetic results must also be considered even though in artificial systems. In this way a better understanding of such systems is guaranteed.

The proposed system of *Papaver somniferum* reflects a growing artificiality in the lower levels of infraspecific classification. For the distinction of varieties, seed- and flower-colours are used. For the seed-colours a domestication trend has alread been pointed out. For the flower-colours a similar trend can be shown but it is not so clear cut. The varieties are principally defined according to DANERT (1958).

The proposed system of classifying *P. somniferum* is shown in table 2. For two

Table 2

Proposed system of classifying *Papaver somniferum*

| | | | | | |
|---|---|---|---|---|---|
| species level | | *Papaver somniferum* L. | | | |
| ssp. level | | ssp. *somniferum* | | ssp. *songaricum* BASIL. | ssp. *setigerum* (DC.) CORB. |
| convar. level | | convar. *somniferum* | – [1] | – [1] | convar. *orientale* DANERT s. l. |
| | | indehiscent capsules | dehiscent capsules | indehiscent capsules | dehiscent capsules |
| var. level | seeds white, yellowish or pink | var. *somniferum* | var. *dinocarpum* ALEF. | var. *albescens* VESS. | var. *orientale* DANERT |
| | | var. *candidum* VESS. | var. *rubrospermum* VESS. | var. *rubicundum* VESS. | var. *apertum* DANERT |
| | | var. *roseolum* VESS. | var. *sanguineum* VESS. | var. *rhodanthum* VESS. | var. *foratum* DANERT |
| | | var. *paeoniflorum* ALEF. | var. *rutilum* DANERT | var. *igneum* DANERT | var. *fulgidum* DANERT |
| | | var. *macrocarpum* COSS. | var. *hussenotii* ALEF. | var. *parmulatum* DANERT | var. *maculosum* DANERT |
| | | var. *papyrinum* DANERT | var. *pictiflorum* DANERT | var. *apiatum* VESS. | var. *hapalanthum* DANERT |
| | | var. *clausum* DANERT | var. *tenerum* DANERT | var. *limboflorum* DANERT | var. *palleolum* DANERT |
| | seeds | var. *haageanum* ALEF. | var. *contrasticum* DANERT | var. *mundum* DANERT | var. *leucomelum* DANERT |
| | light | var. *coerulescens* ROTHM. | var. *pallidum* ROTHM. | var. *livens* VESS. | var. *gaucescens* (ROTHM.) DANERT |
| | grey, light | var. *oculatum* DANERT | var. *madritense* VESS. | var. *halonatum* DANERT | var. *praetextum* DANERT |
| | blue or | var. *nigrum* HAYNE | var. *quassandum* ALEF. | var. *rubidum* VESS. | var. *rotundilobum* DANERT |
| | dark blue | var. *serenum* DANERT | var. *spilanthum* DANERT | var. *sigillatum* DANERT | var. *ocellatum* DANERT |
| | | var. *subgriseum* VESS. | var. *subviolaceum* VESS. | var. *nubeculosum* DANERT | var. *poriferum* DANERT |

[1] Combination not yet available

convars there is no combination available. A formal name and description will be given in a later paper.

The lowest rank in the hierarchy for cultivated plants is the cultivar. As a formal category it may be called an "atomic" unit (JEFFREY 1968) but from the genetical point of view it may consist of a mixture of lines, especially in named land-races. The cultivars can be grouped into varieties. For some exceptional morphological character combinations or freaks, e.g. filled flowers and many capsules per stem (fig. 1), the formal category forma may be used. But it is not advisable to do

Fig. 1 In different botanical varieties of *P. somniferum* forms with many capsules per stem can be found

so, because, if there is a f. *plenum* in one var., then another epithet has to be chosen for every other var. for the formae with the same "flore pleno-character". This system would overload the easily surveyable morphological system of *P. somniferum* as a framework for taxonomic and practical use.

In most of the main crops intensive breeding work is carried out. Therefore, at least in Europe, many highly productive new cultivars are released which rapidly displace the old land-races with lower yields. In crops of only secondary or lesser importance, land-races can often compete with the only sporadically bred new cultivars because the land-races are usually better adapted to the environmental conditions and the peasants are used to them or prefer them by tradition. One good example of such a secondary crop is *P. somniferum* as has been demonstrated by the results of two collecting missions to Czechoslovakia (KÜHN et al. 1976, 1980), two missions to Poland (HAMMER and HANELT 1979, SCHULTZE-MOTEL and HANELT 1979), and a newspaper-based collecting mission in the GDR (HAMMER et al. 1977). The results will now be summarized here.

About 20 percent of the varieties of our accepted classification were found during the collecting trips to parts of central Europe. Forms with indehiscent capsules predominate but there are still some primitive types with shattering capsules as can be seen from table 3. The frequency of races in the collected material is shown but this represents only a small part of the variation because

Table 3
Varieties and their frequency of poppy-landraces collected in parts of central Europe

| ssp. *somniferum* seeds light | ssp. *somniferum* seeds dark | ssp. *songaricum* seeds light | ssp. *songaricum* seeds dark | |
|---|---|---|---|---|
| 19 | 1 | 4 | – | dehiscent capsules |
| 71 | 1 | 12 | – | indehiscent capsules |

| ssp. *somniferum* indehiscent capsules | ssp. *somniferum* dehiscent capsules | ssp. *songaricum* indehiscent capsules | ssp. *songaricum* dehiscent capsules |
|---|---|---|---|
| var. *nigrum* | var. *pallidum* | var. *sigillatum* | var. *glaucescens* |
| var. *coerulescens* | var. *quassandum* | var. *livens* | |
| var. *serenum* | var. *spilanthum* | | |
| var. *papyrinum* | var. *rutilum* | | |

exeptional forms were especially collected. Besides the poppies with shattering capsules, those from the ssp. *songaricum* are especially interesting.

This race could be found in southern parts of Slovakia, mostly. Only one sample has been collected from the foreland of the Lower Beskids in Poland indicating that there may be a migration from south to north for this race.

*P. somniferum* is only facultatively autogamous and there is a considerable degree of outcrossing. Therefore the flowers have to be isolated prior to anthesis when the collected material is propagated in the fields in Gatersleben. The variable populations of land-races have to be traced back to single morphologically characterized entities to guarantee the maintenance of all morphological traits, as has already been shown. The formal classification has turned out to be very useful as a framework for the selection of morphological types as well as for the quick demonstration of the collected material (table 3). In further investigations on the collected material supplementary information will be added to this framework, e.g. concerning seed colour, seed morphology (Fritsch 1979), leaf shape, plant height, capsule size, thus forming a sufficient data bank of taxonomical information as proposed by Hanelt (1973).

In the collecting areas some small-flowered forms of poppies with shattering capsules were found in beet fields or gardens. These were occasionally cropped, but generally regarded as weeds. These types are descendants of old land-races which became weeds or ruderal plants. Most of the reports about ssp. *setigerum* in central Europe are of this type. They cross freely with the cultivated poppies and the character of dehiscent capsules may derive to a considerable extent from those weedy forms. The question arises, as to where the weedy and ruderal forms fit into the formal system of classification of *P. somniferum*. They can be described according to the accepted morphological sytem and then the central European forms will mostly belong to var. *pallidum*. During a trip through Spain (Lehmann and Hammer 1979) a weedy form belonging to var. *contrasticum* was found (fig. 2). If these races are determined according to the present system, this determination should be supplemented by the informal remark weedy or ruderal race.

In the material collected from the GDR poppies were found with small shat-

Fig. 2 Ruderal race of the var. *contrasticum* collected in Spain (right) and land-race from the GDR with acute lobes of the stigmatic disc (left)

tering capsules and acute lobes of the stigmatic disc, resembling the *P. orientale* L.-group for these characters (fig. 2). Until now this form is an exceptional finding in *P. somniferum* and will be studied with regard to possible introgression.

These are only some interesting examples from recently collected European poppy land-races which will be maintained in the Gatersleben gene-bank for further taxonomical and evolutionary studies as well as for their use in practical breeding programmes.

## Zusammenfassung

Probleme der Klassifikation von *Papaver somniferum* und einige Bemerkungen zu unlängst gesammelten europäischen Mohn-Landsorten

Grundzüge der formalen Klassifikation von Kulturpflanzen unter besonderer Berücksichtigung von *Papaver somniferum* werden diskutiert. Diese Art kann in drei Subspecies aufgegliedert werden – die wilde Ausgangssippe ssp. *setigerum* (DC.) CORB. und die beiden Kultursippen ssp. *somniferum* und ssp. *songaricum* BASIL., die sich durch ihre geografische Verbreitung und Merkmale der Narbenscheiben voneinander abgrenzen lassen. In beiden Kultursippen werden je zwei Convarietäten unter Zugrundelegung ihres Schließ- bzw. Schüttmohncharakters unterschieden. Varietäten werden nach DANERT (1958) anhand der Samen- und Blütenfarben definiert.

Die in jüngster Zeit in einigen Teilen Zentraleuropas gesammelten Mohne werden nach dem vorgeschlagenen System klassifiziert. Im Gegensatz zu anderen Kulturpflanzen können in diesen Gebieten noch sehr variable Mohn-Landsorten gefunden werden. Ungefähr 20% der im System aufgeführten Varietäten konnten gesammelt werden, einschließlich einiger Vertreter der ssp. *songaricum* und von Schüttmohnsippen. Die Bearbeitung und Erhaltung dieses Materials in der Genbank Gatersleben wird kurz diskutiert.

## Краткое содержание

Проблемы классификации *Papaver somniferum* и некоторые замечания по поводу недавно собранных европейских местных сортов мака

Рассматриваются основны формальной классификации культурных растений, причём особое внимание уделяется *Papaver somniferum*. Этот вид может быть расчленён на три подвида: дикорастущая исходная форма – ssp. *setigerum* (DC.) CORB. и обе культурных формы – ssp. *somniferum* и ssp. *songaricum* BASIL., которые разграничены их географическим распространением и признаками лопастей рыльца. Внутри обеих культурных форм можно выделить по две группы разновидностей (convarietas) на основании характера их коробочки – закрытой или вскрывающейся. Разновидности определяются, как это предложил ДАНЕРТ (1958) по цвету семян и цветков.

Собранные в последнее время, в некоторых частях Центральной Европы, формы мака классифицируются по предложенной системе. В противоположность другим культурным растениям, в этих областях можно найти ещё очень изменчивые местные сорта мака. Удалось собрать около 20% разновидностей, приведённых в системе, включая некоторых представителей ssp. *songaricum* и формы высыпающихся маков.

Кратко обсуждается обработка и сохранение этого материала в генном банке Гатерслебена.

## Literature


BASILEVSKAJA, N. A., 1928: On the races of Opium poppy, growing in Semiretchic, and the origin of their culture (Russ., Engl. summary). – Trudy prikl. bot., gen. i sel. **19** (2) 95.

BURTT, B. L.; 1970: Infraspecific categories in flowering plants. – Biol. J. Linn. Soc. **2**, 233–238.

DANERT, S., 1958: Zur Systematik von *Papaver somniferum* L. – Kulturpflanze **6**, 61–88.

–, 1963: Zur Systematik von *Nicotiana rustica* L. – Kulturpflanze **11**, 535–562.

DAVIS, P. H., 1978: The moving staircase: a discussion on taxonomic rank and affinity. – Notes Roy. Bot. Gard. Edinb. **36**, 325–340.

DU RIETZ, E., 1930: The fundamental units of biological taxonomy. – Svensk Bot. Tidskr. **24**, 333–428.

FRITSCH, R., 1979: Zur Samenmorphologie des Kulturmohns (*Papaver somniferum* L.). – Kulturpflanze **27**, 217–227.

HAMMER, K., 1980: Zur Taxonomie und Nomenklatur der Gattung *Aegilops* L. – Feddes Rep. **91**, 225–258.

– und R. FRITSCH, 1977: Zur Frage nach der Ursprungsart des Kulturmohns (*Papaver somniferum* L.). – Kulturpflanze **25**, 113–124.

– und P. HANELT, 1979: Botanische Ergebnisse einer Reise in die VR Polen 1976 zur Sammlung autochthoner Landsorten von Kulturpflanzen. – Kulturpflanze **27**, 109–149.

–, – und C. TITTEL, 1977: Sammlung autochthoner Kulturpflanzen auf dem Gebiet der DDR. – Kulturpflanze **25**, 89–99.

HANELT, P., 1972: Die infraspezifische Variabilität von *Vicia faba* L. und ihre Gliederung. – Kulturpflanze **20**, 75–128.

–, 1973: Merkmalsvariabilität bei *Vicia faba* L. I. Künstliche morphologische Systeme bei Kulturpflanzen-Arten. – Kulturpflanze **21**, 55–60.

Hanelt, P. und J. Schultze-Motel, 1979: Bericht über die Reise in die VR Polen zur Sammlung autochthoner Sippen von Kulturpflanzen im Jahre 1978. – Kulturpflanze **27**, 151–163.

Harlan, J. R. and J. M. J. de Wet, 1971: Toward a rational classification of cultivated plants. – Taxon **20**, 509–517.

Helm, J., 1957: Versuch einer morphologisch-systematischen Gliederung der Art *Beta vulgaris* L. – Züchter **27**, 203–222.

–, 1963: Morphologisch-taxonomische Gliederung der Kultursippen von *Brassica oleracea* L. – Kulturpflanze **11**, 92–210.

–, 1964: 100 Jahre Kulturpflanzen-Taxonomie. Dr. Friedrich Alefeld zum Gedächtnis. – Kulturpflanze **12**, 75–92.

Jeffrey, C., 1968: Systematic categories for cultivated plants. – Taxon **17**, 109–114.

Jirásek, V., 1966: Systematika kulturních rostlin a jegí třídicí kategorie. – Preslia **38**, 267–284.

Kühn, F., K. Hammer und P. Hanelt, 1976: Botanische Ergebnisse einer Reise in die ČSSR 1974 zur Sammlung autochthoner Landsorten von Kulturpflanzen. – Kulturpflanze **24**, 282–347.

–, – und –, 1980: Botanische Ergebnisse einer Reise in die ČSSR 1977 zur Sammlung autochthoner Landsorten von Kulturpflanzen. – Kulturpflanze **28**, 183–226.

Kulpa, W. und S. Danert, 1962: Zur Systematik von *Linum usitatissimum* L. – Kulturpflanze, Beih. **3**, 341–388.

Lehmann, C. O., 1954: Das morphologische System der Saaterbsen (*Pisum sativum* L. em. Gov. ssp. *sativum*). – Züchter **24**, 316–337.

–, 1955: Das morphologische System der Kulturtomaten (*Lycopersicon esculentum* Mill.). – Züchter, 3. Sonderh., 64 pp.

–, 1962: Ein Beitrag zur Systematik der Sojabohnen (*Glycine max* (L.) Merr.). – Züchter **32**, 229–249.

– und K. Hammer, 1979: Bericht über die Reise nach Südwest-Spanien zur Sammlung kultivierter und wildwachsender Leguminosen im Jahre 1978. – Kulturpflanze **27**, 97–108.

– und R. Mansfeld, 1957: Zur Technik der Sortimentserhaltung. – Kulturpflanze **5**, 108–138.

Mansfeld, R., 1950: Das morphologische System der Saatgerste, *Hordeum vulgare* L. s. l. – Züchter **20**, 8–24.

–, 1951: das morphologische System des Saatweizens, *Triticum aestivum* L. s. l. – Züchter **21**, 41–60.

–, 1953: Zur allgemeinen Systematik der Kulturpflanzen I. – Kulturpflanze **1**, 138–155.

–, 1954: Zur allgemeinen Systematik der Kulturpflanzen II. – Kulturpflanze **2**, 130–142.

Mowat, A. B. and S. M. Walters, 1964: *Papaver* L. – In: Tutin, T. G. et al., Flora Europaea **1**, 247–250. – Cambridge.

Parker, P. F., 1978: The classification of crop plants. – In: H. E. Street (Editor), Essays in Plant Taxonomy. – Acad. Press, London, New York, San Francisco. – pp. 97–124.

Raven, P. H. 1974: Proposals for the simplification of infraspecific terminology. – Taxon **23**, 828–831.

Rothmaler, W., 1949: Notulae systematicae. 4. Papaveres. – Index Seminum Inst. Gaterslebense, 42.

Schultze-Motel, J., 1979: Die urgeschichtlichen Reste des Schlafmohns (*Papaver somniferum* L.) und die Entstehung der Art. – Kulturpflanze **27**, 207–215.

Stace, C. A., 1976: The study of infraspecific variation. – Curr. Adv. Plant Sci. **23**, 513–523.

Valentine, D. H., 1975: The taxonomic treatment of polymorphic variation. – Watsonia **10**, 385–390.

Dr. K. Hammer
Zentralinstitut für Genetik und
Kulturpflanzenforschung
DDR 4325 Gatersleben

Kulturpflanze XXIX · 1981 · S. 297–300

# Intraspecific classification of water-melon under cultivation

T. B. FURSA (Leningrad, USSR)


## Summary

The ecogeographical classification of water-melon eliciting the principles governing the development of the above crop is presented. 10 ecogeographical groups were singled out differing in morphologic-anatomical characters and physiological properties of varieties. Based on statistical analysis the groups are included into 3 geographical types: the Russian, the Asiatic and the Eastern types. The idea is suggested on the initial position occupied by the Asiatic type in the evolution of water-melon.


## Introduction

This paper deals with the study of ecogeographical classification in cultivated plants. The famous botanist and geographer N. I. VAVILOV, first director of the Institute for Plant Industry in Leningrad (VIR), was the first to use this method in the Soviet Union which is presented here.

The proposed ecogeographical classification of cultivated water-melons is no taxonomic classification but only a complement to it and helps to study collections for practical purposes.

Till recently water-melon, cultivated on rather large areas of the USSR, was scarcely studied botanically. A many-years work with the collection contributed to our suggesting a new botanical classification of water-melon, new ecological differentiation of varieties being added and presented in the paper.

The genus *Citrullus* comprises 4 species of which in culture only one, *C. lanatus* (THUNB.) MATS. et NAKAI, is present.

It is most polymorphic among the other species and according to our classification (FURSA 1972) it comprises 3 subspecies: subsp. *lanatus*, subsp. *vulgaris* (SCHRAD.) FURSA, and subsp. *mucosospermus* FURSA. The subsp. *vulgaris* has two varieties: var. *vulgaris*, the water-melon, and var. *cordophanus*.

All the diversity of water-melon varieties pertains to the former varieties. In other words, the water-melon assortment is botanically very uniform and therefore difficult to systematize. It is no coincidence that a well known Soviet botanist K. J. PANGALLO, who developed in details the intraspecific systematics of the melon, left the idea to differentiate the water-melon.

More than 2000 water-melon varieties collected from 55 countries are available in the VIR collection. With the object of systematizing them, the ecogeographical principle was used by us. 10 ecogeographical groups were singled out: 1–the Russian, 2–of Asia Minor, 3–the West-European, 4–the Transcaucasian, 5–the Central Asiatic, 6–the Indian, 7–the Afghani, 8–the American, 9–the East-Asiatic, 10–the Far Eastern (FURSA 1965).

When including into the groups the following criteria were considered: 1) the provenance and regions of variety cultivation, 2) morphology and anatomical leaf structure, 3) physiological indices–water capacity and water retaining capacity–in the leaf, 4) sexual type of plants.

Application of statistical analysis methods contributed to including the above groups into three large geographical types: the Russian type comprising the 1, 2, 3 groups; the Asiatic consisting of the 4, 5, 6, 7 groups and the Eastern one–the groups 8, 9, 10 (FURSA and KORNEJČUK 1976).

All of them differ in the xeromorphic or mesomorphic degree of the external plant habit, sex type, flesh colour and in some economically valuable characters. Distinct differences in sex have to be stressed, the latter is defined in water-melon by humidity conditions in the region of growing and is controlled by natural selection.

Water-melon varieties belonging to the Russian type differ, as a rule, in xeromorphy. Their leaves are very dissected with a large number of stomata in the inferior epidermis, the leaf palisade parenchyma is well developed. The leaves possess a high water holding capacity up to 18 atm. By their sexual types the varieties are andromonoecious–i.e. male and bisexual flowers (♂ and ⚥) are developed on the same plant. Pink and crimson tints are predominant in the flesh colour.

The Russian group is the largest ecogeographical group of this type. It comprises the USSR water-melon varieties grown in the Lower Volga area, the northern Caucasus, the Ukraine; the varieties have been created, in the main, by folk's breeding. The fruits are with good flavour qualities and high sugar content.

The group of Asia Minor consists of Turkey varieties presented mainly by local populations. The group seemed to be initial material for the assortment of Russian water-melons.

The West-European group includes water-melons from Bulgaria, Hungary, Romania, Yugoslavia; is presented chiefly by bred varieties, developed partly on the base of the Russian set of varieties.

Water-melon varieties pertaining to the eastern type contrast by their ecological characters with those of the Russian type. They are characterized by the mesomorphy of the external plant habit–poor or moderate dissection of the leaf, little number of stomata in the inferior epidermis, less developed palisade parenchyma, lower water holding capacity of the leaf (up to 12 atm.). Regarding their sex the varieties are typical monoecious; male and female flowers (♂ and ♀) are developed on the same plant; red tints are predominant in the flesh colour, yellow and orange tints may occur.

The most numerous ecogeographical group of this type is the American group. It comprises water-melon varieties raised in the USA but widely spread over the whole of the world. The group is presented by high-bred varieties only noted for their uniformity, high sugar content, resistance to diseases; an elongated form is characteristic for the fruits.

The East-Asiatic group includes the varieties from Japan and eastern China distinguished by high sugar content; the forms with yellow and orange flesh occur rather often.

The Far-Eastern group consists of varieties cultivated in the Primorye area of the USSR. It is presented only by very early local forms.

The Asiatic type according to all the characters studied occupies the intermediate position between the Russian and the Eastern types. Within the type xeromorphic and mesomorphic forms occur, varieties of the monoecious and andromonoecious sexual type diverse as regards flesh colour and sugar content in their fruits, semi-cultivated forms are not seldom met with.

Heterogeneity of the Asiatic type and an abundance of populations–hardly involved into breeding–gives ground to assume its initial position in the evolution of the water-melon. This assumption is in agreement with data of linguistics. The word "arbuz" is of Turkish provenance. With little alterations it is found in many Asiatic dialects and by its roots it can be traced back to the Sanscrit language.

The Asiatic type includes 4 ecogeographical groups. The largest among these is the Central Asiatic group, it is composed of varieties from Uzbekistan, Turkmenistan, Tajikistan, Kasakhstan and Western China. The group is highly polymorphous and is mainly presented by variegated populations.

The Afghani group is presented by water-melons from Iran and Afghanistan. It has large fruits, sometimes of unusual form and coloration and is characterized by local varieties.

The Indian group comprises water-melons from northern India and Pakistan, which have been formed under the conditions of the monsoon climate. The fruits are normal as regards their consistence and flesh colour; it is presented by local varieties.

The Transcaucasian group includes varieties from Armenia, Georgia, Azerbaijan, which are variegated populations. It occupies the intermediate position between the Russian and Asiatic types.

The ecogeographical classification of the water-melon suggested by us contains the whole complex of the natural structure discovered in the crop. The objective of the classification is to show the influence of ecological conditions and of controlled breeding on the development of different types of varieties. It contributes to establish affinity among them and to trace the ways of water-melon distribution in the countries of the globe. The differentiation into the groups gives ground to thoroughly study the biology of the crop, to establish general principles governing ecological variability and select initial material for various purposes of breeding.

With the view to work conveniently with the collection, water-melon varieties were also grouped in variety groups (concultivars), singled out on the base of their distinguishing features–form, colour and fruit pattern (FURSA 1977). 32 variety groups were found out in the collection. Lightgreen and green fruits with narrow prickly stripes were discovered to be most widely common. Similar varietal types occur in all ecogeographical groups, the law of parallel variability of characters being confirmed. However it should be mentioned, the phenotypical similarity of water-melon varieties ascribed to the same concultivar does not elicit their genetic affinity, though in certain cases the latter may be observed.

The classification of water-melon diversity by morphology of fruit-characters seems to be of value, it helps to determine new material, to find out undesired admixtures and to identify the variety.

Thus, to the botanical classification of the species *C. lanatus* elaborated by us were added both ecogeographical differentiation of water-melon embracing general principles governing its evolution and distribution and the grouping of varieties by variety groups according to external fruit characters.

## Zusammenfassung

Die intraspezifische Klassifikation der kultivierten Wassermelonen

Die ökogeographische Klassifikation der Wassermelonen wird anhand der Evolutionsprinzipien dieser Kulturart dargelegt. 10 ökogeographische Gruppen werden aufgeführt, die sich durch morphologisch-anatomische Merkmale und physiologische Eigenschaften ihrer Sorten unterscheiden. Auf der Grundlage statistischer Analysen werden die Gruppen drei geographischen Typen zugeordnet: russischer, asiatischer und östlicher Typ. Es wird angenommen, daß vom asiatischen Typ die Evolution der Wassermelonen ausging.

## Краткое содержание

Внутривидовая классификация столового арбуза

Дается эколого-географическая классификация столового арбуза на основе принципов эволюции этой культуры. Приводятся 10 эколого-географических групп, различающихся морфологически-анатомическими признаками и физиологическими особенностями, входящих в них сортов. Пользуясь статистическим анализом, эти группы можно разбить на три географических типа: русский, азиатский и восточный. Можно предположить, что эволюция столового арбуза ведёт своё начало от азиатского типа.

## Literature

Fursa, T. B., 1965: Ekotipy arbuzov. – Trudy prikl. bot., genet. selekc. **37**, vyp. 2, 65–75.
–, 1972: K sistematike roda *Citrullus* Schrad. – Bot. Žur. **57**, 31–41.
–, 1977: Klassifikacija stolovogo arbuza. – Trudy prikl. bot., genet. selekc. **61**, vyp. 1, 3–10.
–, i V. A. Kornejčuk, 1976: Ispol'zovanie statističeskich metodov dlja ekologo-geografičeskoj differenciacii stolovogo arbuza. – Bot. Žur. **61**, 516–525.

Dr. T. B. Fursa, N. I. Vavilov All-Union Institute of Plant Industry,
190000, Leningrad, USSR

Kulturpflanze XXIX · 1981 · S. 301–310

# The evolution of weed beet in sugar beet crops

ALICE EVANS[1] and JAQUELINE WEIR (Cambridge, United Kingdom)

## Summary

The extent of the weet beet problem in Eastern England is described and the factors to the evolution of weed beet are outlined. Some preliminary experiments designed to distinguish between wild, weed and cultivated populations of beet are described. The response of salt (sodium chloride) solution is a potentially useful discriminating character whereas variation in morphometric characters was not so conclusive. Preliminary results on variation in cytological characters suggest that chiasma frequency is lower in weed than in wild populations and cultivars of *Beta vulgaris*. Finally the possible role of crop-related weeds in breeding is discussed.

## I. Introduction

In recent years there has been a dramatic increase in the occurrence of *Beta vulgaris* forms observed as weeds in a range of crops, including sugar beet. These weed forms have been described as volunteer beet, annual or wild beet and as hybrids between wild and cultivated beet. ARNOLD (1977) has suggested that the term 'weed beet' be used to describe any beet plants, regardless of origin, found in an unintended place in agriculture.

The presence of weed beet has escalated to the point where it is the biggest single problem facing sugar beet growers in Britain. A survey carried out in 1977 (MAUGHAN 1978) on 2.5 per cent of the national sugar crop estimated that on average 15 per cent of the crop was infested by weed beet; in some areas, however, 30 per cent of the crop was affected. In the areas with the most serious infestations some farmers are taking heavily infested fields out of the crop rotation in an attempt to eradicate the large weed beet populations (HOLLOWELL 1979).

Weed beet is also widespread in parts of Europe and the USA; the problem is particularly severe in France and Belgium where a short crop rotation is used (BOITEAU and CHRISTMANN 1977).

## II. Factors contributing to the evolution of weed beet

A number of changes have taken place in agronomic practice and in sugar beet breeding and these have interacted to create a weed beet problem.

[1] We are deeply impressed by the untimely death of Dr. A. Evans. The editors.

*A. Changes in agronomic practice*

Over the last 20 years or so there has been a trend towards earlier sowing of the sugar beet crop. Early sowing effectively lengthens the growing season and increases yield but it also increases the probability of vernalization requirements being met. This has led to an increase in the number of bolters; plants in which flowering is initiated during the first year of growth. For example trials carried out in 1951 over 12 sites with sowing dates ranging from 11 March to 10 April showed that in the early sowing 11.3 per cent of the crop bolted (LONGDEN, SCOTT and TYLDESLEY 1975).

A further change has been away from hand hoeing to the use of selective herbicides. This trend occurred with the increased use of monogerm varieties where seed could be drilled to the plant population stand required and thinning was no longer necessary. The absence of hand-hoeing meant that volunteer beet seedlings in the crop, often off the row, were no longer removed in the spring and could therefore produce seed.

The use of mechnical harvesters has meant that a number of roots or parts of roots have been left in the ground at harvest. Some of these 'ground-keepers' overwinter and produce seed in the following year.

Furthermore, in spite of recommendations that the seed crop should be grown away from the root crop there is always the temptation to grow sugar beet for seed on the better land where the root crop is grown. Seed-stock contamination could therefore occur if bolters from a root crop shed seed in a field that was subsequently used for seed production.

A further trend has been to move the seed production areas to a drier climate. BOSEMARK (1970) and O'CONNOR (1970) have shown that as seed production has progressively moved from Sweden to Italy, bolting in the subsequent root crop was reduced. But in the seed producing areas of Southern Europe wild annual forms of *Beta* occur; hybridization can take place and the seed crop becomes contamined by seed of the annual beet and this is subsequently drilled with the commercial root crop.

All these agronomic changes have contributed substantially to the evolution of the annual weed beet.

*B. Changes in breeding procedures*

Monogerm varieties were introduced in the 1950's and many of the early monogerm varieties from the USA, bolted badly and so contributed to the production of annual weed beet.

In the 1960's anisoploid ('polyploid') multigerm synthetics were widely used and these were in turn superseded by monogerm varieties many of which were triploid. The latter are produced by harvesting seed from a diploid, male-sterile monogerm pollinated by a tetraploid multigerm. SCOTT and LONGDEN (1970) have demonstrated that diploid plants release their pollen at a relative humidity of approximately 95 per cent whereas tetraploid plants do not do so until the relative humidity has dropped to 75 per cent. This results in a two hour period in the morning when the stigmata of the female parent are receptive but the tetraploid pollinators are not yet producing an effective pollen cloud. It is a this time that mother plants are most likely to be pollinated by diploid wild forms of *B. vulgaris* subsp. *maritima* which are found around the coast of Britain.

Thus weed beet could have arisen from the following sources:

1. from bolters within the root crop that cross pollinated to produce viable seed.
2. from bolters growing in root and seed crops near the coast which could have been pollinated by wild beet *Beta vulgaris* subsp. *maritima*.
3. from seed produced in Southern Europa which was contaminated by hybridization with annual beet growing locally.
4. from sugar beet roots left in the field after harvest (ground keepers) and producing seed the next year.

Highly variable populations can be produced from all these sources. Natural selection will favour those types that are most fertile and thus progressing most rapidly towards the annual habit; in other words, types with the greatest adaptation to survive as weeds would be selected.

## III. Investigations on the origin of weed beet

Relatively little work has been done to determine the possible origin of weed beet. HORNSEY and ARNOLD (1979) collected samples from two weed beet populations and made test crosses with monogerm 0-type (non-restorer) lines. From the presence or absence of the monogerm character in the progenies they concluded that one of the populations had built up gradually over several years and had evolved from bolters in normal sugar beet varieties whereas the other had evolved recently through contamination by hybridization of a monogerm variety.

The objective of the present investigation is to attempt to distinguish between wild, weed and cultivated populations of beet, and between weed populations of possibly differing origins.

Firstly the growth of seedlings of wild, weed and varietal populations of *B. vul-*

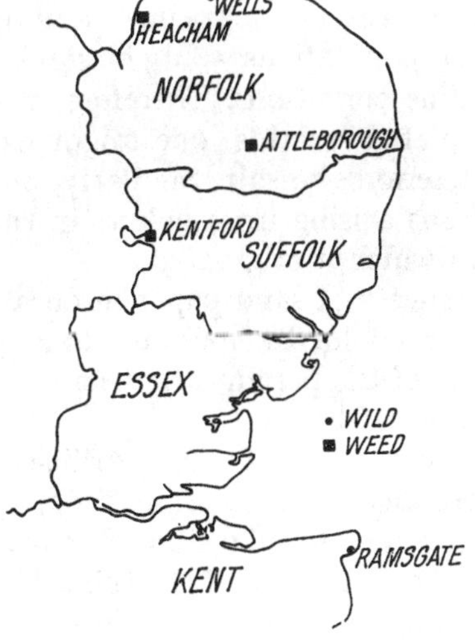


Fig. 1 Sites of wild and weed populations in Eastern England

*garis* were studied in the presence of salt (sodium chloride) solution and secondly a study was carried out to ascertain if there are morphological or cytological characters which can be used to distinguish these populations.

*A. Sources of material*

Samples from the following wild, weed and varietal populations of *Beta vulgaris* were used and the locations from which the wild and weedy types originated is given in figure 1.

Wild beet populations. *B. vulgaris* subsp. *maritima* (L.) THELL.

1. Ramsgate:–a seed sample collected from a small group of plants. (Received from Mr. F. DALE, Department of Plant Biology, University of Birmingham).
2. Wells:–a seed sample collected in October 1978 from a population growing on the seawall at Wells.

*Weed populations*

3. Heacham:–a seed sample collected from an infested field near Heacham; the field had in previous years been growing a polyploid sugar beet variety.
4. Kentford:–this sample originated from cultivated plants left in the ground when part of the field was used for a new motorway.
5. Attleborough:–a seed sample originating from a weed population near Attleborough. The field had a long history of cultivation with a triploid monogerm variety.
   (Seed of 4 and 5 were received from Dr P. C. LONGDEN, Brooms Barn Experimental Station, Higham, Bury St. Edmunds, Suffolk).
6. Sugar beet cultivar. *Beta vulgaris* subsp. *vulgaris* L. A diploid monogerm variety: (seed was obtained from the National Institute of Agricultural Botany (NIAB), Cambridge.)

*B. The response of samples of the populations to salt solution*

The response to salt was chosen as a potentially interesting character because *B. vulgaris* subsp. *maritima* is known to be relatively tolerant to high levels of chlorine in the aerial spray of the sea and in the soil water of maritime habitats and it was thought that the tolerance of the varietal populations would be significantly lower because they are not exposed to saline conditions. Therefore, if a weed population had developed from wild×cultivated hybrids, one might expect such a population to have an intermediate tolerance to salt. Similarly, one might expect the response of weed beet populations arising from bolters in the crop to be similar to the varietal populations from which it originated.

Fruits of all populations were rubbed between sheets of sand paper in an attempt to remove the corky perianth and then washed for 24 hours in running water to remove the water soluble toxic substances of the perianth that have an inhibitory effect on germination. (STOUT and TOLMAN 1940).

Seeds were germinated in petri dishes in the dark at 22 °C. When a sufficient number of seedlings of each populations had produced a root approximately 3 cm long, each seedling was carefully transferred to a glass tube (25 mm long × 4 mm diameter) so that the cotyledons protruded from the top of the tube. The tubes were suspended in boxes containing KNOP's nutrient solution. After five days had elapsed 0.25 M salt solution was added and the seedlings were also sprayed daily with the salt solution. The five-day delay before treatment allowed the waxy cuticles on the wild populations to develop.

The number of seedlings surviving the treatment with salt solution was scored daily up to seven days and the results are presented in figure 2.

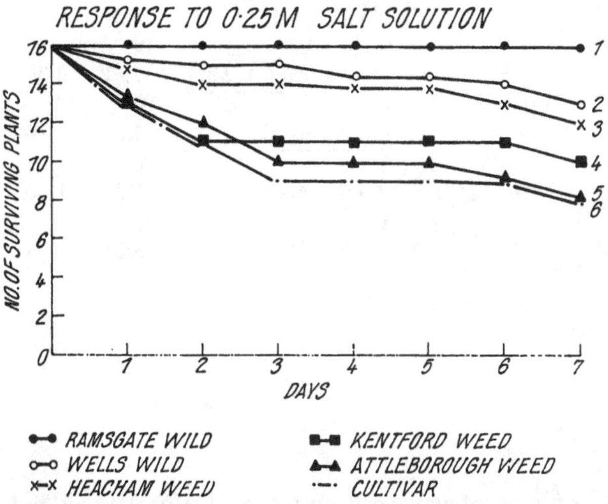


Fig. 2 No. of surviving plants over a period of seven days' treatment with salt

All seedlings of the Ramsgate wild population survived the salt treatment while there were a few deaths in the Wells wild population and the Heacham weed population. The Wells population had introgressed with neighbouring sugar beet crops whereas the Ramsgate population was isolated with no sugar beet growing in the vicinity. The Heacham population was also probably a result of hybridization between the Heacham sugar beet crop and wild material growing nearby.

The Kentford and Attleborough weeds responded in a similar way showing nearly 50% death of the seedlings after seven days treatment with salt. These populations probably developed as a result of hybridization between bolters in the sugar beet crops from which they developed and their response is close to that of the cultivar which showed the lowest survival as was expected.

Although this experiment is the result of a preliminary study of the response of seedlings to salt solution it appears that this technique may be useful in determining the origins of weed populations and in discriminating between wild, weed and cultivated populations.

It is possible that the differences between the populations could be detected without exposing samples of the populations to salt, hence the next experiment was to determine if any characters measured could be used to distinguish the weed population from the wild or varietal populations.

## C. *Variation in morphometric characters of plants from five populations*

Twenty plants from each population were grown in pots in the glasshouse and arranged in a completely randomized block. After 12 weeks the following characters were measured:–root fresh weight and dry weight; total leaf area; leaf width and leaf number per plant.

Table 1

The classification of five populations of *B. vulgaris* for the five characters measured (using DUNCAN's multiple range test)

| Character | | | | | | | | |
|---|---|---|---|---|---|---|---|---|
| Leaf number | S, 21.9 | M, 21.6 | B 23.8 | A 31.0 | | | W 38.3 | S.E. 1.95 |
| Leaf width (cm) | M 11.96 | | | S, 10.8 | A, 10.4 | B 10.3 | W 7.82 | 0.342 |
| Leaf Area ($cm^2$) | M 144.9 | | | B, 122.5 | S, 117.8 | A 114.4 | W 79.8 | 7.44 |
| Root Fresh Weight (gm) | S, 76.2 | M 74.0 | | A 45.6 | B 43.3 | | W 16.3 | 4.60 |
| Root Dry Weight (gm) | S, 23.6 | M 22.8 | | A, 13.5 | B 10.8 | | W 5.05 | 1.35 |

Key:
S 'Sharpes kleine Monobeet' cultivar
M 'Monotri' cultivar
B Bolters
A Attleborough Weed
W *B. maritima* from Wells

The data were subjected to analysis of variance and the DUNCAN's multiple range test was used to group similar populations together and to show the significant differences between the others. Table 1 gives the classification of the five populations for the five characters scored.

The wild population from Wells was distinct for all the characters measured. The Attleborough weed and the population from the bolters could be distinguished on the basis of some of the characters.

Although the experiment can be used to separate some of the populations the results are not very conclusive. A further experiment not reported here, on the response of adult plants to salt solution showed that such a treatment could be used to accentuate the differences between wild weed and cultivated populations of beet.

## *D. Variation in cytological characters*

The method described by BOSEMARK and BARMATOV (1971) was used to study the mitotic chromosomes. All the wild, weed and cultivated plants had the diploid chromosome number of $2n = 2\times = 18$.

Preparations of meiotic chromosomes were obtained from anthers fixed in 6 : 3 : 1 methanol, chloroform and propionic acid and stained in lacto-propionic orcein for one minute before squashing. The chiasmata of bivalents from 10 pollen mother cells per plant from each population were recorded. All preparations were viewed under phase-contrast microscopy. The results suggest that the chiasma frequency is lower in weed populations than in the wild or cultivated populations. Fig. 3 shows the meiotic chromosomes from a plant of a weed population and a plant of a cultivated population.

The implications of these preliminary results will now be considered. The low

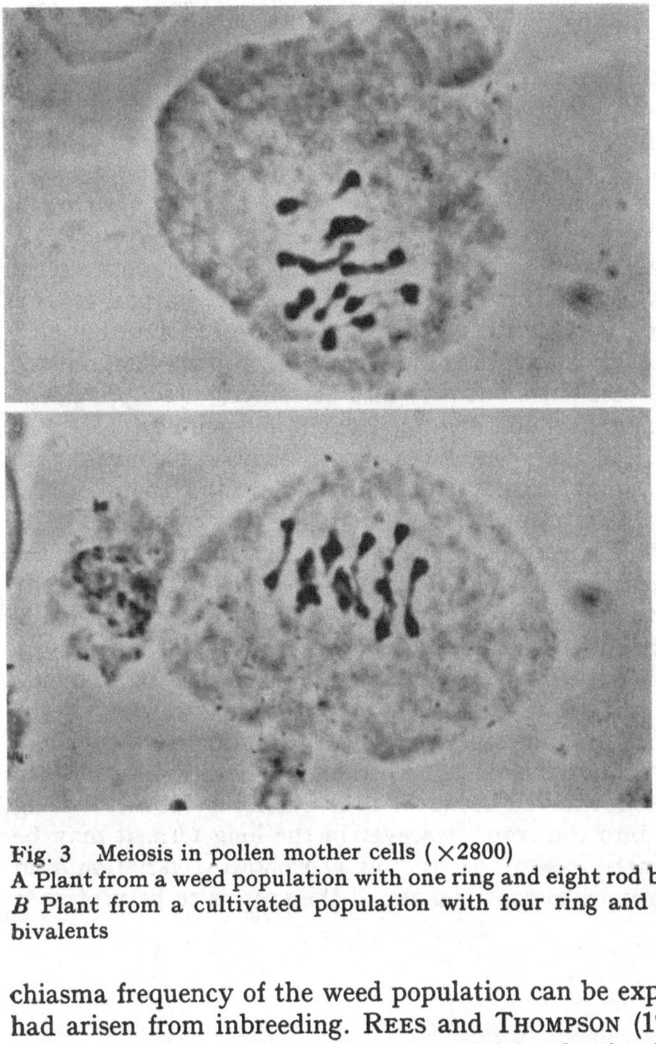

Fig. 3 Meiosis in pollen mother cells (×2800)
A Plant from a weed population with one ring and eight rod bivalents
*B* Plant from a cultivated population with four ring and five rod bivalents

chiasma frequency of the weed population can be explained if these populations had arisen from inbreeding. REES and THOMPSON (1956) demonstrated that in inbred lines of rye there was a substantial reduction in chiasma frequency when compared with outpollinated genotypes. The higher chiasma frequencies of the wild and cultivated populations can be explained because of their greater variability.

## IV. The possible role of crop-related weeds in breeding

Weed beet populations are now widespread in the sugar beet growing areas of Western Europe and in recent years plant breeders and growers have concentrated on trying to eradicate the problem. Sugar beet is a relatively young crop in that it has been grown in Europe for less than 200 years. Yet its genetic base is probably narrower than that of most other cross-fertilizing crop species. Hybridization between sugar beet and *B. maritima* and other wild *Beta* species could provide the plant breeder with new and valuable sources of genetic variation. This type of hybrid almost certainly contributed to the populations from which the early sugar beet varieties were selected. However, most attempts along these lines have not been pursued intensively and have been of little practical value (COONS 1975). One notable exception is the work of MUNERATI (1932). He was able to produce *Cercospora*-resistant sugar beet lines by repeated selection from resistant *B. maritima*×cultivated sugar beet crosses. Most of the leaf spot resistent varieties in Europe have been derived from MUNERATI's selections.

The advantages of natural crosses over artificially produced crosses between the wild and cultivated populations of a crop plant may be that the natural cross is one in which there is greater hybrid fertility. Many natural crosses occur but only the most fertile are observed. Such crosses in the sugar beet crop may not only be sources of disease and virus resistance but may also be sources of adaptive tolerance to a range of environmental conditions e.g. drought and salinity.

It is generally recognised that the weedy races of crop plants can contribute to the genetic variability of the crop. The mechanisms that allow this to happen have been outlined by HARLAN (1965). In several crops, for example rice (CHANG 1978) and sorghum (DOGGETT 1978) the weedy races have played a significant role in the continued evolution and improvement of the crop. The development of weedy races in rice is very similar to the situation in weed beet; in both crops, perennial and annual types have contributed to the weed problem.

The genetic basis of sugar beet has decreased since the introduction of monogerm varieties (BOSEMARK 1979) and it is therefore questionable whether the sugar beet breeder is correct to eradicate weed beet populations. Obviously it would be dangerous in a highly developed temperature agricultural system to allow weed beet to be left uncontrolled because of the possibility of introducing the dominant annual gene into the crop. However in the long term it may be beneficial to broaden the genetic base of the crop by incorporating variation from weed beet if breeding progress is to continue and if the crop is to be protected against hazards.

## Zusammenfassung


Die Evolution der Unkrautrüben in Zuckerrübenbeständen

Es wird die Verbreitung des Problems der Unkrautrüben im östlichen England beschrieben und die Faktoren, die die Evolution der Unkrautrübenpopulation beeinflussen, werden umrissen. Einige vorläufige Versuche zur Unterscheidung zwischen Wild-, Unkraut- und kultivierten Populationen werden beschrieben. Die Reaktion auf Salzlösung (Natriumchlorid) zeigte sich als eine brauchbare Trennmethode, während die Variation in der morphometrischen Charakterisierung nicht so überzeugend war. Vorläufige Ergebnisse über die Unterschiede in der cytologischen Charakterisierung weisen darauf hin, daß die Chiasmafrequenz in Unkrautrüben geringer ist, als in Wildpopulationen und Kultursorten von *Beta vulgaris*. Abschließend wird die mögliche Bedeutung der den Kulturpflanzen verwandten Unkräuter für die Züchtung diskutiert.

**Краткое содержание**

**Эволюция сорных форм *Beta* L. в посевах сахарной свёклы**

Рассматривается проблема распространения сорных форм *Beta* в посевах культурной сахарной свёклы в восточной Англии и факторы, которые способствуют эволюции популяций этих сорняков. Описываются некоторые предварительные опыты, на основании которых можно установить различия между дикорастущими, сорняковыми и культурными популяциями. Реакция на соляной раствор (хлористый натрий) оказалась приемлемым признаком для выявления различий, в то время как морфометрический метод давал мало убедительные результаты. Предварительные цитологические исследования показали, что частота хиазм у сорной свёклы оказалась ниже, чем у дикорастущих популяций и у культурных сортов *Beta vulgaris*. В заключение обсуждаются связи и возможности использования в селекции сорняков, являющихся сородичами культурных растений.

## Literature

ARNOLD, M. H., 1977: Weed beet and bolters. – British Sugar Review **45**, 12–13.

BOITEAU, R. and J. CHRISTMAN, 1977: Weed beets: present position and importance of the problem. – In: Proceedings of the 40th Winter Congress of the International Institute of Sugar Beet Research, Brussels, pp. 157–165.

BOSEMARK, N. O., 1970: Influence of seed crop environment to root crop characteristics. – Journal of the International Institute of Sugar Beet Research **4**, 193.

–, 1979: Genetic poperty of the sugar beet in Europe. – In: Proceedings of the Conference for the Broadening of the Genetic Base in Crops, Wageningen 1978 (Edit. A. C. ZEVEN and A. M. VAN HARTEN) pp. 29–35.

– and V. E. BORMOTOV, 1971: Chromosome morphology in a homozygous line of sugar beet. – Hereditas **69**, 205–212.

CHANG, T. T., 1976: Rice. – In: The evolution of crop plants (Edit. N. W. SIMMONDS) pp. 98–104. Longman Group Limited, London.

COONS, G. H., 1975: Interspecific hybrids between *Beta vulgaris* L. and the wild species of *Beta*. – Journal of the American Society of Sugar Beet Technologists **18**, 281–306.

DOGGETT, H., 1976: Sorghum. – In: The evolution of crop plants (Edit. N. W. SIMMONDS) pp. 112–117. Longman Group Limited, London.

HARLAN, J. R., 1965: The possible role of weedy races in the evolution of cultivated plants. – Euphytica **14**, 173–176.

HOLLOWELL, W., 1979: The weed beet menace. – British Sugar Review **47**, No. 2, 5–16.

HORNSEY, K. G. and M. H. ARNOLD, 1979: The origins of weed beet. – Annals of Applied Biology **92**, 279–285.

LONGDEN, P. C., R. K. SCOTT and J. B. TYLDESLEY, 1975: Bolting of sugar beet grown in England. – Outlook on Agriculture **8**, 188–193.

MAUGHAN, A., 1978: Rothamsted Experimental Station Annual Report for 1977.

MUNERATI, O., 1932: Sull' incrocio della barba-bietola coltivata con la beta selvaggia della costa adriatica. – Industria saccar ital. **25**, 303–304. Cited in BOSEMARK, N. O. 1979 above.

O'CONNOR, L. J., 1970: Environmental influence during beet seed production on bolting and quality characteristics of the subsequent root crop. – Journal of the International Institute of Sugar Beet Research **4**, 207.

REES, H. and J. B. THOMPSON, 1956: Genotypic control of chromosome behaviour in rye. III. Chiasma frequency in homozygotes and heterozygotes. – Heredity **10**, 409–424.

SCOTT, R. K. and P. C. LONGDEN, 1970: Pollen release by diploid and tetraploid sugar beet plants. – Annals of Applied Biology **66**, 129–135.
STOUT, M. and B. TOLMAN, 1940: Toxic effects on germinating sugar beet seed of water soluble substances in the seed ball. – Journal of Agricultural Research **61**, 817–830.

*Acknowledgements*

One of us (JW) acknowledges the financial support of the Department of Agriculture and Fisheries for Scotland.

DR. ALICE EVANS
Department of Applied Biology,
University of Cambridge,
Pembroke Stret,
Cambridge CB2 3DX
United Kingdom

Kulturpflanze XXIX · 1981 · S. 311–319

# Contribution to the taxonomy of *Vicia* species belonging to the section *Faba*

PIETRO PERRINO and DOMENICO PIGNONE (Bari, Italy)

## Summary

*Vicia faba* belongs to the section *Faba*. While the other species of the section *V. narbonensis, V. johannis* and *V. bithynica* cross enough well each other, *V. faba* seems to be strongly isolated from them. In addition the chromosome morphology of *V. faba* differs from the other species of the section which among themselves are enough homogenous in this respect.

To contribute to a wider knowledge on the taxonomical position of *V. faba* in its section, electrophoretical and karyological analysis of 500 individuals belonging to 50 populations of different origin were carried out.

The analyzed species were: *V. faba, V. narbonensis, V. serratifolia, V. bithynica, V. galilaea* and *V. johannis*.

## Introduction

The cultivation of *V. faba* is spreading, also to countries different from the Mediterranean and S. W. Asia regions where its cultivation is as old as agriculture itself. The high protein content of seeds, the ability to improve the ground quality, the ability to organicate the atmospherical nitrogen, the resistance to weeds after sowing, are characteristics which are promoting interest to this crop, even in those countries, as Italy, where its cultivation decreased in the past.

Genetic and breeding research is particularly devoted to search for possible wild relatives that can offer genes useful for modifying the plant habit and/or carrying resistances, as to orobanche, aphids, etc.

Studies made on species relationships came out with the hypothesis that *V. faba* was originated from an extinct progenitor (SCHÄFER 1973). These and other studies were concentrated on species morphologically more similar to *V. faba* (LADIZINSKY 1975a, 1975b, ABDALLA and GÜNZEL 1979). *V. bithynica*, though belonging to the same section, because of its different morphology and because it seems to be absent from S. W. Asia was always excluded.

With the aim to contribute to a wider knowledge on this topic, a cytological and electrophoretical study on species of sect. *Faba* including *V. bithynica* have been undertaken.

## Material and methods

31 samples belonging to 6 species of section *Faba* were utilized in the present study. A list of the material and its origin is given in tables 1 and 2.

Cytological observation were conducted on 10 slides per sample prepared and stained with Quinacrine HCl as described by VOSA and MARCHI (1972) and PIGNONE and ATTOLICO (1980).

Since the benzimide derivate H33258 sometimes displays a different banding pattern in comparison with Quinacrine (VOSA 1973; GATTI et al. 1976), five additional slides per sample were stained with this chemical. Dry slides were soaked 5 min. in M/15 SÖRENSEN's buffer at pH 7, stained with 0.002 mg/ml of H33258 dissolved in the same buffer for 10 min. and differentiated in buffer for 1 min. Slides were then mounted in glicerol: pH 7 buffer (1 : 1) and observed with a fluorescence microscope.

For electrophoresis experiments seed proteins were extracted following the method of PERRINO et al. (1977). Freeze-dried extracts were separated by disc-electrophoresis method of DAVIS and ORNSTEIN (1961) with a current of 5 mA per tube.

Table 1

Species, variety, line number and origin of *Vicia* samples analyzed by electrophoresis

| Gel No. | M.G. No. | Species | Variety | Origin |
|---|---|---|---|---|
| 1 | 106742 | *faba* | *major* | Tunisia |
| 2 | 106466 | „ | „ | Ethiopia |
| 3 | 106362 | „ | „ | Algeria |
| 4 | 106568 | „ | *equina* | Italy |
| 5 | 106995 | „ | „ | Greece |
| 6 | 106596 | „ | *minor* | Italy |
| 7 | 109613 | „ | „ | Lebanon |
| 8 | 107273 | „ | *paucijuga* | Spain |
| 9 | 107272 | „ | „ | Spain |
| 10 | 105846 | *narbonensis* | | Malta |
| 10 | 106293 | „ | | USSR |
| 10 | 105854 | „ | | DDR |
| 10 | 105799 | „ | | BDR |
| 11 | 106288 | „ | | Crete |
| 11 | 106107 | „ | | Hungary |
| 11 | 105786 | „ | | Sweden |
| 11 | 105982 | „ | | Australia |
| 11 | 105751 | „ | | Sicily |
| 12 | 106225 | „ | | Japan |
| 13 | 106204 | *galilaea* | | Israel |
| 14 | 106290 | *johannis* | | USSR |
| 15 | 106215 | „ | | Turkey |
| 16 | 106214 | „ | | Turkey |
| 17 | 106293 | „ | | USSR |
| 18 | 106251 | *serratifolia* | | France |
| 19 | 105849 | „ | | Malta |
| 20 | 106252 | „ | | Turkey |
| 21 | 106181 | *bithynica* | | Greece |
| 22 | 105866 | „ | | DDR |
| 23 | 105681 | „ | | Italy |
| 23 | 105291 | „ | | Italy |

Table 2
Species, number of samples and origin of material analyzed cytologically

| Species | No. of samples | Origin |
|---|---|---|
| *Vicia faba* | 9 | Spain, Greece, Italy, Tunisia, Algeria, Ethiopia, Lebanon. |
| *Vicia narbonensis* | 10 | Sicily, Sweden, Crete, DDR, Malta, Australia, Japan, USSR, BDR, Hungary. |
| *Vicia serratifolia* | 3 | Turkey, France, Malta. |
| *Vicia bithynica* | 4 | Italy, Greece, DDR. |
| *Vicia johannis* | 4 | Turkey, USSR. |
| *Vicia galilaea* | 1 | Israel. |

## Results

### *Cytology*

#### *Vicia faba* ($2n=12$)

The banding pattern produced by H33258 is similar to that produced by Quinacrine. This consists in centromeric $Q^+$ bands on all the chromosomes, except chromosome 6, while after H33258 staining, all the centromeric bands on the telocentric chromosomes are more intense than those produced by Quinacrine; in addition two chromosomes show a more complicated pattern: chromosome 4 ex-

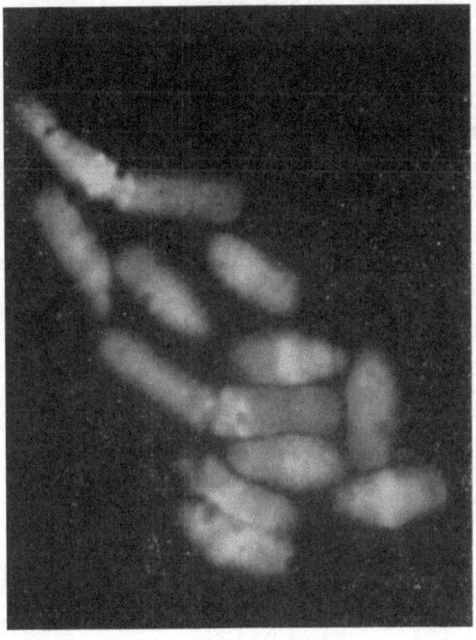

Fig. 1 H-33258 banded metaphase of *Vicia faba*

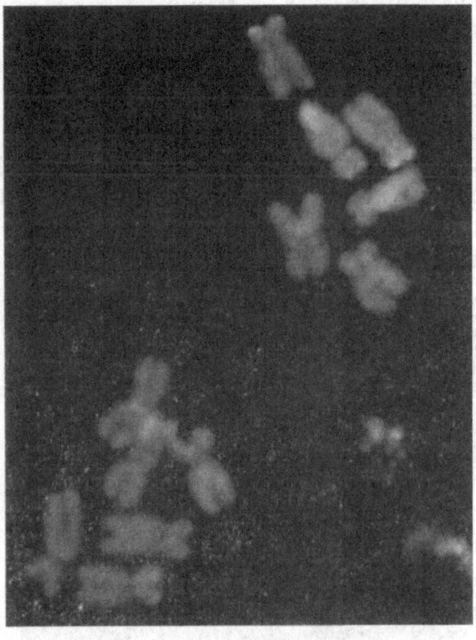

Fig. 2 *Q*-banded metaphase of *Vicia narbonensis*

hibits a very intense band in the middle of the long arm, and chromosome 6 occasionally displays two or three faint bands between the centromere and the middle of the long arm (Fig. 1). This general pattern is highly repetitive in all the examined material.

*Vicia narbonensis*

It is possible to identify three caryotypes in this species (all 2n = 14): these caryotypes differ for little rearrangements of the chromosomes; the chromosomes are all metacentric or sub-metacentric.

Results are in agreement with those obtained by SCHÄFER (1973). With respect to banding the chromosomes exhibited only telomeric bands fainter than those of *Vicia faba* when stained with both the fluorochromes; one intercalary band is present on the secondary constriction (Fig. 2).

Both the fluorochromes gave identical pattern; H33258 gave a more intense fluorescence than Quinacrine. Position and number of bands did not allow specific chromosomal identification.

*Vicia serratifolia*

Results confirm the karyotype proposed by SCHÄFER (1973) (2n = 14). The karyotype is similar to those of *V. narbonensis* and differ for the position of the satellite. Quinacrine and H33258 patterns indicated the presence of telomeric bands as in *Vicia narbonensis*. Fluorescence did not help in the identification of individual chromosomes.

*Vicia galilaea* and *Vicia johannis*

These two species possess almost identical karyotypes similar to the karyotype A of *V. narbonensis* (SCHÄFER 1973). The banding patterns could not help in distinguishing them from each other and from *Vicia narbonensis*.

*Vicia bithynica*

The karyotype of *V. bithynica* (2n = 14) differs from those of all other species in this section. The karyotype is similar to that recently described by D'AMATO et al. (1978). Six pairs of chromosomes are telocentric or subtelocentric and one pair, the one carrying the satellite is sub-metacentric. Four intense bands appeared on three chromosomes by Q and H33258 staining. Two intercalary bands are present on the long arm of chromosome 2, one in the middle and the other on the distal part. A third band is visible in the middle of the long arm of the chromosome

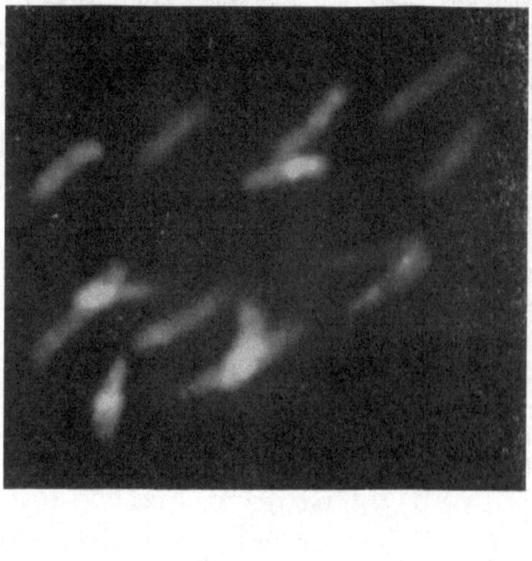

Fig. 3 *Q*-banded metaphase of *Vicia bithynica*

5; the last one is at the terminal quarter of the long arm of the chromosome 7 (Fig. 3). It is worth noticing that these bands occupy approximately the same position as the four largest C-bands described by D'AMATO et al. (1978) on the chromosome B, D and E respectively.

*Description of profiles*

*Vicia faba* L. var. *major* (profiles: 1, 2, 3).

Three accessions with three different origins (Tunisia, Ethiopia and Algeria) showed very similar profiles. The accessions from Ethiopia (2) and Algeria (3) can be distinguished from the Tunisian (1) material for a medium slow moving albumine that is present in the former and absent in the latter and for two also medium fast albumines present in the latter and absent in the former.

Another difference arises about the position of the most fast moving band: this is slower in the Tunisian line and faster in the Ethiopian and Algerian ones. The Ethiopian and Algerian lines though more similar to each other than the Tunisian one show some quanti-qualitative differences for two fast moving albumines.

*Vicia faba* L. var. *equina* (profiles: 4 and 5).

The profile 4 of the two lines from Italy showed two more fast moving albumines than the profile 5 corresponding to the Greek line. The Greek line of *V. faba equina* (5) differs from *V. faba major* from Tunisia (1) for not showing one fast moving albumine. Other differences are merely quantitative.

*Vicia faba* L. var. *minor* (profiles: 6 and 7).

The profiles of *V. faba minor* from Italy (6) differ from the Libyan line (7) for two albumines reciprocally present or absent. The profiles of this variety of *V. faba* are closer to those of the variety *equina* than to those of the variety *major*; however the two lines analyzed can be distinguished from the varieties *major* and *equina* for showing two instead of one very fast moving albumine.

*Vicia faba* var. *paucijuga* (profiles: 8 and 9).

The two lines from Spain showed two very similar profiles. They differ to each

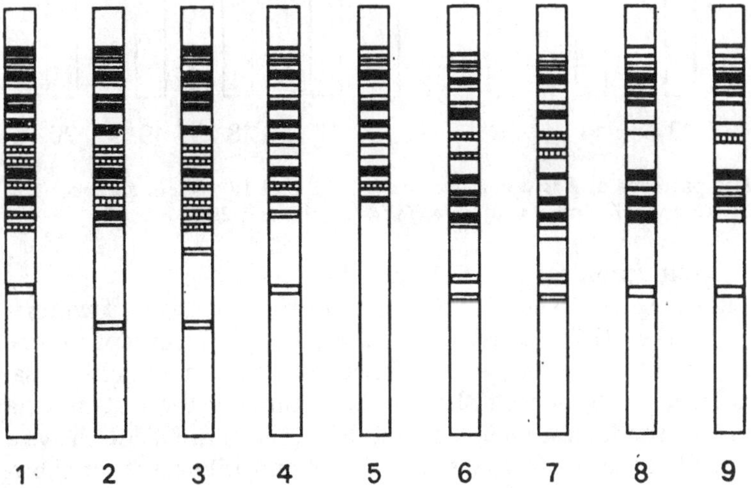


Fig. 4 Electrophoretic patterns of *Vicia faba*

other on a quantitative base only. Both profiles are practically identical to *V. faba equina* from Italy. The main differences arise when quantitative observations are made.

*Vicia narbonensis* L. (profiles: 10, 11 and 12).

The ten lines analyzed in this study showed not more than three very similar profiles. Lines from Malta, USSR, DDR and BRD showed a profile (10) with more bands than those lines from Crimea, Hungary, Uppsala, Australia and Sicily (11). The profile of the latter (11) does not show three medium fast and two fast moving albumines that are present in the profile of the former (10). The Japanese line showed a profile (12) distinguishable from the two previous ones (10 and 11) for two more medium fast moving albumines besides some other quantitative differences.

*Vicia galilaea* PLITM. et ZOH. (profile: 13).

This species shows a pattern very similar to that of the *V. narbonensis* lines analyzed in this study. Its profile, apart from few quantitative differences against to the *narbonensis* group, can be distinguished from the Japanese line (12) for not showing one band (the sixth from the last) and from the other lines of different origin (10 and 11) for showing at least one band more (the seventh from the last).

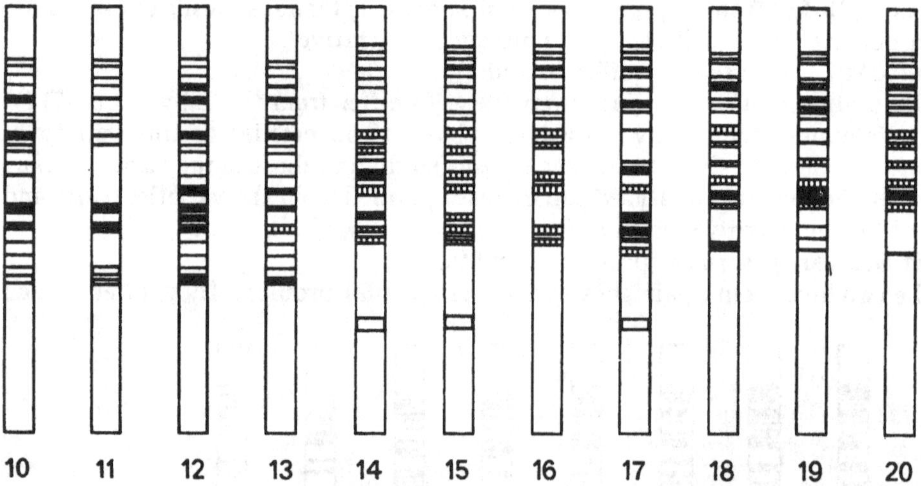


Fig. 5 Electrophoretic patterns of *Vicia narbonensis* (10, 11 and 12), *Vicia galilaea* (13), *Vicia johannis* (14, 15, 16 and 17) and *Vicia serratifolia* (18, 19 and 20)

*Vicia johannis* TAMANSCH. (profiles: 14, 15, 16 and 17).

Four similar profiles were observed for four lines of different origin. Two lines from USSR (14) and Turkey (15) showed almost the same profile except for one very faint albumine present in the former and absent in the latter and vice versa. A second line from Turkey (16) doesn't show at least three of the fast moving albumines and above all the fastest one. A second line (17) from USSR shows a pattern practically identical to the former two lines (14 and 15) except for showing one more fast moving band and for not showing one or two less medium fast moving albumines.

*Vicia serratifolia* Jacq. (profiles: 18, 19 and 20).

Three lines from three different origins showed three very similar patterns. The two lines from France (18) and Turkey (20) are closer to each other than to the line from Malta (19). The main difference between the two former lines consists in one more slow moving albumine for the Turkish line (20). The line from Malta, apart for the absence of this slow moving albumine, shows two more fast moving albumines which do not appear in the other two lines of different origin.

*Vicia bithynica* (L.) L. (profiles: 21, 22 and 23).

Two main profiles were observed for four lines. The two lines from Greece

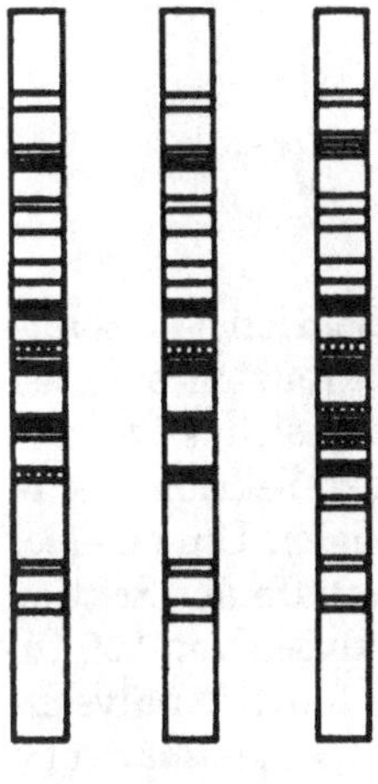

21 22 23 Fig. 6 Electrophoretic patterns of *Vicia bithynica*

and DDR (21 and 22) differ for a very faint fast moving albumine only. The two lines from Italy show practically the same profile (23) and though similar can be distinguished from the Greek and German lines for at least three more almost fast moving albumines. Both profiles especially that corresponding to the Italian material (23) show similarities with the profile of *Vicia faba* var. *minor* (6). These two different species show to be very similar at least for the fast moving albumines.

## Discussion

Chromosome morphology and banding patterns allow the subdivision of the section *Faba* into two main groups of species:

1) species with metacentric chromosomes and faint telomeric $Q^+$ bands (*V. narbonensis, V. serratifolia, V. galilaea, V. johannis*);
2) species with mainly telocentric chromosomes and intense $Q^+$ intercalary bands (*V. faba, V. bithynica*).

All the species of the first group possess caryotypes similar for shape and dimension of the chromosomes and for nuclear DNA content (Chooi, 1971); caryotypes can derive from the same one by single translocations (Schäfer 1973).

Conversely, in the second group, the two species show caryotypes similar in shape and banding pattern, although different in size; the nuclear DNA content of *V. faba* is nearly three times that of *V. bithynica* (Chooi 1971).

A similar conclusion can be drawn from albumine profiles. Caryological and electrophoretical results suggest that, within the analyzed species, *V. bithynica* is the closest to *V. faba*. However, since this similarity was not expected, this study must be understood as a very preliminary one. Further research on a wider proportion of germplasm of the two species is needed before to engage a deep discussion on the mentioned similarity between two morphologically very different species of *Vicia*.

## Zusammenfassung


Beitrag zur Taxonomie der *Vicia*-Arten der Sektion *Faba*

*Vicia faba* gehört zur sect. *Faba*. Während andere Arten der Sektion (*V. narbonensis, V. johannis* und *V. bithynica*) untereinander relativ gut kreuzbar sind, ist *V. faba* von ihnen deutlich isoliert. Darüberhinaus unterscheidet sich *V. faba* in der Morphologie seiner Chromosomen von den anderen Arten der Sektion, die in dieser Hinsicht eine relativ einheitliche Merkmalsausprägung zeigen. Um zu einer besseren Kenntnis der systematischen Stellung von *V. faba* innerhalb der Sektion beizutragen, wurden chromatographische und karyologische Analysen von 500 Individuen aus 50 Populationen verschiedener Herkunft durchgeführt. Analysiert wurden *V. faba, V. narbonensis, V. serratifolia, V. bithynica, V. galilaea* und *V. johannis*.


## Краткое содержание


О таксономии видов *Vicia*, относящихся к секции *Faba*

*Vicia faba* относится к секции *Faba*. В то время как другие виды секции (*V. narbonensis*, *V. johannis*, *V. bithynica*) относительно хорошо скрещиваются между собой, *V. faba* от них чётко изолирована. Кроме того, *V. faba* отличается морфологией своих хромосом от прочих видов секции, показывающих в этом отношении относительно единообразное выражение признаков. Чтобы лучше выявить систематическое положение *V. faba* внутри секции, были проведены хроматографические и кариологические анализы 500 индивидуумов из 50 популяций различного происхождения. Анализировались: *V. faba*, *V. narbonensis*, *V. serratifolia*, *V. bithynica*, *V. galilaea*, *V. johannis*.


## Literature


Abdalla, M. M. F. and Günzel, G., 1979: Protein content and electrophoresis of seed protein of certain *Vicia faba* L. stocks and their assumed ancestors. Z. Pflanzenzüchtg. **83**, 148–154.

CHOOI, W. Y., 1971: Variation in nuclear DNA content in the genus *Vicia*. Genetics **68**, 195–211.
D'AMATO, G., G. BIANCHI, R. CAPINERI, and P. MARCHI, 1978: Localizzazione delle bande C nel cariotipo di cinque specie del genere *Vicia* e due del genere *Lathyrus*. Annali di Botanica **37**, 189–201.
DAVIS, B. J. and L. T. ORNSTEIN, 1961: "Disc-electrophoresis" Distillation Product Industries. – Rochester, U.S.A.
GATTI, M., S. PIMPINELLI and G. SANTINI, 1976: Characterization of *Drosophila* heterochromatin: I. Staining and decondensation with Hoechst 33258 and Quinacrine. Chromosoma **57**, 351–375.
LADIZINSKY, G., 1975a: On the origin of the broad bean *Vicia faba* L. Isr. J. Bot. **24**, 80–88.
–, 1975b: Seed protein electrophoresis of wild and cultivated species of section *Faba* of *Vicia*. Euphytica **24**, 758–788.
PERRINO, P., I. MAELLARO, and A. BLANCO, 1977: Patterns di electroforesi in specie del genere *Vicia*. Genet. Agraria **31**, 121–130.
PIGNONE, D., and M. ATTOLICO, 1980: Chromosome banding in four groups of *Vicia faba* L. Caryologia **32**, 283–288.
SCHÄFER, H. I., 1973: Zur Taxonomie der *Vicia narbonensis*-Gruppe. Kulturpflanze **21**, 211–273.
VOSA, C. G., 1973: Heterochromatin recognition and analysis of chromosome variation in *Scilla sibirica*. Chromosoma **43**, 269–278.
– and P. MARCHI, 1972: On the Quinacrine and GIEMSA patterns of chromosomes of *Vicia faba*. Giornale Bot. Ital. **106**, 151–159.

## *Acknowledgements*

We wish to thank Mr. COLAPRICO C. and SPLENDIDO R. for their help and Dr. LAFIANDRA D. for his advises during electrophoretic analyses.

Dr. P. PERRINO and Dr. D. PIGNONE
Istituto del Germoplasma,
C.N.R., Via G. Amendola 165/A
I-70 126 Bari, Italy

Kulturpflanze XXIX · 1981 · S. 321–323

# Historical background and taxonomy of cultivated large-flowered *Clematis* in Europe[1]

WILLEM A. BRANDENBURG (Wageningen, the Netherlands)

## Summary

The development of large-flowered *Clematis* hybrids is briefly surveyed. The need for a botanic classification of these hybrids has been discussed.

## Outlinear history of cultivated large-flowered *Clematis* in Europe

*Clematis* species have been grown by man since the late Middle Ages. At first they were cultivated as medicinal plants as e.g. NIJLANDT (1682) mentioned.

In the 16th century several European species, especially *C. viticella* L., became known as ornamental. From *C. viticella* several selections (cultivars) were raised, which were vegetatively propagated.

At the end of the 18th century and especially in the 19th century *Clematis* species from China and Japan were introduced into Europe. Among these, *C. florida* THUNB., *C. lanuginosa* LINDL. et PAXT., and *C. patens* MORR. et DECNE. are said to be involved in hybridizations with *C. viticella* (MOORE and JACKMAN 1872; LAVALLEE 1884). This has given rise to many large-flowered, vegetatively propagated *Clematis* hybrids (cultivars), which are grown this very day.

At the end of the 19th century species from North America, especially *C. texensis* BUCKL., were crossed in the mentioned hybrids. Before that time there were only hybrids with open or flat flowers. The new crosses, however, resulted in hybrids with large, bell-shaped flowers.

In the 20th century several noteworthy hybrids were raised. For the extension of the variation in hybrid material, however, it will be necessary to look at other species to cross.

## Classification of large-flowered Clematis hybrids

MOORE and JACKMAN (1872) have proposed a classification for the cultivated *Clematis*, known at that time.

They did not base their classification on botanic characters. Their mode of classification 'will be found to bring together all those species and varieties which are similar in habit and character, and will, moreover, assist us in arranging, in some

1 Presented as poster

intelligible order, the instructions we shall have to offer regarding the cultivation of the different types of *Clematis*' (page 21).

With regard to large-flowered hybrids they made groups of cultivars, naming these groups after the species, which was most similar in habit and growth. With slight extensions, made by SPINGARN (1935), their classification is followed this very day (JOUIN 1907; MARKHAM 1951; FISK 1975; KRÜSSMANN 1976; LLOYD 1978) as to large-flowered hybrids. MOORE and JACKMAN (1872), however, have looked at *Clematis* from a horticulturist's point of view. So, they have put hybrid cultivars and species with a totally different genetic background into one group only because of similarity in habit and especially in growth.

*Clematis* species from China and Japan were introduced in different, European countries. They were described by several taxonomists and horticulturists (see MOORE and JACKMAN 1872; LAVALLEE 1884; KUNTZE 1885). This situation has resulted in a wide range of synonym names. The descriptions, however, were not based on natural populations, but on introduced plants, grown in gardens. Consequently differences between these plants, considered as to be different species, could be caused by a normal variation pattern of natural populations in China and Japan.

To make a botanic classification of all known large-flowered *Clematis* hybrids, biosystematic investigations and cytotaxonomic data of species and hybrids will be needed, as it is not certain, which role each of the earlier mentioned species has played in hybridizations. It seems to be important to start from seeds, collected from wild plants, as the taxonomic status of the parental species is dubious.

## Acknowledgements

I am much indebted to Dr. IR. J. HEYTING, Mr. J. G. VAN DE VOOREN and Mr. E. H. OOST for their helpful remarks and stimulating discussions, and to Miss M. DE GEUS for the aquarelles of *Clematis* cultivars in the presented poster.

## Zusammenfassung

Historischer Hintergrund und Taxonomie der kultivierten großblütigen *Clematis* in Europa.

Die Entstehung großblütiger *Clematis*-Bastarde wird in kurzer Form umrissen. Die Notwendigkeit einer botanischen Klassifikation dieser Hybriden wird diskutiert.

## Краткое содержание

Исторический фон и таксономия возделываемых крупноцветковых форм *Clematis* в Европе

Вкратце описывается история возникновения гибридных форм *Clematis* с крупными цветками. Обсуждаются возможности выяснения происхождения и ботанического классифицирования этих гибридов.

## Literature

FISK, J., 1975: The queen of climbers. – Fisk, Westleton.

JOUIN, E., 1907: Die in Deutschland kultivierten, winterharten *Clematis*. – Mitt. Deutsch. Dendr. Ges. (**1907**), 228–238.

KRÜSSMANN, G., 1976: Handbuch der Laubgehölze, Bd. **I**, 2nd. ed. – Paul Parey, Hamburg, Berlin.

KUNTZE, O., 1885: Monographie der Gattung *Clematis*. – Abh. Bot. Ver. Prov. Brandenburg **26**, 83–202.

LAVALLEE, A., 1884: Les Clématites à grandes fleurs. – J. B. Baillière et fils, Paris.

LLOYD, CH., 1978: *Clematis*. – Collins, London.

MARKHAM, E., 1951: The large and small flowered *Clematis* and their culture in the open air, 3rd. ed. London.

MOORE, T. and G. JACKMAN, 1872: *Clematis* as a garden flower. – J. Murray, London.

NIJLANDT, P., 1682: De Nederlandtse Herbarius of Kruidt-boeck. – Amsterdam (facsimile, 1976, Interbook, Schiedam).

SPINGARN, J. E., 1935: The large-flowered *Clematis* hybrids, a tentative checklist. – Nat. Hort. Mag. (**1935**), 64–94.

Ir. W. A. BRANDENBURG
Department of Taxonomy of
Cultivated Plants,
Agricultural University
Haagsteeg 3,
NL-6708 PM Wageningen
The Netherlands

## Section III:

**Biosystematics of cultivated plants and breeding research**

Chairmen:

A. C. Zeven
J. Mac Key

Kulturpflanze XXIX · 1981 · S. 327–335

# Biosystematic studies of cultivated plants as an aid to breeding research and plant breeding

JOHN G. HAWKES (Birmingham, England)


## Summary

The value of biosystematics to the breeder is shown by reference to examples taken from potato breeding.


## Introduction

Various types of taxonomic research have been developed during the 20th century, some of which are of greater value and interest to plant breeders than others. Of particular value is the discipline known as biosystematics, which seeks to investigate experimentally the cytological and genetical relationships between species, their reproductive biology and their patterns of variation. Reproductive barriers and the extent of actual or potential gene flow between and within species throw light on evolutionary relationships. Such ideas and information are of obvious value to plant breeders, particularly when seeking to make wide crosses, using wild species that may at first sight bear little relationship with the crop which the breeder is trying to improve.

Taxonomists from the times of the Greek and Roman writers until the beginning of this century have concentrated their attentions almost entirely on morphology–the outward form of plants and animals. In the latter half of the 19th century the geographical distribution of plants was found to be significantly correlated with their patterns of variation. However, the morphogeographical approach whilst valuable in giving the breeder a general concept of the total range of variability and its distribution, does not answer the questions he really needs to know.

As N. I. VAVILOV (1940) pointed out, in an attempt to relate studies of cultivated plants to plant breeding objectives:

"The agronomist is more interested in biological and physiological characters, in the relation of varieties to different diseases, in their behaviour with respect to drought, cold, etc. Morphological characters represent only the first approach to knowledge." We might add also, that the plant breeder is interested to know how such characters may be transferred into the primary gene pool of his crop species, and thus what are the breeding barriers that may restrict gene transfer.

Thus, the discipline of biosystematics, linked to that of chemotaxonomy, whereby chemical variation is analysed within and between species, has been of great value to breeders. It has led to much living material being collected in the field and has broadened the genetic base of the gene pool known to breeders.

## The biosystematics of potatoes

Let us now see how biosystematic studies have thrown light on the evolutionary relationships of potato species and have by so doing helped potato breeders in their choice of materials.

(i) Cultivated species (see fig. 1)

JUZEPCZUK and BUKASOV (1929) demonstrated the presence of a polyploid series in cultivated potatoes, ranging from diploids (2n = 24) to triploids (2n = 36), tetraploids (2n = 48) and pentaploids (2n = 60) All these possess their appropriate distribution pattern in the Andes and adjacent regions of South America.

Biosystematic studies have shown that three of the diploids (*S. stenotomum*, *S. phureja*, *S. goniocalyx*) are so closely related that they may perhaps be regarded

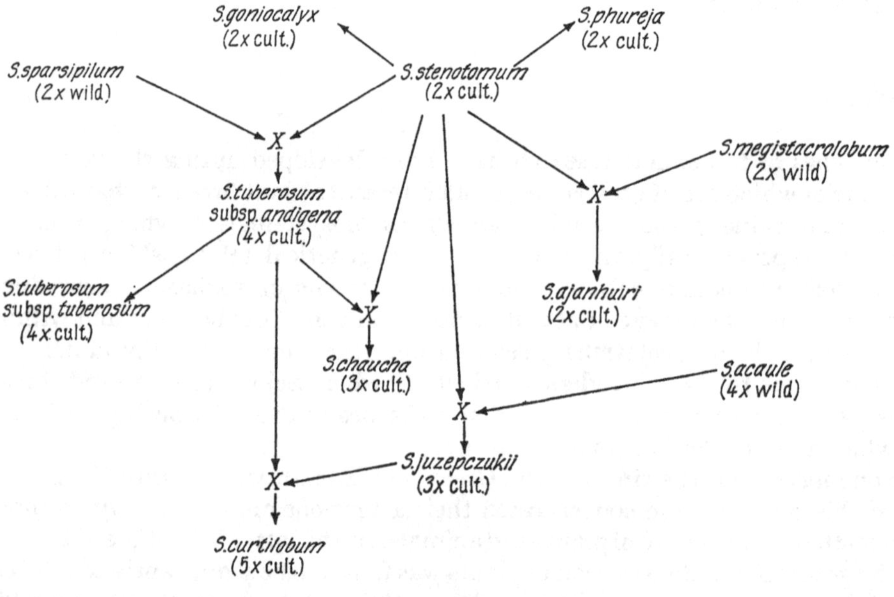


**Fig. 1 Evolutionary relationships of cultivated potatoes**

as no more than subspecies of a single species (TAY 1980, using numerical and biochemical techniques). The fourth diploid species, *S. ajanhuiri*, has now been clearly shown to be a natural hybrid of *S. stenotomum* by a wild species, *S. megistacrolobum* (HUAMAN 1975). Since the latter is highly frost resistant and transmits this character to its hybrid, *S. ajanhuiri*, this analysis clearly indicates a possible new route for breeding frost-resistance into potatoes at the diploid level, using a much wider range of diploid cultivated material and various wild accessions, than occurred when the hybrid was originally formed on the cold windswept plateaux of Bolivia.

The odd-numbered polyploids, triploids and pentaploids, reproduce entirely vegetatively and are obviously very different in their reproductive biology from the sexually outbreeding diploids and tetraploids. The purists would assert that these triploids and pentaploids ought not to be called species or given latin binomials. Yet they exist, are widely grown, and for reasons of convenience and by convention they might as well continue to be given such status.

There are two triploid species, *S. juzepczukii* and *S. chaucha*. The first-named was shown by HAWKES (1962a) to be a natural hybrid of a wild frost-resistant tetraploid species, *S. acaule*, crossed with the cultivated diploid *S. stenotomum*. This was confirmed later by SCHMIEDICHE (1977), who showed that the bitter alkaloids in *S. juzepczukii* might be reduced in a plant breeding programme through selecting *S. acaule* lines with low alkaloid content and *S. stenotomum* with good combining ability. Here, then, is another example of biosystematic studies pointing the way to a totally new breeding programme for frost resistance. It shows that parental selection rather than progeny selection could be used in the production of triploid potato varieties, just as it could with triploid bananas.

The other triploid species, *S. chaucha*, was shown to be a natural hybrid of *S. tuberosum* subsp. *andigena* (tetraploid) crossed with *S. stenotomum* (diploid) (JACKSON 1975; JACKSON et al. 1977, 1978).

The artificial cross produces such a small number of seeds per berry (about 1.5 on average) that there is certainly a barrier to gene flow, though this is not complete. The reciprocal is completely unsuccessful. In addition to triploids a certain number of tetraploid seedlings are produced in the $F_1$, the number of seeds and the proportions of triploids to tetraploids depending entirely on the parental material chosen. Triploids themselves are partly fertile and can be crossed back to the parents, though the number of seedlings produced is again very small. This work demonstrates that characters found in the diploid cultivated potatoes may very likely have flowed into the tetraploids (and vice-versa) over many thousands of years by means of these partially fertile triploid hybrids.

The true nature of the tetraploid potato, *S. tuberosum*, with its Andean subspecies, *andigena*, has also been established (CRIBB 1972) (Fig. 2). This is an amphidiploid hybrid of the cultivated diploid species *S. stenotomum* and the diploid weed species, *S. sparsipilum*. Since little genome evolution seems to have taken place in the South American potatoes (with some exceptions in other taxonomic series), the chromosomes in the tetraploid show multivalent formation, and this has given rise to the assertion that the tetraploid potato is an autopolyploid. In fact the biochemical and morphological data indicate very clearly the hybrid origin of *S. tuberosum*. The weed species *S. sparsipilum* is of value in its own right also, as a source of resistance to root-knot nematodes, and it has endowed the tetra-

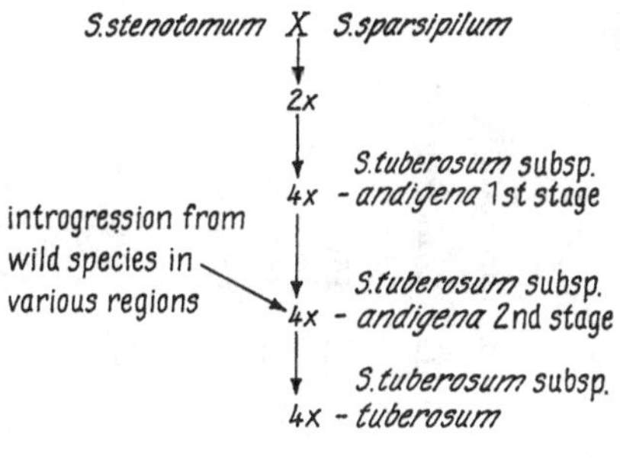


Fig. 2 Origin of *Solanum tuberosum* (CRIBB 1972)

Table 1

Classification and chromosome number of the more important wild and cultivated potato species

| *Series* | | *Species arranged according to chromosome number* ($x=12$) | | | | |
|---|---|---|---|---|---|---|
| | | *2x* | *3x* | *4x* | *5x* | *6x* |
| I | Juglandifolia | *S. juglandifolium*<br>*S. lycopersicoides* | | | | |
| II | Etuberosa | *S. brevidens*<br>*S. etuberosa* | | | | |
| III | Morelliformia | *S. morelliforme* | | | | |
| IV | Bulbocastana | *S. bulbocastanum*<br>*S. clarum* | *S. bulbocastanum* | | | |
| V | Pinnatisecta | *S. cardiophyllum*<br>*S. jamesii*<br>*S. pinnatisectum*<br>*S. trifidum* | *S. cardiophyllum*<br>*S. jamesii* | | | |
| VI | Commersoniana | *S. chacoense*<br>*S. commersonii*<br>*S. tarijense* | *S. calvescens*<br>*S. commersonii* | | | |
| VII | Circaeifolia | *S. capsicibaccatum* | | | | |
| VIII | Conicibaccata | *S. chomatophilum*<br>*S. violaceimarmoratum* | | *S. colombianum*<br>*S. oxycarpum* | | *S. moscopanum* |
| IX | Piurana | *S. piurae* | | *S. tuquerrense* | | |
| X | Acaulia | | | *S. acaule* | *(S. acaule)* | *S. acaule* |
| XI | Demissa | | | | *S. ×edinense*<br>*S. ×semidemissum* | *S. brachycarpum*<br>*S. demissum* |
| XII | Longipedicellata | | *S. ×vallis-mexici* | *S. fendleri*<br>*S. polytrichon*<br>*S. stoloniferum* | | |
| XIII | Polyadenia | *S. polyadenium*<br>*S. lesteri* | | | | |
| XIV | Cuneoalata | *S. infundibuliforme* | | | | |

| | | | | | | |
|---|---|---|---|---|---|---|
| XV | Megistacroloba | *S. boliviense* | | | | |
| | | *S. megistacrolobum* | | | | |
| | | *S. raphanifolium* | | | | |
| | | *S. sanctae-rosae* | | | | |
| | | *S. toralapanum* | | | | |
| XVI | Ingaefolia | *S. rachialatum* | | | | |
| XVII | Olmosiana | *S. olmosense* | | | | |
| XVIII a | Tuberosa (wild) | *S. ×berthaultii* | | | | |
| | | *S. bukasovii* | | | | |
| | | *S. canasense* | | | | |
| | | *S. gourlayi* | | *S. gourlayi* | | |
| | | *S. kurtzianum* | | | | |
| | | *S. leptophyes* | | | | |
| | | *S. maglia* | *S. maglia* | | | |
| | | *S. microdontum* | *S. microdontum* | | | |
| | | *S. oplocense* | | *S. oplocense* | | *S. oplocense* |
| | | *S. sparsipilum* | | | | |
| | | *S. spegazzinii* | | *S. sucrense* | | |
| | | *S. vernei* | | | | |
| | | *S. verrucosum* | | | | |
| XVIII b | Tuberosa (cultivated) | *S. ×ajanhuiri* | *S. ×chaucha* | *S. tuberosum* subsp. *tuberosum* | *S. ×curtilobum* | |
| | | *S. goniocalyx* | *S. ×juzepczukii* | *S. tuberosum* subsp. *andigena* | | |
| | | *S. phureja* | | | | |
| | | *S. stenotomum* | | | | |

ploid cultigen with a greater adaptation to a wider range of environmental conditions than existed in the original diploid cultigen.

There is only one pentaploid species, *S. curtilobum*, which was shown by HAWKES (1962a) to be a cross between *S. juzepczukii* (where an "unreduced" gamete functioned) and *S. tuberosum* subsp. *andigena*.

SCHMIEDICHE's work (1977) indicated that all the *S. curtilobum* lines known to us may well have come from the same original cross. Whilst HAWKES was able to re-synthesis the pentaploid, SCHMIEDICHE tried successfully to form the tetraploid *S. juzepczukii* × *S. goniocalyx* cross. This could again be a new way of breeding for frost resistance by crossing this tetraploid (never seen as such in nature) with *S. tuberosum* varieties, thus keeping to the tetraploid level.

This brief review of biosystematic studies on cultivated potatoes shows how their modes of origin and relationships to one another can be elucidated and clearly indicates lines of advance in plant breeding.

(ii) Wild species

There are probably over 150 species of wild potato, and these have been classified into 18 series (Table 1). Crossability studies have shown that although most of the South American and some of the Mexican species will cross readily with *S. tuberosum*, there are four Mexican series that will not do so, except in one or two special cases. These are also immunologically distinct from the cultivated species (GELL et al. 1956, 1960). However, it was found, for instance, that *S. acaule* could be used as a bridge species to cross *S. bulbocastanum* with *S. tuberosum* (DIONNE 1963; HERMSEN 1966; HERMSEN and RAMANNA 1969). The work also includes certain other species in the Mexican group, all of which are of value to breeders in respect or resistance to late blight (*Phytophthora infestans*).

The taxonomically very widely separated non-tuberiferous series *Etuberosa* has attracted notice recently because of its virus resistance (particularly to potato leaf roll virus). HERMSEN and TAYLOR (1980) have shown that *Etuberosa* species can be crossed, though with extreme difficulty, with tuber-bearing species. Work on in vitro fertilization and protoplast fusion is now in progress to see whether better chances of hybridisation can be obtained by such methods rather than the conventional ones.

Apart from the incompatibility barriers between species at the diploid level, there is also evidence for a certain degreee of genome evolution in wild potatoes that may be valuable to breeders.

This genome evolution is seen in the series *Longipedicellata* and *Demissa* in Mexico and the series *Conicibacata* which occurs both in Mexico and South America. For instance, crosses between *S. stoloniferum* (Mexico, 4×) and *S. chacoense* (S. American 2×) tend to give $12_{II}+12_{I}$ at meiosis (MATSUBAYASHI 1955; KAWAKAMI and MATSUBAYASHI 1956, 1958). They can then be postulated as possessing the genome $A_4$ B (see Table 2). In the hexaploid species of series Demissa there is a tendency in the hybrids to the formation of $24_{II}+24_{I}$. The genome formula A, $A_4$ (B, C, D etc.) would show that $A_1$ and $A_4$ had their homologues present in the hybrid, but the B and C sets could not pair (MARKS 1955, 1965). Where these came from we do not yet know, but it is assumed that at least some of the *Longipicellata* tetraploids are incorporated in the *Demissa* hexaploids. This matter is described in greater detail elsewhere (HAWKES 1978).

Table 2

Suggested genome formulae for certain potato series

| Genome | Taxonomic series |
|---|---|
| $A_1$ | Series Tuberosa, Commersoniana, Cuneoalata, Megistacroloba |
| $A_2$ $A_3$ | Series Acaulia |
| $A_4$ B | Series Longipedicellata (at least in part) |
| $A_1$ $A_4$ (B, C, D, etc.) | Series Demissa |

Recent work by LOPEZ (1979) has shown similar genome differentiation in the series *Conicibaccata,* which also contains hexaploid species as well as diploids and tetraploids.

As another example, I should like to mention the study of the Bolivian weed potato species *S. sucrense* (ASTLEY 1979; ASTLEY and HAWKES 1979). Biosystematic studies have indicated that this occurred as a series of natural hybrids between *S. oplocense* (a tetraploid wild species with cyst nematode resistance) and *S. tuberosum* subsp. *andigena* (also tetraploid). *S. sucrense* retains the nematode resistance of its wild ancestor and apparently backcrosses with *andigena* also. Certain *andigena* varieties have been collected in central Bolivia (department of Potosí) with the same sort of extreme resistance that is found in *S. oplocense,* and the biosystematic studies reported here seem to provide the explanation by indicating the presence of backcrossing to the *andigena* parent. Obviously, more collections were needed from this region, and these have indeed been made recently by HONDELMANN and ASTLEY.

Evidence of introgression in wild potatoes can be seen from studies of the Argentine wild diploid species *S. chacoense* and *S. microdontum* (HAWKES 1962b). These have different altitude requirements but they overlap in a small altitudinal zone and exchange genes by introgression. It has been shown that the virus Y resistance allele is found in both species (HAWKES and HJERTING 1969, p. 219), and further morphological analysis also confirms the introgression hypothesis. Probably more examples could be found in other species after careful study.

## Conclusions

Classical taxonomy and even modern numerical taxonomy can only indicate relationships based on phenotype similarities. These may be interesting and indeed highly significant, but, as VAVILOV (1940) pithily comments "Phenotypes are not genotypes" and only by cyclic crossings is it possible to determine the best combinations. The breeders are interested in taxonomic work which elucidates evolutionary relationships, based on experiment, particularly on the results of crosses, fertility of $F_1$ and later generations, the evolution of polyploidy and the identification and inheritence of resistance genes. Here, biosystematics forms strong links with cytogenetics on the one hand and chemotaxonomy on the other. We still know all too little about the relationships of our crop plants to related wild, weed and ancestral species. Biosystematics, in common with cytogenetics, is attempting to throw light on these problems so as to make wild and primitive material more available to breeders and to provide a groundwork of knowledge that breeders will find valuable.

## Zusammenfassung


Biosystematische Studien an Kulturpflanzen als Hilfsmittel für die Züchtungsforschung und Pflanzenzüchtung

Der Wert der Biosystematik für den Züchter wird an Beispielen aus der Kartoffelzüchtung dargestellt.


## Краткое содержание


Биосистематическое изучение культурных растений как вспомогательное средство для теоретической и практической селекции

Значение биосистематики для селекционера излагается на примерах селекции картофеля.


## Literature


Astley, D., 1979: *Solanum sucrense* Hawkes – a clarification. In: Proc. Conf. Broadening Genet. Base Crops, Wageningen, 1978 (Edits. A. C. Zeven, and A. M. van Harten) pp. 233–235. Pudoc, Wageningen, Netherlands.

–, and J. G. Hawkes, 1979: The nature of the Bolivian weed potato species *Solanum sucrense* Hawkes. – Euphytica **28**, 685–696.

Cribb, P. J., 1972: Studies on the origin of *Solanum tuberosum* L. subspecies *andigena* (Juz. et Buk.) Hawkes – the cultivated tetraploid potato of South America. – Ph. D. thesis, University of Birmingham. Birmingham, England.

Dionne, L. A., 1963: Studies on the use of *Solanum acaule* as a bridge between *Solanum tuberosum* and species in the series *Bulbocastana*, *Cardiophylla* and *Pinnatisecta*. – Euphytica **12**, 263–269.

Gell, P. G. H., S. T. C. Wright, and J. G. Hawkes, 1956: Immunological methods in plant taxonomy. – Nature, Lond., **177**, 573.

–, J. G. Hawkes, and S. T. C. Wright, 1960: The application of immunological methods to the taxonomy of species within the genus *Solanum*. – Proc. Roy. Soc. B. **151**, 364–383.

Hawkes, J. G., 1962a: The origin of *Solanum juzepczukii* Buk. and *S. curtilobum* Juz. et Buk. – Z. Pflanzenz. **47**, 1–14.

–, 1962b: Introgression in certain wild potato species. – Euphytica **11**, 26–35.

–, 1978: History of the potato. In: The Potato Crop: The Scientific Basis for Improvement (Edit. P. M. Harris). Chapman and Hall, London.

–, and J. P. Hjerting, 1969: The potatoes of Argentina, Brazil, Paraguay and Uruguay – a biosystematic study. – Oxford University Press, Oxford.

Hermsen, J. G. Th., 1966: Crossability, fertility and cytogenetic studies in *Solanum acaule* × *S. bulbocastanum*. – Euphytica **15**, 149–155.

–, and M. S. Ramanna, 1969: Meiosis in different $F_1$-hybrids of *Solanum acaule* Bitt. and *S. bulbocastanum* Dun. and its bearing on genome relationships, fertility and breeding behaviour. – Euphytica **18**, 27–35.

Hermsen, J. G. Th. and L. M. Taylor, 1979: Successful hybridization of non-tuberous *Solanum etuberosum* Lind. and tuber-bearing *S. pinnatisectum* Dun. – Euphytica **28**, 1–7.

HUAMÁN, Z., 1975: The origin and nature of *Solanum ajanhuiri* JUZ. et BUK. – a South American cultivated diploid potato. – Ph. D. thesis, University of Birmingham. Birmingham, England.

JACKSON, M. T., 1975: The evolutionary significance of the triploid cultivated potato, *Solanum* ×*chaucha* JUZ. et BUK. – Ph. D. thesis, University of Birmingham, Birmingham, England.

–, J. G. HAWKES, and P. R. ROWE, 1977: The nature of *Solanum* ×*chaucha* JUZ. et BUK., a triploid cultivated potato of the South American Andes. – Euphytica **26**, 775–783.

–, P. R. ROWE, and J. G. HAWKES, 1978: Crossability relationships of Andean potato varieties of three ploidy levels. – Euphytica **27**, 541–555.

JUZEPCZUK, S. W., and S. M. BUKASOV, 1929: (A contribution to the question of the origin of the potato). – Proc. U.S.S.R. Congr. Genet. Pl. and Animal Breed., **3**, 593–611.

KAWAKAMI, K., and M. MATSUBAYASHI, 1956: Studies on the species differentiation in the section *Tuberarium* of *Solanum*. II. Meiotic behaviour in the triploid $F_1$-hybrid from *S. longipedicellatum* × *S. chacoense*, with some regards to chromosome affinity between the parent species. – Rep. Hyogo Univ. Agric., Ser. Agric. **2**, 143–148.

–, –, 1958: Studies on the species differentiation in the section *Tuberarium* of *Solanum*. VI. Meiotic behaviour of amphiploids from *S. longipedicellatum* × *S. chacoense* and its bearing on genomic affinity between the parent species. – Rep. Hyogo Univ. Agric., Ser. Agric. **3**, 124–130

LÓPEZ, L. E., 1979: A biosystematic study of the series *Conicibaccata* of the genus *Solanum* L. – Ph. D. thesis, University of Birmingham. Birmingham, England.

MARKS, G. E., 1955: Cytogenetic studies in tuberous *Solanum* species. I. Genomic differentiation in the group *Demissa*. – J. Genetics **53**, 262–269.

–, 1965: Cytogenetic studies in tuberous *Solanum* species. III. Species relationships in some South and Central American species. – New Phytologist **64**, 293–306.

MATSUBAYASHI, M., 1955: Studies on the species differentiation in the section *Tuberarium* of *Solanum*. III. Behaviour of meiotic chromosomes in $F_1$-hybrid between *S. longipedicellatum* and *S. schickii* in relation to its parent species. – Rep. Hyogo Univ. Agric., Ser. Agric. **2**, 25–31.

SCHMIEDICHE, P. E., 1977: Biosystematic studies on the cultivated frost-resistant potato species *Solanum* ×*juzepczukii* BUK. and *Solanum curtilobum* JUZ. et BUK. – Ph. D. thesis, University of Birmingham. Birmingham, England.

TAY, C. S., 1980: Evolutionary studies on the cultivated diploid potatoes – *Solanum stenotomum* JUZ. et BUK., *S. goniocalyx* JUZ. et BUK. and *S. phureja* JUZ. et BUK. – Ph. D. thesis, University of Birmingham. Birmingham, England.

VAVILOV, N. I., 1940: The new systematics of cultivated plants. – In: The New Systematics (Edit. J. HUXLEY) pp. 549–566. – Oxford.

Prof. Dr. J. G. HAWKES
Department of Plant Biology
The University of Birmingham
P. O. Box 363
Birmingham B15 2TT, England U.K.

Kulturpflanze XXIX · 1981 · S. 337–352

# Alloenzyme variation and relationships among Spanish land-races of *Cucumis melo* L.

José T. Esquinas (FAO/IBPGR, Rome, Italy)

## Summary

Electrophoresis variability was examined in 66 cultivars of *Cucumis melo* representing 35 Spanish provinces to determine the genetic variation within cultivars, genetic differentiation between cultivars and relationships between provinces. The survey defined 15 enzyme-coding *loci* among which the total number of alloenzymes was 22. Progeny tests conducted for 6 polymorphic *loci* confirmed their monogenic nature.

Genetic variation within cultivars was measured by the proportion of polymorphic *loci*, which ranges from 0–0.27, and by the frequency of heterozygosity per *locus*, which ranges from 0–0.10. Genetic differentiation between provinces was estimated by Nei's genetic distance, D, and the range is from 0.01 to 0.26. The average value of D for all pairwise comparisons is $0.07 \pm 0.01$; meaning that on average about one allelic substitution per every 14 *loci* exist between any two Spanish provinces.

The results show that:

(a) almost 75% of the genome remains constant in all Spanish cultivars studied, and these provided a good genetic base for common features of all Spanish melons;
(b) melons from populations which differ in alloenzyme composition usually have distinctive morphological features;
(c) the variant alloenzymes are widespread throughout the country;
(d) the distribution of these variants is not random and may reflect some regional specialization;
(e) on average, more polymorphism is present along the coast than in the interior.

The Spanish cultivars seem to have a common origin and distinctive characteristics, without evidence of recent introgression with foreign melons despite considerable opportunities for interchange within Iberia. Limited regional specializations are also present. In this work, collections from some provinces have been grouped together according to their alloenzyme composition.

## Introduction

Although no mention was found in the available literature until the 15th century (Columbus 1494), there are good reasons to believe that melons were introduced into Spain much earlier. Probably they were known before the Arab domination in the 7th century since the Spanish name for melon comes from Latin and not from Arabic, as happened in the case of watermelon. This argument could be invalidated if the Arabs who invaded Spain did not know this crop, which is however very unlikely.

The above hypothesis is supported by a personal communication from an inhabitant of Lazagurria (Navarra) who found local church records from the 5th century showing that the parishoners paid the tithe in melons. No evidence has been found that may lead to believe that melons from this region are older than other Spanish cultivars.

Melons were probably brought to the Iberian peninsula during the Roman domination or through the barbarian invasions; the fact that current melons from

Fig. 1 upper row – left: "Melo escrit" from Mahon, Baleares. Early season melon
right: "Amarillo rayado temprano" from Esparragalejo, Badajoz. Mid-season melon
lower row – left: "Pinyonet" from a seed company. Mid-season melon
right: "Tendral amarillo" from Elche, Alicante. Late seson melon

Spain have very little in common with either Italian or French cultivars may favour the second hypothesis.

The distinctive characteristics of Spanish melons suggest a common origin for all of them. Unpublished research carried out by the author points towards India as their possible remote source. From a botanical point of view, all Spanish cultivars may be classified into two of NAUDIN's botanical varieties: var. *inodorus* for the late season melons and *reticulatus* for those of earlier production (Figure 1). Melons with intermediate characteristics are also present. There is a third NAUDIN botanical variety, *cantaloupensis*, grown in Spain, but it is used only for exportation. Several attempts to introduce foreign varieties in Spanish markets have soundly failed.

The Spanish consumer has a traditionally specialized taste for melons that has probably prevented their cultivars from introgressing with melons from elsewhere. This may explain the clear-cut delineation among the characteristics of Spanish melons and those from neighbouring countries. Some common features of most Spanish cultivars are: andromonoecious plants, white flesh, orange placenta and cavity, low pH (5 to 5.5), absence of ribs, etc. Distinctive characteristics of winter Spanish melons are: very wrinkled rind, good storage quality, crisp flesh texture, etc.

In spite of their national isolation and the similarities mentioned above, the Spanish cultivars have an amazing amount of organoleptic and morphological variation. This variability may be a product of regional specialization. The organoleptic diversity was already stressed in the first book on Spanish agriculture written by HERRERA (1513) in a chauvinistic comparison of Spanish melons with women: "if they are good, they are the best of things and they cannot be surpassed, but if they are bad, they are one of the most disgusting things that exist". The array of morphological differences was described by ESQUINAS (1972) and is shown in Table 1.

Table 1

Limits of variation of some morphological characters within Spanish melons. The study was carried out in Aranjuez (Spain) with more than 350 cultivars from all over the country (ESQUINAS 1972)

| Characters* | Limits of Variation | |
|---|---|---|
| | Maximum | Minimum |
| Number of fruits per plant | 9.0 | 1.0 |
| Weight of the fruit (gms) | 10,000.0 | 250.0 |
| Length of the fruit (cms) | 50.0 | 10.0 |
| Diameter of the fruit (cms) | 40.0 | 10.0 |
| Thickness of the rind (cms) | 2.5 | 0.1 |
| Content of soluble solids (%)** | 18.0 | 4.0 |

Also, the rind varies in colour from white, through yellow, orange, red, brown, green, grey, until almost black.

* All cultivars have been studied in the same environmental conditions.

** Content in soluble solids was measured with a refractometer.

A major problem with such data is the difficulty of separating the genetic from the environmental components of melon variability, particularly since much of this variability is quantitively, not monogenetically, inherited. Development of electrophoretic techniques to separate isoenzymes has provided us with a powerful tool to identify numerous monogenetic traits in either homozygotic or heterozygotic states.

The nature of isoenzyme variation in 66 cultivars of Spanish melons was studied not only for the abovementioned purpose, but also to investigate variation within populations, the extent of genetic differentiation between populations and to compare the zymograms of cultivated melons from 35 different Spanish provinces.

## Materials and methods

### *A. Materials*

66 local cultivars representing the 35 provinces where melons are grown in Spain were employed in this study. The seeds were kindly provided by local farmers. The average number of plants analysed per cultivar was 73.

The identification number, source, origin, sex type and other information concerning each tested population are listed in table 2.

Table 2

Identification number, source, accession site, sex type and local names for each tested cultivar.

| Identification Number | Accession Site | Sex[1] | Local Name |
|---|---|---|---|
| 1 | Alborea, Albacete | A | "Amarillo" |
| 2 | Elche, Alicante | A | "Tendral blanco" |
| 3 | Elche, Alicante | A | "Tendral amarillo" |
| 4 | Elche, Alicante | A | "Tendral negro" |
| 5 | Elche, Alicante | A | "Pinyonet" (piel de sapo) |
| 6 | Huercalovera, Almeria | A | "Melon de olor" |
| 7 | Guareña, Badajoz | A | "Pequeño amarillo" |
| 8 | Guareña, Badajoz | A | "Gordo amarillo" |
| 9 | La Albuera, Badajoz | A | "Melonas amarillas" |
| 10 | Esparragalejo, Badajoz | A | "Banda de Godoi" |
| 11 | Esparragalejo, Badajoz | A | "Amarillo rayado temprano" |
| 12 | Mahon, Baleares | A | "Melo escrit" |
| 13 | Igualada, Barcelona | A | "Pinyonet" |
| 14 | V. de Alcantara, Caceres | A | "Pipa de trigo" |
| 15 | Rota, Cadiz | A | "Amarillo grande" |
| 16 | San Mateo, Castellon | A | "Tendral" |
| 17 | Valdepeñas, Cuidad Real | A | "Fino redondo" |
| 18 | Calzada de Calatrava, Cuidad Real | A | |
| 19 | Calzada de Calatrava, Cuidad Real | A | |
| 20 | Calzada de Calatrava, Cuidad Real | A | "Averolo" |
| 21 | Fernan Nuñez, Cordoba | A | "Cama caliente" |

Table 2 (Cont.)

| Identification Number | Accession Site | Sex[1] | Local Name |
|---|---|---|---|
| 22 | Fernan Nuñez, Cordoba | A | "Cocas" |
| 23 | Puentegenil, Cordoba | A | "Lopereño o maduro" |
| 24 | Villa del Rio, Cordoba | A | "Madeno temprano escrito" |
| 25 | Bolliga, Cuenca | A/M | "Aguadierte" |
| 26 | Tarancon, Cuenca | A | "Dorado" |
| 27 | San Clemente, Cuenca | A | |
| 28 | Figueras, Gerona | A | "Cabanas o ronos" |
| 29 | Baza, Granada | A | |
| 30 | Baza, Granada | A | |
| 31 | Baza, Granada | A | "De la marina" |
| 32 | Baza, Granada | A | "Blancos lisos" |
| 33 | Atienza, Guadalajara | A | |
| 34 | Moguer, Huelva | A | "Cartoyero" |
| 35 | Monzon, Huesca | A | "Tendral negro" |
| 36 | Monzon, Huesca | A | "Tendral amarillo" |
| 37 | Alcaudete, Jaen | A | "Loperano" |
| 38 | Arjona, Jaen | A | "Variedad agostizo" |
| 39 | Torredonjimeno, Jaen | A | "Caña dulce" |
| 40 | V. de D. Juan, Leon | A | "Mentolado" |
| 41 | Mollerusa, Lerida | A | "Tendral Temprano" |
| 42 | Calahorra, Logroño | A | "Pipa de trigo" |
| 43 | Calahorra, Logroño | A | "Tempranillo" |
| 44 | Calahorra, Logroño | A | "Tendral tardio de agua" |
| 45 | Aranjuez, Madrid | A | "Puchero" |
| 46 | Aranjuez, Madrid | A | "Mochuelo" |
| 47 | V. Salvanes, Madrid | A | "Mochuelos" |
| 48 | A. de Henares, Madrid | A | |
| 49 | Almargen, Malaga | A | "Invernizo" |
| 50 | Santomera, Murcia | A | "Escrito" |
| 51 | Murchante, Navarra | A | |
| 52 | Vitigudino, Salamanca | M | "Pedroso" |
| 53 | Cuidad Rodrigo, Salamanca | A | "Invernizo" |
| 54 | Coca, Segovia | A | |
| 55 | Ecija, Sevilla | A | "Coca japonesa" |
| 56 | Vendrelli, Tarragona | A | "Invern o invierno" |
| 57 | Mora, de Ebro, Tarragona | A | |
| 58 | Valls, Tarragona | A | "De invierno" |
| 59 | Alacon, Teruel | A | "Comun" |
| 60 | Alcañiz, Teruel | A | "Puionete" |
| 61 | Belvis de Jara, Toledo | A | "Bolas" |
| 62 | Belvis de Jara, Toledo | A | "Negros" |
| 63 | Foyos, Valencia | A | |
| 64 | Tordesillas, Valladolid | A | "Blanco amarillo" |
| 65 | Toro, Zamora | A | |
| 66 | Alagon, Zaragoza | A | "Mochuelo" |

[1] Sex: M = monoecious
A = andromonoecious

## *B. Methods*

Gel and electrode buffers: Two buffer systems were used: 1. a gel buffer: 0.15 M Tris and 0.0004 M citrate and electrode buffer: 0.3 M borate, pH 7.8 (Rick and Fobes 1976) and 2. a gel buffer: 0.009 M Tris and 0.003 M citric acid, pH 7.1 and electrode buffer: 0.135 M Tris and 0.045 M citric acid, pH 7.1 (Ayala et al. 1974a).

Gel preparation: 250 mls of gel buffer solution were measured into a 250 ml volumetric flask. 30 gr of hydrolyzed potato starch with about one-third of the gel buffer solution were put into a 500 ml suction flask and the mixture shaken until the starch was in suspension. The remaining gel buffer solution was brought to a boil and added at once to the starch suspension while agitating the latter rapidly. A vacuum was applied immediately to remove all air bubbles from the gel mixture.

Sample preparation: A crude extract was obtained by crushing root and hypocotyl of 4–6 week seedlings. This extract was absorbed onto two 7×5 mm pieces of filter paper. The paper containing the samples was inserted along a slot in the gel made about 5 cm from the edge. The gel was placed horizontally between two buffer trays with electrodes. Sponge wicks were used to establish contact between the gel and the electrode buffer in the trays. An electric field at 160 V was applied for an initial period of 20 minutes, after which pieces of paper with the sample extracts were removed. Electrophoresis was then continued at a constant power of 11 W for 230 minutes. Apparatus and methodology for constant power and gel cooling are described by Fobes and Paige (1975). After the run was completed, the two ends of the gel, which were in contact with the sponge wicks, were cut off. The rest of the gel was sliced horizontally into 2 mm thick slices with a taut piano wire and 2 mm thick plexiglass guides. The top slice was discarded and the other three were used for three different enzyme assays. Since the crude extract was absorbed with two pieces of filter paper and each one placed in a different gel, as many as six enzyme assays could be obtained from an individual seedling.

Enzyme assays: Methods were taken from Shaw and Koen (1968), Brewer (1970), Rick et al. (1974) or Ayala et al. (1972) with or without modification.

Table 3 lists the six enzymes assayed, the abbreviations used to designate the enzymes, the buffer system used, and the number of *loci* scored.

Gel fixation: After the enzyme bands appear, the reaction was stopped by washing the gel with water and fixing with glycerine diluted 1 : 1 in water. Antibiotic was then added. After eight hours the gel could be wrapped and sealed with Saran wrap for storage. Most gels keep their appearance for many months but with dyes using NBT, the bands start fading after a few weeks (Ayala et al. 1972).

Table 3

Enzymes assayed and procedures used

| Enzyme | Abbreviation | Buffer system | Number of *loci* scored |
|---|---|---|---|
| Peroxidase | *PRX* | 1 | 4 |
| Acid Phosphatase | *APS* | 1 | 2 |
| Glutamate Oxalacetate Transaminase | *GOT* | 1 | 4 |
| Phosphoglucomutase | *PGM* | 2* | 1 |
| Phosphoglucoisomerase | *PGI* | 2 | 2 |
| 6-Phosphoglucodehydrogenase | *6-PDGH* | 2 | 2 |

* Different zone of activity of this enzyme was detected using buffer 1, although for practical reasons it was only scored in the zone of activity of buffer 2.

*C. Calculations*

Genetic differences between provinces were measured using the statistics: genetic similarity (I) and genetic distance (D), following NEI's procedure (1972). If X and Y are two populations, the mean genetic similarity per *locus* between them is: $I = I_{xy}/(I_x \cdot I_y)^{1/2}$, where $I_{xy}$, $I_x$, and $I_y$ are the arithmetic means, over all *loci* $x_i y_i$, $x_i^2$ and $y_i^2$; with $x_i$ and $y_i$ being the frequencies of the *i*th allele in populations X and Y, respectively. Genetic similarity may be roughly defined as the proportion of *loci* at which the allelic constitution of two populations are essentially identical. Genetic distance ($D = -\log I$) between two populations can be considered a measure of the average number of electrophoretically detectable allelic substitutions per *locus* which have accumulated since the two populations separated from a common ancestor, if it is assumed that the number of allelic substitutions per *locus* follows a POISSON distribution and the codon substitutions within a *locus* occur independently from each other (AYALA *et al.* 1974b; AYALA 1975).

## Results and discussion

*A. Characterization of the alloenzymes*

The nomenclatural system adopted consists of a few letters, a hyphenated number and a superscript. The letters designate the enzyme system; the hyphenated number, the *locus* and the superscript, the allele; *n* symbolizing a null allele. The allele prevailing for each *locus* in both wild and cultivated forms is assumed to be the normal or standard. The superscript assigned to the standard alleles is always *1*. The standard zymotype is defined as the one with only standard alleles.

Figure 2 illustrates diagrammatically all the melon variants found for the six enzyme systems tested. Fifteen *loci* were defined in this survey and the total

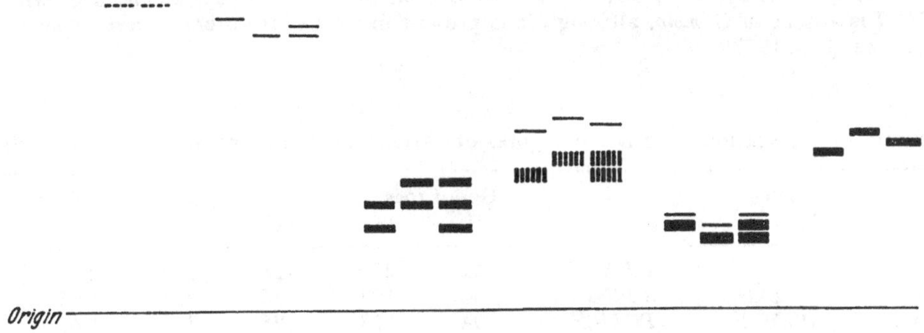


Fig. 2 Diagram showing the segregation of the six polymorphic loci of *C. melo*, the inheritance of which has been studied in this work

number of alloenzymes represented among the Spanish cultivars at all *loci* were 22. In some instances, *6-PGDH-2* for example, a simple band corresponds to a single *locus*; for others, *PRX-1* for example, one gene regulates a group of adjacent bands.

Eleven *loci* proved to be monomorphic, while the remaining four are polymorphic, for each of which two alleles are known; the rare alleles are not included in this classification. Except for the null alleles, the two contrasting homozygotes can be distinguished from each other and from the corresponding heterozygote. Non-Spanish cultivars and wild melons were used to carry out progeny tests for those *loci* that were fixed among the Spanish cultivars but show variability within the species. $F_2$ segregation (Figure 3) verified the monogenic nature and confirmed the identification of most polymorphic genotypes (Table 4); the identifi-

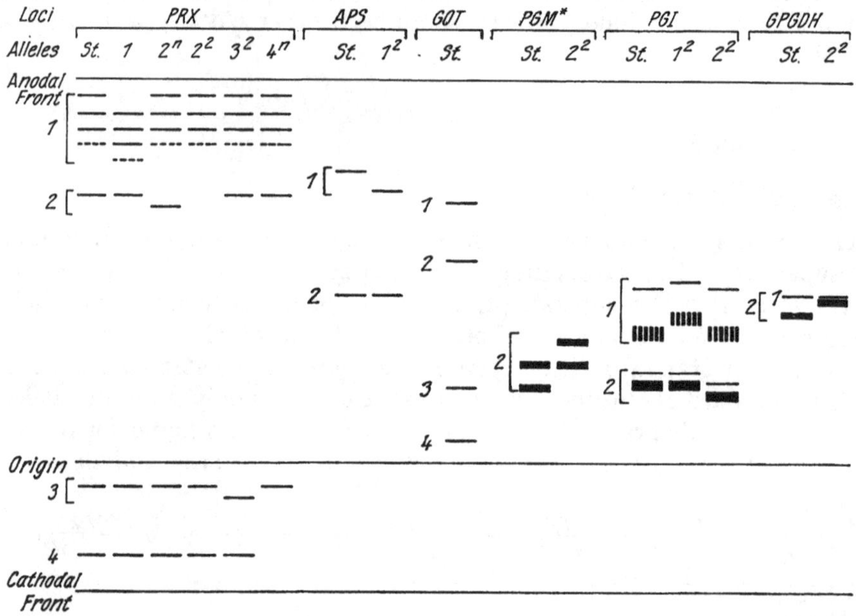


Fig. 3 Diagram of zymotypes for all alleles studied for six enzyme systems in *C. melo*. **PGM-1* is absent in *C. melo*, although it is present in most of the other *Cucumis* species. (Esquinas, J. T. 1977)

Table 4

Summary of segregation from heterozygotes of various combinations of alleles of *C. melo*

| Locus | Alleles 1 | Alleles 2 | Genotypes 1/1 | 1/2 | 2/2 | n | $X^2$ |
|---|---|---|---|---|---|---|---|
| *PRX-1* | $PRX\text{-}1^1$ | $PRX\text{-}1^2$ | 122 | 289 | 128 | 539 | 2.91 |
| *APS-1* | $APS\text{-}1^1$ | $APS\text{-}1^2$ | 43 | 107 | 56 | 206 | 1.95 |
| *PGM-2* | $PGM\text{-}2^1$ | $PGM\text{-}2^2$ | 74 | 173 | 78 | 325 | 1.46 |
| *PGI-1* | $PGI\text{-}1^1$ | $PGI\text{-}1^2$ | 69 | 171 | 76 | 316 | 2.45 |
| *PGI-2* | $PGI\text{-}2^1$ | $PGI\text{-}2^2$ | 91 | 167 | 84 | 342 | 0.47 |
| *6-PGDH-2* | $6\text{-}PGDH\text{-}2^1$ | $6\text{-}PGDH\text{-}2^2$ | 35 | 98 | 54 | 187 | 3.85 |

cation of other monomorphic genotypes is based on comparisons with other *Cucumis* species (ESQUINAS 1977).

*APS-1*$^2$ and *PG1-1*$^1$ interact with the wild allele, but not all expected banks were shown in the tissues tested. Numerous examples of such asymmetric enzyme patterns in heterozygous condition are described and studied by HARRIS (1975); a number of alternative explanations for this phenomenon are also provided in the second edition of his book.

*B. Distribution of the alleles*

*PRX-1*$^2$, *PRX-2*$^n$, *APS-1*$^2$ and *PGI-1*$^2$ are the only four non-standard alloenzymes that were found, with the exception of rare alleles. Figure 4 shows the geographic distribution of these variants. They are widespread in Spain with some cases of regional restrictions but without any obvious replacement tendencies.

Rare variants are: *PRZ-2*$^2$, *PRX-3*$^2$ and *PRX-4*$^n$. *PRX-2*$^2$ seems to appear randomly in several cultivars, although their frequency rarely exceeds 1% and never 4%. *PRX-3*$^2$ was only found in two cultivars in frequencies of 1 and 4%. *PRX-4*$^n$ was shown in two plants of another cultivar; however, the fact that it is a null allele, not detectable in heterozygotes, and the fact that progeny tests were not performed leave doubts about its genetic origin.

*C. Genetic variation within populations*

The number of *loci* studied, the sample size, the proportion of polymeric *loci* (two criteria) and the heterozygosity per *locus* for the 15 *loci* in each one of the 66 cultivars studied are listed in Table 5.

Table 5

Sample size, proportion of polymorphic loci and estimated heterozygosity. Each cultivar has been tested for 15 loci

| ID Number | Average sample size per locus | Proportion of polymorphic loci Criterion 1* 5% | Criterion 2** 1% | Proportion of genome heterozygosity per individual |
|---|---|---|---|---|
| 1 | 86 | 0 | 0 | 0 |
| 2 | 85 | 0 | 0.07 | 0.01 |
| 3 | 72 | 0.20 | 0.20 | 0.02 |
| 4 | 61 | 0 | 0.13 | 0.01 |
| 5 | 82 | 0.07 | 0.20 | 0.04 |
| 6 | 79 | 0.13 | 0.13 | 0.04 |
| 7 | 77 | 0.13 | 0.13 | 0.06 |
| 8 | 88 | 0.07 | 0.13 | 0.01 |
| 9 | 73 | 0 | 0.13 | 0.01 |
| 10 | 77 | 0.13 | 0.27 | 0.05 |
| 11 | 74 | 0.07 | 0.07 | 0.05 |
| 12 | 76 | 0.07 | 0.07 | 0.03 |
| 13 | 74 | 0.07 | 0.07 | 0.03 |
| 14 | 77 | 0.13 | 0.07 | 0.04 |
| 15 | 86 | 0 | 0 | 0 |
| 16 | 89 | 0.13 | 0.20 | 0.06 |
| 17 | 64 | 0 | 0 | 0 |
| 18 | 70 | 0 | 0 | 0 |

Table 5 (Cont.)

| ID Number | Average sample size per locus | Proportion of polymorphic loci Criterion 1* 5% | Criterion 2** 1% | Proportion of genome heterozygosity per individual |
|---|---|---|---|---|
| 19 | 76 | 0.13 | 0.13 | 0.02 |
| 20 | 61 | 0.20 | 0.20 | 0.06 |
| 21 | 64 | 0.27 | 0.27 | 0.06 |
| 22 | 73 | 0.13 | 0.13 | 0.01 |
| 23 | 94 | 0.27 | 0.27 | 0.03 |
| 24 | 82 | 0.20 | 0.27 | 0.05 |
| 25 | 77 | 0.13 | 0.13 | 0.07 |
| 26 | 64 | 0.07 | 0.07 | 0.04 |
| 27 | 61 | 0.07 | 0.20 | 0.01 |
| 28 | 73 | 0.13 | 0.20 | 0.04 |
| 29 | 68 | 0 | 0 | 0 |
| 30 | 73 | 0.13 | 0.13 | 0.05 |
| 31 | 63 | 0 | 0 | 0 |
| 32 | 88 | 0 | 0 | 0 |
| 33 | 59 | 0.07 | 0.07 | 0.02 |
| 34 | 77 | 0 | 0 | 0 |
| 35 | 86 | 0.07 | 0.13 | 0.03 |
| 36 | 89 | 0.20 | 0.20 | 0.07 |
| 37 | 64 | 0.07 | 0.07 | 0.01 |
| 38 | 75 | 0 | 0 | 0 |
| 39 | 79 | 0 | 0 | 0 |
| 40 | 63 | 0 | 0.07 | 0.01 |
| 41 | 70 | 0 | 0 | 0 |
| 42 | 80 | 0 | 0 | 0 |
| 43 | 79 | 0.13 | 0.13 | 0.06 |
| 44 | 89 | 0.13 | 0.13 | 0.13 |
| 45 | 78 | 0 | 0 | 0 |
| 46 | 76 | 0.07 | 0.07 | 0.03 |
| 47 | 64 | 0 | 0 | 0 |
| 48 | 51 | 0 | 0 | 0 |
| 49 | 59 | 0.13 | 0.13 | 0.07 |
| 50 | 96 | 0.13 | 0.13 | 0.07 |
| 51 | 84 | 0.13 | 0.13 | 0.02 |
| 52 | 71 | 0.20 | 0.20 | 0.09 |
| 53 | 71 | 0.07 | 0.07 | 0.01 |
| 54 | 65 | 0.07 | 0.07 | 0.01 |
| 55 | 60 | 0.07 | 0.07 | 0 |
| 56 | 68 | 0.13 | 0.13 | 0.03 |
| 57 | 43 | 0.07 | 0.07 | 0.07 |
| 58 | 62 | 0.07 | 0.07 | 0.04 |
| 59 | 61 | 0 | 0 | 0 |
| 60 | 67 | 0 | 0.07 | 0.01 |
| 61 | 71 | 0.13 | 0.13 | 0.06 |
| 62 | 77 | 0.13 | 0.13 | 0.03 |
| 63 | 61 | 0.13 | 0.20 | 0.04 |
| 64 | 93 | 0.07 | 0.07 | 0.03 |
| 65 | 61 | 0.20 | 0.20 | 0.10 |
| 66 | 74 | 0 | 0 | 0 |

* Criterion 1: The frequency of the most common allele is $\leqq .95$
** Criterion 2: The frequency of the most common allele is $\leqq .99$

Frequencies of cultivars with standard vs. non-standard alleles of cultivated melons in each of thirty-five Spanish provinces*

| Provinces & Identification no. of cultivars studied | No. of cultivars | Frequency of cultivars w/ fixed standard zymotype | $PRX\text{-}1^2$ | $PRX\text{-}2^n$ | $APS\text{-}1^2$ | $PGI\text{-}1^2$. | Others (Rare) |
|---|---|---|---|---|---|---|---|
| Albacete (1) | 1 | 0 | 0 | 0 | 1.00 | 1.00 | |
| Alicante (2, 3, 4, 5) | 4 | 0 | 0.10 | 0.51 | 0.50 | 0.65 | $PRX\text{-}2^2$ = *v.r.*, $PRX\text{-}3^2$ = 0.01 |
| Almeria (6) | 1 | 0 | 1.00 | 0.37 | 0 | 0 | $PRX\text{-}2^2$ = *v.r.*, $PRX\text{-}4^n$ = 0.10 |
| Badajoz (7, 8, 9, 10, 11) | 5 | 0 | *v.r.* | 0 | 0.43 | 0.86 | $PRX\text{-}2^2$ = *v.r.* |
| Baleares (12) | 1 | 0 | 0 | 0 | 0.35 | 0 | |
| Barcelona (13) | 1 | 0 | 0 | 0 | 1.00 | 0.64 | |
| Caceres (14) | 1 | 0 | 0.89 | 0.33 | 1.00 | 1.00 | |
| Cadiz (15) | 1 | 0 | 0 | 0 | 1.00 | 0 | |
| Castellon (16) | 1 | 0 | 0.02 | 0.71 | 0.41 | 0 | |
| Ciudad-Real (17, 18, 19, 20) | 4 | 0 | 0 | 0.27 | 0.58 | 0.41 | |
| Cordoba (21, 22, 23, 24) | 4 | 0 | 0.15 | 0.67 | 0.66 | 0.89 | |
| Cuenca (25, 26, 27) | 3 | 0 | 0.13 | 0.14 | 1.00 | 0.49 | |
| Gerona (28) | 1 | 0 | 0.75 | 0.57 | 1.00 | 1.00 | $PRX\text{-}2^2$ = *v.r.*, $PRX\text{-}3^2$ = 0.10 |
| Granada (29, 30, 31, 32) | 4 | 0 | 0.20 | 0.25 | 0.62 | 0.75 | |
| Guadalajara (33) | 1 | 0 | 0 | 1.00 | 0.75 | 1.00 | |
| Huelva (34) | 1 | 0 | 0 | 0 | 1.00 | 0 | |
| Huesca (35, 36) | 2 | 0 | 0.16 | 0.35 | 0.46 | 0.37 | |
| Jaen (37, 38, 39) | 3 | 0 | 0.31 | 0.33 | 1.00 | 1.00 | |
| Leon (40) | 1 | 0 | 0 | 0.97 | 1.00 | 0 | |
| Lerida (41) | 1 | 1.00 | 0 | 0 | 0 | 0 | |
| Logroño (42, 43, 44) | 3 | 0 | 0.08 | 0 | 0.84 | 0.32 | |
| Madrid (45, 46, 47, 48) | 4 | 0 | 0 | 0.42 | 0.75 | 0 | |
| Malaga (49) | 1 | 0 | 0 | 0.51 | 0.54 | 0 | |
| Murcia (50) | 1 | 0 | 0.47 | 0.62 | 1.00 | 1.00 | |
| Navarra (51) | 1 | 0 | 0.06 | 0.86 | 1.00 | 1.00 | |
| Salamanca (52, 53) | 2 | 0 | 0.14 | 0 | 0.80 | 0.53 | |
| Segovia (54) | 1 | 0 | 0 | 1.00 | 0.05 | 0 | |
| Sevilla (55) | 1 | 0 | 0.57 | 0.56 | 1.00 | 0 | |
| Tarragona (56, 57, 58) | 3 | 0.33 | 0.20 | 0.11 | 0.02 | 0 | |
| Teruel (59, 60) | 2 | 0 | 0 | 1.00 | 0.52 | 0 | |
| Toledo (61, 62) | 2 | 0 | 0.38 | 0.33 | 1.00 | 1.00 | $PRX\text{-}2^2$ = *v.r.* |
| Valencia (63) | 1 | 0 | 0.04 | 0.20 | 0.11 | 0 | |
| Valladolid (64) | 1 | 0 | 0 | 0 | 1.00 | 0.77 | |
| Zamora (65) | 1 | 0 | 0.42 | 0.28 | 0.33 | 0 | |
| Zaragoza (66) | 1 | 0 | 0 | 1.00 | 1.00 | 0 | |

* The frequency of null alleles has been calculated considering the Hardy Weinberg equilibrium. *v.r.* = very rare.

Little or no allelic variation was found at most of the *loci* in any population. If a *locus* is considered polymorphic when the most common allele has a frequency no greater than 0.95, the number of polymorphic *loci* ranges from zero (0%) for twenty-two populations to 4 (27%) for two populations. If a *locus* is considered polymorphic when the most common allele has a frequency no greater than 0.99, the number of polymorphic *loci* ranges from zero (0%) for seventeen populations to 4 (27%) for four populations. The mean frequency of heterozygous individuals per *locus* ranges from 0% for eighteen populations to 13% for population number 44.

Little significance, however, should be attached to the different levels of heterozygosity within populations, since they are cultivars and also many of them have been restricted by small population size for several generations; furthermore, the confidence of the estimates is low since so few *loci* contribute to the heterozygosity estimate.

*D. Characterization of provinces*

A summary of the average zymotype per province and a list of their populations is shown in Table 6.

*E. Genetic variation between provinces*

Genetic distances and genetic similarities have been calculated according to the method of Nei (1972) for all population pairs.

Table 7 shows the genetic distances and similarities between cultivated melons from all province pairs. They range from D=0.01 (several cases) to D=0.26 (Almeria vs. Leon). The average value of D for all pairwise comparisons is 0.07±0.01, meaning that on average 0.07 substitutions per *locus* exist between cultivars from any two Spanish provinces; that is, a substitution has occurred for every 14 *loci*.

## Conclusions

Electrophoretic data are compatible with available data from other sources. Table 7 and Figure 4 can be interpreted in the light of information shown in the introduction of this paper, as follows: 1. About 75% (11 out of 15 *loci*) of the genome remain constant in all Spanish cultivars studied and provided a good genetic base for the common features of all Spanish melons; 2. The four variant alloenzymes are widespread throughout the country without any obvious replacement tendency, probably due to the good opportunities for interchange that they have had within Iberia; 3. However, the distribution of these variants is not random and may reflect some regional specialization.

Regarding the geographical distribution of variants some points deserve to be noted: 1. On average more polymorphism is present along the coast than in the interior. 2. Provinces like Almeria, in the south, and Gerona, in the north, clearly differ in the alloenzyme composition from their neighbouring provinces. In both cases their cultivars represented in this work, "Melon de olor" in Almeria and "Cabañas" or "Roñas" in Gerona, also show distinctive morphological features.

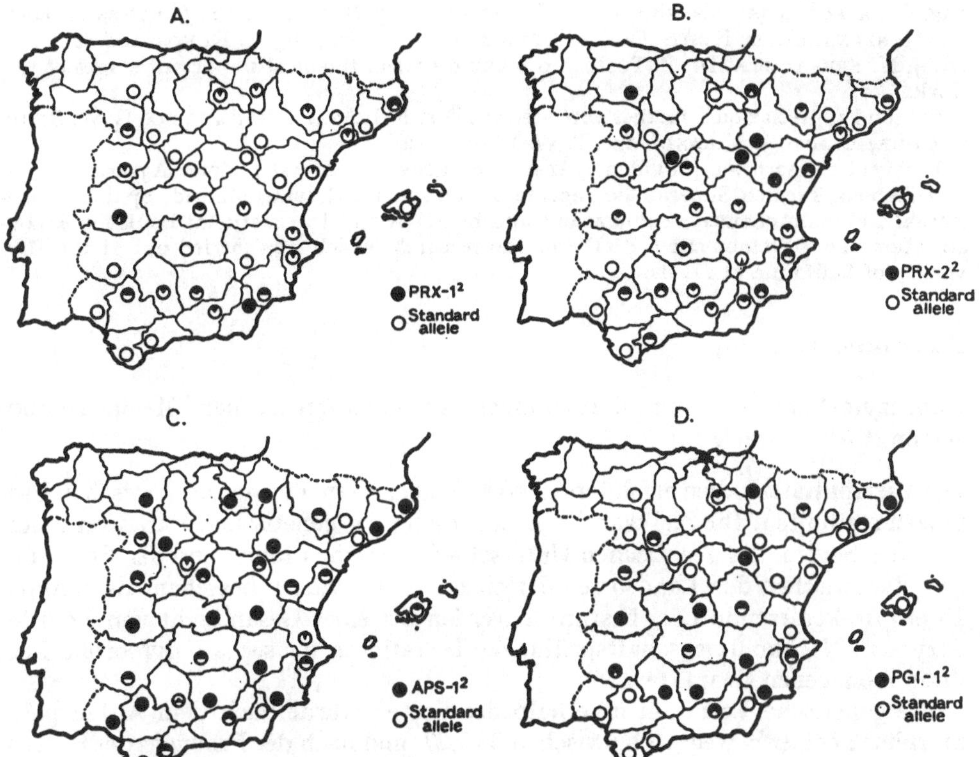


Fig. 4 Geographical distribution of *PRX-1*, *PRX-2*, *APS-1* and *PGI-1* among cultivated melons in Spain

3. On the contrary, many neighbouring provinces may be grouped together according to their alloenzyme composition. Some of these groups are:

(a) Huesca, Lerida, Tarragona, Castellon, Valencia and Baleares; (b) Teruel and Taragoza; (c) Valladolid and Salamanca; (d) Caceres, Toledo, Cuenca and Albacete; (e) Badajoz, Ciudad-Real, Cordoba, Jaen and Granada; (f) Huelva, Sevilla, Cadiz and Malaga.

In summary, the Spanish cultivars seem to have a common origin and distinctive characteristics, without evidence of recent introgression with foreign melons despite considerable opportunities for interchange within Iberia. Limited regional specializations are also present.

Unfortunately, the great source of germplasm present in Spain until very recently is steadily decaying due to the standardization of products needed for harvesting and commercialisation. This is one of the tributes that developing countries are paying in order to compete in a technological world.

*Acknowledgements*

I wish to express my sincere gratitude to Dr. C. M. Rick for his guidance, encouragement and continued assistance. Also to Dr. E. Sanchez-Monge for his continued support during collection, multiplication and morphological characterization of cultivars.

This research would not have been possible without the previous collaboration of

Ing. C. CASANOVA and his wife, Dr. M. L. DETONI, Ing. N. DURAN, Dr. J. FOBES, C. GARCIA-MANZANARES, D. HUNT, Dr. P. GURLICK, C. NEWMAN, Ing. C. RAMOS and his wife, Dr. J. M. SERRADILLA, Dr. M. TÒSIC and many others in the most important stages of this work.

Grateful appreciation is further extended to: Drs. F. J. AYALA, R. BARRERA, G. W. BOHN, C. GOMEZ-CAMPO, J. MIRANDA and T. W. WHITAKER.

Financal support was provided by Instituto Nacional de Investigaciones Agrarias (INIA) and Escuela Tecnica Superior de Ingenieros Agronomos (ETSIA), Madrid, Spain, for collection and maintenance of populations and by INIA and International Bank for Reconstruction and Development (IBRD) for the research, which was carried out at the University of California at Davis.

## Zusammenfassung

Alloenzym-Variation und -Beziehungen zwischen spanischen Melonen-Landsorten (*Cucumis melo* L.)

Die Variabilitätsstudien erfolgten an 66 Melonensorten, die aus 35 spanischen Provinzen stammten. Ihr Ziel war die Bestimmung der genetischen Variation innerhalb der Sorten, der genetischen Unterschiede zwischen ihnen und der Beziehungen, die zwischen den Landsorten der einzelnen Provinzen bestehen. Es konnten 15 enzym-kodierende Loci bestimmt werden bei einer Gesamtzahl von 22 Alloenzymen. Nachkommenschaftsprüfungen bestätigten für sechs polymorphe Loci ihren monogenen Charakter.

Die genetische Variation innerhalb der Sorten wurde nach dem Anteil polymorpher Loci, er bewegt sich zwischen 0–0,27, und nach der Frequenz der Heterozygotie je Locus, sie erstreckt sich von 0–0,10, veranschlagt. Die zwischen den Provinzen bestehenden genetischen Unterschiede der Landsorten wurden mit Hilfe von NEI's genetischer Distanz, D, geschätzt; ihr Wert bewegt sich zwischen 0,01 und 0,26. Für alle paarweisen Vergleiche beträgt der Durchschnittswert von D $0{,}07 \pm 0{,}01$. Das bedeutet, daß zwischen zwei Provinzen im Durchschnitt ungefähr eine allele Substitution für jeden der 14 Loci besteht.

Die Ergebnisse zeigen, daß
(a) in allen untersuchten spanischen Sorten fast 75% des Genoms konstant bleiben, diese ergeben eine gute Grundlage für gemeinsame Merkmale aller spanischen Melonen;
(b) Melonen aus Populationen, die sich in ihrer Alloenzymzusammensetzung unterscheiden, gewöhnlich charakteristische Merkmale besitzen;
(c) die varianten Alloenzyme im Lande weit verbreitet sind;
(d) die Verteilung dieser Varianten nicht zufällig ist; sie kann eine gewisse regionale Spezialisierung widerspiegeln;
(e) im Durchschnitt der Polymorphismus entlang der Küste stärker als im Landesinnern vertreten ist.

Die spanischen Sorten scheinen eine gemeinsame Herkunft und unterscheidende Merkmale zu haben. Es gibt keinen Beweis für rezente Introgression mit ausländischen Melonen, obwohl dafür erhebliche Möglichkeiten auf der Iberischen Halbinsel bestehen. Begrenzte regionale Spezialisierungen sind festzustellen. Sammlungen aus einigen Provinzen werden entsprechend ihrer Alloenzymzusammensetzung in Gruppen zusammengefaßt.

## Краткое содержание

Изменчивость аллоэнзимов и родственные связи испанских местных сортов дыни (*Cucumis melo* L.)

Изменчивость изучалась на 66 сортах дыни, происходящих из 35 провинций Испании. Целью работы было выявление генетической изменчивости внутри сортов, генетических различий между ними и связей, существующих между местными сортами отдельных провинций. Найдено 15 локусов, кодирующих энзимы; общее число энзимов составляло 22. Расщепление потомств выявило моногенный характер 6 локусов.

Генетическая изменчивость внутри сортов оценивалась по доле полиморфных локусов (в пределах от 0 до 0,27) и по частотам гетерозиготности на локус (между 0 и 0,10). Генетические различия между местными сортами из различных провинций оценивались при помощи генетической дистанции D (Nei, 1972), колебавшейся между 0,01 и 0,26. Для всех парных сравнений средняя величина D составляла $0,07 \pm 0,01$. Это значит, что в среднем между двумя провинциями, приблизительно, происходит одно аллельное замещение на каждый из 14 локусов.

Результаты показали, что
(а) во всех исследованных испанских сортах почти 75% генома остаётся постоянным; на основании этого можно принять значительную общность признаков всех испанских дынь;
(б) популяции дынь, различающихся по составу их аллоэнзимов, обыкновенно обладают и характерными морфологическими признаками;
(в) изменчивость аллоэнзимов широко распространена по всей стране;
(г) распределение этой изменчивости не случайно; оно может отражать некоторую региональную специализацию;
(д) в среднем, полиморфизм чаще встречается вдоль берегов, чем внутри страны.

Испанские сорта, очевидно, имеют общее происхождение, но обладают характерными признаками. Ничто не указывает на современную интрогрессию иностранных форм дынь, хотя на Иберийском полуострове для этого существуют значительные возможности. Наблюдается ограниченная региональная специализация. В этой работе сделана группировка форм из разных провинций, в соответствии с составом аллоэнзимов, возделываемых в них дынь.


## Literature


Ayala, F. J., 1975: Genetic differentiation during the speciation process. – In: Evolutionary Biology, vol. **8** (Eds. T. Dobzhanski, M. K. Hecht and W. C. Steere), pp. 1–78. Plenum Press, New York.

–, J. R. Powell, M. L. Tracey, C. A. Mourão and S. Pérez-Salas, 1972: Enzyme variability in the *Drosophila willistoni* group. IV. Genetic variation in natural populations of *Drosophila willistoni*. – Genetics **70**, 113–139.

–, J. W. Valentine, L. G. Barr and G. S. Zumwalt, 1974a: Genetic variability in a temperate intertidal Phoronid, *Phoronopsis viridis*. – Biochemical Genetics **11**, 413–427.

–, M. L. TRACEY, D. HEDGECOCK and R. C. RICHMOND, 1974b: Genetic differentiation during the speciation process in *Drosophila*. – Evolution **28**, 576–592.
BREWER, G. J., 1970: An introduction to isozyme techniques. – Academic Press, New York.
COLUMBUS, C., 1494: Letters of Christopher Columbus (Ed. MAJOR, R. H.). – London 1847.
ESQUINAS, J. T., 1972: Estudio de variedades indigenas de melon con vistas a la mejora de esta planta. –.IX Jornadas de Genetica Lusoespañolas. (Communication) October 1972; pp. 47–48.
–, 1977: Allozyme variation and relationships in the genus *Cucumis*. – Ph. D. Thesis, University of California, Davis.
FOBES, J. F. and D. F. PAIGE, 1975: Refinements in starch gel electrophoresis for genetic screening of horticultural material: application to nematode testing of *Lycopersicon esculentum*. – HortScience **10**, 313.
HARRIS, H., 1975: The principles of human biochemical genetics. – North Holland Publishing Company, Amsterdam-Oxford. Frontiers of Biology. (Ed. A. NEUBERGER and E. L. TATUM) 473 pp.
HERRERA, G. A. DE, 1513: Obra de Agricultura. – Alcalá de Henares. 369 pp.
NEI, M., 1972: Genetic distance between populations. – Amer. Natur. **106**, 283–292.
RICK, C. M. and J. F. FOBES, 1976: Peroxidase complex with concomitant anodal and cathodal variation in red-fruited tomato species. – Proc. Nat. Acad. Sci. USA **73**, No. 3, 900–904.
–, R. W. ZOBEL and J. F. FOBES, 1974: Four peroxidase loci in red-fruited tomato species: genetics and geographic distribution. – Proc. Nat. Acad. Sci. USA **71**, No. 3, 835–839.
SHAW, C. R. and A. L. KOEN, 1968: Starch gel zone electrophoresis of enzymes. In: Chromatographic and electrophoretic techniques. Vol. 2, 2nd ed. (Ed. I. SMITH). – John Wiley, New York.

Dr. JOSÉ T. ESQUINAS
AGPE/IBPGR, Food and Agriculture Organization of the United Nations
Via delle Terme di Caracalla
I-00100 Roma, Italia

Kulturpflanze XXIX · 1981 · S. 353–360

# Investigation into chromosome morphology of sugar beet and related wild species

HANS DE JONG (Amsterdam, Netherland)


## Summary

In the genus *Beta*, chromosomes at pachytene are about 10 times longer than in mitotic metaphase and show a conspicuous differentiation of chromomeres along their arms. It is shown that in sugar beet (*Beta vulgaris*) these chromomeres are mainly restricted to well defined blocks adjacent to the centromere regions. In the wild species *B. procumbens*, the chromomeres are found not only in the centromere regions, but also occur intercalary. In both species, conspicuous telomeres have not been observed. Tentative FEULGEN-DNA measurements revealed that the difference in chromomere distribution between these species is probably not caused by significant differences in DNA amounts. As the patterns observed in sugar beet also occur in other members of the section *Vulgares* and those observed in *B. procumbens* in the other two members of the section *Patellares*, it is hypothesized that the heterochromatin distributions are representative for the sections mentioned.


## Introduction

The genus *Beta* comprises four sections: *Vulgares* TR., *Corollinae* TR., *Nanae* ULTRICH and *Patellares* TR. (COONS 1975). The number of species in this genus is 10–14, depending on the classification of some species of the *Vulgares* section. Sugar beet and other cultivated *Beta vulgaris* varieties have often been hybridized with wild species belonging to the *Corollinae* and the *Patellares* sections in order to transfer desirable genes to the beet genome (BOSEMARK 1969). Although the production of $F_1$-hybrids and their backcross generations have several times been claimed to be successful, the exchange of genes between homoeologous chromosomes is rare. To understand this phenomenon, several chromosome studies of interspecific hybrids and their derivatives have been made (NAGL 1969, WALIA 1971).

In the Netherlands, at the Instituut voor Rationele Suikerproduktie, Bergen op Zoom, a cytogenetic study has been started to support the breeding program for beet cyst nematode resistance in sugar beet. Emphasis was put on chromosome identification of *Beta vulgaris* and the nematode resistant wild species of the *Patellares* section, in order to facilitate the study of pairing and further meiotic behaviour of the alien chromosome in resistant monosomic additions to *Beta vulgaris*.

It has been shown earlier (BOSEMARK and BORMOTOV 1971, DE JONG and DE BOCK 1978) that individual mitotic chromosomes of *Beta vulgaris* cannot be identified morphologically; they are small and rather similar and C-bands as obtained with the BSG-technique are mainly restricted to the centromere regions. During pachytene, chromosomes are about 10 times longer and show differentiation along their length. This has enabled us to identify some of the bivalents and to study variation in chromosome morphology within and between species. In this report the results of a comparative study on pachytene chromosomes of *Beta vulgaris* and *B. procumbens* are given. As the chromomere distribution in pachytene chromosomes generally corresponds to the C-banding pattern of metaphase chromosomes, we have compaired the locations of constitutive heterochromatin in root tip preparations of both species. Variation in distribution and size of chromomeres within a group of related species proves often to be based on variation in constitutive heterochromatin content and consequently, in the total amount of DNA per 2 C genome. To evaluate this correlation in the genus *Beta*, FEULGEN-DNA contents of three species belonging to the *Vulgares* section and all three species of the *Patellares* section were determined.

## Materials and methods

Flower bud and root tip material were obtained from young plants of *Beta vulgaris* (2n=18 and 36), *B. maritima* (2n=18) and *B. atriplicifolia* (2n=18), all belonging to the *Vulgares* section; of *B. procumbens* (2n=18), *B. webbiana* (2n=18) and *B. patellaris* (2n=36), all belonging to the *Patellares* section and of interspecific hybrids between *B. vulgaris* and *B. procumbens*. Pachytene nuclei were stained in an ethanol-hydrochloric acid-carmine mixture and poststained in GIEMSA. The complete technique was described previously (DE JONG 1978). In order to preclude variation in staining intensity, all important steps of the procedure were carefully standardized. Mitotic chromosomes were C-banded with the BSG-technique according to DE JONG and DE BOCK (1978). For the cytophotometrical determination of the DNA content, root tips were fixed in ethanol-acetic acid (3 : 1). Hydrolysis was followed in normal hydrochloric acid during 6 minutes at 60 °C. The material was stained in SCHIFF's reagent for 2 hours in the dark and subsequently squashed in acetic acid 45%. Mounting was in fluormount. The preparations were stored at −20 °C until use. Cytophotometric determinations of telophase nuclei were performed according to TEMPELAAR (1980). All root tips were treated at the same time under stricktly identical circumstances, so that the FEULGEN-DNA values from different preparations of the same plant were fairly similar.

## Results

At pachytene the nine bivalents of *Beta vulgaris* have a length of 20–50 μm. Their centromere position is subterminal to median with arm ratio's (=s/1+s) of 0.16–0.48. Sometimes considerable deviations occur as a result of local stretching or asynchronous condensation of chromosome arms. One bivalent is associated with the nucleolus. Pachytene chromosomes of *Beta* have been found to show a characteristic pattern of chromomeres. All distinct chromomeres are located in the centromere regions and the NOR. The centromeres themselves are often rather long and easy to locate. Striking telomeres are not visible. A careful comparison of the chromomere patterns between *B. vulgaris* and *B. procumbens* has revealed some interesting differences. In *B. vulgaris*, the proximal regions are composed of

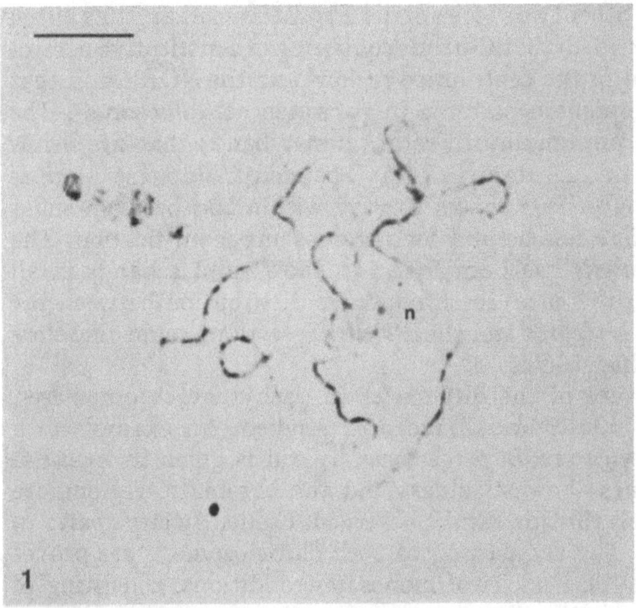


Fig. 1 Pachytene nucleus of *Beta vulgaris*. The proximal segments of the bivalents are deeply stained in contrast to the distal portions which are almost achromatic. *n* = nucleolus. Bar equals 10 μm in all figures

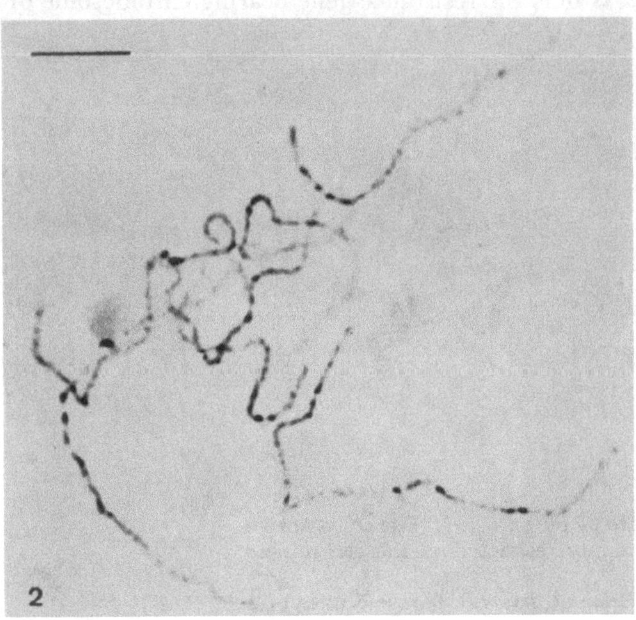


Fig. 2 Pachytene nucleus of *B. procumbens*. Around the centromeres, small distinct chromomeres are visible. Note the conspicuous intercalary chromomeres on some bivalents

several densely clustered chromomeres. This is in contrast to the distal parts of the bivalents which are rather achromatic and show only very few small chromomeres. An example is given in figure 1. This type of chromomere distribution also occurs in *Beta maritima* and *B. atriplicifolia*, both species belonging to the *Vulgares* section. In *B. procumbens* the proximal chromatic zone is densely stained but it is significantly smaller than in *Beta vulgaris*. Most bivalents also show some intercalary chromomeres which prove to be helpful markers for chromosome indentification. An example of this chromomere pattern is depicted in figure 2. This pattern has also been found in the other two species of the *Patellares* section. It has been shown (DE JONG and DE BOCK 1978) that C-bands consisting of constitutive heterochromatin are mainly located in the centromere regions and the NOR's and that they correspond to the chromomere patterns in pachytene chromosomes. The C-bands are very small and presumably there are minor bands that are below optical microscopic resolution. An example of a C-banded metaphase plate is shown in figure 3. C-bands have been shown to vary within and between sugar beet varieties and are therefore not helpful for chromosome identification. The C-band pattern of *B. procumbens* has been found to show similar bands in all chromosomes. The observed differences in chromomere distribution between the investigated members of the *Vulgares* and the *Patellares* sections could therefore not be supported by C-banding studies.

In order to study the nature of the differences in pachytene chromomeres, pachytene nuclei of interspecific hybrids have been studied. An example of a pachytene nucleus of a *B. vulgaris* × *B. procumbens* hybrid is given in figure 4. The bivalents or chromosomes show stickiness and the chromatic regions are clustered. Loops and unpaired threads can be observed. Evidently large parts of the chromosome fail to pair. The comparison of both chromosome types proves to be more succesful in pachytene nuclei of monosomic additions, consisting of the nine bivalents of *B. vulgaris* plus the resistance-gene bearing chromosome of

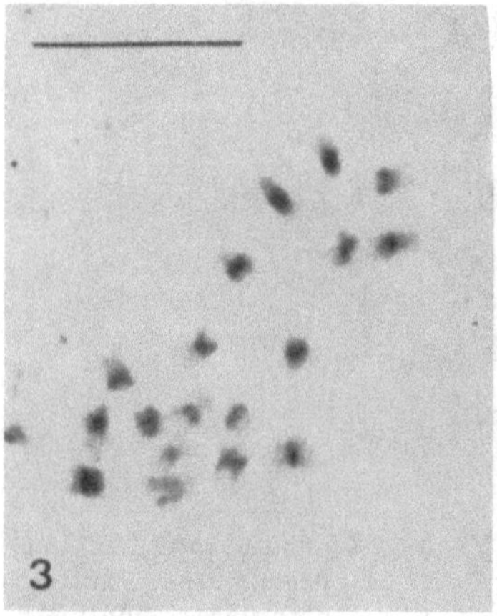


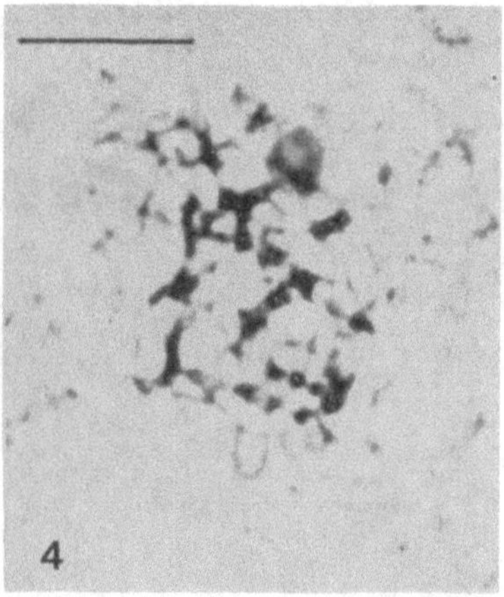


Fig. 3 C-banded metaphase-plate of *Beta vulgaris*. The constitutive heterochromatic segments are mainly restricted to the centromere regions
Fig. 4 Pachytene nucleus of a *Beta vulgaris* × *B. procumbens* hybrid

*B. procumbens* (SAVITSKY 1975). The difference in chromomere pattern is quite easy to demonstrate in these plants (DE JONG et al., in preparation). To answer the question as to whether the chromomere distribution can be ascribed to differences in DNA content, cytophotometrical measurements of FEULGEN-DNA-values have been performed. Fig. 5 presents the values as they have been obtained from telophase nuclei of the six *Beta* species. An analysis of variance is summarized in table 1. Comparison of the FEULGEN-DNA measurements have revealed that the distributions of 2 C values of *B. vulgaris* and *B. maritima* show striking

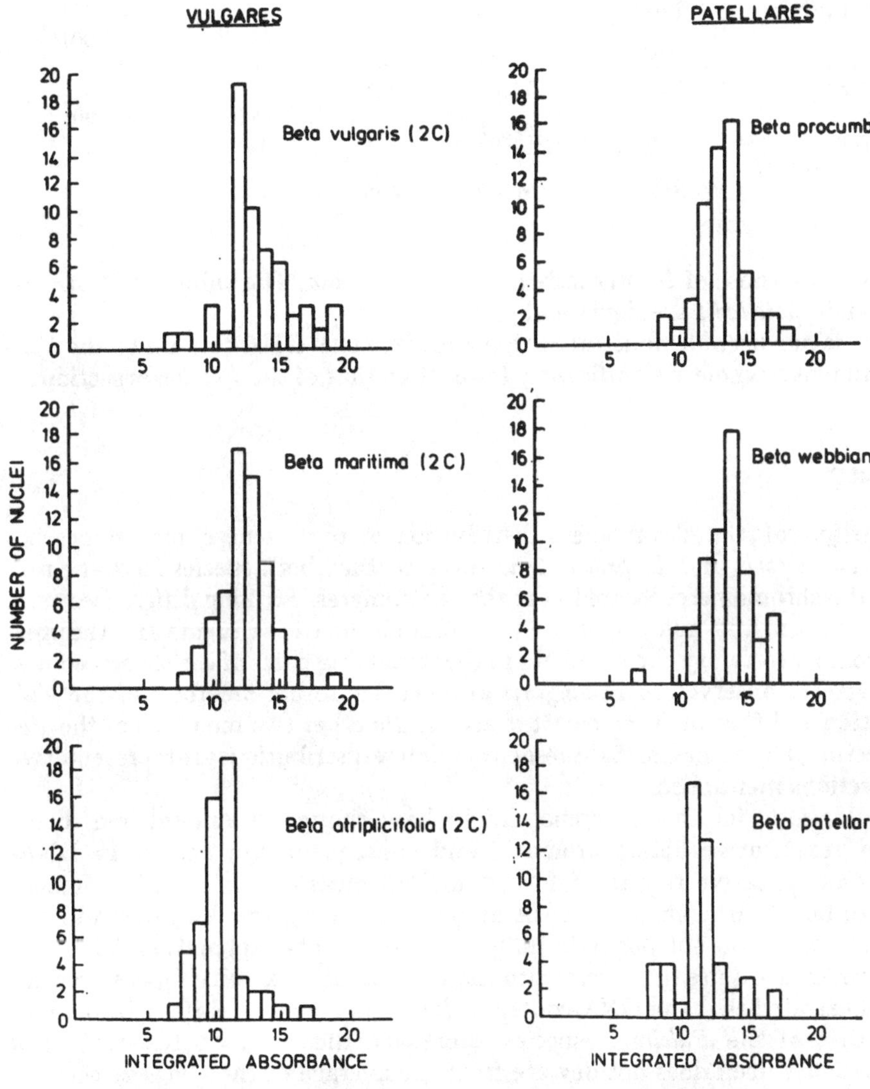


Fig. 5 Comparative histograms of relative FEULGEN-DNA contents of 2C nuclei in root tip meristems of *Beta vulgaris*, *B. maritima*, *B. atriplicifolia* (three species of the *Vulgares* section) and *B. procumbens*, *B. webbiana* and *B. patellaris* (together the *Patellares* section)

Table 1

Analysis of variance of the DNA-FEULGEN contents of 2C nuclei of *Beta vulgaris*, *B. maritima*, *B. atriplicifolia* (section *Vulgares*) and *B. procumbens*, *B. webbiana* and of $^{1}/_{2}$ the value of 2C nuclei of *Beta patellaris* (section *Patellares*)

| Source of variation | df | SS | MS | F | P |
|---|---|---|---|---|---|
| Between all species | 5 | 32684 | 6536 | – | – |
| Within all species | 340 | 128657 | 378 | – | – |
| | | | | 17.27 | <0.001 |
| Between all *Vulgares* | 2 | 18222 | 9111 | – | – |
| Within all *Vulgares* | 174 | 70799 | 406 | – | – |
| | | | | 22.39 | <0.001 |
| Between all *Patellares* | 2 | 11094 | 5547 | – | – |
| Within all *Patellares* | 166 | 57858 | 348 | – | – |
| | | | | 15.93 | <0.001 |
| Between the *Vulgares* and the *Patellares* | 1 | 3367 | 3367 | – | – |
| Total | 345 | 161342 | 468 | – | – |
| | | | | 7.19 | <0.01 |

similarity as do those of *B. procumbens* and *B. webbiana*. The values of *B. atriplicifolia* and *B. patellaris* are deviant (fig. 5).

Finally, it has been demonstrated that the FEULGEN-DNA content of the *Vulgares* section as a whole is significantly lower than that of the *Patellares* section.

## Discussion

A comparison of the chromomere distribution of pachytene chromosomes between *Beta vulgaris* and *B. procumbens* revealed that both species have the majority of the chromomeres located near the centromeres. Striking differences exist in respect to size and number of the proximal chromatic segments and the presence of some intercalary chromomeres in the remaining parts of the chromosomes. As the patterns observed in *B. vulgaris* also occur in other members of the *Vulgares* section and that of *B. procumbens* also in the other two members of the *Patellares* section, I hypothesize that these chromomere distributions are representative for the sections mentioned.

An explanation for this difference might be a change in amount and distribution of constitutive heterochromatin and consequently, a change in DNA-content. Our tentative results of FEULGEN-DNA-measurements, which demonstrate significant differences in DNA-amounts between the *Vulgares* and the *Patellares* sections do not preclude the possibility of this explanation. These results, however, are quite the reverse situation to that of NAKASHIMA (pers. comm.) who demonstrated that the DNA-contents of the *Vulgares* species are equal or higher than that of the *Patellares* species. Moreover, the average DNA-content of the *B. patellaris* itself does not deviate from the average of the *Vulgares* section. It can be concluded from these results that the differences of the DNA-content within and between the two sections show too much variations for the conclusions that the chromomere differences are related to differences in DNA-content.

Alternatively, one may assume that large differences in the level of the rearrangements of DNA in different segments of the chromosomes underly the two chromomere types. Finally, they may arise as a result of differential contraction of whole chromosomes or specific segments.

Differences in pachytene morphology have been described earlier by BROWN (1966) between tomato and petunia, both belonging to the *Solanaceae*. In tomato, the heterochromatin is located in well defined proximal blocks and some conspicuous telomeres, whereas in petunia it is distributed in many chromomeres mainly in the centromere regions and gradually decreasing in size towards the chromosome ends. Conspicuous telomeres are absent in petunia. It is reasonable to believe that the differences in chromomere pattern or amount of heterochromatin affect the pairing between homoeologous chromosomes in hybrids between species of the *Vulgares* and the *Patellares* sections and so might be one reason why chromosome exchange between the alien chromosome and one of the sugar beet chromosome in monosomic additions occurs only in very small frequencies.

*Acknowledgements*

I am thankful to Mrs. H. BLOM and Mr. F. SCHURING for assistance. Drs. W. HEIJBROEK of the Instituut voor Rationele Suikerproduktie, Bergen op Zoom and Mr. T. S. M. DE BOCK of the Foundation of Plantbreeding (SVP), Wageningen, are acknowledged for providing material. I am also very indepted to Dr. L. PIJNACKER and Dr. M. J. TEMPELAAR for using their cytophotometer equipment. This study was a part of a project on cytogenetics of interspecific hybrids resistant to beet cyst nematode, supported by the Instituut voor Rationele Suikerproduktie, Bergen op Zoom.

## Zusammenfassung

Untersuchung zur Chromosomenmorphologie der Zuckerrübe und verwandter Wildarten

In der Gattung *Beta* sind die Chromosomen im Pachytän etwa 10mal so lang wie in der mitotischen Metaphase und zeigen eine deutliche Differenzierung der Chromomeren entlang ihrer Arme. Es wird gezeigt, daß diese Chromomeren bei der Zuckerrübe (*Beta vulgaris*) hauptsächlich auf gut definierte, den Centromerregionen benachbarte Regionen beschränkt sind. Bei der Wildart *B. procumbens* werden die Chromomeren nicht nur in den Centromerregionen gefunden, sondern sie kommen auch interkalar vor. Bei beiden Arten sind deutliche Telomeren nicht beobachtet worden. Probemessungen von FEULGEN-DNA ergaben, daß der Unterschied in der Chromomerenverteilung zwischen diesen Arten wahrscheinlich nicht durch signifikante Unterschiede im DNA-Gehalt verursacht wird. Da die bei der Zuckerrübe beobachteten Muster auch bei anderen Vertretern der Sektion *Vulgares* und die bei *B. procumbens* beobachteten bei den anderen beiden Vertretern der Sektion *Patellares* auftreten, wird angenommen, daß die Heterochromatinverteilungen typisch für die erwähnten Sektionen sind.

## Краткое содержание

**Исследования по морфологии хромосом сахарной свёклы и её дикорастущих сородичей**

У растений рода *Beta* хромосомы в пахитене приблизительно в 10 раз длинее, чем в метафазе мейоза и проявляют чёткую дифференциацию хромомер вдоль их плечей. Показано, что эти хромомеры у сахарной свёклы (*Beta vulgaris*) находятся, главным образом, на хорошо ограниченных участках, лежащих рядом с центромером. У дикорастущего вида *B. procumbens* хромомеры находятся не только на центромерном участке, но встречаются и интеркалярно. Теломеры не наблюдались чётко ни у культурного ни у дикого вида. Пробные измерения ДНК показали, что различия между видами в распределении хромомер, по всей видимости, не обусловлены достоверными различиями в содержании ДНК. Так как структура, наблюдавшаяся у сахарной свёклы, проявляется и у других представителей секции *Vulgares*, а, наблюдавшаяся у *B. procumbens* – и у других представителей секции *Patellares*, то можно считать, что распределение гетерохроматина является типичным для названных секций.

## Literature

BOSEMARK, N. O., 1969: Interspecific hybridization in *Beta* L.; prospects and value in sugar beet breeding I.I.R.B. Report **4**, 112–119.

–, and V. E. BORMOTOV, 1971: Chromosome morphology in a homozygous line of sugar beet. – Hereditas **69**, 205–212.

BROWN, S. W., 1966: Heterochromatin. – Science **151**, 417–425.

COONS, G. H., 1975: Interspecific hybrids between *Beta vulgaris* L. and the wild species of *Beta*.-J. Am. Soc. Sugar beet Technol. **11**, 281–306.

JONG, J. H. DE, 1978: A carmine-GIEMSA technic for meiotic prophase chromosome of the genus *Beta* L. – Stain Technol. **53**, 169–172.

–, and T. S. M. DE BOCK, 1978: Use of haploids of *Beta vulgaris* L. for the study of orcein and GIEMSA stained chromosomes. – Euphytica **27**, 41–47.

NAGL, W., 1969: The course of the first meiotic prophase in *Beta procumbens* and in the $F_1$ between *B. vnlgaris* and *B. procumbens*. – Theor. Appl. Genet. **39**, 356–360.

SAVITSKY, H., 1975: Hybridization between *Beta vulgaris* and *B. procumbens* and transmission of nematode (*Heterodera Schachtii*) resistance to sugar beet. – Can. J. Genet. Cytol. **17**, 197–209.

TEMPELAAR, M. J., 1980: DNA-content of isolated nuclei of postembryonic stages of progeny from normal and irradiated males of *Tetranychus urticae* KOCH (Acari, Tetranychidae). – Chromosoma (Berl.) **77**, 359–371.

WALIA, K., 1971: Meiotic prophase in the genus *Beta* (*B. vulgaris* 2x and 4x, *B. webbiana* and *B. patellaris*). – Z. Pflanzenzücht. **65**, 141–150.

DR. H. DE JONG
Institute of Genetics,
University of Amsterdam
Kruislaan 318
NL-1098 SM Amsterdam,
The Netherlands

Kulturpflanze XXIX · 1981 · S. 361–367

# New botanical variants of *Triticum sphaerococcum* Perc. and *Triticum spelta* L. developed by hybridization of species from the genus *Aegilops* L. with *Triticum aestivum* L.

Bogdan Bochev (Sofia, Bulgaria)

## Summary

As a result of the hybridization of *Aegilops ovata* auct. non L. with the *Triticum aestivum* L. cultivar Bezostaja 1, new biological variants of *T. sphaerococcum* Perc. have been obtained. Besides the basic morphological and biological characters, they possess also some new useful features, such as better expressed winter hardiness, resistance to fungal diseases, and higher content of protein. New botanical varieties of *T. spelta* L. are a result of hybridization between *Ae. crassa* Boiss. ssp. *vavilovii* Zhuk. (2n=42) and different cultivars of *T. aestivum*. Some of them have useful biological and agricultural characters and are considered to be of interest to further utilization in plant breeding. Thus, on the basis of all these results, new suggestions about the origin of the genus *Triticum* are made. At the same time, the new biological variants of both of these species increase the possibilities of their practical utilization in wheat breeding.

## Introduction

*Triticum sphaerococcum* Perc. is an endemic species of India and Pakistan. It is most widely extended in the state of Punjab, India. The species is slightly polymorphic and is distinguished by its considerable morphological uniformity. Up to now only the six varieties described by Percival (1921) were known. Later 11 more new botanical varieties developed by the use of various methods (Singh 1959, Nigmatullin 1972, Stankov and Tsikov 1973, Bochev 1974, Udačin and Šachmedov 1974, and Stankov 1977) were included in this species. Thus a new interest in *T. sphaerococcum* was aroused concerning its origin, evolution, taxonomy and the possibilities of using it in bread wheat breeding.

Contrary to *T. sphaerococcum*, the species *T. spelta* L. is of the remote past. It was of interest for agriculture in Egypt, Greece and Rome. *T. spelta* is a polymorphic species. In the opinion of Dorofeev (1972), it is represented by 33 varieties most of which belong to the Asian subspecies *kuckuckianum* Gökg., and is the result of hybridization between tetraploid wheats and *T. aestivum* L. and supplementary spontaneous mutations affecting the alleles of the gene Q. *T. spelta* has been the object of numerous and diverse studies. As a result of these studies, a number of questions concerning its primary origin, its secondary origin from *T. aestivum*, the localization of the gene q controlling the speltoid-type of the spike, etc., was explained. It is of particular interest that a dominant gene is present in *T. spelta* which restores the fertility of *T. aestivum* having *T. timopheevi* ZHUK. cytoplasm. The importance of *T. spelta* for phylogenetical investigations grew considerably after the successful experimental synthesis of the species by crossing *T. dicoccoides* KÖRN. with *Ae. squarrosa* auct. non L. and subsequent doubling of the number of chromosomes (MCFADDEN and SEARS 1946). The number of valuable biological and economic traits of *T. spelta* and particularly the high protein and lysine content of its grain and the complex resistance to fungal diseases of some of its forms contribute very much to the great interest in this species.

A large-scale research programme aiming at the investigation of the interrelations between the species of the genera *Aegilops* L. and *Triticum* L. by hybridization of their genome and chromosome homeologies and the existing possibilities for using the genus *Aegilops* in bread wheat improvement is carried out at the Institute of Genetics of the Bulgarian Academy of Sciences. As a result of hybridization between definite *Aegilops* species and various *T. aestivum* cultivars, the first new biological variants of the species *T. sphaerococcum*, *T. spelta* and *T. dicoccon* SCHRANK were developed.

Object of the present study are some of the main morphological and biological characteristics of the developed new botanical varieties of *T. sphaerococcum* and *T. spelta*.

## Materials and methods

The study includes five new varieties of *T. sphaerococcum* and seven of *T. spelta*. The basic morphological characteristics distinguishing them from the existing varieties, the level of their winter or spring habit, their photoperiodic reaction and their grain protein content were observed. The level of winter or spring habit was determined by the length of the vernalization stage. Moistened, germinating seeds were vernalized for 0, 10, 20 and 30 days, respectively, at a temperature of 0–2 °C. The photoperiodic reaction of plants from the various vernalization treatments was assessed at day lengths of 10 and 15 h. The grain protein content was determined after KJELDAHL. Plants of the various experimental treatments were grown under controlled conditions in glasshouses ensuring daily air temperature above 15 °C during the vegetation period so that no supplementary vernalization could take place.

## Results and discussion

### *Triticum sphaerococcum* PERC.

The new botanical *T. sphaerococcum* varieties were developed by crossing *Ae. ovata* auct. non L. with cv. Bezostaja 1 of *T. aestivum*. They were selected in the course of the large-scale form developing process in the $F_3$ to $F_5$ hybrid generations. On the basis of their main taxonomic, morphological and biological characteristics they were named as follows:

1. *Triticum sphaerococcum* PERC., variety "Pseudo albopubescens". Spike awned. Awns black. Glumes white, pubescent. Caryopsis reddish. The variant was selected from the $F_5$ of the cross *Ae. ovata* × *T. aestivum* cv. Bezostaja 1.

2. *Triticum sphaerococcum* PERC., variety "Pseudo albo glabrum". Spike awned. Awns black. Caryopsis reddish. The variant was selected from the $F_5$ of the cross *Ae. ovata* × *T. aestivum* cv. Bezostaja 1.

3. *Triticum sphaerococcum* PERC., variety "Pseudo echinatum". Spike awned. Awns black. Glumes white, glabrous. Caryopsis white. The variant was selected from the $F_5$ of the cross *Ae. ovata* × *T. aestivum* cv. Bezostaja 1.

4. *Triticum sphaerococcum* PERC., variety "Pseudo rubroaristatum". Spike awned. Awns black. Glumes red, pubescent. Caryopsis reddish. The variant was selected from the $F_8$ of the cross *Ae. ovata* × *T. aestivum* cv. Bezostaja 1.

5. *Triticum sphaerococcum* PERC., variety "Pseudo rubrum". Spike awnless. Glumes short, pubescent. Floral glumes short, oval, pubescent with short to medium long (at the tip of the spike up to 5 cm long) hairs. Selected from the $F_3$ of the cross *Ae. ovata* × *T. aestivum* cv. Bezostaja 1.

An entirely novel and particularly valuable biological character of the new *T. sphaerococcum* variants is their pronounced winter habit expressed by their low temperature (0–2 °C) requirements during the vernalization stage, by its duration (table 1), and by their field cold resistance in case of autumn sowing. This character is not inherent to *T. sphaerococcum* which is a typical spring wheat type. Along with this new character the variants possess, and are distinguished for, their outstanding vigour, resistance to fungal diseases and different day-length reaction (table 1).

Data presented in table 2 show that the variants have high grain protein content. With higher protein content than that of the varieties *globosum* PERC. and *rubiginosum* PERC., the variety "Pseudo-alboglabrum" and "Pseudo rubrum" are distinguished.

Taking into consideration the fact that the "*sphaerococcum* forms" and the sphaerococcoides are developed only from crosses between *Ae. ovata* and the cultivars Bezostaja 1 and Rannjaja 12, it can be assumed that the cytoplasmic-nuclear interrelations existing in the hybrids are the decisive factor for the changed allelic state of the gene S, originating from *T. aestivum*.

Table 1

Effect of vernalization stage and day length on the period from plant emergence to heading (in days)

| Varieties of *Triticum sphaerococcum* | Day length | | | | | | | |
|---|---|---|---|---|---|---|---|---|
| | 15 h | | | | 10 h | | | |
| | Vernalization stage, days | | | | | | | |
| | 30 | 20 | 10 | 0 | 30 | 20 | 10 | 0 |
| 1. "Pseudo-albopubescens" | 65 | 68 | 76 | 76 | 127 | –[1] | – | – |
| 2. "Pseudo-alboglabrum" | 66 | 78 | 96 | 94 | 83 | 123 | 125 | 125 |
| 3. "Pseudo-echinatum" | 59 | 58 | 69 | 81 | 70 | 75 | 89 | 120 |
| 4. "Pseudo-rubroaristatum" | 68 | 88 | 116 | – | 117 | 120 | – | – |

[1] – "no heading"

Table 2

Grain protein content of varieties of *T. sphaerococcum*

| Varieties of *Triticum sphaerococcum* | dry matter % | Protein content % (dry matter) |
|---|---|---|
| 1. "Pseudo-albopubescens" | 88.53 | 16.5 |
| 2. "Pseudo-alboglabrum" | 88.51 | 27.3 |
| 3. "Pseudo-echinatum" | 88.45 | 15.1 |
| 4. "Pseudo-rubroaristatum" | 88.58 | 15.7 |
| 5. "Pseudo-rubrum" | 88.51 | 23.3 |
| 6. var. *globosum* | 88.61 | 19.9 |
| 7. var. *rubiginosum* | 88.57 | 18.1 |

*Triticum spelta* L.

The new botanical *T. spelta* variants were developed as a result of hybridization of *Ae. crassa* Boiss. ssp. *vavilovii* Zhuk. (2n=42) with the *T. aestivum* cultivars Rusalka, Bezostaja 1, Rannjaja 12 and IBO 1828, they were selected from the $F_5$ to $F_8$ hybrid generations. On the basis of their main taxonomic, morphological and biological characters they were named as follows:

1. *Triticum spelta* L., variety "Arista inflata". Spike partly composed, awned, glumes glabrous, grayish-brown, brighter to white at the edge. Awns black at base and curved. Caryopsis reddish. Differs from its close var. *samuricum* Dorof. by its tendency to complicate the spike. The variant was selected from the $F_7$ of the cross *Ae. crassa* ssp. *vavilovii* × *T. aestivum* cv. Rusalka.

2. *Triticum spelta* L., variety "Inermis composita". Spike awnless, partly composed. Glumes glabrous, dark grey to almost black, brighter to white towards the edge. Caryopsis reddish. Differs from var. *jakubzineri* Dorof. by its tendency to complicate the spike. The variant was selected from the $F_7$ of the cross *Ae. crassa* ssp. *vavilovii* × *T. aestivum* cv. Rusalka.

3. *Triticum spelta* L., variety "Inermis maculata". Spike awnless. Glumes glabrous, reddish with black dots. Caryopsis reddish. The variant was selected from the $F_8$ of the cross *Ae. crassa* ssp. *vavilovii* × *T. aestivum* cv. Bezostaja 1.

4. *Triticum spelta* L., variety "Arista brevis". Spike awnless. Glumes grayish black, reddish towards the edge, pubescent. Lemma shortly awned. Caryopsis reddish. Differs from its close var. *alefeldii* Körn. by its short lemma awns. The variant was selected from the $F_8$ of the cross *Ae. crassa* ssp. *vavilovii* × *T. aestivum* cv. Bezostaja 1.

5. *Triticum spelta* L., variety "Spica fragilis". Spike grayish-black, awnless. Glumes glabrous with black spots, reddish towards the edge. Caryopsis reddish. Intermediate of *T. spelta* and *T. dicoccum*. The variant was selected from the $F_8$ of the cross *Ae. crassa* ssp. *vavilovii* × *T. aestivum* cv. Rannjaja 12.

6. *Triticum spelta* L., variety "Albido punctata". Spike pure white, awnless. Glumes glabrous, pure white black dots. Caryopsis reddish. The variant was selected from the $F_8$ of the cross *Ae. crassa* ssp. *vavilovii* × *T. aestivum* cv. IBO 1828.

7. *Triticum spelta* L. ssp. *spelta*, variety "Nigro pubescens". Spike awnless. Glumes blackish-red, shortly pubescent. Caryopsis reddish. Differs from its close

var. *neglecta* Körn. by its blackish glumes. The variant was selected from the $F_8$ of the cross *Ae. crassa* ssp. *vavilovii* × *T. aestivum* cv. Rannjaja 12.

Along with these botanical varieties selected after the hybridization of *Ae. crassa* ssp. *vavilovii* with various *T. aestivum* cultivars, almost all varieties of *T. spelta* described by Dorofeev (1972) were found in the later (after $F_5$) generations. In the opinion of this author some of them are typical only for the Asian subspecies *kuckuckianum*, and deviate from it only in the intensity of expression of the individual taxonomic, morphological and biological characters. These synthetic forms as well as the new biological varieties differ from each other considerably in their winter or spring habit, which is manifested by their various temperature requirements during the vernalization stage, by its duration and by their photoperiodic reaction to different day lengths (table 3).

Considerable differences are observed between the individual varieties and var. *duhamelianum* Mazz. in respect to protein content of the grain (table 4).

Table 3

Effect of vernalization stage and day length on the period from plant emergence to heading (in days)

| Varieties of *Triticum spelta* | Day length | | | | | | | |
|---|---|---|---|---|---|---|---|---|
| | 15 h | | | | 10 h | | | |
| | Vernalization stage, days | | | | | | | |
| | 30 | 20 | 10 | 0 | 30 | 20 | 10 | 0 |
| 1. "Arista-inflata" | 98 | 93 | 93 | 109 | 168 | 168 | –[1] | – |
| 2. "Inermis-composita" | 72 | 79 | 158 | 159 | 84 | 114 | 154 | – |
| 3. "Inermis-maculata" | 74 | 79 | 79 | 82 | 157 | 157 | 157 | 157 |
| 4. "Arista-brevis" | 72 | 73 | 83 | 83 | 94 | 96 | 109 | 114 |
| 5. "Spica-fragilis" | 69 | 87 | 166 | – | 167 | 165 | – | – |
| 6. "Albido-punctata" | 84 | 124 | 157 | 158 | 165 | 165 | – | – |
| 7. ssp. *spelta* "Nigro-pubescens" | 69 | 87 | 157 | 157 | 106 | 138 | 157 | 158 |

[1] – "no heading"

Table 4

Grain protein content of varieties of *Triticum spelta*

| Varieties of *Triticum spelta* | dry matter % | Protein content % (dry matter) |
|---|---|---|
| 1. "Arista-inflata" | 88.36 | 22.84 |
| 2. "Inermis-composita" | 88.46 | 23.07 |
| 3. "Inermis-maculata" | 88.75 | 21.71 |
| 4. "Arista-brevis" | 88.91 | 22.78 |
| 5. "Albido-punctata" | 88.72 | 17.77 |
| 6. ssp. *spelta* "Nigro-pubescens" | 88.75 | 21.71 |
| 7. "Brevi-duhamelianum" | 88.14 | 21.69 |
| 8. var. *duhamelianum* | 88.66 | 18.64 |

Data presented in table 4 show clearly that among the studied samples the following represent definite breeding interest: "Inermis composita", "Arista inflata", "Arista brevis", and "Inermis maculata".

The new synthetic forms of *T. spelta*, besides the above indicated valuable characters, are also distinguished by their higher earliness and resistance to diseases and lodging.

## Conclusions

New biological variants of *T. sphaerococcum* and *T. spelta* were developed as a result of hybridization between species of the genus *Aegilops* and *T. aestivum* cultivars. The first winter and winterspring forms of the species *T. sphaerococcum* have been selected. These new varieties contribute to the morphological diversity of the species and widen the scope of their further use in bread wheat breeding.

Results obtained in the conducted investigations reveal one of the ways in which the existing species and varieties of the genus *Triticum* originated, were reproduced and enriched on the basis of their hybridization with the genus *Aegilops*. These results represent an important theoretical prerequisite for a new approach and direction of the future investigations concerning the evolution and the interrelations of the genera *Triticum* and *Aegilops*. From the standpoint of wheat breeding it is evident that species of the genus *Aegilops* should be more widely included in the breeding programmes.


## Zusammenfassung

Neue botanische Varietäten von *Triticum sphaerococcum* PERC. und *Triticum spelta* L. aus Kreuzungen von Arten der Gattung *Aegilops* L. mit *Triticum aestivum* L.

Durch Kreuzungen von *Aegilops ovata* auct. non L. mit der Sorte 'Bezostaja 1' von *Triticum aestivum* L. wurden neue biologische Varianten von *T. sphaerococcum* PERC. erhalten. Neben den morphologischen und biologischen Merkmalen besitzen sie auch einige neue nützliche Eigenschaften, wie bessere Winterfestigkeit, Resistenz gegenüber Pilzkrankheiten, und einen höheren Proteingehalt. Neue botanische Varianten von *T. spelta* L. sind Ergebnis von Kreuzungen zwischen *Ae. crassa* BOISS. ssp. *vavilovii* ZHUK. ($2n = 42$) und verschiedenen Sorten von *T. aestivum*. Einige von ihnen weisen nützliche biologische und pflanzenbauliche Eigenschaften auf und können daher für eine künftige Anwendung in der Pflanzenzüchtung interessant sein. Auf der Grundlage dieser Ergebnisse werden neue Vermutungen über den Ursprung der Gattung *Triticum* angestellt. Die neuen biologischen Varianten beider Arten erweitern die Möglichkeiten ihrer praktischen Nutzung in der Weizenzüchtung.

## Краткое содержание

Новые ботанические разновидности *Triticum sphaerococcum* Perc. и *T. spelta* L. от скрещивания видов рода *Aegilops* L. с *T. aestivum* L.

Путём скрещивания *Aegilops ovata* auct. non. L. с сортом ‘Безостая 1’ (*Triticum aestivum* L.) были получены новые биологические варианты *T. sphaerococcum* PERC. Кроме морфологических и биологических признаков они имеют и некоторые новые полезные качества, как напр. лучшую зимостойкость, сопротивляемость грибным заболеваниям, повышенное содержание протеина. Новые ботанические расновидности, *T. spelta* L. получены в результате скрещиваний *Ae. crassa* BOISS. ssp. *vavilovii* ZHUK. (2n = 42) с различными сортами *T. aestivum*. Некоторые из них обнаружили полезные биологические и хозяйственные свойства и могут быть интересны для применения в селекционных целях. На основании этих результатов высказываются новые предположения о происхождении рода *Triticum*. Новые биологические варианты обоих видов расширяют возможности их практического использования в селекции пшеницы.

## Literature

BOCHEV, B., 1974: *Triticum sphaerococcum* PERC. var. *pseudorubrum* BOCH. – C. R. Acad. Agric. G. Dimitrov, Sofia, **7**, 3, 9–11.

DOROFEEV, V. F., 1972: Pšenicy Zakavkaz'ja. – Leningrad.

MCFADDEN, E. S., and E. R. SEARS, 1946: The origin of *T. spelta* and its free-threshing hexaploid relatives. – J. Hered. **37**, 81–89, 107–116.

NIGMATULIN, F. G., 1972: Novaja raznovidnost' šarozernoj pšenicy. – Doklady AN Tadž. SSR. **15**, 10, 58.

PERCIVAL, I., 1921: The Wheat Plant. – Duckworth Co., London.

SINGH, R. D., 1959: Indian J. Genet. Pl. Breed. **19**, 199.

STANKOV, I., 1977: C. R. Acad. Bulg. Sci. **30**, 1.

STANKOV, I. Ts., and D. K. TSIKOV, 1973: Two new botanical varieties of *Triticum sphaerococcum* PERC. – C. R. Acad. Bulg. Sci. **26**, 6, 807–810.

UDAČIN, R. A., and I. Š. ŠACHMEDOV, 1974: K poznaniju vida *Triticum sphaerococcum* PERC. – Bjul. VNII Rastenievodstva **46**, 19–25.

Dr. B. BOCHEV
Institute of Genetics
BG - 1113 Sofia, Bulgaria

According to the author's request the term "variety" was used for the hybrid derivatives.

Editors

Kulturpflanze XXIX · 1981 · S. 369–375

# Possible relationships between wild and cultivated carrots (*Daucus carota* L.) in the Netherlands

WILLEM A. BRANDENBURG (Wageningen, the Netherlands)

## Summary

A theory about the descent and development of modern carrot cultivars in Europe formulated by BANGA (1963) is discussed. According to this author, the cultivated carrot originates from Afghan carrot material. Furthermore, the origin of carrots with fleshy, orange roots is suggested to lay in the Netherlands in the 17th century.

However, it is shown that wild populations of the carrot with white coloured roots have occurred in Europe since prehistoric times. During the late Middle Ages and Renaissance these carrots were used as medicinal plants.

It is suggested that these indigenous carrots have played a minor role in the development of modern cultivated carrots.

## Introduction

In a study of the possible relationships between wild and cultivated populations of a species in the same area, the following questions are important:
- What is the taxonomic status of both the wild and the cultivated population;
- Is the wild population endemic to the area;
- What has been the influence of domestication on the cultivated population in the same area;
- Are there links between wild and cultivated populations in that particular area.

As there have been both wild and cultivated carrots in the Netherlands for at least the past three centuries, the above stated questions are starting-points for a research project about the development of cultivated carrots in the Netherlands.

## Taxonomy

Both the wild and the cultivated carrot belong to the species *Daucus carota* L. This very polymorphic species is the type species of the genus *Daucus* L., which belongs to the tribe *Caucalideae* BENTH. of the *Umbelliferae*. Because of its great variation several subspecies are recognized, mainly based on characters of the inflorescence, the infrutescence and the fruit (HEYWOOD and DAKSHINI 1971; SAENZ DE RIVAS and HEYWOOD 1974; OKEKE 1978).

In the Netherlands we are dealing with two subspecies:
– *D. carota* L. subsp. *carota* wild carrot;
– *D. carota* subsp. *sativus* (HOFFM.) ARCANGELI cultivated carrot.
The latter subspecies is difficult to describe.

Data collected by MACKEVIČ (1929, 1932) indicate that its centre of origin is probably Afghanistan. THELLUNG (1927) suggested that it is the result of hybridizations of mainly *D. carota* subsp. *carota* and *D. maximus* DESF. in Mediterranean regions. The most distinct characters of *D. carota* subsp. *sativus* are a thick, fleshy, sweet, edible root, variably coloured, and a well developed whorl of barbs at the top of the spines on the vallecular ridges of the mericarp, whereas *D. carota* susp. *carota* has a thin, though, bitter, non-edible, white root, a poorly developed central umbellet with variable coloured flower(s). Further descriptions can be found in LINNAEUS (1753), HOFFMAN (1791), DE CANDOLLE (1830), BENTHAM (1867), LANGE (1874), ARCANGELI (1882), CALESTANI (1905), THELLUNG (1926a, 1926b), ONNO (1936), MARGRUDER et al. (1940), BANGA (1963), HEYWOOD (1968a, 1968b) and SMALL (1978).

## Wild carrots in the Netherlands

It is unknown whether wild carrots are endemic to the Netherlands or not. Data reported by BERTSCH und BERTSCH (1947) and HELMQUIST (1979) show that the wild carrot occurs in respectively Southern Germany and Sweden from prehistoric times.

Fossil seeds were found near human settlements. However, it is not clear whether the carrot came there as a more or less domesticated plant or whether it occurred there independent of human activities.

Although VAN ZEIST (1968) did not mention the carrot to occur in the Netherlands in prehistoric times, he (VAN ZEIST 1980) stated that at that time it possibly had come here from Southern Europe.

It was not until the late Middle Ages and Renaissance that several herbal books (FUCHS 1543; DODONAEUS 1554; TABERNAEMONTANUS und BAUHIN 1613; NIJLANDT 1682) mentioned the wild carrot as a medicinal plant. All plant parts were used. There is, however, little evidence that people grew the wild carrot in herbal gardens (FUCHS 1543). Usually, parts of naturally grown plants were collected.

At present, according to WESTHOFF en DEN HELD (1969), the wild carrot grows in the Netherlands in the following plant associations:
– Arrhenateretum elatioris, BR.-BL., 1919;
– Melilotetum albo-officinalis, SISS., 1950.

So, it can be found in large parts of the country, including most of the arable areas. Wild carrot populations in the Netherlands show a wide range of variation, mainly in flower and umbel characters. Gynodioecy was observed by BEYERINCK (1894). BRAAK and KHO (1958) have studied sex differences between flowers within umbels per plant. They, as did ZEVEN (1975, pers. com.), observed also male sterility. In a preliminar study I observed different types of male sterility (petaloidy and brown anther type) in several wild populations in the Netherlands

(BRANDENBURG 1977). Besides this, I found that umbel size, form and colour, and the life span were extremely variable characters. It is remarkable that in these wild populations characters (male sterility) occur that are important for breeding purposes. It is unknown whether the cause of this is introgressive hybridization with cultivated carrots, or the reverse, or whether these characters have developed in both wild and cultivated carrots independent of each other.

From the above mentioned variation it can be learned that because of the lack of a representative sample of Dutch wild carrots, seeds must be collected from several plants at different sites of the country.

## The cultivated carrot in the Netherlands

On the strength of data by THELLUNG (1926b, 1927) MACKEVIČ (1929, 1932), and ZAGORODSKICH (1939), and ethnobotanic sources and paintings, BANGA (1963) formulated a theory about the descent and development of modern carrot cultivars in Europe. In the 12th century carrot material with purple and yellow roots was brought by Moors and Arabs from Afghanistan into Spain. From there it was distributed all over Europe. These introduced populations were considered as the only source from which the European cultivated carrot has been developed.

For two reasons, according to BANGA (1963), the origin of carrots with fleshy, orange roots has been laid in the Netherlands in the 17th century. Firstly, there are Dutch paintings from the 17th century depicting for the first time both orange and yellow carrot roots. Secondly, in the 18th century cultivars with orange roots were described by Dutch horticulturists (e.g. KNOOP 1752, 1769) as derived from carrots with yellow roots.

The first reason, however, is a doubtful one, because the colours of the meant paintings (e.g. WTTEWAAL, "Woman selling vegetables", Centraal Museum, Utrecht, Neth.) are not always true to type. The painter used colour effects in arranging his subjects. So, in paintings, the differences between yellow and orange roots are due to artistical features rather than to differences between cultivars. Paintings from the 16th and 17th century (e.g. PIETER AERTSEN, "Fruits and vegetables stand", Museum Boymans-Van Beuningen, Rotterdam, Neth.; NICOLAES MAES, "A market scene at Dordrecht, Rijksmuseum, Amsterdam, Neth.) show clearly differences between purple/red and yellow/orange carrot roots. KNOOP (1752, 1769) recorded that two types of orange carrot roots occurred in the 18th century:

- big, long, (pale) orange carrots;
- finer, intensively orange coloured carrots (the so called 'Horn' cultivars).

It is suggested that the fine, orange carrots were raised in the Netherlands in the 17th century, because Dutch painters were not depicting them until that time (e.g. GERARD DOU, "The Quacksalver", Museum Boymans-Van Beuningen, Rotterdam, Neth.).

BANGA (1963) mentioned that cultivars with fleshy, white coloured roots were recorded in the 18th century.

GIBAULT (1912) and SANGERS (1952) mentioned medieval documents that refer to white carrot roots. These records are dubious, for at that time carrots were largely confused with parsnips (*Pastinaca sativa* L.). ZEVEN and ZHUKOVSKY (1975) suggested that carrots having fleshy, white roots are possibly mutants, derived from populations with yellow or purple roots. In the Netherlands nowadays there are only cultivated carrots with orange coloured roots. There is no division in groups of cultivars but in season of cultivation (Rassenlijst voor Groentegewassen 1980). They are used as a food crop for man (vegetable, fresh or tinned).

In the earlier days there were carrot cultivars, which were mainly used as fodder crops (SCHEYGROND 1953; WIJBRANS 1953). They had red, yellow, pale yellow, or white roots. So, it can be concluded, that uniformity of root colour is a recent feature of cultivated carrots in the Netherlands.

## Links between wild and cultivated carrots

The occurrence of cultivated carrots with fleshy, differently coloured roots and of wild carrots with though, white roots has been discussed. Although VILMORIN (1886) stated that he had bred cultivated carrots from wild ones, his experiment was never repeated.

DALE (1975), however, mentioned introgressive hybridization from wild into cultivated carrots, McCOLLUM (1975) reported hybridizations between wild and cultivated carrot accessions (*D. carota* × *D. capillifolius* GILLI), which would be readily crossed. DUDITS et al. (1978) have studied somatic hybrids of *D. carota* × *D. capillifolius*, which have fleshy, white roots. The $F_1$ hybrids between wild and cultivated carrots are mostly annuals. Since cultivated carrots are biennals, the hybrids can be easily observed the first year and removed from commercial seed-growing fields. In the second year, however, segregating $F_2$ generations are flowering together with the cultivated carrots. Therefore, it will be difficult to prevent contamination of wild and cultivated carrots.

From genetic experiments on the inheritance of colour of carrot roots BUISHAND and GABELMAN (1979) point out that the character is controlled by a few major genes. Furthermore, it is influenced by several minor genes and environmental factors. They state, that white roots have two dominant alleles. So, hybrids between wild and cultivated carrots can easily be detected also in this way. The occurence of cultivated carrots with fleshy, white roots should be an indication that wild populations have played a minor role in the development of modern cultivated carrots in the Netherlands. Firstly, the fact that BANGA (1963) disregarding the occurrence of wild carrots, has laid the origin of orange carrot roots in the Netherlands. Secondly, the results of the above mentioned genetic experiments. Numerical analysis of the *Daucus carota* complex as used by SMALL (1978) will be helpful in interpreting the results.

*Acknowledgements*

I wish to express my appreciation to Dr. Ir. A. C. ZEVEN, Mr. E. H. OOST, Mr. J. G. VAN DE VOOREN and Prof. Dr. Ir. J. H. VAN DER VEEN for their helpful remarks and stimulating discussions.

I also thank Mrs. C. DE BRUYN of the State Office for Art Documentation for her assistance, and Mr. J. S. DE BLOCK, who kindly corrected the English manuscript.

## Zusammenfassung

Verwandtschaftliche Beziehungen zwischen wilden und kultivierten Möhren (*Daucus carota* L.) in den Niederlanden

Die von BANGA (1963) formulierte Abstammungs- und Entwicklungstheorie über die rezenten Möhrensorten wird diskutiert. Nach seiner Meinung sind die kultivierten Möhren aus afghanischen Formen entstanden. Darüber hinaus wird vermutet, daß Möhren mit fleischigen, orangefarbigen Wurzeln in den Niederlanden im 17. Jahrhundert entwickelt worden sind.

Es wird jedoch gezeigt, daß es schon seit prähistorischen Zeiten in Europa Wildpopulationen von Möhren mit weißfarbigen Wurzeln gegeben hat. Diese Möhren wurden im späten Mittelalter und in der Renaissance als Heilpflanzen verwendet.

In vorliegender Arbeit wird darauf hingewiesen, daß diese heimischen Möhrenformen bei der Entstehung unserer heutigen Möhrensorten eine nur geringe Rolle gespielt haben dürften.

## Краткое содержание

Родственные связи между дикорастущими и культурными формами моркови (*Daucus carota* L.) в Нидерландах

Обсуждается теория происхождения и развития современных сортов моркови (БАНГА, 1963). Согласно этой теории культурная морковь возникла из афганских форм. Кроме того, Банга предполагает, что морковь с мясистым оранжевым корнем была выведена в Нидерландах в 17 веке.

Однако в дальнейшем показано, что уже в доисторические времена в Европе существовали дикорастущие популяции моркови с белёсыми корнями. Эта морковь в позднем Средневековье и в эпоху Возрождения использовалась как лекарственное растение.

В настоящей работе указывается, что эти местные формы, очевидно играли только незначительную роль в возникновении наших современных сортов моркови.

## Literature

ARCANGELI, G., 1882: Compendio della flora Italiana, 2nd ed. – D. Loescher, Torino.

BANGA, O., 1963: Western carotene carrot, Maintypes and Origin. – Tjeenk Willink, Zwolle.

BENTHAM, G., 1867: *Daucus*, in G. BENTHAM and J. D. HOOKER, Genera Plantarum, vol. 1, 928.

BERTSCH, K., und F. BERTSCH, 1947: Geschichte unserer Kulturpflanzen. – Stuttgart.

BEYERINCK, M. W., 1894: Gynodioecie bei *D. carota* L. – Ned. Kruidk. Arch. **2** (4), 345.

BRAAK, J. P., and Y. O. KHO, 1958: Some observations on the floral biology of the carrot (*Daucus carota* L.). – Euphytica **7**, 131–139.

BRANDENBURG, W. A., 1977: Botanisch onderzoek naar wilde en cultuurpeen. (Stud. Res. Rep.) – Instituut voor Plantenveredeling, Landbouwhogeschool, Wageningen.

BUISHAND, J. G., and W. H. GABELMAN, 1979: Investigations on the inheritance of color and carotene content in phloem and xylem of carrot roots (*Daucus carota* L.). Euphytica **28**, 611–632.

CALESTANI, V., 1905: Contributo alla sistematica delle Ombrellifere d'Europa. – Webbia **1**, 89–280.

DE CANDOLLE, A. P., 1830: *Daucus*. – Prod. Syst. Nat. Regni Veg. **4**, 209–215.

DODONAEUS, R., 1554: Cruydeboeck. – Jan van der Loe, Antwerpen.

DUDITS, D., G. HADLACZKY, E. LEVI, O. FEJER, Z. HAYDU and G. LAZAR, 1977: Somatic hybridisation of *Daucus carota* and *D. capillifolius* by protoplast fusion. – Theor. Appl. Genet. **51**, 127–132.

FUCHS, L., 1543: Kreuterbuch. – Basel.

GIBAULT, M. G., 1912: Histoire des Légumes. – Librairie Horticole, Paris.

HELMQUIST, H., 1979: Beiträge zur Kenntnis der prähistorischen Nutzpflanzen in Sweden. – Op. Bot. **47**.

HEYWOOD, V. H., 1968a: *Daucus*, in Flora Europaea, vol. **2**, 373–375. – University Press, Cambridge.

–, 1968b: The *Daucus carota – gingidium* complex (Flora Europaea Notulae Systematicae ad Floram Europaeam spectantes No. 7). – Feddes. Rep. **79**, 65–68.

– and K. M. DAKSHINI, 1971: Fruit structure in the *Umbelliferae – Caucalidae*, in V. H. HEYWOOD (ed.), The biology and chemistry of the *Umbelliferae*: 215–232. – Academic Press, London, New York.

HOFFMANN, G. F., 1791: Deutschlands Flora. – J. J. Palm, Erlangen.

KNOOP, J. H., 1752: De beknopte huishoudelijke Hovenier, vol. **1**, 309–310. – Noordbeek, Leeuwarden.

–, 1769: Beschrijving van de moes – en keukentuin. – Ferwerda and Tresling, Leeuwarden.

LANGE, J., 1874: *Daucus*, in M. WILLKOMM and J. LANGE, Prodr. Fl. Hisp. **3**, 18–24.

LINNAEUS, C., 1753: Species Plantarum, 1 st. ed. – Holmiae.

MAGRUDER, R., V. R. BOSWELL, S. L. ENSWELLER, J. C. MILLER, A. E. HUTCHINS, J. F. WOOD, M. M. PARKER and H. H. ZEMMERLEY, 1940: Descriptions of types of principal American varieties of orange-fleshed carrots. – U.S.D.A. Misc. Publ. **361**.

MACKEVIČ, V. I., 1929: The carrot of Afghanistan. – Bull. Appl. Bot. **20**, 517–562.

–, 1932: Ovoščnye rastenija Turcii. – Leningrad.

MCCOLLUM, G. D., 1975: Interspecific hybrids *Daucus carota* ×*D. capillifolius*. – Bot. Gaz. **136**, 201–206.

NIJLANDT, P., 1682: De Nederlandtse Herbarius of Kruidt – boeck. – Amsterdam (facsimile, 1976, Interbook, Schiedam).

OKEKE, S. E., 1978: Morphological variation of bracts, bracteolules and fruits in *Daucus* L. – Actes du 2e Symposium International sur les Ombellifères (Perpignan, 1977), 161–174.

ONNO, M., 1936: Die Wildformen von *Daucus* sect. *Carota*. – Beih. Bot. Centralbl. **56** (B), 83–136.

Rassenlijst voor Groentegewassen (Vollegrondsgroenten), 1980. – RIVRO, Wageningen.

SAENZ DE RIVAS, C., e V. H. HEYWOOD, 1974: Estudio preliminar sobre los *Daucus* de la Espana Peninsular. – Anal. Inst. Bot. Cavanilles **31**, 97–118.

SANGERS, W. J., 1952: De ontwikkeling van de Nederlandse Tuinbouw. – Tjeenk Willink, Zwolle.

SCHEYGROND, W., 1953: Voederwortelen II. De teelt, de waardebepalende hoedanigheden en enkele voor de veredeling belangrijke correlaties. – Euphytica **2**, 157–160.

SMALL, E., 1978: A numerical taxonomic analysis of the *Daucus carota* complex. – Can. J. Bot. **56**, 248–276.

TABERNAEMONTANUS, J. TH., und C. BAUHIN, 1613: Neuw.vollkommentlich Kreuterbuch. – Frankfurt a. M.

THELLUNG, A., 1926a: Die Linnéschen *Daucus* – Arten im Lichte der Original – Herbarexemplare. – Repert. Spec. Nov. Regni Veg. **22**, 305–315

–, 1926b: *Daucus* in G. HEGI, (ed.), Illustrierte Flora von Mittel-Europa, Vol. **5**, 1501–1526. – J. F. Lehmann, München.

–, 1927: Die Abstammung der Gartenmöhre und des Gartenrettichs. – Feddes Repert. Spec. Nov. Beih. **46**, 1–7.

van Zeist, W., 1968 (1970): Prehistoric and early historic food plants in the Netherlands. – Palaeohistoria **14**, 41–173.
–, 1980: Prehistorische Cultuurplanten. – Intermediair **16** (3), 35–47.
Vilmorin, M. P. P. A. Levêque de, 1886: Notice sur l'amélioriation de la carotte sauvage. – Paris.
Westhoff, V., en A. J. den Held, 1969: Plantengemeenschappen van Nederland. – Thieme, Zutphen.
Wijbrans, J. R., 1953: Voederwortelen I. De indeling en de beschrijving van het in Nederland geteelde sortiment. – Euphytica **2**, 149–156.
Zagorodskich, P., 1939: New data on the origin and taxonomy of the cultivated carrot. – Compt. Rend. (Dokl.) Acad. Sci. URSS **25** (6), 520–523.
Zeven, A. C., and P. M. Zhukovsky, 1975: Dictionary of cultivated plants and their centres of diversity. – Pudoc, Wageningen. 219 p.

Ir. W. A. Brandenburg
Department of Taxonomy of
Cultivated Plants,
Agricultural University
Haagsteeg 3,
NL-6708 PM Wageningen
The Netherlands

Kulturpflanze XXIX · 1981 · S. 377–388

# Biosystematics of crop-weed complexes

BARBARA PICKERSGILL (Reading, England)

## Summary

Many crops have conspecific or congeneric weedy relatives with which they are sympatric in part at least of their range. The crop may have been domesticated from plants similar to the modern weedy types; or the modern weeds may be derived from the crop (often through natural hybridisation); or crop and weed may have diverged simultaneously from a shared wild ancestor. Gene flow between the crop and its companion weed may be limited; or considerable but predominantly in one direction; or extensive in both directions, crop to weed and weed to crop. Phylogenetic relationships and the extent of gene exchange between crop and weed affect the amount of genetic diversity and the likelihood that weed populations will provide a source of useful characters for improvement of the crop, hence are relevant in collection and conservation of crop genetic resources. Gene exchange and the strength of the selection pressures operating to keep crop and weed distinct also affect the probability that genes of adaptive value to crop and weed respectively will be linked, which can greatly hamper use of weedy relatives in plant breeding programmes.

## Introduction

Many crop plants, in part at least of their range, grow sympatrically with weedy races to which they are closely related. Because crop and weed often differ strikingly in morphology, they may be placed in different species, but they can, and frequently do, exchange genes readily. HARLAN (1965, 1970) considers such weedy races to be important reservoirs of genetic diversity for crop improvement. Genes may be transferred from weed to crop by natural hybridisation, and this partially offsets the trend under domestication to greater uniformity, hence greater vulnerability to pests, diseases and adverse environmental conditions. Plant breeders also have exploited the diversity present in wild or weedy relatives of crops. Resistances to nematodes, blight and various virus diseases have been transferred from wild species of *Solanum* to the cultivated potato (HAWKES 1977); wild peas may be a source of adaptation to warmer and drier conditions (HARLAN 1976); wild oat has been used to introduce seed dormancy into cultivated oats bred for autumn sowing in Canada (THOMAS and JONES 1976). These and other examples have increased awareness of the role that wild and weedy relatives may play in the future improvement of crops and such races are thus now routinely collected and conserved in genetic resource collections.

However, the wild and weedy relatives of crop plants are not always considered an asset. PARKER (1977) points out that natural hybridization between modern cultivars and their sympatric weedy relatives may give rise to new and exceedingly troublesome weeds, hence should be prevented as far as possible. Induced mutations are sometimes claimed to be a more efficient source of genetic diversity than wild species because mutants can be induced in the genotype of the crop itself, thus avoiding problems of desirable characters being linked with undesirable wild-type traits. Such problems can be very considerable, and have made some plant breeders reluctant to use wild species in crop improvement programmes. HAWKES (1977) cites the example of transfer of nematode resistance from *Lycopersicon peruvianum* to the cultivated tomato. The interspecific hybrid was made in 1944 but it was not until 1956 that the tight linkage between resistance and undesirable fruit characters was broken.

These different views on the role and usefulness of weedy relatives of crop plants are not as contradictory as they may appear. Crop-weed complexes may represent biosystematically very different situations, with very different consequences for the amounts of diversity present in weedy populations and for the genetic organisation of that diversity. These in turn affect the usefulness of weed races as a genetic resource. The factors of greatest importance are the phylogenetic relationships between crop and weed, and the extent and direction of gene flow between crop and weed.

## Phylogenetic relationships in crop-weed complexes

Three different evolutionary relationships between crops and their weedy relatives are possible. Firstly, the weed may have been domesticated to produce the crop. Secondly, the weedy races may be derived from the crop. Thirdly, crop and weed may have diverged simultaneously from a common ancestral wild population.

### *1. Weedy races are ancestral to the crop*

Most of the species of chili pepper cultivated in Latin America have close weedy relatives. The most intensively studied of these crop-weed pairs is ***Capsicum annuum***, and the most convincing evidence that the domesticate is derived from the weedy races comes from the mitotic chromosomes (PICKERSGILL 1971, 1977). Most accessions of weedy *C. **annuum*** have one pair of acrocentric chromosomes, as do most other species in the genus. Domesticated *C. **annuum*** has two pairs of acrocentric chromosomes. This derived karyotype has not been found in any other species, but it does occur also in weedy *C. **annuum***, though it is not common and is virtually confined to central Mexico. The earliest archaeological specimens also come from central Mexico, date back to about 7000 B.C., and have seeds comparable in size to modern weedy *C. **annuum***. Large seeds, which must have come from domesticated peppers, occur much later, well after the domestication of other crops. These data suggest that *C. **annuum*** first became a tolerated weed of cultivation, then was domesticated in central Mexico from a small sample of local wild peppers carrying the derived karyotype.

Other examples of crops domesticated from ancestors adapted to growing in open primary or disturbed secondary habitats (weeds in the terminology developed after the invention of agriculture) include barley and einkorn wheat (HARLAN and ZOHARY 1966) and possibly potato (HAWKES, this symposium).

*2. Weedy races are derived from the crop*

The wild sea beet (*Beta vulgaris* ssp. *maritima*) of Britain and northern Europe lacks sufficient competitive ability to survive as a weed (LONGDEN 1974). Nevertheless, weed beet now infests over 20% of the sugar beet crop in England and in France has caused some land to be taken out of beet production (LONGDEN 1976). Many modern sugar beet cultivars are triploid monogerm (recessive mm) produced by using a selected tetraploid to pollinate a cytoplasmically male-sterile diploid. Tetraploid beets shed pollen at lower relative humidity, hence later in the day, than diploid beets. If commercial seed is produced in an area where wild diploid beets occur, the male sterile diploid seed parent is likely to be pollinated by pollen from the wild diploids before pollen from the tetraploid intended as the male parent becomes available. Seed sold to the growers will thus contain some cultivated×wild $F_1$ hybrids which, since the annual habit is dominant, will develop into bolters or weed beet, reducing both yield and quality of the crop.

HORNSEY and ARNOLD (1979) investigated whether contaminated seed was the sole source of the British weed beet problem. They test-crossed weed beets from two infested fields with a male-sterile non-restorer diploid monogerm line. In one field the weed beets carried alleles for the monogerm trait and/or male sterility, hence originated from contaminated commercial seed. In the other field the weeds did not carry monogerm or male sterility alleles, hence had evolved locally from older multigerm cultivars in response to changes in agricultural practice, particularly earlier sowing of the crop (with the attendant risk that the seedlings will receive sufficient vernalisation to cause bolting) and replacement of hand-thinning and weeding by drilling and chemical weed control so that volunteer beet seedlings are no longer rogued out in spring. Nevertheless, despite their different origins, these weed beets are clearly all derived from the crop and not ancestral to it.

In the United States, use of male sterility to produce $F_1$ hybrid sorghum seed facilitates crossing between male-sterile sorghum and the sympatric tetraploid weed, *Sorghum halepense*. Some commercially distributed seed thus grows into sterile triploids which do not contribute to crop yield, are difficult to control chemically, and persist as vigorous perennial weeds (BAKER 1972). Commercial carrot seed is no longer produced in parts of Canada where the weedy wild carrot occurs because domesticated and wild carrots cross so readily (SMALL 1978). Examples such as these explain why PARKER (1977) views wild relatives of crops as potential sources of new weed problems.

Other derived weeds are of less concern to farmers. 6-row wild barley is a sporadic weed in various localities but seldom persists for long. It is sometimes treated as a distinct species, *Hordeum agricrrithon*, but has been shown to arise from natural crossing between 6-row domesticated barley and wild or weedy 2-row

*H. spontaneum* (ZOHARY 1959). In California, similar natural crossing between the domesticated radish, *Raphanus sativus*, and weedy *R. raphanistrum* has produced distinct weed races of *R. sativus* (PANETSOS and BAKER 1967).

In all these examples, successful evolution of weedy races derived from the crop has involved hybridization between the crop and related wild or weedy taxa. Many crops have been so modified by human selection that they cannot survive for long without man. For example, to succeed as weeds, the derived weed races of *Raphanus sativus* had to lose the biennial habit, soft fruits, swollen tap root and restricted lateral roots of the domesticated radish. Near-simultaneous back-mutation in the independent genetic loci governing all these traits is extremely unlikely. Instead, the necessary changes were achieved in one step by introduction of the necessary wild-type alleles from *R. raphanistrum*. There are very few instances of derived weedy races evolving from crops without some form of hybridization, and these few examples are from crops which have not themselves been much altered by human selection.

### *3. Crop and weed diverge simultaneously from a common ancestral wild population*

The classic example of disruptive selection leading to simultaneous evolution of crop and weed races is sorghum (DOGGETT and MAJISU 1968), but two less-familiar examples will be discussed here. The grain chenopods of Mexico and the Andes (*Chenopodium nuttalliae* and *C. quinoa* respectively) each have a companion weed race. The two domesticates evolved under human selection and developed compact terminal inflorescences, synchronous anthesis, loss of seed dispersal and seed dormancy, and pale seeds with thin testas. The two weedy forms evolved under natural selection and retained diffuse inflorescences, an extended flowering period, and easily-dispersed seeds with thick dark testas. The ancestor of the entire group was probably similar to modern *C. berlandieri* var. *zschackei*, which grows wild in the western United States (WILSON and HEISER 1979). This taxon is morphologically variable and highly polymorphic for at least one of the two loci coding for the enzyme leucine aminopeptidase. The two crop-weed complexes are virtually monomorphic, but each is fixed for a different allele. Each cropweed pair thus incorporates a small, but different, portion of the variation present in the common ancestor. Superimposed on this common heredity are the striking differences between crop and weed within each complex, brought about by disruptive selection for adaptation to different evolutionary niches.

If domesticated maize evolved from teosinte (*Zea mexicana*), then Jutiapa teosinte, which occurs in swampy areas of Guatemala out of genetic contact with domesticated maize, may represent the ancestral wild population from which both maize and the Mexican weed teosintes arose (GALINAT 1973). The weed teosintes grow as maize-mimics in cultivated cornfields and have retained the wild-type characters of hard rachis and glumes, seeds sunk in deep cupules in the rachis for protection from birds, fragile rachis which breaks up into single-seeded dispersal units, and several female inflorescences per tiller. On the other hand, human selection for yield and palatability in domesticated maize has reduced the depth of the cupules and the induration of rachis and glumes, so grains can be easily shelled from the cob, and has produced a rachis which does not disarticulate at maturity, an increase in both size of grains and number of grains per inflorescence, but a reduced number of inflorescences, often only one per tiller. Again disruptive selection for adaptation to crop and weed niches respectively maintains two distinct phenotypes in the population despite interbreeding between the crop and its companion weed.

## Effects on the genetic diversity of weedy populations

Where a crop has been domesticated from types similar to modern weedy populations, founder effects are likely to result in only a small part of the diversity of the weedy races being included in the gene pool of the crop. This is apparent in *Capsicum annuum*, where the weedy races show much greater variation in chromosome morphology, particularly number, size and position of satellites, than the domesticate (PICKERSGILL 1971). Screening and exploitation of the weedy chili peppers for useful characters are only just beginning, but already it appears that weedy *C. annuum* contains useful alleles for hypersensitivity to tobacco mosaic virus, partial resistance to cucumber mosaic virus and immune-type resistance to potato virus Y which are not present in domesticated *C. annuum* (POCHARD, personal communication).

A more dramatic example of the decline in variation which may accompany domestication comes from barley. Wild barley (*Hordeum spontaneum*) is widespread throughout much of the Middle East. Israeli populations alone were found to contain 105 alleles at 28 different loci coding for various enzymes. These same 28 loci were assayed in 100 plants from the $F_{17}$ generation of a composite cross involving 6200 lines of spring barley from the world collection and only 48 different alleles were found. Israeli wild barley also proved to contain slightly more variation at 4 esterase loci than was present in the entire world collection of cultivated barleys (NEVO et al. 1979).

Where a crop is derived from a small sample of ancestral weedy forms, it is thus evident that extant weedy races may represent a rich source of diversity not already present in the crop. Furthermore, where weedy races are ancestral to the crop they are, by definition, closer than the crop to the common stock from which other species in the genus diverged. They may therefore provide a useful bridge for transfer of desirable characters from more distantly related species to the crop gene pool. For example, *Capsicum baccatum* is reported to be resistant to cucumber mosaic virus, potato virus Y, possibly tobacco mosaic virus, and *Phytophthora capsici*, so various workers are attempting to obtain hybrids between *C. baccatum* and domesticated *C. annuum*, with only limited success (DUMAS DE VAULX and PITRAT 1977). I have never obtained viable seeds of *C. baccatum* × domesticated *C. annuum*, but I have raised several different $F_1$ combinations of *C. baccatum* × weedy *C. annuum*, suggesting that this is a possible route for introducing disease resistance into domesticated *C. annuum*.

Where weedy races are derived from the crop, as in the case of the weed beets, founder effects have operated twice, once in the origin of the crop and again in the origin of the derived weeds. The latter are thus likely to be more genetically depauperate than the crop. On theoretical grounds, it is difficult to see such derived weed races as a source of any unique characters. However, if the weeds originated from cultivars which have been superseded and which are not themseves represented in genetic resource collections, it may be useful to collect and conserve the derived weeds. Otherwise, maximum capture of genetic diversity should be achieved by collecting appropriate samples from the crop and the wild relatives implicated in the origin of the derived weeds.

Where weed and crop populations have diverged simultaneously, under the action of disruptive selection, the weed populations may initially have incorporated a somewhat different sample of genetic diversity from that present in the crop. Whether the weeds still contain different alleles will depend on the extent of gene exchange between crop and weed. The most valuable sources of alleles not represented in the crop are however, likely to be the original wild populations. Jutiapa teosinte, perhaps the original wild ancestor of both maize and weedy teosinte, carries genes for resistance to both maize mosaic virus and *Puccinia sorghi*, whereas weedy teosinte and mainland Mesoamerican races of maize are susceptible. Brewbaker (1979) interprets this as indicating that Jutiapa teosinte originated through recent introgression from Caribbean maize, in which mosaic resistance is widespread, but the chromosome knob data (Galinat 1971) do not support this suggestion.

## Gene flow in crop-weed complexes

Gene exchange between crops and their companion weeds may be very limited, or it may be extensive but predominantly in one direction, or it may be extensive in both directions, crop to weed and weed to crop.

### 1. *Limited gene flow*

Many crop plants are facultative inbreeders. In wheat, barley, lentils and peas, for example, gene exchange with sympatric weedy relatives is limited by habitual self-pollination in both crop and weed. In *Capsicum annuum*, outcrossing under field conditions may vary from 1% to nearly 70% (Quagliotti 1979). Hybrids between weedy and domesticated peppers may occur wherever the two grow together, but the hybrids are never frequent and $F_2$ or backcross swarms have not been reported. Gene exchange is limited not only by predominant self-pollination but also by the small size of the crop population. Since one or two bushes provide an ample supply of fruits for an entire household, each plant will contribute very few progeny to the next generation and the chances of hybrids or their derivatives establishing successfully are therefore small.

### 2. *Predominantly unidirectional gene flow*

In the grain chenopods of Mexico and the Andes, and also in the very similar grain amaranths, human selection restricts gene flow from weed to crop. The grain of the domesticates is pale in colour while that of the associated weeds is dark. Natural crossing between weed and crop is frequent, but the dark grain of the weeds is dominant and farmers select only pale grain for sowing. Although some $F_1$ crop×weed seed may therefore be harvested and sown along with the crop, human selection discriminates successfully against later hybrid generations and gene flow is thus predominantly from crop to weed. This may enhance the ability of the weeds to survive by mimicking the crop. Certainly in Mexico the weed amaranths show certain traits derived from the domesticated amaranths, while the crop remains essentially uncontaminated by genes derived from the weeds (SAUER 1967).

In the chuño potatoes of the high Andes, on the other hand, gene flow is predominantly from weed to crop. At altitudes of around 3500 m, where night frosts are a hazard, cultivated potatoes grow together with the weedy tetraploid *Solanum acaule* and cross freely with it (HAWKES 1962). The morphological effects of this hybridisation can be seen in the short stems and crinkled leaves of the chuño potatoes. More significantly, the crop has acquired cold tolerance from the weed and also the wild-type character of bitter tubers, which necessitates elaborate processing by soaking, repeated freezing and thawing, and finally trampling to express the juice and produce the form of freeze-dried tuber known as chuño. *Solanum acaule*, on the other hand, has not been conspicuously modified by genes from the cultivated potato. Presumably natural selection restricts gene flow in this direction.

*3. Extensive gene flow*

The classic example of massive gene flow between crop and weed is maize and its weedy relative teosinte. Both are wind-pollinated plants adapted for cross-pollination by monoecy combined with strong protandry. Where they grow together, hybridisation between crop and weed is frequent and gene exchange proceeds in both directions, to the advantage of both crop and weed (WILKES 1967, 1977). Teosinte survives as a weed of maize fields by mimicking maize so effectively that it cannot be rogued out until the inflorescences develop. Gene flow from maize to teosinte increases the effectiveness of teosinte as a maize mimic, by enabling it to pick up the pigmentation patterns and leaf characters of the various cultivars amongst which it grows. Gene flow from teosinte to maize produces a heterotic effect so marked that some Mexican farmers claim presence of the weed improves the crop. However, GALINAT (1974) has suggested that teosinte has also had a braking effect on the evolution of domesticated maize and that extreme forms such as the large-kernelled race Cuzco were able to develop only outside the range of teosinte.

## Consequences of gene flow for genetic diversity

Where gene exchange between crop and weed is limited, there is a good chance of finding genetic variants in the weed which are not represented in the crop, for example the resistances to various virus diseases present in weedy *Capsicum annuum*. Where there is unidirectional gene flow from crop to weed, the weed may act as a reservoir of genetic diversity in HARLAN's sense (HARLAN 1965), and may retain some variation from crosses with older cultivars which are no longer grown.

Such weed races are thus worth collecting and conserving as sources of diversity now absent from the crop. However, where there is extensive gene exchange between weed and crop, or where gene flow is predominantly from weed to crop, the weed races are unlikely to contain any alleles not present in the crop, and it will then be easier to use the appropriate races of the crop as sources of desirable characters in breeding programmes.

## Effects of gene flow on linkage

As has already been discussed, when crops and their sympatric weeds hybridise, disruptive selection will usually operate against the hybrids and maintain the fitnesses of crop and weed populations to their respective niches despite gene flow between them. This is likely to lead either to increased isolation between the crop and its companion weed, or to tighter linkage between genes of adaptive value in the two populations (PICKERSGILL and HEISER 1976). Linkage between genes useful for crop improvement and genes governing undesirable wild-type characters creates considerable problems for plant breeders and may affect their willingness to use wild material in breeding programmes.

Where predominant self-pollination restricts gene flow between crop and weed, few hybrids are formed and disruptive selection consequently operates only weakly. In *Capsicum*, its effects are further diminished because, at least in subsistence agriculture, hybrids between crop and weed are not at any marked disadvantage once they succeed in establishing. Hot peppers are often picked green, so it does not matter that the hybrids have non-persistent wild-type fruit when mature. Small fruits are not discriminated against because yield is rarely important, a few bushes more than providing for the needs of an entire household. There is consequently little pressure to evolve tight linkages between adaptive combinations of alleles, and it is not surprising that no linkage has so far been detected between genes governing different wild-type traits in *Capsicum*. Fruit colour, pungency and persistence all assort independently (PICKERSGILL 1977). Polygenes governing differences in fruit size and other quantitative characters are presumably also scattered throughout the chromosome complement. The breeder is thus likely to find that genes for disease resistance from the weedy peppers will each be linked with different but undesirable wild-type characters.

On the other hand, in crops such as maize where extensive gene exchange between the crop and its weedy relative is combined with strong disruptive selection for adaptation to crop versus weed niches, there is likely to be tight linkage between genes of adaptive value in the two populations. In wild Jutiapa teosinte the genes governing morphology of the female inflorescence are scattered on several different chromosomes, but in the Mexican weed teosintes these genes are linked into a complex carried on chromosome 4. Recombination between genes on this chromosome segment and its homologue in domesticated maize is further suppressed by an inversion (GALINAT 1971, 1973). A more subtle interaction has also been reported. In both maize and teosinte, heterozygosity for chromosome knobs reduces crossing-over in the vicinity of the knob. GALINAT (1977) quotes examples of maize-teosinte complexes in which the maize partner is essentially knobless while the teosinte partner has many chromosome knobs. Hybrids between such maize-teosinte pairs will therefore show reduced recombination for segments of many different chromosomes and adaptive combinations of alleles will again be conserved.

Similar linkages between genes controlling adaptive differences between crops and weeds have been reported in radish which, like maize, is an outbreeder (PANETSOS and BAKER 1967) and in oats, which is an inbreeder (BAUM 1968), but in which there is strong disruptive selection against plants morphologically intermediate between weed and crop.

It thus appears that tight linkages between adaptive wild-type alleles, which can ease the plant breeder's task of transferring other useful alleles from weed to crop without serious problems from undesirable linkages, are in general most likely to occur where extensive gene exchange between crop and weed has left little unique variation in the weed populations for the breeder to exploit. Conversely, useful alleles for the plant breeder are most likely to occur in situations where the evolutionary history and biosystematic interactions of crop and weed make it very probable that any useful characters will be linked with undesirable wild-type traits.

## Conclusions

Biosystematic aspects of crop-weed complexes are complicated, subtle and have seldom been investigated in depth. Phylogenetic relationships between weed and crop and the extent and direction of any gene exchange between them influence the amount of unique genetic diversity to be found in weedy populations, hence the likelihood that weedy populations will constitute reservoirs of useful genes for crop improvement. Consideration of the biosystematic interactions in particular crop-weed complexes may help collectors in planning strategies for field work and priorities for conservation. In general, ancient weeds such as the weedy chili peppers and weeds which do not exchange genes freely with their companion crops are more likely to contain characters not represented in the crop-gene pool than are derived weeds such as the weedy beets or weeds which hybridise extensively with their associated crops. However, problems of utilisation are likely to be correspondingly greater in weeds which are genetically isolated from the crops to which they are related and amongst which they grow.

## Zusammenfassung


Biosystematik von Kulturpflanze-Unkraut-Komplexen

Viele Kulturpflanzen haben Unkrautverwandte derselben Art oder Gattung, mit denen sie zumindest in einem Teil ihres Areals sympatrisch verbreitet sind. Die Kulturpflanze kann durch die Domestikation von Pflanzen entstanden sein, die den modernen Unkrauttypen ähnlich sind, oder die modernen Unkräuter können sich von der Kulturpflanze abgeleitet haben (häufig durch natürliche Hybridisation), oder die Kulturpflanze und das Unkraut können sich gleichzeitig aus einem gemeinsamen wilden Vorfahren entwickelt haben. Der Genfluß zwischen der Kulturpflanze und seinem zugehörigen Unkraut kann begrenzt oder beträchtlich, aber in einer Richtung vorherrschend, oder in beiden Richtungen, von der Kulturpflanze zum Unkraut und vom Unkraut zur Kulturpflanze, bedeutend sein. Die phylogenetische Verwandtschaft und die Größe des Genaustausches zwischen der Kulturpflanze und dem Unkraut beeinflussen den Umfang der genetischen Vielfalt sowie die Wahrscheinlichkeit, daß Unkrautpopulationen eine Quelle von nützlichen Merkmalen für die Verbesserung der Kulturpflanze sein werden, also wichtig für die Sammlung und Konservierung der genetischen Ressourcen von Kulturpflanzen sind. Genaustausch und Stärke des Selektionsdruckes, die die Trennung von Kulturpflanze und Unkraut bewirken, beeinflussen die Wahrscheinlichkeit, mit der Gene mit Anpassungswert in der Kulturpopulation bzw. in der Unkrautpopulation gekoppelt werden, was im letzten Falle die Verwendung von Unkrautverwandten in Kreuzungsprogrammen beträchtlich behindern kann.

**Краткое содержание**

Биосистематика комплексов « культурное растение – сорняк »

Многие культурные растения имеют сородичей того же вида или рода, являющихся сорняками, которые, по крайней мере в некоторых частях ареала, встречаются симпатрически. Культурное растение, путём одомашнивания, могло произойти от растений, сходных с современными сорняковыми типами. С другой стороны, современный сорняк мог произойти от культурного растения (часто путём естественной гибридизации). И наконец, культурное и сорное растения могли возникнуть оба от какого-то общего дикорастущего предка. Циркуляция генов между культурными растениями и родственными сорняками может быть органичена или значительна, преобладая или в одном направлении или в обоих, т. е. от культурного растения к сорняку и от сорняка к культурному растению. Филогенетическое родство и масштаб генного обмена между культурным растением и сорняком влияют на размах генетического многообразия и на вероятность того, что сорняковая популяция может стать источником полезных признаков для улучшения культурного растения. Тогда эта популяция оказывается ценной для сборов и сохранения генетических ресурсов. Генный обмен и сила селекционного давления, которые обуславливают расхождение культурной и сорной популяций, влияют на вероятность сцепления генов, имеющих адаптационную ценность, или в культурной, или в сорной популяциях, что может, в последнем случае, значительно усложнить использование родственных сорных растений в селекционных программах.

## Literature

BAKER, H. G., 1972: Migrations of weeds. – In: Taxonomy, phytogeography and evolution (Edit. D. H. VALENTINE), pp. 327–347. Academic Press, New York and London.

BAUM, B. R., 1968: On some relationships between *Avena sativa* and *A. fatua* (Gramineae) as studied from Canadian material. – Can. J. Bot. **46**, 1013–1024.

Brewbaker, J. L., 1979: Diseases of maize in the wet lowland tropics and the collapse of the Classic Maya civilisation. – Econ. Bot. **33**, 101–118.
Doggett, H. and B. N. Majisu, 1968: Disruptive selection in crop development. – Heredity **23**, 1–23.
Dumas de Vaulx, R. and M. Pitrat, 1977: Croisement interspecifique entre *C. annuum* et *C. baccatum*. – In: Capsicum 77. C.R. 3me Congrès EUCARPIA Piment (Edit. E. Pochard) pp. 75–81. Institut National de la Recherche Agronomique, Montfavet-Avignon.
Galinat, W. C., 1971: The origin of maize. – Ann. Rev. Genetics **5**, 447–478.
–, 1973: Preserve Guatemalan teosinte, a relict link in corn's evolution. – Science **180**, 323.
–, 1974: The domestication and genetic erosion of maize. – Econ. Bot. **28**, 31–37.
–, 1977: The origin of corn. – In: Corn and Corn Improvement, 2nd edition (Edit. G. F. Sprague) pp. 1–47. American Society of Agronomy, Madison.
Harlan, J. R., 1965: The possible role of weed races in the evolution of cultivated plants. – Euphytica **14**, 173–176.
–, 1970: Evolution of cultivated plants. – In: Genetic resources in plants – their exploration and conservation (Edit. O. H. Frankel and E. Bennett) pp. 19–32. International Biological Programme, London.
–, 1976: Genetic resources in wild relatives of cultivated crops. – Crop Science **16**, 329–333.
– and D. Zohary, 1966: Distribution of wild wheats and barley. – Science **153**, 1074–1080.
Hawkes, J. G., 1962: The origin of *Solanum juzepczukii* Buk. and *S. curtilobum* Juz. et Buk. – Zeitschr. für Pflanzenzüchtung **47**, 1–14.
–, 1977: The importance of wild germplasm in plant breeding. – Euphytica **26**, 615–621.
Hornsey, K. G. and M. H. Arnold, 1974: The origin of weed beet. – Ann. appl. Biol. **92**, 279–285.
Longden, P. C., 1974: Sugar beet as a weed. – Proc. 12th Brit. Weed Control Conf. **1**, 301–308.
–, 1976: Annual beet: problems and prospects. – Pestic. Sci. **7**, 422–425.
Nevo, E., D. Zohary, A. H. D. Brown and M. Haber, 1979: Genetic diversity and environmental associations of wild barley, *Hordeum spontaneum*, in Israel. – Evolution **33**, 815–833.
Panetsos, C. A. and H. G. Baker, 1967: The origin of variation in "wild" *Raphanus sativus* (Cruciferae) in California. – Genetica **38**, 243–274.
Parker, C., 1977: Prediction of new weed problems, especially in the developing world. – In: Origins of pest, parasite, disease and weed problems (Edit. J. M. Cherrett and G. R. Sagar) pp. 249–264. Blackwell Scientific Publications, Oxford.
Pickersgill, B., 1971: Relationships between weedy and cultivated forms in some species of chili peppers (genus *Capsicum*). – Evolution **25**, 683–691.
–, 1977: Chromosomes and evolution in *Capsicum*. – In: Capsicum 77. C.R. 3me Congrès EUCARPIA Piment (Edit. E. Pochard) pp. 27–37. Institut National de la Recherche Agronomique, Montfavet-Avignon.
– and C. B. Heiser, 1976: Cytogenetics and evolutionary change under domestication. – Phil. Trans. R. Soc. Lond. B. **275**, 55–69.
Quagliotti, L., 1979: Floral biology of *Capsicum* and *Solanum melongena*. – In: The biology and taxonomy of the Solanaceae (Edit. J. G. Hawkes, R. N. Lester and A. D. Skelding) pp. 399–419. Academic Press, London.
Sauer, J. D., 1967: The grain amaranths and their relatives – a revised taxonomic and geographic survey. – Ann. Mo. Bot. Gard. **54**, 103–137.
Small, E., 1978: A numerical taxonomic analysis of the *Daucus carota* complex. – Can. J. Bot. **56**, 248–276.
Thomas, H. and I. T. Jones, 1976: Origins and identification of weed species of *Avena*. – In: Wild oats in world agriculture (Edit. D. Price Jones) pp. 1–18. Agricultural Research Council, London.
Wilkes, H. G., 1967: Teosinte: the closest relative of maize. – Bussey Inst., Harvard University, Cambridge.
–, 1977: Hybridisation of maize and teosinte in Mexico and Guatemala and the improvement of maize. – Econ. Bot. **31**, 254–293.

WILSON, H. D., 1976: Genetic control and distribution of leucine aminopeptidase in the cultivated chenopods (*Chenopodium*). – Biochem. Gen. **14**, 913–919.
– and C. B. HEISER, 1979: The origin and evolutionary relationships of 'huauzontle' (*Chenopodium nuttalliae* SAFFORD), domesticated chenopod of Mexico. – Amer. J. Bot. **66**, 198–206.
ZOHARY, D., 1959: Is *Hordeum agriocrithon* the ancestor of six-rowed cultivated barley? – Evolution **13**, 279–280.

Dr. B. PICKERSGILL,
Department of Agricultural Botany,
Plant Science Laboratories,
University of Reading,
Whiteknights,
Reading, RG6 2AS,
Berkshire, England.

Kulturpflanze XXIX · 1981 · S. 389

# Activity of r-RNA cistrons as a tool for the identification of genomes in the genus *Triticum* L.[1]

SEVDALIN GEORGIEV (Sadovo, Bulgaria), MARLEN KRISTEVA (Sofia, Bulgaria) and HRISTO NICOLOFF (Sofia, Bulgaria)

**(Summary, text not received)**

18 different ploidy (2x, 4x, 6x) representatives of the *Aegilops* genus and 15 of the *Triticum* genus were analysed cytologically and by means of the nucleolar test.

As a result of these researches, availability of different satellite chromosome numbers among the representatives with a same ploidy of the *Triticum* and *Aegilops* genus was demonstrated in a certain way. Thus among the diploid (2x) representatives of the *Aegilops* genus 2 groups were determined in regard to the satellite chromosomes: a) representatives possessing one intermediary SAT chromosome pair, b) representatives which have had one intermediary SAT chromosome pair, one micro SAT chromosome pair. Among the diploid representatives of the *Triticum* genus 2 micro SAT chromosomes pairs were found only. Among the 4x representatives of the *Aegilops* genus 3 groups were observed: a) with one macro SAT chromosome pair, b) with one macro and one intermediary SAT chromosome pair, and c) one macro and one intermediary and 2 micro SAT chromosome pairs. Among the 4x representatives of the *Triticum* genus only 2 macro SAT chromosome pairs were observed to be available.

In this respect the group of the 6x representatives proves to be very polymorphic and in which the following was established in the *Aegilops* genus: a) with 1 macro SAT pair, 1 intermediary SAT pair and 2 micro SAT pairs, b) representatives with 2 macro SAT pairs and 1 micro SAT pair, which coincides with 6x representatives of the *Triticum* genus possessing 2 macro SAT pairs and 1 micro SAT chromosome pair. The variety 'Kavkaz', where 1 macro SAT, 1 intermediary SAT and 1 micro SAT chromosomes were observed formed an exception in this regard.

As a result of the data obtained by means of the nucleolar test and the karyological analysis, the karyotypes of the studied 2x, 4x, 6x representatives of *Aegilops* and *Triticum* have been precised and corrected. These results have contributed to the elucidation and precision of some problems related to the phylogeny of the *Triticum* genus.

The results obtained by us show that nowadays 4x and 6x wheat species have evidently polyphyletic origin, along with that both in 2x, 4x, and 6x species of the *Triticum* and *Aegilops* genus a very characteristic genome heterogenity is available.

1 Presented as poster

Dr. S. GEORGIEV, Institute of Plant Introduction and Genetic Resources, Sadovo, Bulgaria

Kulturpflanze XXIX · 1981 · S. 391–399

# Isoenzyme variation within the wild *Secale* L. species[1]

JERZY PUCHALSKI, and BOGUSŁAW MOLSKI (Warszawa, Poland)

## Summary

The degree of inter- and intra-specific variation in the populations of allogamous and autogamous *Secale* L. species was estimated in support of isoenzyme markers. Totally, 11 different species of *Secale* were analysed including perennial species: *S. kuprijanovii* GROSSH., *S. chaldicum* FED., *S. montanum* GUSS., *S. anatolicum* BOISS., and annual species: *S. silvestre* HOST., *S. vavilovii* GROSSH., *S. ancestrale* ZHUK., *S. dighoricum* VAV., *S. afghanicum* VAV., *S. segetale* ROSHEV.; as well as the cultivated rye *S. cereale* L. cv. 'Dańkowskie Złote'. Isoenzymes were analysed in the first leaf of single young seedlings grown in controlled conditions using starch-gel slab electrophoresis technique and the electrophoretic forms of two different enzyme systems were detected: nonspecific esterases (E.C. 3.1.1) and peroxidase (E.C. 1.11.1.7). It was found that the breeding system of rye species has a distinct affect on the isoenzyme variation in their populations. The self-fertile annual species, *S. silvestre*, was monomorphic for both enzyme systems. A very low esterases polymorphism was also found in other autogamous species, *S. vavilovii*, however a similar degree of polymorphism in allogamous annual species, *S. segetale* and *S. dighoricum*, was observed. Generally, the perennial *Secale* species were more polymorphic than the annual ones, and the isoenzyme variation in the population of cultivated rye *S. cereale* was similar to some other wild annual species. It was observed that the intra-specific isoenzyme variation in *Secale* was bigger than the inter-specific. Esterases electrophoretic forms showed to be very sensitive markers of variation in rye species.

## Introduction

The phylogenetic relations in the genus *Secale* L., though continually revised, are not completely clear and need more investigation. For this study, it seems that many advantages are offered by the new technique of electrophoretic analysis of isoenzymes. Isoenzymes were widely applied for studies on many plant species, as well in taxonomy (GOTTLIEB 1977), in genetics (JACOBS 1975) and especially in population genetics and evolutionary studies (BROWN 1978). Isoenzymes are very valuable markers for estimating the variation between, and within, plant species and can supplement other taxonomic traits. Isoenzymes were also analysed in the

1 Presented as poster

populations of genus *Secale*: in rye cultivars (PUCHALSKI and MOLSKI 1975), inbred lines (SCHMIDT-STOHN 1979) and different species (JAASKA 1975, 1979), but the degree of intra-specific variation was not yet estimated. The aim of this study was to compare the degree of isoenzyme polymorphism in the populations of autogamous and allogamous *Secale* species to conclude about the breeding system in rye and phylogenetic affinities.

## Materials

For our investigations, we used the seed samples of *Secale* species reproduced in the collection of the Botanical Garden of Polish Academy of Science, harvested from the plots isolated with linen tents during the flowering period. Totally, 11 different *Secale* species, 10 wild species and 1 cultivated variety 'Dańkowskie Złote' were analysed. Most of the seed stocks were obtained from the Vavilov All-Union Institute of Plant Industry (VIPI) in Leningrad. The source of analysed seeds is given in brackets. The list of analysed species contained:

A – perennial species: 1. *Secale kuprijanovii* GROSSH. (Agricultural University, Szcecin, Poland), 2. *S. chaldicum* FED. (VIPI), 3. *S. montanum* GUSS. (Bot. Garden Nancy, France), 4. *S. anatolicum* BOISS. (VIPI); B – annual species: 5. *S. silvestre* HOST. (VIPI), 6. *S. vavilovii* GROSSH. (VIPI), 7. *S. dighoricum* VAV. (VIPI), 8. *S. ancestrale* ZHUK. (VIPI), 9. *S. afghanicum* VAV. (VIPI), 10. *S. segetale* ROSHEV. (VIPI), 11. *S. cereale* cv. 'Dańkowskie Złote' (Plant Breeding Station Rogaczewo, Poland).

For simplification, all analysed samples were treated as separate species, however some of them are classified as subspecies or botanical varieties.

## Methods

The rye plants for analysis were grown in the growth chamber in strictly controlled conditions. Extraction of enzyme proteins was made from the whole first leaves of the 7-day old green seedlings. Isoenzymes in support of starch-gel slab electrophoresis technique were separated (KRISTJANSSON 1963). After electrophoresis, the specific enzyme activities were histochemically stained. Non-specific esterases (E.C. 3.1.1) were stained in the anodic part of the gel slab according to SHAW and KOEN (1968) with α-naphtyl acetate (Koch-Light Lab.) and Fast Garnet GBC salt (Sigma). Activity of peroxidase (E.C.1.11.1.7) was detected according to GRAHAM et al. (1965) technique with SHAW's and PRASAD's (1970) modification and the 3-amino-9-ethylocarbazole (Sigma) as the hydrogen donator was used. Peroxidase electrophoretic forms were stained only in the cathodic part of the gel slab. The other details concerning preparation of plant materials, extraction, electrophoresis, and enzyme activity detection, were described previously (PUCHALSKI and MOLSKI 1975). The analytical populations of several *Secale* species contained from 81 to 132 single plants.

The individual variation within the species populations in support of polymorphic index formula was calculated (MARSHALL and ALLARD 1970).

$$P_I = \frac{\sum_{i=1}^{n} p_i (1-p_i)}{n} \quad 0 < P_I < 0{,}25$$

$P_I$ – population polymorphic index
$p_i$ – frequency of particultural isoenzyme band in the population
n – number of isoenzyme bands detected in the species population

## Results and discussion

Both analysed enzyme systems, esterase and peroxidase, were very heterogeneous in the *Secale* species and consisted of many multiple electrophoretic forms (fig. 1). In the esterase species, patterns totalling more than 20 electrophoretic forms were

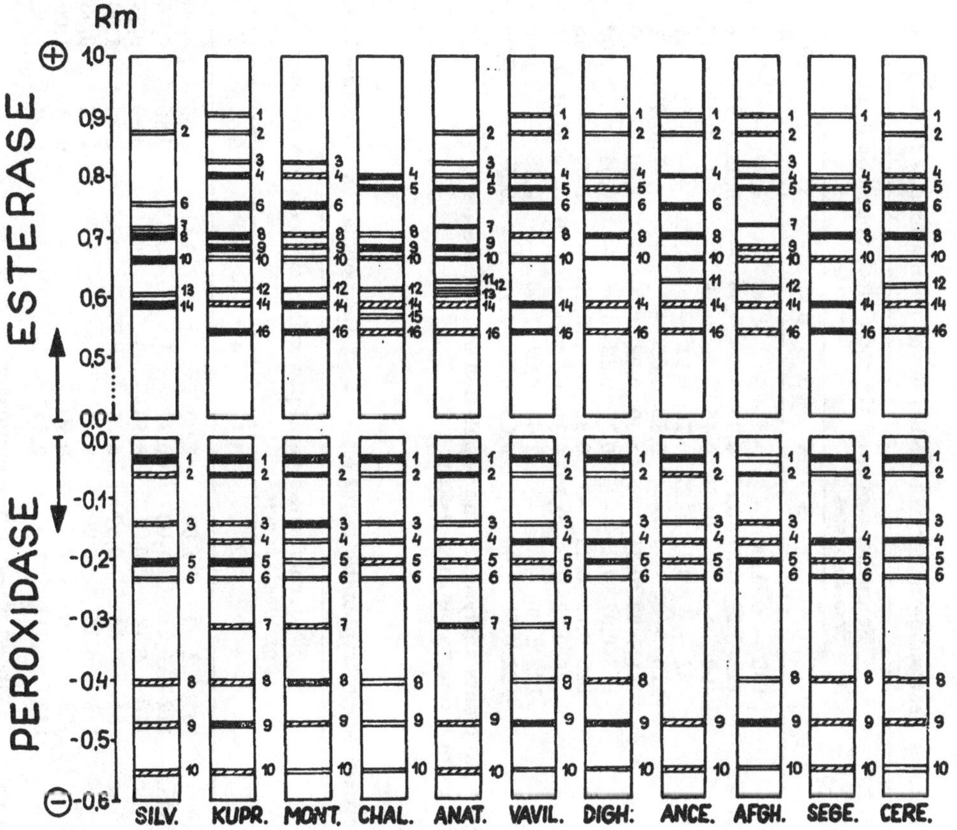


Fig. 1 Starch gel electrophoretic patterns of anodic esterase forms and cathodic peroxidase forms in the different *Secale* L. species. Abbreviations: SILV – *S. silvestre* Host., KUPR – *S. kuprijanovii* Grossh., MONT – *S. montanum* Guss., CHAL – *S. chaldicum* Fed., ANAT – *S. anatolicum* Boiss., VAVIL – *S. vavilovii* Grossh., DIGH – *S. dighoricum* Vav., ANCE – *S. ancestrale* Zhuk., AFGH – *S. afghanicum* Vav., SEGE – *S. segetale* Roshev., CERE – *S. cereale* L. cv. 'Dańkowskie Złote', Rm – relative electrophoretic mobility

Table 1

Polymorphism of esterases electrophoretic forms (anodic) in the populations of *Secale* L. species

| Species | N | Frequencies of several anodic electrophoretic forms of esterases in populations of *Secale* species | | | | | | | | | | | | | | | | PI |
|---|---|---|---|---|---|---|---|---|---|---|---|---|---|---|---|---|---|---|
| | | 1 | 2 | 3 | 4 | 5 | 6 | 7 | 8 | 9 | 10 | 11 | 12 | 13 | 14 | 15 | 16 | |
| *S. silvestre* HOST. | 100 | – | 1,00 | – | – | – | 1,00 | 1,00 | 1,00 | – | 1,00 | – | – | 1,00 | 1,00 | – | – | 0,000 |
| *S. kuprijanovii* GROSSH. | 112 | 0,29 | 0,37 | 0,85 | 0,95 | – | 0,99 | – | 0,93 | 0,97 | 0,54 | – | 0,82 | – | 1,00 | – | 1,00 | 0,113 |
| *S. montanum* GUSS. | 90 | – | – | 0,33 | 0,46 | – | 0,58 | – | 0,93 | 0,79 | 0,77 | – | 0,16 | – | 1,00 | – | 0,96 | 0,141 |
| *S. anatolicum* BOISS. | 82 | – | 0,27 | 0,27 | 0,76 | 0,51 | – | 0,29 | – | 0,88 | 0,94 | 0,21 | 0,40 | 0,11 | 0,88 | – | 0,81 | 0,163 |
| *S. chaldicum* FED. | 110 | – | – | – | 0,85 | 0,65 | – | – | 0,77 | 0,96 | 0,81 | – | 0,66 | – | 1,00 | 0,36 | 1,00 | 0,131 |
| *S. vavilovii* GROSSH. | 81 | 0,59 | 0,51 | – | 1,00 | 0,02 | 1,00 | – | 1,00 | – | 1,00 | – | – | – | 1,00 | – | 1,00 | 0,072 |
| *S. dighoricum* VAV. | 118 | 0,07 | 0,95 | – | 0,79 | 0,32 | 0,96 | – | 0,94 | – | 1,00 | – | – | – | 1,00 | – | 1,00 | 0,082 |
| *S. ancestrale* ZHUK. | 116 | 0,15 | 0,85 | – | 0,81 | – | 0,89 | – | 1,00 | – | 0,68 | 0,63 | – | – | 0,75 | – | 0,88 | 0,139 |
| *S. afghanicum* VAV. | 85 | 0,72 | 0,51 | 0,52 | 0,36 | 0,53 | – | 1,00 | – | 0,91 | 0,73 | – | 0,80 | – | 0,96 | – | 0,80 | 0,165 |
| *S. segetale* ROSHEV. | 102 | 1,00 | – | – | 1,00 | 0,02 | 1,00 | – | 0,98 | – | 1,00 | – | – | – | 1,00 | – | 0,98 | 0,025 |
| *S. cereale* L. | 132 | 0,60 | 1,00 | – | 0,63 | 0,54 | 0,99 | – | 0,39 | – | 0,95 | – | 0,66 | – | 0,99 | – | 0,98 | 0,127 |

N – number of individual plants analysed in the population of each species
PI – polymorphic index of esterases for each species

Table 2

Polymorphism of peroxidase electrophoretic forms (catodic) in the populations of *Secale* L. species

| Species | N | Frequencies of several peroxidase cathodic electrophoretic forms in the populations of *Secale* species | | | | | | | | | | PI |
|---|---|---|---|---|---|---|---|---|---|---|---|---|
| | | 1 | 2 | 3 | 4 | 5 | 6 | 7 | 8 | 9 | 10 | |
| *S. silvestre* HOST. | 100 | 1,00 | 1,00 | 1,00 | – | 1,00 | 1,00 | – | 1,00 | 1,00 | 1,00 | 0,000 |
| *S. kuprijanovii* GROSSH. | 112 | 0,21 | 1,00 | 0,22 | 1,00 | 0,65 | 0,86 | 0,96 | 0,86 | 0,96 | 1,00 | 0,088 |
| *S. chaldicum* FED. | 110 | 0,14 | 0,71 | 0,40 | – | 0,19 | 0,89 | 0,83 | 0,63 | 0,38 | 1,00 | 0,159 |
| *S. vavilovii* GROSSH. | 81 | 0,35 | 0,69 | 0,26 | 0,27 | 0,65 | 0,72 | 1,00 | 0,76 | 0,28 | 1,00 | 0,164 |
| *S. dighoricum* VAV. | 118 | 0,25 | 0,99 | 0,60 | – | 0,99 | 1,00 | 0,98 | 0,73 | 0,49 | 0,99 | 0,103 |
| *S. afghanicum* VAV. | 85 | 0,16 | 0,79 | 0,59 | – | – | 0,61 | 1,00 | 1,00 | 0,45 | 1,00 | 0,128 |
| *S. cereale* L. | 132 | 0,41 | 1,00 | 0,34 | – | 0,25 | 1,00 | 1,00 | 0,39 | 0,20 | 1,00 | 0,117 |

N – number of individual plants analysed in the population of each species
PI – polymorphic index of peroxidase for each species

detected. In the zone Rm 0,53–0,91, containing the fastest bands, 16 different esterase bands in all species occured but only 7 to 12 bands in the populations of single species were found (fig. 1). Some of the esterase bands were specific for particular species; for example bands Nos. 11 and 13 for *S. anatolicum*, band No. 15 for *S. chaldicum* and No. 7 for *S. silvestre*. Bands designated as 10 and 14 were common for all *Secale* species.

Peroxidase multiple forms were moving mainly into the cathode and in the cathodic part of the gel slab–10 different bands in all species were detected (fig. 1). Most of them occured in all analysed species; for example bands designated with numbers 1, 2, 4, 5, 9 and 10 for all species were common. The peroxidase band No. 7 was specific for perennial species and *S. vavilovii*.

The bigger inter-specific differences, in comparison with the number of bands, for the frequencies of particular electrophoretic enzyme forms were noted (table 1, table 2). Some bands in several species with very high frequencies occured, but in others only in a few plants were found. The biggest inter-specific differences for esterase bands designated as 1, 2, 4, 8 or 10 could be observed (table 1).

A similar phenomenon was also noted for the frequencies of peroxidase electrophoretic forms and the biggest inter-specific differences for bands designated with number 2, 3, 6, 8, and 10 were found, however, the slowest peroxidase band No. 1 in every plant of all species appeared (table 2). The populations of particular species also contained other monomorphic bands, esterase as well as peroxidase. The population of autogamous species *S. silvestre* was monomorphic for all assayed esterase and peroxidase bands and only small differences in the intensity of several bands in the single plants patterns could be observed (fig. 2a). In *S. vavilovii* 2/3 of all esterase bands were monomorphic and *S. segetale* contained as much as 87,5% of monomorphic esterase bands (table 1). In comparison, all esterase bands occuring in perennial species *S. anatolicum* were polymorphic.

The analysed *Secale* species showed very different degrees of intra-specific variation. This phenomenon is distincly visible on the individual plants enzymatic patterns separated in the starch gel (fig. 2). The annual self-fertile species, *S. silvestre*, was completely uniform for both enzyme systems (fig. 2a), yet a very high level of enzymatic variation in the perennial *Secale* species was noted, for example in *S. anatolicum* (fig. 2c). The population polymorphism was estimated in all *Secale* species using the polymorphic index (PI) formula (see: Methods) on the basis of the frequencies for all electrophoretic isoenzyme forms (table 1 and table 2). For esterase bands, the highest PI values attained all perennial species and annual *S. afghanicum*. On the contrary, the low polymorphism of esterase in other annual species was estimated: in autogamous *S. vavilovii* and allogamous species *S. segetale* and *S. dighoricum* (table 1). Smaller differences in polymorphism degree, in comparison to esterase, for peroxidase PI values were observed (table 2) and the lowest polymorphism in *S. kuprijanovii* and *S. dighoricum* was found. The degree of polymorphism in the population of cultivated rye was similar to other annual wild species of *Secale*, especially *S. ancestrale*.

The investigation showed that in rye species the distinct isoenzymatic polymorphism appeared. Very high isoenzyme variation was also found within the rye cultivars (PUCHALSKI and MOLSKI 1975), even in rye inbred lines (JOHANSSON

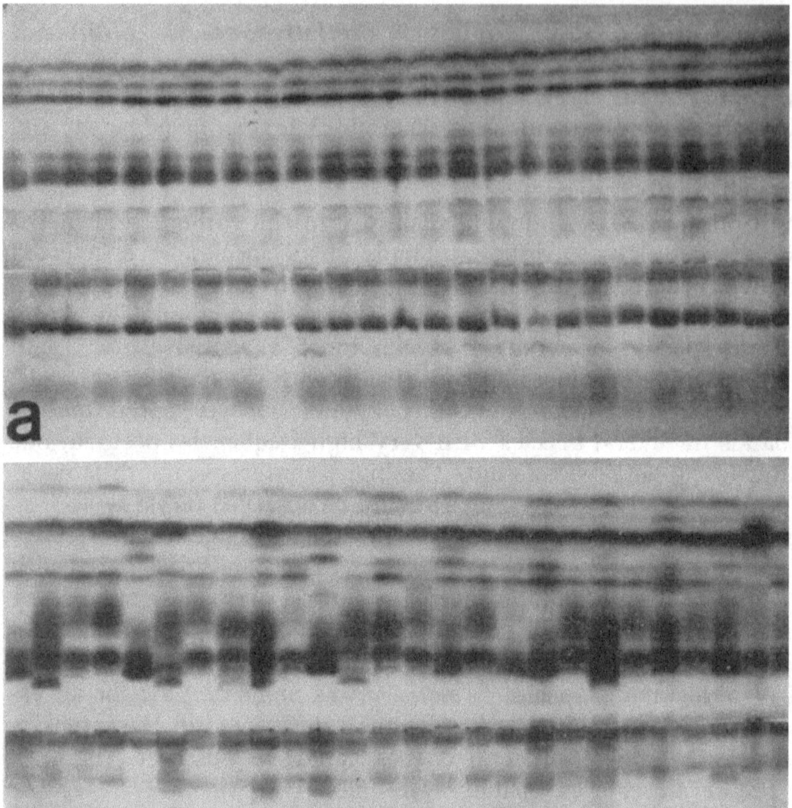

Fig. 2 Intra-specific variation of isoenzymes in the populations of different *Secale* L. species. Upper slice – anodic isoesterases, Bottom slice – cathodic isoperoxidases. *a* – *Secale silvestre* HOST., *b* – *S. vavilovii* GROSSH., *c* – *S. anatolicum* BOISS. The presented species showed very different degrees of individual variation, self-fertile *S. silvestre* was monomorphic for both enzyme systems, *S. vavilovii* was low polymorphic, and perennial *S. anatolicum* exhibited very high degree of the polymorphism.

1969; SCHMIDT-STOHN 1979). JAASKA (1975) also estimated the variation within the wild *Secale* species and found that autogamous species, *S. silvestre*, *S. vavilovii*, and *S. africanum*, were monomorphic for their isoenzyme patterns. In our studies, the self-fertile species, *S. silvestre*, didn't show any variation within population, esterase as well as peroxidase. It seems that the breeding system in rye has a large affect on the level of intra-specific variation, however in our studies, the cleistogamous species *Secale vavilovii* showed a low polymorphism for esterase and yet a rather high polymorphism for peroxidase. ARROYO (1975) analysed the populations of allogamous species, *Limnanthes alba* and autogamous *L. floccosa*, and didn't find the predicted association between the level of autogamy and the number of monomorphic isoenzyme loci in populations. MARSHALL and ALLARD (1970) also estimated 54% and 31% polymorphic isoenzyme loci in highly self-pollinating species *Avena fatua* and *A. barbata*. The small heterozygosity at enzyme loci was also found by BABBEL and WAIN (1977) in self-fertile species *Hordeum jubatum*, resulting from duplication and polyploidy. The heterozygosity of enzyme loci after gene duplication was proven by GOTTLIEB (1974) on the basis of his studies on alcohol dehydrogenase patterns in *Clarkia franciscana*. However, the inbreeding species usually show less enzyme variation than the outbreeding species, the differences being smaller than predicted. BROWN (1979) explained this phenomenon on the basis of the evolution of the mating system in plants.

Our studies showed that the perennial rye species showed a higher level of esterase polymorphism than annual species. Similar results were obtained by MORISHIMA and OKA (1970) in their studies on isoenzyme variation in the wild and cultivated *Oryza* species. It seems that the level of the enzyme polymorphism is correlated with the adaptation ability for the environmental conditions. SHAHI et al. (1969) found that widely distributed species of *Oryza*, *O. perennis*, *O. officinalis* and *O. latifolia*, were more enzymatically variable than the cultivated *O. sativa* and *O. glaberrima* and the wild *O. brachyantha*, distributed in limited areas. The association between isozyme phenotypes and the type of the environment was observed by KAHLER et al. (1980) in the populations of slender wild oat *Avena barbata* occupying different sites in Israel. The higher esterase polymorphism of perennial rye species than in annual rye also seems to be connected with their wider distribution and better environmental adaptability.

## Zusammenfassung


Isoenzym-Variation bei Wildarten der Gattung *Secale* L.

Der Grad der inter- und intraspezifischen Variation in Populationen von allogamen und autogamen Arten der Gattung *Secale* L. wurde mit Hilfe von Isoenzymmerkmalen geschätzt. Insgesamt wurden 11 verschiedene Arten von *Secale* einschließlich der mehrjährigen Arten analysiert: *S. kuprijanovii* GROSSH., *S. chaldicum* FED., *S. montanum* GUSS., *S. anatolicum* BOISS. und die einjährigen Arten: *S. silvestre* HOST., *S. vavilovii* GROSSH., *S. ancestrale* ZHUK., *S. dighoricum* VAV., *S. afghanicum* VAV., *S. segetale* ROSHEV., wie auch des Kulturroggens *S. cereale* L. cv. Dankowskie Złote. Die Isoenzyme wurden am ersten Blatt junger Einzelsämlinge, die unter kontrollierten Bedingungen angezogen wurden, unter Anwendung der Stärkegel-Plattenelektrophorese-Technik geprüft, und die elektrophoretischen Formen von zwei verschiedenen Enzymsystemen wurden nachgewiesen: nichtspezifische Esterasen (E. C. 3. 1. 1.) und Peroxidase (E. C. 1. 11. 1. 7.). Es wurde gefunden, daß das Fortpflanzungssystem der Roggen-Arten eine bestimmte Wirkung auf die Isoenzymvariation in ihren Populationen hat. Die selbstfertile einjährige Art, *Secale silvestris*, war hinsichtlich beider Enzymsysteme monomorph. Ein sehr schwach ausgeprägter Esterasenpolymorphismus wurde in einer anderen autogamen Art, *S. vavilovii*, gefunden, jedoch wurde ein ähnlicher Grad von Polymorphismus in allogamen einjährigen Arten, *S. segetale* und *S. dighoricum*, beobachtet. Grundsätzlich waren die mehrjährigen *Secale*-Arten polymorpher als die einjährigen, und die Isoenzym-Variation der Populationen des kultivierten Roggens *S. cereale* war ähnlich der einiger anderer einjähriger Wildarten. Es wurde beobachtet, daß die intraspezifische Enzym-Variation von *Secale* größer war als die interspezifische. Die esterase-elektrophoretischen Formen erwiesen sich als sehr empfindliche Indikatoren für die Variation der Roggenarten.

**Краткое содержание**

Изменчивость изоэнзимов у дикорастущих видов рода *Secale* L.

Степень межвидовой и внутривидовой изменчивости в популяциях аллогамных и автогамных видов рода *Secale* L. оценивалась при помощи признаков изоэнзимов. Всего анализировалось 11 различных видов; многолетние: *S. kuprijanovii* Grossh., *S. chaldicum* Fed., *S. montanum* Guss., *S. anatolicum* Boiss.; однолетние: *S. silvestre* Host, *S. vavilovii* Grossh., *S. ancestrale* Zhuk., *S. dighoricum* Vav., *S. afghanicum* Vav., *S. segetale* Roshev. и культурная рожь *S. cereale* L., сорт 'Даньковские Злоте'. Изоэнзимы исследовались в первом листе отдельных молодых сеянцев, выращенных в контролированных условиях, с применением методики электрофореза (крахмал – гель). Были выявлены электрофоретические формы двух различных систем энзимов: неспецифические эстеразы (E. C. 3. 1. 1.) и пероксидазы (E. C. 1. 11. 1. 7.). Оказалось, что система размножения видов ржи оказывает определённое влияние на изменчивость изоэнзимов в их популяциях. Самофертильный однолетний вид *S. silvestre* был мономорфен в отношении обоих систем энзимов. Очень слабо выраженный полиморфизм эстераз был найден у другого явтогамного вида, *S. vavilovii*, однако сходная степень полиморфизма наблюдалась и у аллогамных однолетних видов, *S. segetale* и *S. dighoricum*. В общем, многолетние виды ржи оказались более полиморфными, чем однолетние, а изменчивость изоэнзимов популяций культурной ржи была сходна с таковой некоторых других однолетних дикорастущих видов. Внутривидовая изменчивость энзимов у рода *Secale* оказалась сильнее, чем межвидовая. Электроферетические эстеразные формы проявили себя как очень чувствительные индикаторы изменчивости видов ржи.

## Literature

ARROYO, M. T. K. DE, 1975: Electrophoretic studies of genetic variation in natural populations of allogamous *Limnanthes alba* and autogamous *Limnanthes floccosa* (Limnanthaceae). – Heredity **35**, 153–164.

BABBEL, G. R. and R. P. WAIN, 1977: Genetic structure of *Hordeum jubatum*. I. Outcrossing rates and heterozygosity levels. – Can. J. Genet. Cytol. **19**, 143–152.

BROWN, A. H. D., 1978: Isozymes, plant population genetic structure and genetic conservation. – Theor. Appl. Genet. **52**, 145–157.

–, 1979: Enzyme polymorphism in plant populations. – Theor. Popul. Biol. **15**, 1–42.

GOTTLIEB, L. D., 1974: Gene duplication and fixed heterozygosity for alcohol dehydrogenase in the diploid plant *Clarkia franciscana*. – Proc. Nat. Acad. Sci. USA **71**, 1816–1818.

–, 1977: Electrophoretic evidence and plant systematics. – Ann. Mo. Bot. Gard. **64**, 161–180.

GRAHAM, R. C., U. LUNDHOLM, and M. J. KARNOVSKY, 1965: Cytochemical demonstration of peroxidase activity with 3-amino-9-ethylcarbazole. – J. Histochem. Cytochem. **13**, 150.

JAASKA, V., 1975: Evolutionary variation of enzymes and phylogenetic relationships in the genus *Secale* L. (in Russ.). – Eesti NSV Tead. Akad. Toim. Biol. **24**, 179–198.

–, 1979: Genetic polymorphism of acid phosphatase in population of rye, *Secale cereale* L. s.l. – Eesti NSV Tead. Akad. Toim. Biol. **28**, 185–193.

JACOBS, M., 1975: Isozymes and a strategy for their utilization in plant genetics. II. Isozymes as a tool in plant genetics. – In: Genetic manipulations with plant material (Edit. L. LEDOUX) pp. 379–389. Plenum Press, New York.

JOHANSSON, H., 1969: Multipla enzymformer i råg. – Nord. Jordbr. Forskn. **51**, 123–124.

KAHLER, A. L., R. W. ALLARD, M. KRZAKOWA, C. F. WEHRHAHN, and E. NEVO, 1980: Associations between isozyme phenotypes and environment in the slender wild oat (*Avena barbata*) in Israel. – Theor. Appl. Genet. **56**, 31–47.

KRISTJANSSON, F. K., 1963: Genetic control of two pre-albumins in pigs. – Genetics **18**, 1059–1063.

MARSHALL, D. R., and R. W. ALLARD, 1970: Isozyme polymorphism in natural populations of *Avena fatua* and *A. barbata*. – Heredity **25**, 373–382.

MORISHIMA, H., and H.-J. OKA, 1970: A survey of genetic variations in the populations of wild *Oryza* species and their cultivated relatives. – Jap. J. Genet. **45**, 371–385.

PUCHALSKI, J., and B. MOLSKI, 1975: Esterases variability within some Polish rye cultivars. – Hod. Rośl. Aklim. Nasien. **19**, 479–485.

SCHMIDT-STOHN, G., 1979: Genetische Analysen von Esterase-Loci beim Roggen (*Secale cereale* L.) mit Hilfe der isoelektrischen Fokussierung in Flachgelen. – Z. Pflanzenzüchtg. **83**, 155–162.

SHAHI, B. B., H. MORISHIMA, and H.-J. OKA, 1969: A survey of variations in peroxidase, acid phosphatase and esterase isozymes of wild and cultivated *Oryza* species. – Jap. J. Genet. **44**, 303–319.

SHAW, C. R., and A. L. KOEN, 1968: Starch gel zone electrophoresis of enzymes. – In: Chromatographic and electrophoretic techniques. – Vol. **II**. Zone electrophoresis (Edit. I. SMITH) pp. 325–364. Heinemann Med. Books, London.

–, and R. PRASAD, 1970: Starch gel electrophoresis of enzymes – a compilation of recipes. – Biochem. Genet. **4**, 297–320.

Dr. J. PUCHALSKI and Dr. B. MOLSKI
Botanical Garden of the Polish Academy of Sciences
ul. Prawdziwka 2
00 - 973 Warszawa p-84, Poland

# II. Originalarbeiten aus dem Institut

Kulturpflanze XXIX · 1981 · S. 403–415

# Variabilität der Keimwurzelzahl bei Gerste

Peter Apel, Achim Bergmann, Hans-Wolfgang Jank und Christian O. Lehmann

(Eingegangen am 16. Dezember 1980)


## Zusammenfassung

Es wurde die Keimwurzelzahl von 957 Varietäten der Saatgerste (*Hordeum vulgare* L.) bestimmt. Dieser Wert variiert zwischen 3,70 und 8,47 Keimwurzeln pro Karyopse. Aus anderen Untersuchungen waren Werte für die Kornmasse, die Pflanzenhöhe zur Reife, den Proteingehalt und den Lysingehalt der untersuchten Varietäten verfügbar. Zwischen diesen Parametern und der Keimwurzelzahl gibt es einige signifikante Korrelationen. Die physiologische und evtl. züchterische Bedeutung dieser Zusammenhänge wird diskutiert.


## 1. Einleitung

Wasser ist nicht nur in semiariden, sondern auch in humiden Getreideanbaugebieten zumindest zeitweise ein ertragsbegrenzender Faktor (Monteith 1977, Lieth 1976). Die Ausnutzung des verfügbaren Bodenwassers zur Erzeugung von Biomasse hängt zum einen ab vom Transpirationskoeffizienten, d. h. vom Verhältnis des durch Transpiration abgegebenen Wassers zum photosynthetisch gebundenen $CO_2$, zum anderen von der Fähigkeit des Wurzelsystems, das verfügbare Bodenwasser aufzunehmen.

Im Zusammenhang mit Überlegungen über eine optimale Bestandesentwicklung bei Getreide, die letztlich auf die Forderung nach einer möglichst raschen Jungpflanzenentwicklung hinauslaufen, muß Wassermangel in jungen Entwicklungsstadien als besonders gravierend eingeschätzt werden. Es ist daher im Zusammenhang mit der Züchtung auf Trockenheitsresistenz von Interesse zu wissen, wie groß die genetisch bedingte Variabilität bezüglich der Wurzelentwicklung insbesondere in jungen Entwicklungsphasen ist.

Die quantitative Erfassung des Wurzelwachstums unter Standortsbedingungen im Freiland ist aus methodischen Gründen schwierig, so daß darüber bisher nur wenig vergleichende Werte vorliegen. Es darf angenommen werden, daß bei Getreide die Keimwurzel für die Wasserversorgung in frühen Entwicklungsstadien die dominierende Rolle spielen. Aus Untersuchungen an Weizen (Robertson et al. 1979, Fritsch 1977, dort auch ausführliche Literaturangaben) ist bekannt, daß die Zahl der Keimwurzeln art- und sortenspezifisch erhebliche Unterschiede aufweisen kann. Außerdem verdeutlichen diese Untersuchungen, daß zwischen Keimwurzelzahl und Kornmasse eine positive Korrelation besteht und moderne Saatweizensorten eine besonders hohe Keimwurzelzahl aufweisen; das könnte auf einen indirekten Selektionseffekt durch Züchtung auf hohe Kornmasse hindeuten.

Es war das Ziel dieser Untersuchungen, die genetische Variabilität der Keimwurzelzahl bei Gerste zu ermitteln. Außerdem sollte mit diesen Ermittlungen eine Basis für weiterführende Experimente geschaffen werden, die darauf abzielen, nach Beziehungen zwischen Keimwurzelzahl und Jungpflanzenentwicklung unter Streßbedingungen zu suchen. Es gibt einen weiteren Aspekt, der Entwicklung der Primärwurzeln besondere Aufmerksamkeit zuzuwenden. Es gibt Hinweise darauf, daß vorzugweise in den Wurzeln gebildete Phytohormone die meristematische Aktivität des Sprosses und damit dessen quantitative Entwicklung beeinflussen (WENT 1938, MURRAY 1968, DE ROPP 1946). Obwohl unsere Untersuchungen hierzu keine Aussagen liefern, darf aus der Analyse des Gesamtmaterials erwartet werden, daß hierdurch interessantes Ausgangsmaterial für weiterführende Untersuchungen über das Sproß-Wurzel-Verhältnis v. a. im jungen Entwicklungsstadium anfallen wird. Da für das nach dem Zufallsprinzip aus dem Gesamtsortiment von H. vulgare ausgewählte Material aus der ständigen Bearbeitung des Kulturpflanzensortimentes in Gatersleben einige weitere quantitative Analysenwerte (TKM, Höhe, Protein- und Lysingehalt) für die einzelnen Sippen vorlagen, bot es sich an, auch diese in die statistische Bearbeitung mit einzubeziehen, obwohl die dabei erhaltenen Korrelationskoeffizienten nicht mit der ursprünglichen Fragestellung in Beziehung stehen.

## 2. Material und Methoden

Zur Untersuchung gelangten 957 Land- und Zuchtsorten der Saatgerste (*Hordeum vulgare* L.) aus dem Gaterslebener Kulturpflanzenweltsortiment. Zu Vergleichszwecken wurden Herkünfte (nach Anbauort und Erntejahr) einer modernen Hochzuchtsorte untersucht.

Zur Ermittlung der Keimwurzelzahl wurden die Karyopsen in Keimrollen bei 20 °C zur Keimung gebracht und die Keimwurzeln nach 5 Tagen ausgezählt. Jede Stichprobe (Varietät, Herkunft) umfaßt 50 Keimlinge. Die Werte für die Tausendkornmasse, die Pflanzenhöhe sowie den Protein- und Lysingehalt stammen aus der Dokumentation des Kulturpflanzenweltsortimentes Gatersleben.

Alle statistischen Berechnungen sowie die Bildung von Gruppen nach morphologischen Merkmalen oder geographischer Herkunft erfolgten über eine EDVA (KRS 4201). Zum Test auf Normalverteilung wurde der Kolmogorow-Smirnow-Test benutzt.

## 3. Ergebnisse

### 3.1. *Genetische Stabilität der Keimwurzelzahl*

Untersuchungen über die genetische Stabilität des Merkmals Keimwurzelzahl wurden nicht durchgeführt. Ein Vergleich der Keimwurzelzahlen der Sorte Trumpf von Saatgut aus verschiedenen Anbauorten und Erntejahren (Tab. 1) zeigt jedoch, daß der Einfluß der Umwelt während der Samenreife auf die Ausbildung dieses Merkmals gering, seine genetische Determiniertheit offenbar stark ausgeprägt ist. Dies stützt ähnliche Schlüsse, die ROBERTSON (ROBERTSON et al. 1979) aus Befunden an Nachkommenschaftsuntersuchungen bei Weizen zog.

Tabelle 1

Keimwurzelzahlen der Sorte 'Trumpf'. Saatgut von unterschiedlichen Anbausorten und -jahren. (Bei den Anbauorten handelt es sich um verschiedene Saatzuchtstationen in der DDR

| Anbauort | Anbaujahr | Keimwurzelzahl $\bar{x}$ (n = 50) | $\pm s_{\bar{x}}$ |
|---|---|---|---|
| A | 1976 | 6,04 | (0,1) |
| B | 1976 | 6,28 | (0,1) |
| C | 1976 | 6,32 | (0,1) |
| D | 1976 | 6,47 | (0,1) |
| E | 1976 | 6,32 | (0,1) |
| F | 1976 | 6,57 | (0,1) |
| G | 1976 | 6,73 | (0,1) |
| H | 1976 | 6,27 | (0,1) |
| I | 1976 | 6,24 | (0,1) |
| K | 1973 | 6,22 | (0,1) |
| K | 1974 | 6,48 | (0,1) |
| K | 1976 | 6,03 | (0,3) |

### *3.2. Merkmalsmittelwerte innerhalb morphologischer Gruppen*

Der Mittelwert von 6,10 Keimwurzeln für das gesamte Material von 957 Sorten liegt deutlich über den für Saatweizen (*Triticum aestivum* L.) ermittelten Werten, die zwischen 3,7 und 5,8 liegen (Robertson et al. 1979, Apel, unveröffentlicht).

Ein signifikanter Unterschied besteht zwischen den zwei- und mehrzeiligen Gersten. Bei den zweizeiligen Formen ist die Keimwurzelzahl höher ($\bar{x} = 6,54$) als bei den mehrzeiligen ($\bar{x} = 5,78$). Innerhalb dieser beiden Hauptgruppen werden die Untergruppen nach den Merkmalen nackte oder bespelzte Karyopse gebildet und bei den mehrzeiligen als weitere kleine Untergruppe die Winterformen getrennt betrachtet.

Hierbei zeigte sich, daß sich die Mittelwerte für die Untergruppen nicht signifikant von denen der Hauptgruppen unterscheiden, d. h. das Merkmal mehrzeilig oder zweizeilig ist für die Keimwurzelzahl von größerer Bedeutung als die Merkmale „nackt" oder „bespelzt", bzw. „Winter-" oder „Sommerform". Im Mittel weisen die zweizeiligen Formen eine etwas höhere Kornmasse auf als die mehrzeiligen, was wohl Ausdruck der allgemein negativen Korrelation zwischen Kornmasse und Kornzahl/Ähre ist. Eine weitere Ursache der geringeren TKM der mehrzeiligen Gerten ist möglicherweise in der schwächeren Ausbildung der Karyopsen der Seitenährchen zu sehen.

Der Unterschied in der Kornmasse zwischen bespelzten und nackten Formen ist bei den zweizeiligen Gersten weitaus geringer als dies bei den mehrzeiligen der Fall ist. Bezüglich der Pflanzenhöhe sind gruppenspezifische Unterschiede angedeutet, jedoch nicht statistisch signifikant.

Die Ergebnisse der Protein- und Lysinuntersuchungen sind ausführlich bei Lehmann et al. (1978) dargestellt und werden deshalb hier lediglich im Zusammenhang mit Korrelationsbeziehungen zu anderen Merkmalen betrachtet.

Der Test auf Normalverteilung nach Kolmogorow-Smirnow wies aus: Keines

Tabelle 2

Merkmalsmittelwerte in morphologisch unterschiedlichen Gruppen der Saatgerste (*Hordeum vulgare* L.)
K = Keimwurzelzahl; TKM = Tausendkornmasse (g); H = Pflanzenhöhe zur Reife (cm); P = Rohprotein der Körner (%); L = Lysingehalt des Rohproteins (%); n = Anzahl der untersuchten Varietäten
Die Zahlen in Klammer sind die Standardabweichungen des Mittelwerts; die Zahlen darunter stellen den kleinsten und größten Wert in der untersuchten Gruppe dar.

| | n | K | TKM | H | P | L |
|---|---|---|---|---|---|---|
| gesamtes Material | 957 | 6,10 (0,697)<br>3,70 –8,47 | 44,5 (0,76)<br>21,3 –66,7 | 99 (14,6)<br>41 –152 | 14,7 (1,67)<br>10,4 –20,7 | 3,09 (0,36)<br>1,53 –4,23 |
| Mehrzeilig | 537 | 5,78 (0,563)<br>3,70 –7,63 | 42,7 (7,78)<br>21,3 –63,3 | 97 (15,5)<br>41 –152 | 14,7 (1,63)<br>10,7 –20,3 | 3,08 (0,37)<br>1,53 –4,21 |
| Mehrzeilig bespelzt | 457 | 5,81 (0,517)<br>3,70 –7,63 | 44,2 (0,69)<br>26,7 –63,3 | 98 (14,3)<br>43 –152 | 14,5 (1,48)<br>10,7 –20,3 | 3,10 (0,37)<br>1,53 –4,21 |
| Mehrzeilig nackt | 80 | 5,58 (0,749)<br>3,76 –6,90 | 33,9 (6,41)<br>21,3 –58,3 | 90 (19,9)<br>41 –132 | 16,1 (1,71)<br>12,8 –20,2 | 2,96 (0,31)<br>2,24 –3,68 |
| Mehrzeilig Sommer | 522 | 5,78 (0,568)<br>3,70 –7,63 | 42,6 (7,85)<br>21,3 –63,3 | 96 (15,3)<br>41 –152 | 14,8 (1,62)<br>10,7 –20,3 | 3,08 (0,37)<br>1,53 –4,21 |
| Mehrzeilig Winter | 15 | 5,86 (0,334)<br>5,04 –6,32 | 44,4 (4,59)<br>36,2 –52,3 | 114 (15,6)<br>75 –135 | 13,1 (0,98)<br>11,1 –14,7 | 3,19 (0,27)<br>2,81 –3,67 |
| Zweizeilig | 379 | 6,54 (0,639)<br>4,74 –8,47 | 47,1 (6,80)<br>30,7 –66,7 | 102 (13,0)<br>60 –140 | 14,6 (1,74)<br>10,4 –20,7 | 3,10 (0,36)<br>1,84 –4,04 |
| Zweizeilig bespelzt | 355 | 6,56 (0,634)<br>4,74 –8,34 | 47,2 (0,69)<br>30,7 –66,7 | 103 (13,1)<br>60 –140 | 14,5 (1,68)<br>10,4 –19,7 | 3,11 (0,36)<br>1,84 –4,04 |
| Zweizeilig nackt | 24 | 6,31 (0,672)<br>5,49 –8,47 | 45,6 (5,93)<br>33,7 –58,3 | 99 (11,0)<br>76 –121 | 16 1 (1,99)<br>12,8 –20,7 | 2,89 (0,30)<br>2,28 –3,61 |
| Wechselzeilig | 41 | 6 14 (0,519)<br>5,19 –7,37 | 44,5 (3,89)<br>36,0 –52,0 | 102 (9,7)<br>83 –125 | 15,6 (1,25)<br>13,3 –18,1 | 3,20 (0,33)<br>2,58 –4,23 |

der Merkmale ist innerhalb des Gesamtmaterials oder innerhalb der Gruppen, die mit >100 Sippen vertreten waren, normalverteilt. Für Protein und Lysin war dies schon von LEHMANN et al. (1978) an einem wesentlich umfangreicheren Material gefunden worden.

### 3.3. *Korrelationen zwischen den Merkmalen innerhalb morphologischer Gruppen*

Korrelationskoeffizienten (nach BRAVAIS-PEARSON) wurden sowohl für das gesamte Material als auch für die Untergruppen (Tab. 1) berechnet.

Von der Gesamtheit aller Werte sollen hier nur die statistisch signifikanten angeführt werden.

Keimwurzelzahl: TKM

| | r |
|---|---|
| gesamtes Material | +0,224[xx] |
| mehrzeilig | +0,212[xx] |

| | |
|---|---|
| mehrzeilig, Sommer | +0,207xx |
| mehrzeilig, Winter | +0,600x |
| mehrzeilig, nackt | +0,324xx |
| mehrzeilig, bespelzt | +0,130x |

Die signifikant positive Korrelation zwischen Keimwurzelzahl und Kornmasse innerhalb des gesamten Materials ist mit $r = 0,224$ wesentlich geringer als die von ROBERTSON et al. für diploide (r = 0,70), tetraploide (r = 0,76) und hexaploide (r = 0,63) Weizen gefundenen Werte. Das bedeutet, daß die Chance, bei Gerste indirekt mit der Selektion auf hohe Kornmasse auch auf hohe Keimwurzelzahl zu selektieren, wesentlich geringer ist als bei Weizen. Auffällig ist ferner, daß eine solche Beziehung bei den zweizeiligen Gersten überhaupt nicht gefunden wurde, sondern nur bei den mehrzeiligen und deren Untergruppen. Hier ist sie bei den Winterformen mit r = +0,600 am engsten, obwohl dieser Wert auf Grund des geringen Gruppenumfangs nur auf dem 5% Niveau signifikant ist.

Keimwurzelzahl : Pflanzenhöhe

| | r |
|---|---|
| gesamtes Material | +0,240xx |
| mehrzeilige | +0,261xx |
| mehrzeilige, Sommer | +0,258xx |
| mehrzeilige, nackt | +0,525xx |
| mehrzeilige, bespelzt | +0,140xx |

Die positive Korrelation zwischen Keimwurzelzahl und Pflanzenhöhe innerhalb des gesamten Materials ist auf den hohen Anteil an mehrzeiligen Varietäten zurückzuführen. Bei den zweizeiligen fehlt ein solcher Zusammenhang. Da ein kausaler Zusammenhang zwischen beiden Merkmalen kaum angenommen werden kann, ist es am wahrscheinlichsten, daß die ebenfalls nur bei den mehrzeiligen Gersten vorhandene positive Korrelation zwischen TKM und Pflanzenhöhe (s. dort) diesen Effekt bewirkt.

Die Selektion auf Kurzstrohformen könnte daher bei mehrzeiligen Sorten durch diesen Zusammenhang einen nachteiligen Einfluß auf die TKM haben.

Keimwurzelzahl : Protein

| | r |
|---|---|
| gesamtes Material | −0,112xx |
| Wechselgersten | −0,322x |
| mehrzeilig | −0,093x |
| mehrzeilig, Winter | −0,523x |
| mehrzeilig, nackt | −0,442x |
| zweizeilig | −0,146x |
| zweizeilig, bespelzt | −0,121x |

Die im Gesamtmaterial signifikante, wenn auch nur sehr lose negative Korrelation zwischen Keimwurzelzahl und Proteingehalt wird hauptsächlich durch die zweizeiligen Formen und den, zahlenmäßig allerdings kleinen Anteil an mehrzeiligen Winter- und Nacktgersten bewirkt. Zwar ist auch innerhalb der mehrzeiligen dieser Zusammenhang signifikant, aber mit r = 0,093 doch sehr schwach ausgeprägt. Physiologisch ist diese, wenn auch nur schwach, aber regelmäßig in fast allen Gruppen ausgeprägte Beziehung derzeit kaum zu interpretieren, zumal ein indirekter Zusammenhang über Korrelationen zwischen TKM und Protein (s. dort) und TKM und Keimwurzelzahl als Erklärungsmöglichkeit ausscheidet.

Keimwurzelzahl : Lysin
Die einzige signifkante Korrelation zwischen diesen beiden Parametern wurde mit r = 0,234ˣ innerhalb der Gruppe der mehrzeiligen Nacktgersten gefunden.

TKM : Pflanzenhöhe

| | r |
|---|---|
| gesamtes Material | +0,143ˣ |
| mehrzeilig | +0,170ˣˣ |
| mehrzeilig, Sommer | +0,162ˣˣ |
| mehrzeilig, nackt | +0,318ˣˣ |

Die positive Korrelation zwischen Kornmasse und Pflanzenhöhe wird innerhalb des gesamten Materials hauptsächlich durch den Anteil an mehrzeiligen Formen hervorgerufen. Innerhalb der zweizeiligen liegt ein solcher Zusammenhang nicht vor. Insgesamt ist aber auch bei den mehrzeiligen dieser Zusammenhang so lose, daß es nur evtl. bei der Gruppe der mehrzeiligen Nacktgersten Nachteile bei der gleichzeitigen Verfolgung der Zuchtziele Kurzstrohigkeit und hohe Tausendkornmasse bringen könnte.

TKM : Protein

| | r |
|---|---|
| zweizeilig | +0,403ˣˣ |
| zweizeilig, bespelzt | +0,473ˣˣ |
| mehrzeilig | −0,318ˣˣ |
| mehrzeilig, Sommer | −0,316ˣˣ |

Innerhalb des Gesamtmaterials ist kein signifikanter Zusammenhang zwischen beiden Merkmalen nachzuweisen. Dies ist sicher darauf zurückzuführen, daß sich die positive Korrelation bei den zweizeiligen und die negative bei den mehrzeiligen bei der Rechnung über alle Werte aufheben.

TKM : Lysin
Nur in der Gruppe der mehrzeiligen Nacktgersten wurde eine signifikante Korrelation mit r = +0,234ˣ gefunden. Diese dürfte zugleich auch die Korrelation zwischen Keimwurzelzahl und Lysin in dieser Gruppe erklären.

Höhe : Protein

| | r |
|---|---|
| gesamtes Material | −0,162ˣˣ |
| zweizeilig | −0,112ˣ |

| | |
|---|---|
| zweizeilig, bespelzt | −0,127$^{xx}$ |
| mehrzeilig | −0,203$^{xx}$ |
| mehrzeilig, Sommer | −0,178$^{xx}$ |
| mehrzeilig, nackt | −0,317$^{xx}$ |

Die im gesamten Material und in 5 Untergruppen signifikant negative Beziehung zwischen Pflanzenhöhe und Proteingehalt ist physiologisch zwar kaum zu interpretieren, läßt aber eine günstige Prognose zu bezüglich der Zuchtziele Kurzstrohigkeit und Proteingehalt.

Höhe : Lysin

| | r |
|---|---|
| mehrzeilig | +0,194$^{xx}$ |
| mehrzeilig, Sommer | +0,199$^{xx}$ |
| mehrzeilig, nackt | +0,346$^{xx}$ |
| mehrzeilig, bespelzt | +0,140$^{xx}$ |

Die innerhalb der mehrzeiligen Gersten durchweg positive Korrelation zwischen Pflanzenhöhe und Lysin, die bei den mehrzeiligen Nacktgersten am ausgeprägtesten ist, ist wohl in Beziehung zu bringen mit der negativen Beziehung Höhe : Protein einerseits (s. o.) und der nachfolgend angeführten Beziehung Protein : Lysin andererseits. Obwohl die zweizeiligen Formen innerhalb dieses Materials nicht ganz ins Bild passen, scheint es allgemein so zu sein, daß lange Formen proteinarm und proteinarme lysinreich sind. Dieser Zusammenhang zwischen Protein- und Lysingehalt war bereits an dem wesentlich umfangreicheren Material von Lehmann et al. (1978) deutlich geworden.

Protein : Lysin

| | r |
|---|---|
| gesamtes Material | −0,139$^{xx}$ |
| zweizeilig | −0,117$^{x}$ |
| mehrzeilig | −0,164$^{xx}$ |
| mehrzeilig, Sommer | −0,156$^{xx}$ |
| mehrzeilig, nackt | −0,276$^{xx}$ |

### *3.4. Merkmalsmittelwerte innerhalb geographischer Gruppen*

In Tab. 3 sind die Merkmalsmittelwerte und deren Streuungsbereich für Gruppen angegeben, die nach der Herkunft des Materials gebildet wurden. Tabelle 4 weist den Anteil der morphologischen Typen innerhalb der geographischen Gruppen und Untergruppen aus. Der für Europa und Asien vorgenommene Versuch einer stärkeren Untergliederung brachte auf der Basis des vorliegenden Materials keine wesentlich neuen Erkenntnisse bezüglich der Merkmalsausprägung und evtl. korrelativer Beziehungen zwischen diesen. Die Mittelwerte für die Merkmale werden wesentlich durch den morphologischen Typ determiniert, der in der jeweiligen Untergruppe dominierend vertreten ist. Ganz ausgeprägt zeigt sich dies bei der Gruppe „Ostasien“, (Herkünfte aus Japan, China und Tibet), deren niedrige

Tabelle 3

Merkmalsmittelwerte in Gruppen geographisch unterschiedlicher Herkunft der Saatgerste (*Hordeum vulgare* L.)
K = Keimwurzelzahl; TKM = Tausendkornmasse (g); H = Pflanzenhöhe zur Reife (cm); P = Rohprotein der Körner (%); L = Lysingehalt des Rohproteins (%); n = Anzahl der untersuchten Varietäten. Die Zahlen in Klammer sind die Standardabweichungen des Mittelwerts; die Zahlen unter der Klammer stellen den kleinsten und größten Wert in der untersuchten Gruppe dar.

| | n | K | | TKM | | H | | P | | L | |
|---|---|---|---|---|---|---|---|---|---|---|---|
| Europa | 285 | 6,26 | (0,72) | 44,5 | (7,1) | 103 | (14,3) | 13,8 | (1,3) | 3,0 | (0,36) |
| | | 4,3 | –8,5 | 21,3 | –64,0 | 70 | –145 | 10,4 | –18,4 | 1,8 | –4,0 |
| Asien | 221 | 5,81 | (0,72) | 40,8 | (7,7) | 92 | (18,2) | 14,7 | (1,7) | 3,0 | (0,35) |
| | | 3,7 | –7,3 | 21,7 | –58,3 | 41 | –152 | 10,7 | –20,2 | 1,5 | –4,0 |
| Afrika | 407 | 6,18 | (0,60) | 46,7 | (7,1) | 100 | (10,6) | 15,4 | (1,5) | 3,2 | (0,34) |
| | | 4,5 | –8,1 | 28,0 | –66,7 | 70 | –127 | 11,8 | –20,7 | 2,0 | –4,2 |
| Amerika | 39 | 5,69 | (0,75) | 42,4 | (7.1) | 101 | (15,8) | 14,1 | (1,8) | 3,1 | (0,47) |
| | | 4,7 | –7,6 | 28,0 | –55,7 | 70 | –137 | 11,8 | –19,7 | 2,0 | –4,2 |
| Australien | 5 | 5,86 | (0,38) | 46,3 | (5,4) | 103 | (15,0) | 14,6 | (1,6) | 2,9 | (0,35) |
| | | 5,4 | –6,3 | 39,3 | –51,3 | 86 | –126 | 12,8 | –16,9 | 2,4 | –3,3 |
| West- und Mitteleuropa | 70 | 6,55 | (0,59) | 41,4 | (5,9) | 109 | (11,4) | 13,8 | (1,3) | 3,0 | (0,36) |
| | | 5,4 | –8,5 | 26,0 | –58.3 | 88 | –132 | 11,2 | –16,5 | 1,8 | –4,0 |
| Süd- und Südosteuropa | 144 | 6,06 | (0,73) | 47,2 | (6,7) | 100 | (15,4) | 13,7 | (1,1) | 3,0 | (0,37) |
| | | 4,3 | –8,3 | 26,7 | –63,0 | 70 | –145 | 11,1 | –17,1 | 1,9 | –3,8 |
| Vorderer und mittlerer Orient | 139 | 5,93 | (0,63) | 44,1 | (6,1) | 95 | (14,5) | 14,1 | (1,3) | 3,1 | (0,34) |
| | | 4,7 | –7,3 | 23,3 | –58,3 | 57 | –135 | 10,7 | –19,7 | 2,3 | –4,0 |
| Ostasien | 56 | 5,38 | (0,83) | 34,6 | (7,2) | 83 | (24,7) | 16,0 | (1,9) | 3 0 | (0,38) |
| | | 3,7 | –7,0 | 21,7 | –52 | 41 | –152 | 12,1 | –20,2 | 1,5 | –3,6 |
| Aethiopien | 406 | 6,18 | (0,60) | 46,7 | (7,1) | 100 | (10,6) | 15,4 | (1,5) | 3,2 | (0,34) |
| | | 4,5 | –8,1 | 28,0 | –66,7 | 70 | –127 | 11,8 | –20,7 | 2,0 | –4,2 |

Werte für Keimwurzelzahl, TKM und Höhe bei gleichzeitig hohem Proteingehalt auf den hohen Anteil an mehrzeiligen und darunter wiederum vielen Nacktgersten zurückzuführen ist (s. Tab. 4).

### *3.5. Korrelationen zwischen den Merkmalen innerhalb geographischer Gruppen*

Es werden wiederum nur die statistisch signifikanten Koeffizienten angeführt

Keimwurzelzahl : TKM

| | r |
|---|---|
| Asien | +0,308^xx^ |
| Ostasien | +0,632^xx^ |
| Afrika | +0,255^xx^ |

Es fällt auf, daß bei den asiatischen und besonders darunter den ostasiatischen Formen die Korrelation zwischen Keimwurzelzahl und TKM besonders eng ist. Das ist sicher damit in Zusammenhang zu bringen, daß es sich hier überwiegend um mehrzeilige Formen handelt. Bei dem Material aus Europa ist kein Zu-

Tabelle 4

Anzahl der untersuchten Sortiments-Nummern innerhalb der einzelnen nach morphologischen und geographischen Kriterien gebildeten Gruppen

| morphologische Gruppe / geographische Gruppe | Europa | Asien | Afrika | Amerika | Australien | West- und Mitteleuropa | Süd- und Südosteuropa | Vorderer und mittlerer Orient | Ostasien | Nordostafrika | Summe |
|---|---|---|---|---|---|---|---|---|---|---|---|
| mehrzeilig | 144 | 159 | 203 | 27 | 4 | 14 | 114 | 84 | 52 | 202 | 537 |
| mehrzeilig, bespelzt | 128 | 112 | 186 | 27 | 4 | 9 | 109 | 80 | 17 | 185 | 457 |
| mehrzeilig, nackt | 16 | 47 | 17 | 0 | 0 | 5 | 5 | 4 | 35 | 17 | 80 |
| mehrzeilig, Sommer | 135 | 154 | 203 | 26 | 4 | 13 | 106 | 79 | 52 | 202 | 522 |
| mehrzeilig, Winter | 9 | 5 | 0 | 1 | 0 | 1 | 8 | 5 | 0 | 0 | 15 |
| zweizeilig | 141 | 62 | 163 | 12 | 1 | 56 | 30 | 55 | 4 | 163 | 379 |
| zweizeilig, bespelzt | 136 | 58 | 148 | 12 | 1 | 52 | 30 | 53 | 3 | 148 | 355 |
| zweizeilig, nackt | 5 | 4 | 15 | 0 | 0 | 4 | 0 | 2 | 1 | 15 | 24 |
| wechselzeilig | 0 | 0 | 41 | 0 | 0 | 0 | 0 | 0 | 0 | 41 | 41 |
| Summe | 285 | 221 | 407 | 39 | 5 | 70 | 144 | 139 | 56 | 406 | 957 |

sammenhang nachweisbar und somit ist ein indirekter Selektionseffekt bei Züchtung auf hohe TKM bei der Arbeit mit diesem Material wenig wahrscheinlich.

Keimwurzelzahl : Höhe

| | r |
|---|---|
| Ostasien | $+0{,}593^{xx}$ |
| Asien | $+0{,}331^{xx}$ |
| Europa | $+0{,}205^{xx}$ |

Die positive Beziehung zwischen Keimwurzelzahl und Höhe bei den asiatischen Formen überrascht nicht, da bei diesen auch ein positiver Zusammenhang zwischen TKM und Höhe besteht. Dagegen ist die positive Beziehung zwischen Keimwurzelzahl und Höhe bei den europäischen Formen nicht in dieser Weise zu erklären, da bei diesen die Korrelation zwischen TKM und Höhe signifikant negativ ist.

Keimwurzelzahl : Protein

| | r |
|---|---|
| Asien | $-0{,}224^{xx}$ |
| Ostasien | $-0{,}423^{xx}$ |
| Afrika | $-0{,}146^{xx}$ |

Wie schon innerhalb der morphologischen Gruppen sind die statistisch signifikanten Koeffizienten negativ. Am engsten ist diese Beziehung bei den ostasiatischen Formen, wohl wegen ihres überdurchschnittlich hohen Proteingehaltes in Kombination mit einer geringen Kornmasse.

Keimwurzelzahl : Lysin

| | r |
|---|---|
| Ostasien | +0,261ˣ |
| S. + SO-Europa | +0,204ˣ |
| Amerika | −0,407ˣˣ |

War innerhalb der morphologischen Gruppen nur eine signifikant positive Beziehung gefunden worden, so überrascht hier die relativ enge, aber negative Korrelation bei den Herkünften aus Amerika. Rein formal resultiert dies daraus, daß diese Formen zwar einen dem Gesamtmittel entsprechenden Lysingehalt, aber eine unverhältnismäßig niedrige Keimwurzelzahl aufweisen.

TKM : Höhe

| | r |
|---|---|
| Asien | +0,391ˣˣ |
| Ostasien | +0,518ˣˣ |
| Afrika | +0,139ˣˣ |
| Europa | −0,271ˣˣ |
| S. + SO-Europa | −0,377ˣˣ |
| V. + m. Orient | +0,186ˣ |

Waren die Korrelationen zwischen TKM und Höhe für das Gesamtmaterial sowie für die Untergruppen der mehrzeiligen Gersten durchweg positiv, so fällt bei der Gliederung nach Herkunftsgebieten auf, daß diese Beziehung für die europäischen Sorten insgesamt und insbesondere für die süd- und südosteuropäischen Formen signifikant negativ ist. Da es sich hier nur zum geringsten Teil um neuere Formen handelt, kann dies kaum das Ergebnis gleichzeitiger Selektion auf Kurzstrohigkeit und hohe Kornmasse handeln. Vielmehr könnte dieser Befund dafür sprechen, daß süd- und südosteuropäische Herkünfte bei der Verfolgung dieser beiden Zuchtziele günstigeres Ausgangsmaterial darstellen als z. B. ostasiatische Formen.

TKM : Protein

| | r |
|---|---|
| Asien | −0,407ˣˣ |
| Ostasien | −0,504ˣˣ |

Diese signifikant negativen Koeffizienten sind mit Sicherheit auf den hohen Anteil mehrzeiliger Formen in dieser Gruppe zurückzuführen.

TKM : Lysin

| | r |
|---|---|
| Asien | +0,198ˣˣ |
| Europa | −0,271ˣˣ |
| S. + SO-Europa | −0,347ˣˣ |

Die positive Korrelation innerhalb des asiatischen Materials dürfte ebenfalls auf den hohen Anteil mehrzeiliger Gersten zurückzuführen sein. Eine rein formale Erklärung der negativen Beziehung zwischen den beiden Merkmalen innerhalb der europäischen Formen ergibt sich aus deren relativ hohem TKM bei nur durchschnittlichem Lysingehalt.

Höhe : Protein

| | r |
|---|---|
| Afrika | $-0{,}111^{x}$ |
| Amerika | $-0{,}316^{x}$ |
| Asien | $-0{,}329^{xx}$ |
| Ostasien | $-0{,}464^{xx}$ |

Die in den morphologischen Gruppen nachgewiesene negative Beziehung zwischen Höhe und Protein fehlt interessanterweise bei den europäischen Formen und denen aus dem mittleren und vorderen Orient, während sie für die übrigen Herkunftsgebiete wie für die morphologischen Gruppen immer negativ ist.

Höhe : Lysin

| | r |
|---|---|
| Asien | $+0{,}254^{xx}$ |
| Ostasien | $+0{,}386^{xx}$ |
| V + m. Orient | $+0{,}186^{x}$ |

Dieser Befund liefert keine über die für die morphologischen Gruppen hinausgehende Information, da in den hier angeführten Herkunftsgebieten die mehrzeiligen Formen überwiegen.

Protein : Lysin

| | r |
|---|---|
| Afrika | $-0{,}180^{xx}$ |
| Asien | $-0{,}337^{xx}$ |
| Ostasien | $-0{,}397^{xx}$ |
| V. + m. Orient | $-0{,}188^{x}$ |
| Europa | $-0{,}237^{xx}$ |
| W. + M.-Europa | $-0{,}325^{x}$ |

Die vorn getroffene Feststellung einer fast generell negativen Beziehung zwischen Proteingehalt und Lysingehalt trifft auch für die meisten Herkunftsgruppen zu. Daß eine solche Beziehung für die Gruppe aus Amerika nicht nachgewiesen werden konnte, ist möglicherweise auf den relativ geringen Umfang dieser Gruppe zurückzuführen. Die formale Ursache des Fehlens einer solchen Beziehung bei den süd- und südosteuropäischen Gersten ist in ihrem relativ niedrigen Proteingehalt bei gleichzeitig durchschnittlichem Lysingehalt zu suchen.

Es war das Ziel dieser Untersuchungen, geeignetes Ausgangsmaterial für die Analyse eines evtl. Zusammenhanges zwischen Keimwurzelzahl und Resistenz gegen Wassermangel im Jungpflanzenstadium zu finden. Es kann nicht a priori davon ausgegangen werden, daß ein solcher Zusammenhang tatsächlich existiert. Passioura (1972) fand, daß Weizenpflanzen (Sorte 'Gabo') Bodenwasser am ökonomischsten nutzen, wenn die Leitfähigkeit der Wurzeln für Wasser am geringsten ist. Er erreichte dies experimentell dadurch, daß er nur einer Keimwurzel das Wachstum in tiefere, wasserführende Bodenschichten gestattete, während die übrigen in den oberen, rasch austrocknenden Bodenschichten verblieben. Er leitete aus seinen Ergebnissen ab, daß für Regionen mit reichlichem Winter- und Frühjahrsniederschlag und fast fehlenden Niederschlägen während der Hauptentwicklungsphase Genotypen mit hohem Wurzelwiderstand für die Wasserleitung selektiert werden sollten. Eine solche Forderung ist für Gebiete mit mehr oder weniger gleichmäßig verteilten Niederschlägen während der Ontogenese sicher nicht die optimale Variante. Weiterführende Untersuchungen müssen jedoch erst erweisen, ob die Keimwurzelzahl eines Genotyps unter mitteleuropäischen Niederschlagsverhältnissen für die Gesamtentwicklung und insbesondere die Ausbildung der Ertragskomponenten von Bedeutung ist.


## Summary

Variability of seedling root numbers in barley

Seedling root numbers in 957 varieties of barley were estimated. The numbers varied between 3,70 and 8,47 roots per seedling. From other investigations data of kernel dry weight, plant height at ripeness, protein content of the kernel and lysine content of protein were available.

There are several significant correlations between seedling root number and these other parameters. There physiological significance and possible meaning in barley breeding is discussed.



## Краткое содержание

Изменчивость числа зародышевых корешков у ячменя

Определялось число зародышевых корешков у 957 разновидностей ячменя (*Hordeum vulgare* L.). Это число колеблется от 3,70 до 8,47 корешков на зерновку. Были использованы также данные других исследований на этих разновидностях: о массе зерна, высоте растения (в стадии зрелости), содержании протеина и лизина. Между этими параметрами и числом зародышевых корешков были найдены некоторые достоверные корреляции. Обсуждается физиологическое, а иногда и селекционное, значение этих зависимостей.


## Literatur


de Ropp, R. S., 1946: Studies in the physiology of leaf growth. III The influence of roots on the growth of leaves and stems in rye. – Ann. Bot. **10**, 353–359.

Fritsch, R., 1977: Über morphologische Wurzelmerkmale bei *Triticum* L. und *Aegilops* L. (Gramineae). – Kulturpflanze **XXV**, 45–70.

LEHMANN, CHR., O., A. RUDOLPH, K. HAMMER, A. MEISTER, K. MÜNTZ und F. SCHOLZ, 1978: Eiweißuntersuchungen am Getreide- und Leguminosen-Sortiment Gatersleben. Teil 2: Gehalt an Rohprotein und Lysin von Gersten (*Hordeum vulgare* L. s. l.). – Kulturpflanze **XXVI**, 163–181.

LIETH, H., 1976: The use of correlation models to predict primary productivity from precipitation or evapotranspiration. – In: Water and plant life (Edit. O. L. LANGE et al.) pp. 392–407. Ecol. Studies **19**. Springer-Verlag Berlin, Heidelberg, New York.

MONTEITH, J. L., 1977: Climate and the efficiency of crop production in Britain. – Phil. Trans. R. Soc. Lond. B **281**, 277–294.

MURRAY, D., 1968: Light and Leaf Growth in *Phaseolus*. Ph. D. Thesis, University of Edinburgh, zit. nach DALE, J. E., 1976: "Cell division in leaves" in YEOMAN, M. M. (Ed.), 1976: Cell division in higher plants – Acad. Press.

PASSIOURA, J. B., 1972: The effect of root geometry on the yield of wheat growing on stored water. – Austr. J. agric. Res. **23**, 745–752.

ROBERTSON, B. M., J. G. WAINES and B. S. GILL, 1979: Genetic variability for seedling root numbers in wild and domesticated wheats. – Crop Sci. **19**, 843–847.

WENT, F. W., 1938: Transplantation experiments with peas. – Amer. J. Bot. **25**, 44–55.

Dr. P. APEL, Dipl.-Math. A. BERGMANN,
Dipl.-Math. H.-W. JANK und Dr. CHR. O. LEHMANN

Zentralinstitut für Genetik und
Kulturpflanzenforschung
der Akademie der Wissenschaften der DDR
4325 Gatersleben

Kulturpflanze XXIX · 1981 · S. 417–423

# Zur fluorimetrischen Bestimmung von Chlorophyll a und Chlorophyll b

NATALJA JÄKEL

(Eingegangen am 2. Februar 1981)

## Zusammenfassung

Die Bestimmung von Chl a und b im Gemisch in verschiedenen Lösungsmitteln wird mit einer Fluoreszenzmethode durchgeführt. Verwendet wird dazu ein in unserem Labor aus zwei Monochromatoren aufgebautes Fluoreszenz-Spektrophotometer. Die Anregungswellenlänge für die Fluoreszenzemission liegt bei 455 nm. Die Messungen, die bei Zimmertemperatur durchgeführt werden, eröffnen die Möglichkeit, das Chl a/Chl b-Verhältnis bis zu einer Größenordnung von 60–70 zu bestimmen.

## Einleitung

Gewöhnlich wird die Bestimmung der Konzentration von Chlorophyl a (Chl a) und Chlorophyll b (Ch b) in Lösungen spektrophotometrisch durchgeführt (SCHLYK 1968). Diese Methode der Konzentrationsbestimmung von Pigmenten basiert auf der Messung der optischen Dichte bei zwei Wellenlängen, die den Absorptionsmaxima im roten Spektralbereich entsprechen.

Wenn die optische Dichte der Komponenten jedoch gering ($E < 0.1$) und eine der Komponenten in sehr kleinen Mengen vorliegt, ergibt diese Methode nur sehr ungenaue Werte und ist für Chl a/Chl b-Verhältnisse $> 8$ praktisch nicht mehr anwendbar (ZIEGLER und EGLE 1965; OGAWA und SHIBATA 1965). Dann erweist sich die Anwendung der Fluoreszenzmethode für die Ermittlung der Konzentrationen von Chl a und Chl b als günstiger (BOARDMAN und THORNE 1971; ZENKEVICH und LOSEV 1970; DEGREEF et al. 1975). Ziel dieser Arbeit war es, die letztgenannte Methode in unserem Laboratorium zu etablieren.

## Theorie

Wie bekannt, ist die Fluoreszenzintensität eines Stoffes proportional der Intensität des Anregungslichtes ($F_0$), der Absorption ($1-T$) und der Quantenausbeute ($\varphi$):

$$F = F_0\,(1-T)\,\varphi .$$

Bei niedrigen Konzentrationen ist die Absorption einer Lösung proportional der optischen Dichte und folglich der Konzentration des lumineszierenden Stoffes:

$$(1-T) \sim E \sim C$$

Daraus folgt: $F \sim C$
Für Lösungen mit niedrigen Konzentrationen gilt somit:

$$F(\lambda) = k(\lambda) \cdot C$$

(k ist ein von der Anregungswellenlänge abhängiger Proportionalitätskoeffizient, C die Konzentration des Stoffes).

Im Fall von reinen Chl a- und Chl b-Lösungen (BOARDMAN und THORNE 1971) gelten bei Anregung der Fluoreszenz mit zwei verschiedenen Wellenlängen folgende Gleichungen:

$$F(\lambda_1) = k_b(\lambda_1) \cdot C_b$$
$$F(\lambda_2) = k_a(\lambda_2) \cdot C_a \qquad (1)$$

($F(\lambda_2)$ und $F(\lambda_1)$ sind die Amplituden der Fluoreszenz von Chl a und Chl b im Maximum).
Daraus ergibt sich:

$$\frac{C_a}{C_b} = \frac{k_b(\lambda_1)}{k_a(\lambda_2)} \cdot \frac{F(\lambda_2)}{F(\lambda_1)} \qquad (2)$$

Aus dieser Gleichung (2) geht hervor, daß eine linerare Abhängigkeit zwischen dem Verhältnis Chl a/Chl b und dem Verhältnis $F(\lambda_2)/F(\lambda_1)$ zu erwarten ist.

In Gemischen ist eine solche linerare Beziehung jedoch nicht zu beobachten, da sich die Fluoreszenzemissionsspektren von Chl a und Chl b überlappen. Deshalb ist es notwendig, den Anteil jeder Komponente an der Fluoreszenz der anderen zu berücksichtigen. Die Gesamtfluoreszenzintensität von Lösungen, bestehend aus zwei nicht wechselwirkenden lumineszierenden Komponenten, ist

$$F(\lambda) = F_a(\lambda) + F_b(\lambda)$$

Wenn die Fluoreszenz des Gemisches bei einer Wellenlänge angeregt und die Gesamtintensität bei zwei Wellenlängen, die den Maxima der einzelnen Komponenten entsprechen, gemessen wird, dann gelten die nachstehenden Gleichungen:

$$F(\lambda_1) = F_{a1}C_a + F_{b1}C_b$$
$$F(\lambda_2) = F_{a2}C_a + F_{b2}C_b \qquad (3)$$

($F_{a1}$ ist die Fluoreszenzintensität pro Konzentrationseinheit von Chl a im Fluoreszenzmaximum von Chl b; $F_{b2}$ die Fluoreszenzintensität von Chl b pro Konzentrationseinheit im Fluoreszenzmaximum von Chl a, $F_{b1}$ die Fluoreszenzintensität von Chl b im Maximum und $F_{a2}$ die Fluoreszenzintensität von Chl a im Maximum.)

Aus den Gleichungen (3) folgt:

$$\frac{F(\lambda_2)}{F(\lambda_1)} = \frac{F_{a2}C_a + F_{b2}C_b}{F_{a1}C_a + F_{b1}C_b} \qquad (4)$$

Daraus wird das Konzentrationsverhältnis der Komponenten bestimmt:

$$\frac{C_a}{C_b} = \frac{F(\lambda_2)\,F_{b1} - F(\lambda_1)\,F_{b2}}{F(\lambda_1)\,F_{a2} - F(\lambda_2)\,F_{a1}} = \frac{\dfrac{F(\lambda_2)}{F(\lambda_1)} - \dfrac{F_{b2}}{F_{b1}}}{1 - \dfrac{F(\lambda_2)}{F(\lambda_1)} \cdot \dfrac{F_{a1}}{F_{a2}}} \cdot \frac{F_{b1}}{F_{a2}} \tag{5}$$

($F_{b2}/F_{b1}$ ist dabei der Fluoreszenzanteil von Chl b im Fluoreszenzmaximum von Chl a, $F_{a1}/F_{a2}$ der Fluoreszenzanteil von Chl a im Fluoreszenzmaximum von Chl b). Diese Verhältnisse werden dabei aus den Fluoreszenzemissionsspektren von reinem Chl a und Chl b bestimmt (s. Abb. 1). Die Methode zur Bestimmung des Verhältnisses Chl a/Chl b im Gemisch, wobei die Menge von Chl b im Vergleich zu der von Chl a gering ist, basiert auf der schwachen Anregung der Fluoreszenz von Chl a im Fluoreszenzmaximum von Chl b.

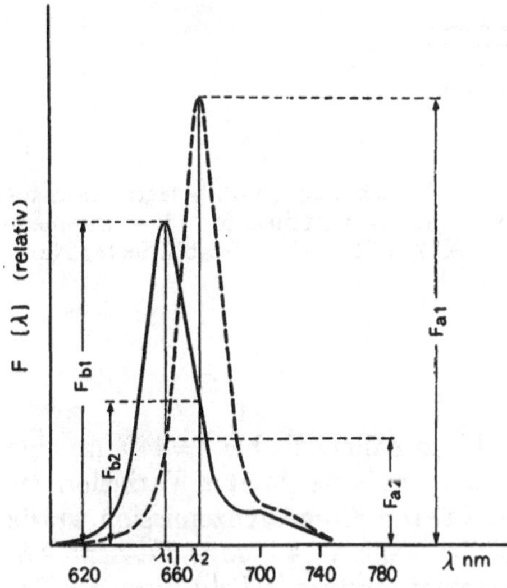


Abb. 1 Fluoreszenzemissionsspektren von Chl a und Chl b in Azeton (unkorrigiert). (———) – Fluoreszenzemissionsspektrum von Chl b (Anregungswellenlänge $\lambda = 455$ nm); (– – –) – Fluoreszenzemissionsspektrum von Chl a (Anregungswellenlänge $\lambda = 429$ nm)

## Material und Methoden

Chl a und Chl b wurden mit Azeton aus Blättern von *Vicia faba* unter Zusatz von geglühtem Quarzsand extrahiert und papierchromatographisch getrennt (Sapozhnikov 1964). Die Pigmente wurden mit Azeton eluiert und rechromatographiert. Aus den eluierten Extrakten wurden reine Chl a- und Chl b-Lösungen bekannter Konzentration hergestellt, wobei als Lösungsmittel Äther, Azeton und Äthanol dienten. Zur Herstellung von Ätherlösungen der Pigmente wurden diese aus Azeton in Äther überführt (Sapozhnikov 1964). Die Konzentrationsbestimmung von Chl a und Chl b erfolgte spektrophotometrisch am UNICAM-SP 800. Zur Berechnung der Konzentrationen wurden spezifische Extinktionskoeffizienten verwendet: für Azetonlösungen die Koeffizienten nach Wettstein (1957), für Ätherlösungen die nach Comar und Zscheile (1942) und für Äthanollösungen die von Boardman und Thorne (1971). Aus Chl a- und Chl b-Lösungen wurden in bestimmtem Verhältnis Gemische hergestellt.

Die Fluoreszenzmessungen erfolgten an einem Spektrofluorimeter, das aus einem Prismenmonochromator SPM 1 (VEB Carl Zeiss Jena, DDR), einem Gittermonochromator (Bausch und Lomb, USA) sowie einem Fluoreszenzmeßeinsatz mit Küvettenraum (VEB Carl Zeiss Jena) aufgebaut wurde. Abb. 2 gibt im Blockschema einen Überblick über das Gerät. Die Fluoreszenzmessungen wurden bei Zimmertemperatur durchgeführt, als Lichtquelle diente eine Xenonlampe. Das Licht vom Anregungsmonochromator wird durch die Bodenfläche der Quarzküvette gestrahlt. Senkrecht hierzu, in horizontaler Richtung, wird die Fluoreszenz vom Emissionsmonochromator mit Hilfe eines Photovervielfachers gemessen, dessen Signal mittels Schreiber aufgezeichnet wurde.

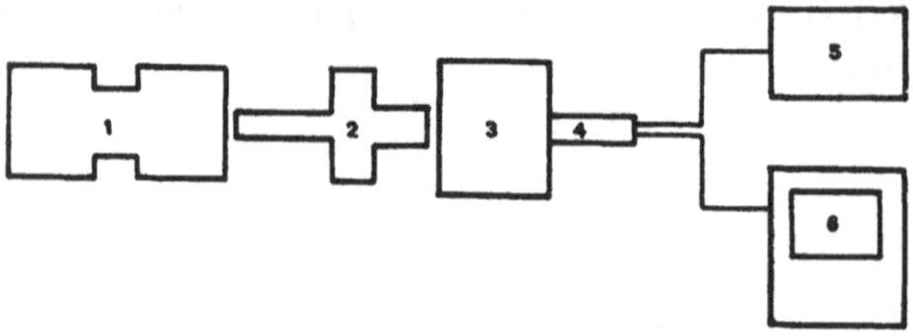


Abb. 2 Blockschema des Spektrofluorimeters. (1) – Bausch und Lomb-Gittermonochromator mit Lichtquelle; (2) – Fluoreszenzmeßeinsatz mit Küvettenraum; (3) – Prismenmonochromator SPM1; (4) – Photovervielfacher M12 FC52; (5) – Stabilisiertes Netzgerät; (6) –Schreiber G1 B1

## Ergebnisse und Diskussion

Aus den Ausgangslösungen von Chl a und Chl b in Äther, die bei $\lambda = 440$ nm eine optische Dichte von 0,1 hatten, wurden Gemische mit einer breiten Variation des Verhältnisses Chl a/Chl b von 70 bis 6 hergestellt. Die Fluoreszenzemission wurde durch monochromatisches Licht mit der Wellenlänge $\lambda = 455$ nm, die dem Absorptionsmaximum von Chl b entspricht, angeregt. Mit einer Erhöhung des Ge-

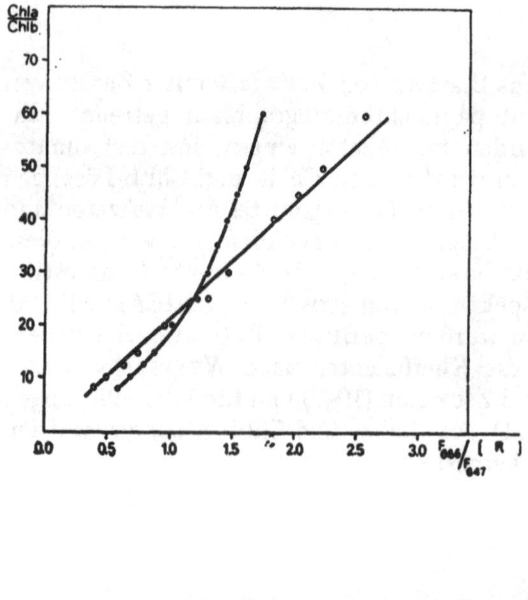


Abb. 3 Zusammenhang zwischen den Verhältnissen von Chl a/Chl b und den Fluoreszenzamplituden für Gemische in Äther. (1) –●—● Verhältnis $F_{666}/F_{647}$ als Abszisse; (2) – ○—○ korrigiertes Verhältnis R als Abszisse

haltes an Chl b konnte im Fluoreszenzspektrum eine Vergrößerung des entsprechenden Intensitätsmaximums und eine relative Reduzierung des Maximums von Chl a beobachtet werden. Aus den Fluoreszenzemissionsspektren wurde das Verhältnis der Amplituden $F(\lambda_2)/F(\lambda_1) = F_{666}/F_{647}$ errechnet. Wie aus Abb. 3 (Kurve 1) ersichtlich, existiert zwischen dem Verhältnis von Chl a/Chl b und dem $F_{666}/F_{647}$-Amplitudenverhältnis keine lineare Abhängigkeit. Um eine solche jedoch zu erhalten, muß, wie schon erwähnt, die Überlappung der Spektren beachtet werden. Zur Berechnung der Korrekturkurve wurde Gleichung (5) in der nachstehenden Form verwendet:

$$\frac{C_a}{C_b} = R\frac{F_{b1}}{F_{a2}}$$

$$\text{Dabei ist: } R = \frac{\dfrac{F(\lambda_2)}{F(\lambda_1)} - k_b}{1 - \dfrac{F(\lambda_2)}{F(\lambda_1)} \cdot k_a} \qquad (6)$$

$$\text{und } k_b = \frac{F_{b2}}{F_{b1}} \qquad k_a = \frac{F_{a1}}{F_{a2}}$$

Verwendet man als Abszisse statt des Verhältnisses $F_{666}/F_{647}$ das entsprechend Gleichung (6) korrigierte Verhältnis R, ergibt sich eine lineare Beziehung (s. Abb. 3, Kurve 2). $F_{b1}/F_{a2}$ ergibt sich dann graphisch als Anstieg der Geraden.

Ähnliche Resultate wurden auch mit Pigmentgemischen in Azeton erhalten. Die optische Dichte der Ausgangslösungen betrug bei $\lambda = 440$ nm 0,15. Angeregt wurde die Fluoreszenz bei $\lambda = 453$ nm und die Berechnungen der Verhältnisse $F(\lambda_2)/F(\lambda_1) = F_{668}/F_{653}$ erfolgten wie beschrieben (s. Abb. 4). Die Werte für $k_a$ und $k_b$ aus Gleichung (6) für Gemische von Chl a und Chl b hängen vom verwendeten Lösungsmittel ab und sind in Tabelle 1 dargestellt. Außerdem wurden für die Kurven 2 aus Abb. 3 und 4 die Korrelationskoeffizienten errechnet (Tabelle 1).

Die Anwendung der dargelegten Methode ist in den Fällen von Nutzen, wo ein relativ geringer Chl b-Anteil auftritt, wie z. B. bei Untersuchungen der Ergrün-

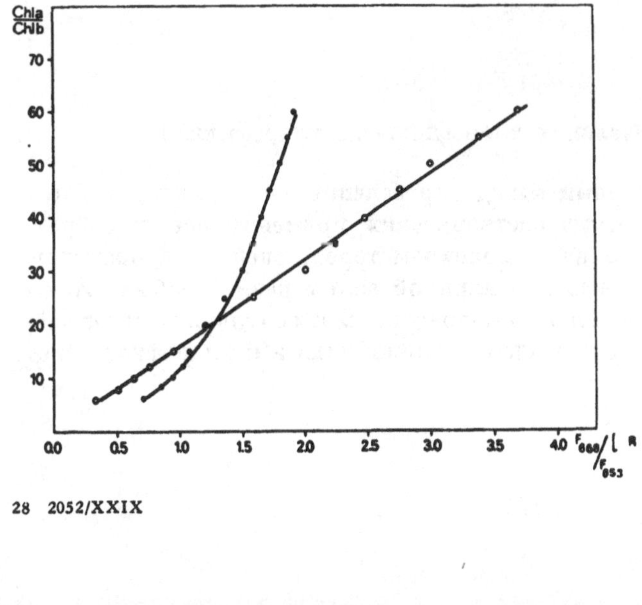


Abb. 4 Zusammenhang zwischen den Verhältnissen Chl a/Chl b und den Fluoreszenzamplitudenverhältnissen für Gemische in Azeton. (1) – (●——●) Verhältnis $F_{669}/F_{653}$ als Abszisse; (2) – (○— —○) korrigiertes Verhältnis R als Abszisse

Tabelle 1

Die Berechnungskoeffizienten $k_a$ und $k_b$ bei Verwendung verschiedener Lösungsmittel

| Lösungsmittel | Koeffizient | | | |
|---|---|---|---|---|
| | $k_a$ | $k_b$ | $F_{b1}/F_{a2}$ | r |
| Äther | 0,25 | 0,27 | 21,70 | 0,993 |
| Äthanol | 0,28 | 0,50 | 14,30 | 0,979 |
| Azeton (100%) | 0,31 | 0,44 | 16,00 | 0,998 |
| Azeton (80%) | 0,38 | 0,44 | 8,33 | 0,973 |

nung von etiolierten Pflanzen. So erlaubt diese Methode, das Vorliegen von Chl b schon in frühen Stadien seiner Synthese festzustellen (THORNE und BOARDMAN 1971; SCHLYK et al. 1970). Für Untersuchungen an Chl b-armen oder -freien Mutanten (MEISTER 1973), bei denen die Chl b-Konzentration im Verhältnis zum Gehalt an Chl a bestimmt werden soll, ist die Fluoreszenzmethode ebenfalls wichtig.

Herrn Dr. A. MEISTER danke ich für die kritischen Hinweise und ständige Unterstützung bei der Durchführung dieser Arbeit.

## Summary

On the fluorimetric estimation of chlorophyll a and chlorophyll b.

The determination of chl a and chl b in mixtures using different solvents was carried out by a fluorimetric method. The used fluorimeter is based on two monochromators and was constructed in our laboratory. The excitation wave length of fluorescence emission was 455 nm. The measurements, which were accomplished at room temperature, reveal the possibility to determine the chl a/chl b ratio up to 60–70.

## Краткое содержание

К флуорометрическому определению хлорофилла a и хлорофилла b

Рассматривается флуоресцентный метод определения хлорофиллов a и b в смеси, в различных органических растворителях, применительно к, собранному в лаборатории на базе двух монохроматоров, спектрофлуорометру. Флуоресценция эмисии возбуждалась длинной волны равной 455 нм. Измерения производились при комнатной температуре. В этих условиях метод даёт достоверные результаты при соотношении хлорофилла a к хлорофиллу b от 60 до 70.

## Literatur

BOARDMAN, N. K. and S. W. THORNE, 1971: Sensitive fluorescence method for the determination of chlorophyll a/chlorophyll b ratios. – Biochim. Biophys. Acta **253**, 222–231.

COMAR, C. L. and F. P. ZSCHEILE, 1942: Analysis of plant extracts for chlorophylls a and b by a photoelectric spectrophotometric method. – Plant Physiol. **17**, 198–209.

DE GREEF, J. A., R. CAUBERGS and R. MERTENS, 1975: Een zeer gevoelige spectrofotofluorometrische methode voor de detectie van kleine concentraties chlorophyl b in aanwezigheid hoge concentraties aan chlorophyl a. – Arch. int. physiol. et biochim. **83**, 955–957.

MEISTER, A., 1973: Untersuchungen zur Zusammensetzung der Blattpigmente an Mutanten von *Lycopersicon esculentum, Lycopericon pimpinellifolium* und *Antirrhinum majus*. – Kulturpflanze **21**, 295–311.

OGAWA, T. and K. SHIBATA, 1965: A sensitive method for determining chlorophyll b in plant extracts. – Photochem. Photobiol. **4**, 193–200.

SAPOZHNIKOV, D. I., 1964: Pigmenty Plastid zelenych Rastenij i Metodika ich Issledovanija. – Nauka, Moskau.

SCHLYK, A. A., 1968: O spektrofotometricheskom opredelenii chlorofillov a i b. – Biochimija **33**, 275–285.

–, A. B. RUDOI and A. YU. VEZITSKII, 1970: Immediate appearance and accumulation of chlorophyll b after short illumination of etiolated maize seedlings. – Photosynthetica 4, 68–77.

THORNE, S. W. and N. K. BOARDMAN, 1971: Formation of chlorophyll b, and the fluorescence properties and photochemical activities of isolated plastids from greening pea seedlings. – Plant Physiol. **47**, 252–261.

WETTSTEIN, D. VON, 1957: Chlorophyll-Letale und der submikroskopische Formwechsel der Plastiden. – Exp. Cell Res. **12**, 427–506.

ZENKEVICH, E. I. und A. P. LOSEV, 1970: Lumineszentnyi metod opredelenija chlorofillov a i b v ich smesi. – Z. prikl. spektr. **13**, 1032–1038.

ZIEGLER, R. und K. EGLE, 1965: Zur quantitativen Analyse der Chloroplastenpigmente. I. Kritische Überprüfung der spektral-photometrischen Chlorophyll-Bestimmung. – Beitr. Biol. Pflanzen **41**, 11–37.

Dipl.-Phys. N. JÄKEL
Zentralinstitut für Genetik und Kulturpflanzenforschung
der AdW der DDR,
DDR - 4325 Gatersleben

Kulturpflanze XXIX · 1981 · S. 425–432

# Zum Ausmaß der Semisterilität bei translokations-heterozygoter Gerste

GOTTFRIED KÜNZEL und HERMANN BRETSCHNEIDER

(Eingegangen am 19. Januar 1981)

## Zusammenfassung

Für 132 verschiedene einfache Translokationen, die auf drei Ausgangsformen zurückgehen, wurde in z. T. zweijährigen Untersuchungen die Semisterilität bei Strukturheterozygotie (gemessen am Kornansatz) ermittelt. Unterschiede im Kornansatz ergaben sich in Abhängigkeit vom Anbaujahr, von der vorliegenden Translokation und von der Ausgangsform. Als Ursachen der beobachteten Variabilität werden Umweltwirkungen sowie cytologische und genetische Faktoren diskutiert.

## Einleitung

Reziproke Translokationen haben sich als effektive Hilfsmittel für die Lösung einer Vielzahl spezieller cytogenetischer Probleme erwiesen, darunter insbesondere für genetische Merkmalsanalysen. Dabei wird in vielen Fällen auf die aufwendigen cytologischen Kontrolluntersuchungen in der Meiose verzichtet und die Semisterilität strukturheterozygoter Individuen als phänotypischer Marker genutzt (BURNHAM 1966). Die aus theoretischen Erwägungen für einfache heterozygote Translokationen zu erwartende Semisterilität ist jedoch im Pflanzenreich nur bei bestimmten Arten tatsächlich realisiert, bei Kulturpflanzen u. a. bei Mais, Reis, Erbse und Ackerbohne (BURNHAM 1956; SJÖDIN 1971). In anderen Fällen ist die Sterilität geringer; für Gerste wird eine Fertilitätsminderung zwischen 30 und 40% angegeben.

Während zum Meioseverhalten heterozygoter Translokationen umfangreiche Untersuchungen vorliegen, beziehen sich Angaben zur Fertilität überwiegend auf nur wenige Translokationslinien und auf relativ summarische experimentelle Daten. Unter Einbeziehung einer größeren Anzahl einfacher Translokationslinien wird daher in der vorliegenden Arbeit untersucht, in welchem Ausmaß die durch Strukturheterozygotie hervorgerufene Fertilitätsreduktion bei Gerste variieren kann.

## Material und Methoden

In die Untersuchungen einbezogen wurden 132 einfache Translokationslinien von Sommergerste (*Hordeum vulgare* L. s. l.), die auf die Sorten 'Elgina', 'Frigga' und einen langstrohigen, zweizeiligen, lockerährigen, der Sorte 'Elgina' sehr ähnlichen Stamm ('Hd. St.') zurückgehen. Alle Translokationen wurden durch Röntgenbestrahlung von Saatgut induziert, nach den bei KÜNZEL und SCHOLZ (1972) sowie KÜNZEL (1976) angegebenen Verfahren isoliert und hinsichtlich der beteiligten Chromosomen definiert.

Zur Erzeugung und Erhaltung strukturheterozygoter Pflanzen erfolgten einmalige Kreuzungen mit der jeweiligen Ausgangsform, wobei für den Nachbau ab $F_2$ auf die zur Hälfte ausspaltenden partiell-sterilen (strukturheterozygoten) Individuen zurückgegriffen wurde. Der Anbau solcher 1:1 spaltenden Populationen sowie der als Kontrollen dienenden Ausgangsformen erfolgte in Freilandparzellen, 1975 mit etwa 70 Pflanzen, 1976 mit rund 100 Pflanzen.

Der Kornansatz wurde an 30–50 Ähren von 10–15 partiell-sterilen Pflanzen aus den spaltenden Populationen bzw. an 100 Ähren der Kontrollvarianten bestimmt, indem die Anzahl voll entwickelter Karyopsen zur Gesamtzahl der Ährchen ins Verhältnis gesetzt wurde. Die verkümmerten Ährchen an der Ährenbasis und -spitze blieben dabei unberücksichtigt.

Zur statistischen Auswertung der Ergebnisse diente der *T*-Test sowie der *U*-Test nach MANN-WHITNEY. Die Irrtumswahrscheinlichkeit wurde mit $\alpha = 5\%$ vorgegeben.

## Ergebnisse

Im Mittel der im Anbaujahr 1975 geprüften 132 heterozygoten Translokationen beträgt der Kornansatz 65,1% und bei den als Kontrollen dienenden Ausgangsformen 97,0%. Die Unterschiede zwischen den auf verschiedene Ausgangsformen zurückgehenden Translokationslinien sind nicht sehr groß; sie betragen für die Mittelwerte weniger als 10% und liegen sowohl für die Minimal- als auch für die

Tabelle 1

Kornansatz pro Ähre (%) heterozygoter Translokationen, Anbau 1975

| Ausgangsform | Anzahl Translokationen | Kornansatz heterozygote Translokat. $\bar{x}$ | $x_{min}$ | $x_{max}$ | Ausgangsform |
|---|---|---|---|---|---|
| 'Elgina' | 42 | 69,5 | 59,7 | 77,4 | 96,1 |
| 'Frigga' | 50 | 61,5 | 52,0 | 69,4 | 97,0 |
| 'Hd. St.' | 40 | 65,1 | 54,6 | 73,5 | 97,8 |
| insges. | 132 | 65,1 | 52,0 | 77,4 | 97,0 |

Tabelle 2

Häufigkeitsverteilung der Kornansatzwerte (%), Anbau 1975

| Ausgangsform | Anzahl Translokationen | 50 | 55 | 60 | 65 | 70 | 75 |
|---|---|---|---|---|---|---|---|
| 'Elgina' | 42 | | 1 | 1 | 10 | 18 | 12 |
| 'Frigga' | 50 | 1 | 9 | 18 | 17 | 5 | |
| 'Hd. St.' | 40 | | 1 | 8 | 18 | 11 | 2 |

Tabelle 3

Zweijähriger Vergleich des Kornansatzes pro Ähre (%) bei ausgewählten heterozygoten Translokationen mit geringerem (I) und höherem (II) Kornansatz im Anbaujahr 1975

| Ausgangsform / Gruppe | | Anzahl Translokationen | 1975 | 1976 | Ø1975/1976 |
|---|---|---|---|---|---|
| 'Elgina' | | | | | |
| I | | 6 | 62,8 | 65,7 | 64,6 |
| II | | 7 | 75,0 | 72,7 | 73,6 |
| | insges. | 13 | 69,4 | 69,4 | 69,4 |
| 'Frigga' | | | | | |
| I | | 5 | 54,2 | 61,4 | 58,6 |
| II | | 7 | 68,4 | 65,7 | 66,7 |
| | insges. | 12 | 62,5 | 63,9 | 63,3 |
| 'Hd. St.' | | | | | |
| I | | 6 | 60,1 | 59,8 | 59,9 |
| II | | 7 | 70,0 | 66,8 | 68,1 |
| | insges. | 13 | 65,4 | 63,6 | 64,3 |

Maximalwerte in etwa der gleichen Größenordnung, wobei die Differenzen zwischen Minimal- und Maximalwerten unabhängig von der Ausgangsform gleichermaßen bei etwa 18% liegen. Lediglich die Linien der Sorte 'Elgina' zeigen im Durchschnitt eine etwas höhere Fertilität (Tab. 1). Dieser Sachverhalt wird durch

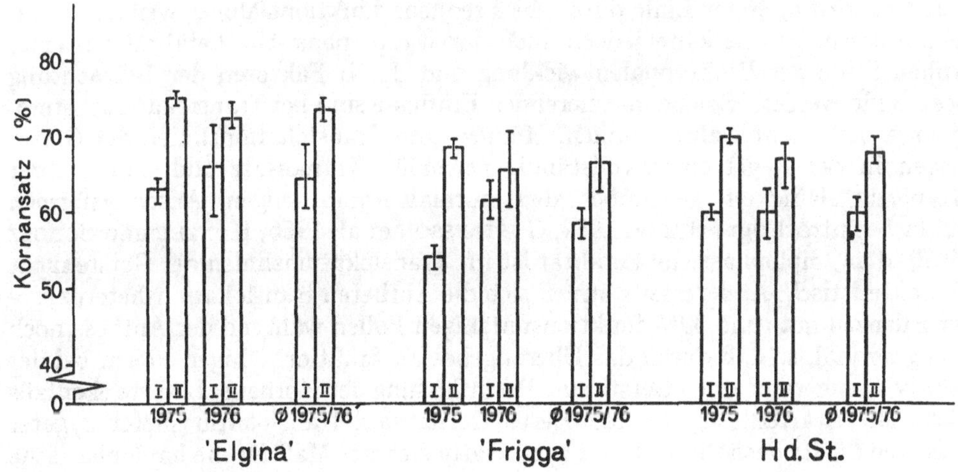


Abb. 1 Zweijähriger Vergleich des Kornansatzes pro Ähre (%) bei ausgewählten heterozygoten Translokationen mit geringerem (I) und höherem (II) Kornansatz im Anbaujahr 1975

Konfidenzintervall für $\alpha = 5\%$

die in Tab. 2 angegebene Häufigkeitsverteilung der Kornansatzwerte noch deutlicher.

Für den Wiederholungsanbau 1976 erfolgte auf der Grundlage der 1975 erzielten Ergebnisse eine Einschränkung auf solche Translokationslinien, bei denen die partielle Sterilität der heterozygoten Pflanzen am stärksten von den Mittelwerten abwich. Auf diese Weise wurde für jede Ausgangsform eine Gruppe von Translokationen mit geringerem (I) und höherem (II) Kornansatz gebildet. Die Differenzen im Kornansatz zwischen diesen willkürlich gebildeten Gruppen lagen 1975 zwischen 10 und 14%; sie bestätigten sich im Wiederholungsanbau 1976, sanken jedoch auf Werte zwischen 4 und 7% (Tab. 3, Abb. 1).

## Diskussion

Bei Sterilitätsphänomenen haplophasischer Natur, die sich gleichermaßen auf weibliche und männliche Gameten beziehen, wie im Falle von Translokationsheterozygotie, wird die Fertilitätsreduktion häufig über Pollenletalitätsbestimmungen mittels Färbeverfahren charakterisiert. Die Färbbarkeit des Pollens ist jedoch nicht immer ein zuverlässiges Kriterium für seine Funktionsfähigkeit. So sind z. B. bei Translokationen mit relativ kurzen ausgetauschten Segmenten die funktionsunfähigen Pollen infolge relativ geringer genetischer Imbalancen noch voll anfärbbar, wie orientierende eigene Untersuchungen ergaben. Sind andererseits die Differenzen zwischen normalen und letalen Pollenkörnern groß, treten in den Farblösungen starke Entmischungserscheinungen ein, die die Erfassung repräsentativer Stichproben sehr erschweren.

Technisch einfacher läßt sich der Kornansatz ermitteln. Die Höhe des Kornansatzes wird in erster Linie durch die Frequenz funktionsfähiger weiblicher Gameten bestimmt; sie kann jedoch auch durch diplophasische Letalität auf einer frühen Stufe der Embryonalentwicklung und durch Faktoren der Befruchtung beeinflußt werden. Solche maskierenden Einflüsse sind bei Translokationsheterozygotie nicht sehr wahrscheinlich. Homozygote Translokationslinien der Gerste zeigen in der Regel einen vollständig normalen Kornansatz und sind in ihrer Kornertragsfähigkeit gegenüber den normalchromosomigen Ausgangsformen kaum beeinträchtigt (Nybom 1954, Gustafsson et al. 1966, Künzel und Scholz 1978), d. h., diplophasische Letalität ist für Translokationslinien der Gerste nicht charakteristisch. Andererseits öffnen sich die Antheren translokationsheterozygoter Pflanzen mit rund 30% funktionsunfähigen Pollen während der Anthese noch völlig normal, so daß durch das Überangebot an funktionsfähigen Pollen bei der relativ streng autogamen Gerste eine Beeinflussung der Kornansatzwerte ebenfalls nicht zu erwarten ist. Der reduzierte Kornansatz translokationsheterozygoter Pflanzen dürfte deshalb bei Gerste ein relativ sicheres Maß für die haplophasische Letalität darstellen.

Die für heterozygote Translokationen charakteristische Semisterilität wird in erster Linie durch die Verteilung der an den multivalenten Paarungsverbänden beteiligten Chromosomen in der Meiose I bestimmt. Ohne Brücksichtigung der bei Gerste ohnehin seltenen Austauschvorgänge in den interstitiellen Segmenten (Abschnitte zwischen Zentromer und Translokationspunkt) können funktionsfähige Gameten mit einem kompletten Genom nur nach alternativen Verteilungen entstehen, da translozierte und normale Chromosomen in den Multivalenten alternieren. Adjazente Verteilungen, bei denen die im Multivalent benachbart angeordneten Chromosomen einem Zellpol zugeteilt werden, führen hingegen zu Genomen mit partiellen Duplikationen und Defizienzen und damit in der Regel zu funktionsunfähigen Gameten. Welche Verteilungen realisiert werden, hängt von der Koorientierung der Zentromere der am Multivalent beteiligten Chromosomen ab. Erfolgt diese zufallsgemäß, sind für einfache Translokationen alternative und adjazente Verteilungen mit gleicher Häufigkeit zu erwarten und nicht im Verhältnis 1 : 2, wie manchmal auch in jüngerer Literatur noch irrtümlich angegeben wird (z. B. DENNHÖFER 1975). In vielen Fällen, darunter auch bei Gerste, erfolgt die Zentromerkoorientierung jedoch nicht zufällig, sondern in dem Sinne gerichtet, daß bevorzugt alternative Verteilungen eintreten. Die der Regulation der Zentromerkoorientierung zugrundeliegenden Prinzipien sind bisher noch weitgehend ungeklärt. Sicher ist jedoch, daß hier ein komplexer Prozeß vorliegt, bei dem zahlreiche chromosomenmorphologisch, genetisch und modifikativ bedingte Einzelfaktoren in Wechselwirkung stehen und letzten Endes zu der jeweils für bestimmte Spezies bzw. für die konkret vorliegende Translokation charakteristischen Relation zwischen alternativen und adjazenten Verteilungen führen.

Bei Gerste werden gewöhnlich 60–70% der Quadrivalente mit alternativer Zentromerkoorientierung angetroffen (u. a. SMITH 1941, MOH und SMITH 1952, NILAN 1964, GRAMATIKOVA 1979), woraus sich ergibt, daß die Fertilitätsreduktion zwischen 30 und 40% liegen sollte. Die bisherigen Befunde, die insgesamt betrachtet auf eine durchschnittliche Fertilitätsminderung von rund 30 bis 40% hinweisen, zeigen eine recht gute Übereinstimmung mit dem bereits genannten Erwartungswert (SMITH 1941, HANSON und KRAMER 1949, BURNHAM et al. 1954, SHIH und SHEBESKI 1960, EKBERG 1969, GRZYWA und NALEPA 1975, YU und HOCKETT 1979). Unsere Ergebnisse, die für die 1975 geprüften 132 heterozygoten Translokationen einen mittleren Kornansatz von 65% ergaben, fügen sich ebenfalls recht gut in dieses Bild ein.

Eine mögliche Einflußgröße auf die Fertilität der strukturheterozygoten Pflanzen ist der konkret vorliegende Strukturumbau selbst. In diesem Zusammenhang werden insbesondere unterschiedliche Bruchpositionen, die die Flexibilität der Multivalente und damit die Zentromerkoorientierung beeinflussen, als Variationsursache diskutiert (BURNHAM 1962, SIMONSEN 1975, COCHRAN 1976). Ist z. B. ein ausgetauschtes Segment kurz, so resultieren vorrangig kettenförmige Quadrivalente, deren Chromosomen stärker zu alternativen Verteilungen neigen als solche aus ringförmigen Quadrivalenten (ÖSTERGREN 1951, BURNHAM 1956, SJÖDIN 1971). Unabhängig von den Ursachen im einzelnen, sollte sich die Wirksamkeit derartiger Faktoren in Fertilitätsdifferenzen zwischen verschiedenen heterozygoten Translokationen äußern, die auf eine gemeinsame Ausgangssorte zurückgehen. Hinweise auf solche Differenzen wurden bereits von BURNHAM et al. (1954) sowie von SHIH und SHEBESKI (1960) gegeben.

Unsere Befunde zeigen ebenfalls gewisse Differenzen im Kornansatz. Bei getrennter Betrachtung der Translokationen einer jeden Ausgangsform betrugen 1975 die maximalen Differenzen in allen drei Fällen im Kornansatz gleichermaßen etwa 18% (Tab. 1). Im Wiederholungsanbau 1976 waren die Differenzen zwischen den Gruppen von Translokationen mit geringerem und höherem Kornansatz jedoch wesentlich geringer (Tab. 3, Abb. 1). Diese Befunde zeigen zunächst den begrenzten Wert einjähriger Ergebnisse und weisen auf eine starke, vom Anbaujahr abhängige Umweltlabilität dieses Merkmals hin. Zwischen beiden Prüfungsjahren können die Differenzen im Kornansatz für einzelne Translokationen bis zu 13% betragen. Insgesamt kann abgeschätzt werden, daß bei dem geprüften Material echte, auf den vorliegenden Strukturumbau zurückgehende Unterschiede im Kornansatz etwa 10% kaum überschreiten dürften. Das schließt jedoch nicht aus, daß in seltenen Einzelfällen dennoch stärkere Abweichungen vorliegen können. Da bei unseren Mutationsexperimenten zur Vorselektion in allen Fällen zunächst ein Screening auf partiell-sterile (translokationsheterozygote) $M_2$-Pflanzen durchgeführt wurde, besteht die Möglichkeit, daß heterozygote Translokationen mit besonders geringer Fertilitätsreduktion nicht erfaßt worden sind.

Eine weitere Einflußgröße auf die Fertilität stellen genetische Faktoren mit Kontrollfunktionen für die Zentromerkoorientierung dar, für deren Wirkung zahlreiche Hinweise bestehen (u. a. LAWRENCE 1958, SIMONSEN 1975, SYBENGA 1975, COCHRAN 1976, LACADENA und CANDELA 1977, LELLEY und GUSTAFSON 1979). Diskutiert werden in diesem Zusammenhang eine Reihe polygen determinierter Merkmale, wie etwa Chiasmapositionen, -frequenz und -terminalisation (BURNHAM 1956, RICKARDS 1964, REES und SUN 1965, SUN und REES 1967, NARANJO und LACADENA 1979), die über Flexibilitätsänderungen der Multivalente auf die der endgültigen Chromosomenverteilung vorangehenden Ko- und Reorientierungsprozesse der Zentromere Einfluß nehmen (SYBENGA 1968, 1975) und auf diese Weise die Relation zwischen alternativen und adjazenten Chromosomenverteilungen verschieben können.

Auf die Wirksamkeit derartiger genetischer Faktoren kann z. B. geschlossen werden, wenn im Durchschnitt für eine größere Anzahl verschiedener Translokationen Fertilitätsunterschiede in Abhängigkeit von der Ausgangsform vorliegen. Unsere Ergebnisse zeigen, daß insbesondere die Translokationen der Sorte 'Elgina' sowohl 1975 (Tab. 1 und 2) als auch bei dem Vergleich der zweijährigen Ergebnisse (Tab. 3) im Durchschnitt einen höheren Kornansatz zeigen als die Translokationen der Ausgangsformen 'Frigga' und 'Hd. St.', wobei ein *U*-Test diese Unterschiede als signifikant ausweist. Diese nicht sehr großen, aber zweifellos echten Differenzen in Abhängigkeit von der Ausgangsform können am zwanglosesten als Hinweis auf die Wirkung unterschiedlicher genetischer Faktoren gedeutet werden.

Insgesamt zeigen unsere Ergebnisse, daß der Kornansatz heterozygoter Translokationen offensichtlich durch komplexe Interaktionen zahlreicher Einzelfaktoren bestimmt wird, an denen relativ große Umweltwirkungen, aber auch chromosomenmorphologische und genetische Faktoren beteiligt sind. Trotzdem erweist sich die im Kornansatz widerspiegelnde Fertilitätsreduktion translokationsheterozygoter Individuen bei Gerste als ein relativ sicherer phänotypischer Marker.

## Summary

Extent of semisterility in barley heterozygous for translocations

For 132 diverse single translocations originating from three varieties, semisterility (measured as seed set) in case of structural heterozygosity was ascertained, partly for two years. Differences in seed set were found depending on the year of cultivation, on the individual translocation and on the original variety. Influences of environment as well as cytological and genetic factors are being discussed as causes of the observed variability.

## Краткое содержание

О размерах снижения фертильности гетерозигот по транслокациям у ячменя

У 132 гетерозиготных простых транслокаций, происходящих от трёх исходных форм, регистрировалось (частично в трёхлетних наблюдениях) образование зёрен. Различия в числе образовавшихся зёрен оказались в зависимости от года возделования, от определённой транслокации и от исходной формы. Как причины, наблюдавшейся изменчивости, обсуждаются влияние среды, а также цитологические и генетические факторы.

## Literatur

Burnham, C. R., 1956: Chromosomal interchanges in plants. – Bot. Review **22**, 419–552.

–, 1962: Discussions in Cytogenetics. – Minneapolis.

–, 1966: Cytogenetics in plant improvement. – In: Plant Breeding Symp. at Iowa State University (Edit. K. J. Frey), Ames, 139–187.

–, F. H. White and R. Livers, 1954: Chromosomal interchanges in barley. – Cytologia **19**, 191–202.

Cochran, D. G., 1976: Disjunction types and their frequencies in two heterozygous reciprocal translocations of *Blattella germanica* (L.). – Chromosoma **59**, 129–135.

Dennhöfer, L., 1975: Inherited preferential segregation in translocation heterozygotes of the mosquito. – Theor. Appl. Genet. **45**, 250–253.

Ekberg, I., 1969: Different types of sterility induced in barley by ionizing radiations and chemical mutagens. – Hereditas **63**, 257–278.

Gramatikova, M., 1979: Genetische und cytologische Analyse gammastrahleninduzierter partieller Sterilität bei Wintergerste (bulg.). – Genetika i Selekcija (Sofia) **12**, 92–98.

Grzywa, M. and S. Nalepa, 1975: Investigation on the possibility of using translocation heterozygotes in the $F_2$ for the localization of genes which cause dwarfness in barley (poln.). – Biul. Inst. Hod. Aklimat. Roślin No. **124/125**, 33–36.

Gustafsson, Å., U. Lundqvist and I. Ekberg, 1966: Yield reactions and rates of origin of chromosome rearrangements and gene mutations in barley. – Hereditas **56**, 200–206.

Hanson, W. D. and H. H. Kramer, 1949: The genetic analysis of two chromosome interchanges in barley from $F_2$ data. – Genetics **34**, 687–700.

Künzel, G., 1976: Indications for a necessary revision of the barley karyogramme by use of translocations. – In: Barley Genetics III (Edit. H. Gaul), Proc. Third Intern. Barley Genet. Symp., Garching 1975, Verl. Thiemig, München, 1976, 275–281.

–, und F. SCHOLZ, 1972: Über Nutzungsmöglichkeiten von induzierten Chromosomenmutationen bei Kulturpflanzen. – Kulturpflanze **20**, 225–262.

–, –, 1978: Development and utilization of multiple translocations in barley. – In: Experimental Mutagenesis in Plants, Proc. Intern. Symp. Varna 1976, Sofia, 1978, 112–118.

LACADENA, J. R. and M. CANDELA, 1977: Centromere co-orientation at metaphase I in interchange heterozygotes of rye, *Secale cereale* L. – Chromosoma **64**, 175–189.

LAWRENCE, C. W., 1958: Genotypic control of chromosome behaviour in rye. IV. Selection for disjunction frequency. – J. Heredity **12**, 127–131.

LELLEY, T. and J. P. GUSTAFSON, 1979: Genotypically controlled alternate distribution of translocated chromosomes in the rye cultivar "Snoopy". – Z. Pflanzenzüchtg. **82**, 306–310.

MOH, C. and L. SMITH, 1952: Three coincidental changes in atombombed barley: a line of barley with all surviving plants heterozygous for an interchange and a gene for white seedlings. – J. Heredity **43**, 183–188.

NARANJO, T. and J. R. LACADENA, 1979: Analysis of centromere co-orientation in a rye-wheat derivative by means of C-banding. – Chromosoma **73**, 227–235.

NILAN, R. A., 1964: The Cytology and Genetics of Barley 1951–1962. – Monographic Supplement No. **3**, Research Studies, Washington State University, Pullman, pp. 278.

NYBOM, N., 1954: Karyotype and viability in barley. – Acta Agric. Scand. **4**, 507–514.

ÖSTERGREN, G., 1951: The mechanism of co-orientation in bivalents and multivalents. – Hereditas **37**, 85–156.

REES, H. and S. SUN, 1965: Chiasma frequency and the disjunction of interchange associations in rye. – Chromosoma **16**, 500–510.

RICKARDS, G. K., 1964: Some theoretical aspects of selective segregation in interchange complexes. – Chromosoma **15**, 140–155.

SHIH, C. and L. H. SHEBESKI, 1960: Sterility in interchange heterozygotes of barley. – Canad. J. Genet. Cytol. **2**, 429–434.

SIMONSEN, Ø., 1975: Cytogenetic investigations in diploid and autotetraploid populations of *Festuca pratensis*. – Hereditas **79**, 73–108.

SJÖDIN, J., 1971: Induced translocations in *Vicia faba* L. – Hereditas **68**, 1–33.

SMITH, L., 1941: An inversion, a reciprocal translocation, trisomics, and tetraploids in barley. – J. Agric. Res. **63**, 741–750.

SUN, S. and H. REES, 1967: Genotypic control of chromosome behaviour in rye. IX. The effect of selection on the disjunction frequency of interchange associations. – J. Heredity **22**, 249–254.

SYBENGA, J., 1968: Orientation of interchange multiples in *Secale cereale*. – J. Heredity **23**, 73–79.

–, 1975: Meiotic Configurations. – Monographs on Theoretical and Applied Genetics I (Edit. R. FRANKEL, M. GROSSMAN, H. F. LINSKENS, D. DE ZEEUW), Verl. Springer, Berlin, Heidelberg, New York, pp. 252.

YU, C. W. and E. A. HOCKETT, 1979: Chromosome behaviour, breeding characteristics and seed set of partially sterile barley, *Hordeum vulgare* L. – Z. Pflanzenzüchtg. **82**, 133–148.

Dr. G. KÜNZEL und Dr. H. BRETSCHNEIDER
Zentralinstitut für Genetik und
Kulturpflanzenforschung
der Akademie der Wissenschaften der DDR
4325 Gatersleben

Kulturpflanze XXIX · 1981 · S. 433—442

# Report of travels to South Italy 1980 for the collection of indigenous material of cultivated plants

PIETRO PERRINO, KARL HAMMER and PETER HANELT

(Eingegangen am 13. Januar 1981)

## Summary

Chronology and results of two collecting missions to South Italy have been shortly described which have been undertaken jointly by the Istituto del Germoplasma, Bari, and the Zentralinstitut für Genetik und Kulturpflanzenforschung, Gatersleben, in early and late summer 1980 in order to study indigenous material of cultivated plants especially of grain legumes. 461 seed samples have been collected among them besides some relic cultigens as einkorn and emmer many variable land-races e. g. of *Vicia faba*, *Phaseolus vulgaris*, and a broad spectrum of local vegetable varieties. Gene erosion has reached in South Italy different levels being dependent on crop species and geographical region.

## Introduction

Within the frame of the official agreement between the Consiglio Nazionale delle Ricerche (C.N.R.) of Italy and the Academy of Sciences of the GDR first joint activities have been developped by the Istituto del Germoplasma in Bari and the Zentralinstitut für Genetik und Kulturpflanzenforschung at Gatersleben in order to study and to collect indigenous material of cultivated plants and related wild species in different parts of South Italy. The Mediterranean region is one of the classic gene centres of the world and a Mediterranean Programme has been initiated by the International Board for Plant Genetic Resources to collect indigenous material in this area (cf. LEHMANN and HAMMER 1979). Being a part of this gene centre Italy is characterized by a considerable variability of its cultivated plants, especially the cereals, grain legumes, vegetable and fruit crops and also by the occurence of many more or less closely related wild taxa of the cultigens. Gene erosion within the last decades has resulted in a decrease of the cultivation of indigenous races in Italy too; therefore there is an urgent necessity for the collection of this local material although for cereals especially wheat this work has largely been done already by the Bari institute (see PORCEDDU 1976, 1979a, 1979b, PORCEDDU and BENNETT 1971, etc.). Thus the collection during the two joint trips in early and late summer 1980 has been concentrated on grain legumes, vegetables and other hitherto neglected crops.

The travels have been organized by the Bari institute, participants have been

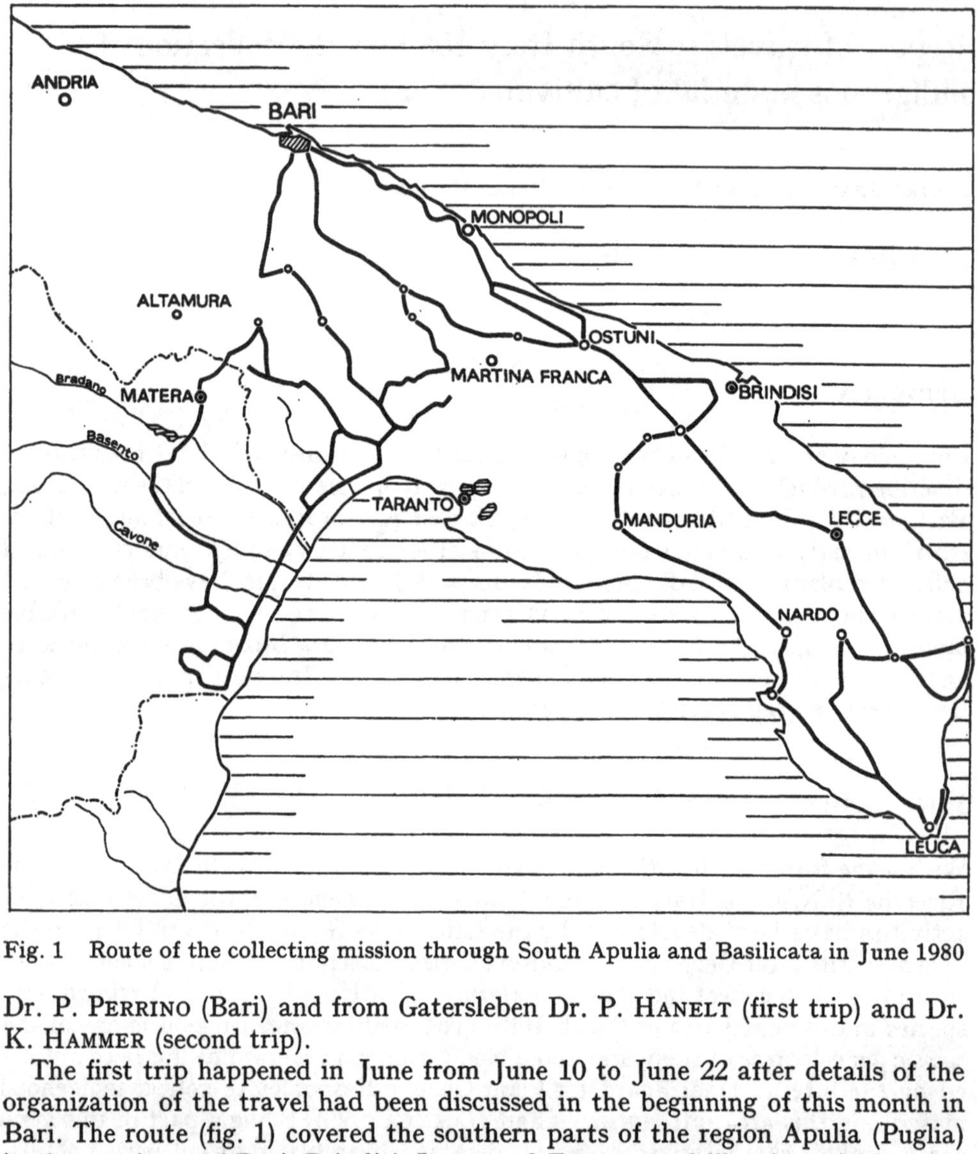


Fig. 1 Route of the collecting mission through South Apulia and Basilicata in June 1980

Dr. P. Perrino (Bari) and from Gatersleben Dr. P. Hanelt (first trip) and Dr. K. Hammer (second trip).

The first trip happened in June from June 10 to June 22 after details of the organization of the travel had been discussed in the beginning of this month in Bari. The route (fig. 1) covered the southern parts of the region Apulia (Puglia) in the provinces of Bari, Brindisi, Lecce and Taranto and likewise neighbouring eastern and north-eastern districts of the region Basilicata. The trip went through all landscape types from the coastal region, the Murge-plateau to the marginal parts of the Apennine mountains. In the following a detailed itinerary will be given:

10. 6. Bari–Casamassima–Turi–Putignano–Alberobello–Cisternino–Ostuni–S. Vito dei Normanni–Brindisi–Mesagne–Lecce; collecting mainly in the region of Cisternino and Mesagne

11. 6. Lecce–Maglie–Otranto–S. Cesarea Terme; visit to the experimental fields of the Istituto Tecnico Agrario "G. Presto" in Lecce, collecting in the region of Maglie and south of Otranto

12. 6. S. Cesarea Terme–Poggiardo–Maglie–Cutrofiano–Galatina–Collepasso–Casarano –Taurisano–S. Maria di Leuca–Leuca; collecting north of S. Cesarea Terme and near Casarano and Galatina
13. 6. Leuca–Gagliano del Capo–Leuca–Ugento–Gallipoli–Conchiglie near S. Maria al Bagno; collecting near Ugento, Gallipoli and between Gallipoli and Nardo
14. 6. S. Maria al Bagno–Nardo–Avetrana–Manduria–Oria–Latiano–Mesagne–S. Vito dei Normanni–Carovigno–Ostuni–Fasano–Bari; collecting in the region of Nardo, Avetrana, S. Vito dei Normanni and Carovigno
15. 6. Bari
16. 6. Bari–Sannicandro di Bari–Acquaviva delle Fonti–Giòia del Colle–Mòttola–Massafra–in direction of Martina Franca and back via Massafra in direction of Metaponto to Lido di Castellaneta; collecting in the region of Massafra and near Lido di Castellaneta
17. 6. Lido di Castellaneta–Metaponto–Policoro–Rotondella–Marina di Nova Siri; collecting mainly near Policoro and in the region of Rotondella
18. 6. Marina di Nova Siri–Policoro–Montalbano Ionico–valleys of Cavone and Basento–Ferrandina–Matera; collecting in Montalbano Ionico, in the valley of the Cavone river and near Matera
19. 6. Matera–Santeramo in Colle–Castellaneta–Lido di Castellaneta; collecting between Matera and Santeramo and near Castellaneta
20. 6. Lido di Castellaneta–Palagiano–Mòttola–direction of Martina Franca–Noci–Putignano–Casamassima–Bari; collecting mainly in the region of Mottola/Massafra and Noci
21. 6. Bari–Fasano–Ostuni–Carovigno–Cisternino–Locorotondo–Alberobello and back to Carovigno; collecting in the Valle d'Itria between Cisternino and Locorotondo
22. 6. Carovigno and back to Bari; collecting near Carovigno

The route of the second mission (fig. 2) went to the northeastern mountainous parts of the region Apulia (Puglia) (Gargano mountains) and to the Apennine mountains of northern Apulia and neighbouring parts of the region Campania from August 29 to September 5. Prior to this mission the organization of the travel has been discussed and some explorating has been done with regard to wild relatives of cultivated plants especially on the sea shore near Bari. The itinerary of the second mission will be given in the following:

29. 8. Bari–Foggia–San Marco in Lamis–Sannicandro–Rodi Garganico; collecting near San Marco in Lamis, Sannicandro, and Cagnano
30. 8. Rodi Garganico–Pèschici–Vieste–Mattinata–Monte S. Angelo; collecting near Pèschici, Vieste, and on the foot of the Monte Sacro
31. 8. Monte S. Angelo–Foresta Umbra–Carpino–S. Giovanni Rotondo; collecting in the surroundings of Mattinata, Monte S. Angelo, Foresta Umbra, and Carpino
1. 9. S. Giovanni Rotondo–Monte Calvo–S. Giovanni Rotondo–S. Marco in Lamis–San Severo–Lucera–S. Bartolomeo in Galdo; collecting especially north of the Monte Calvo, west of S. Marco in Lamis, south of Volturara Appula, and north S. Bartolomeo
2. 9. S. Bartolomeo–Castelveterè–S. Marco di Cavoti–Benevento; collecting in the surroundings of S. Bartolomeo and Castelveterè, north of S. Marco di Cavoti
3. 9. Benevento–S. Giorgio la Molara–Montefalcone V. Fortore–Castelfranco in Misc. –Faeto–Castellùcio V. Maggiore; collecting near S. Giorgio, Montefalcone, and Castelfranco
4. 9. Castellùcio V. Maggiore–Faeto–Castelfranco in Misc.–Miscana river–Castelfranco in Misc.–Vizzino–Savignano Irpino; collecting near Castelfranco
5. 9. Savignano–Savignano Irpino–Savignano Scalo–Greci–Savignano Scalo–Monteleone di Puglia–Accadia–Santa Agata–Candela–Foggia–Bari; collecting near Savignano, Greci, and Monteleone di Puglia

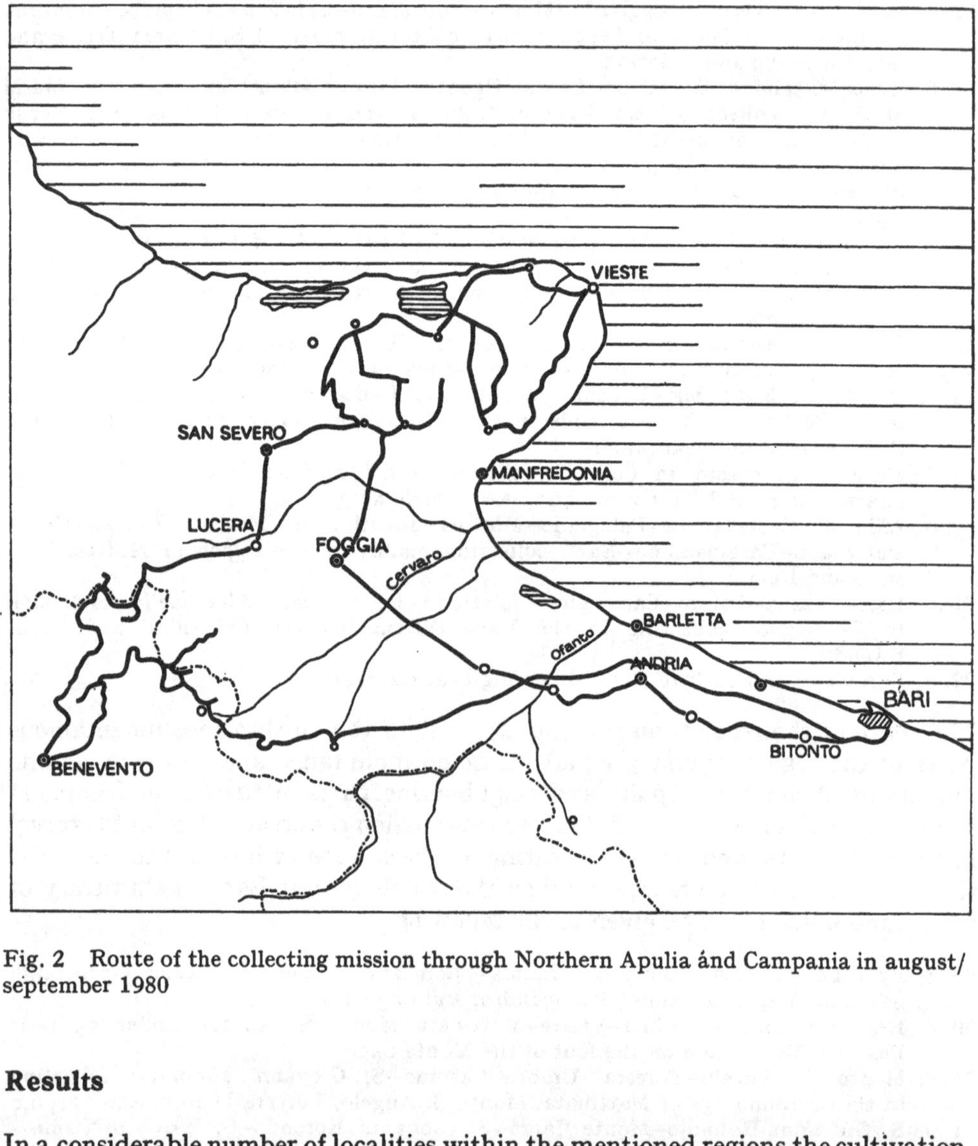


Fig. 2 Route of the collecting mission through Northern Apulia ánd Campania in august/september 1980

## Results

In a considerable number of localities within the mentioned regions the cultivation of local races of cultivated plants is still practised although heavy gene erosion and the loss of former indigenous plant material could be observed in some regions, e.g. in the coastal district of Metaponto, in parts of the Murge plateau, and in the region of Foggia where the agriculture is characterized by a high degree of mechanization and by the use of other modern agricultural methods. Gene erosion is also true for single crops of which nowadays mostly modern varieties are cultivated (e.g. *Triticum durum*) or of which the acreage has been decreased considerably during the last decades (e.g. some grain legumes especially *Lens culinaris*, *Lathyrus sativus*, *Lupinus albus* and some variants of *Cicer arietinum*). The collection of these species nowadays often cultivated only by farmers of the older gene-

Fig. 3 Typical intercropping in a house garden near Avetrana: olives (left) and other fruit trees with different vegetables, some plants of each of the latter crops being grown for the self-made seed production (left *Beta vulgaris*)

Fig. 4 A considerable part of the accessions has been collected from the seed storages of the farmers, as in this foto near Mattinata, Gargano mountains

ration should therefore be enhanced and extended to other districts of South Italy.

By the anomalous late and cool spring it was very often impossible to make field collections in June and one had to rely upon the seed storages of the farmers, the same was true during the late season when the harvest had been already finished almost completely.

The collections contain 461 different accessions nearly the half of them being leguminous plants. Besides them a very variable spectrum of local vegetable taxa could be gathered supposed to be of equal importance for further breeding work.

In the following a detailed list of the collected material will be given:

| Crop | number of accessions first trip | second trip | sum |
|---|---|---|---|
| faba bean, *Vicia faba* | 44 | 28 | 72 |
| chick pea, *Cicer arietinum* | 21 | 15 | 36 |
| common vetch, *Vicia sativa*, and other wild *Vicia* spp. | 19 | 2 | 21 |
| pea, *Pisum sativum* incl. wild type | 17 | 4 | 21 |
| *Phaseolus* beans, *Ph. vulgaris*, *Ph. coccineus*, and cow pea, *Vigna unguiculata* | 14 | 18 | 32 |
| lentils, *Lens culinaris*, and wild *Lens* spp. | 9 | 1 | 10 |
| vetchling, *Lathyrus sativus*, and wild *Lathyrus* spp. | 6 | 3 | 9 |
| white lupin, *Lupinus albus* | 7 | 6 | 13 |
| small-grain legumes (*Trifolium*, *Melilotus*, *Ononis*, *Trigonella*, *Onobrychis*) | 6 | 1 | 7 |
| legumes total sum | 143 | 78 | 221 |
| oat, *Avena* spp., incl. wild taxa | 9 | 5 | 14 |
| barley, *Hordeum vulgare* | 5 | 7 | 12 |
| wheat, *Triticum* spp. | 1 | 29 | 30 |
| maize, *Zea mays* | 2 | 10 | 12 |
| wild grass species (*Aegilops*, *Lolium*, *Briza*, *Haynaldia*) | 6 | 3 | 9 |
| cereals total sum | 23 | 54 | 77 |
| cabbage, mustard, *Brassica* spp., *Sinapis* spp., incl. wild taxa | 16 | 1 | 17 |
| further cruciferous vegetable species (*Eruca*, *Lepidium*, *Raphanus*) | 9 | 2 | 11 |
| salad, endive, chicory (*Lactuca sativa*, *Cichorium*, *Tragopogon*) incl. wild taxa | 25 | 14 | 39 |
| onion, *Allium cepa*, and other *Allium* spp. mostly wild ones | 13 | 9 | 22 |
| cucumber, cucurbits (*Cucumis*, *Cucurbita*, *Lagenaria*) | 8 | 4 | 12 |
| tomato, paprika, egg plant (*Lycopersicon*, *Capsicum*, *Solanum*) | 5 | 5 | 10 |
| carrot, celery, parsley (*Daucus*, *Apium*, *Petroselinum*) incl. wild taxa | 6 | 7 | 13 |

| Crop | number of accessions first trip | second trip | sum |
| --- | --- | --- | --- |
| beet, spinach (*Beta vulgaris, Spinacia oleracea*) incl. wild taxa | 5 | 7 | 12 |
| poppy, linseed (*Papaver, Linum*) incl. wild species | 4 | 3 | 7 |
| further spice or medicinal species | 5 | 15 | 20 |
| vegetables incl. some oil and medicinal plants total sum | 96 | 67 | 163 |
| total sum | 262 | 199 | 461 |

The seed material of these accessions had been divided afterwards in Bari, thus the Bari and Gatersleben gene banks possess identical sets of the collection with the exception of some material collected for special purposes for both the mentioned institutions.

By the isolation of morphologically different lines within rather many of the accessions the real number of the collected strains will increase further, especially the grain legume material contains often different seed types within the same accession number (especially in faba beans, peas, cow peas and garden beans, less so in chick pea, and vetchling).

Herbarium material could be collected from 54 different numbers.

Without going into details the following remarkable observations made during the missions or afterwards in connection with a preliminary examination of the seed material should be mentioned:

– The most variable grain legume seems to be *Vicia faba*, strains with large-sized seeds predominate within the collections; sometimes the populations represent mixtures of small- and large-seeded types, striking seed colour variants being very often present.

– An overwhelming variability show also the seed samples of *Phaseolus vulgaris*. This variability seems to be in part the result of outcrossings. *Phaseolus coccineus* has been found only once during our missions in the Apennine mountains.

– The formerly more widespread dark-coloured seed variants of *Cicer arietinum* are nowadays disappearing in South Italy, they have been collected only once as a uniform local race with black seeds and in a few further cases as rare admixtures of light-seeded races.

– Some new localities of wild lentil species (*Lens ervoides, L. nigricans*) hitherto unknown could be discovered and seeds of them collected; mixed populations of these species have not been observed.

– Different taxa of the *Vicia sativa* aggregate very often grow sympatrically mostly as weeds and sometimes up to 4–6 taxa of this complex can be found within few square-meters (e. g. *V. macrocarpa, V. cordata, V. angustifolia, V. sativa* s. str.) introgressions between them may occur.

– Special interest deserve collections of *Triticum monococcum* and *T. dicoccon* from the Apennine mountains near the village of Castelfranco in Misc. These indigenous cereals were thought to have been lost already some times ago. But some farmers grow them uptillnow as forage for their live-stock. Samples of these wheats

Fig. 5 Typical scenery of the Gargano mountains with predominating types of extensive agriculture (in the foreground fruit-tree of *Castanea sativa*)

from the farmers' seed storages contained also other cereals, pulses, and a lot of interesting weeds (e. g. *Lolium temulentum*, *Agrostemma githago*) which will be further evaluated after reproduction of the material. The details of *T. monococcum* and *T. dicoccon* cultivation in Italy will be published in a separate paper.

– Worth mentioning are also some probably early ripening strains of maize (*Zea mays*) from higher altitudes (more than 600 m above sea level) of the Apennine mountains.

– The chicory, *Cichorium intybus*, exhibits an astonishing spectrum of wholly

different vegetable varieties almost unknown outside South Italy (cf. LARKCOM 1978) and worthwhile a more thorough taxonomic analysis.
– The *Brassica ruvo* described by BAILEY from material of rather uncertain origin in 1940 represents a vegetable typical for South Italian gardens; it must be included within *Br. rapa* as an infraspecific taxon which had been selected as an annual leafy vegetable.
– Variable wild types of beets (*Beta vulgaris*) could be found on some places along the Adriatic sea shore.
– Italy is also a gene centre for fruit trees (cf. MORRETINI 1977). The Gargano mountains proved to be especially rich in cultivars and wild relatives of fruit trees (e.g. *Malus* spp., *Pyrus* spp., *Sorbus domestica, Crataegus azarolus, Arbutus unedo*). This material should be collected and maintained for breeding purposes.

The collections of the two missions will be maintained, reproduced and evaluated in Bari and Gatersleben during the following years according to the procedures practised by these gene banks. A more detailed description of the material will be published afterwards in one of the forthcoming volumes of this journal.

## Zusammenfassung

Bericht über Reisen nach Süd-Italien 1980 zur Sammlung indigenen Materials von Kulturpflanzen

Es wird kurz über Verlauf und Ergebnisse zweier Sammelreisen in Süd-Italien berichtet, die im Früh- und Spätsommer 1980 gemeinsam vom Istituto del Germoplasma, Bari, und dem Zentralinstitut für Genetik und Kulturpflanzenforschung, Gatersleben, zum Studium autochthonen Materials von Kulturpflanzen, besonders Körnerleguminosen, durchgeführt wurden. Es konnten insgesamt 461 Samenproben gesammelt werden, unter denen sich neben Reliktkulturen wie Einkorn und Emmer viele in sich sehr variable Landsorten, z. B. von *Vicia faba, Phaseolus vulgaris,* sowie ein breites Spektrum lokaler Gemüsesorten befinden. Die Gefährdung indigenen Materials durch die Generosion ist nach Kulturart und Region in Süditalien unterschiedlich hoch.

## Краткое содержание

Отчёт о поездках в южную Италию в 1980 г. для сборов автохтонного материала культурных растений

Кратко сообщается о ходе и результатах двух поездок в южную Италию, проведённых в начале и в конце лета 1980 г., совместно Институтом зародышевой плазмы (Бари) и Центральным институтом генетики и культурных растений (Гатерслебен) для обследования автохтонного материала культурных растений, главным образом, зерновых бобовых. Сбор семян составил в общем 461 образец, среди которых находятся реликтовые культуры, как пшеница однозернянка и эммер, многие весьма изменчивые местные сорта *Vicia faba*, *Phaseolus vulgaris* и др., а также – широким спектром – сорта местных овощей. Автохтонный материал подвержен в южной Италии различной степени генной эрозии, зависящей от типа культуры и области возделования.

## Literature

Bailey, L. H., 1940: Certain noteworthy Brassicas. – Gent. herb. **4** (9), 319–330.

Larkcom, J., 1978: Red-leaved chicories. – The Garden **103** (3), 66–69.

Lehmann, Chr. O., und K. Hammer, 1979: Bericht über die Reise nach Südwest-Spanien zur Sammlung kultivierter und wild wachsender Leguminosen im Jahre 1978. – Kulturpflanze **27**, 97–108.

Morretini, A., 1977: Frutticoltura generale e speciale. – Roma.

Porceddu, E., 1976: Laboratorio del germoplasma, Bari. Attivitá scientifica svolta nel 1974. – Ric. sci. (C.N.R., Roma) **45** (1), 49–57.

–, 1979a: Wheat collecting in the Mediterranean region. In: Zeven, A. C., and A. M. Van Harten (ed.), Broadening the genetic base of crops, p. 75–82. – Pudoc, Wageningen.

–, 1979b: Gene bank activities in Italy. In: L. c., p. 123–125.

–, and E. Bennett, 1971: Primi risultati di una spedizione di esploratione e raccolta di vecchie varieta di frumento in Sicilia. – Ecol. agrar. **7** (4), 3–18.

Dr. P. Perrino
Istituto del Germoplasma (C.N.R.)
I–70126 Bari,
Via G. Amendola 165A,
Italy
Dr. K. Hammer and Dr. P. Hanelt,
Zentralinstitut für Genetik und
Kulturpflanzenforschung der AdW der DDR,
DDR - 4325 Gatersleben,
Corrensstr. 3
DDR

Kulturpflanze XXIX · 1981 · S. 443–446

# Hans Helbaeks wissenschaftliches Werk

Jürgen Schultze-Motel

(Eingegangen am 8. April 1981)

Während Hans Helbaeks Tätigkeit als Konservatorassistent (1941) und als Konservator (1948) der Ethnographischen und Altertums-Sammlung am Nationalmuseum in Kopenhagen erkannte der Archäologe Gudmund Hatt dessen Begeisterung für archäologische und naturwissenschaftliche Probleme und gab ihm Unterstützung. Hans Helbaek untersuchte zunächst Abdrücke von Kulturpflanzensamen an Keramik und deren Auswertungsmöglichkeiten, eine Methode, die in Dänemark von Sarauw erstmals angewendet wurde. Er widmete sich außerdem auch der Untersuchung verkohlter Kulturpflanzenreste aus England, Irland und Dänemark. Später dehnte er seine Forschungen auf Material aus Italien, Ägypten und dem Nahen Osten aus. Hier konnte er in dem Fund von Jarmo (Irak) Formen entdecken, die frühe Stadien in der Entwicklung der Getreide darstellen und daher von grundlegender Bedeutung für das Verständnis der Kulturpflanzenentstehung sind.

Seine Forschungen wurden durch Verleihung der Ehrendoktorwürde von Reading/England (1959) und Lund/Schweden (1960) anerkannt. In dieser Zeit nahm Helbaek an archäologischen Expeditionen teil, die vom Oriental Institute Chicago nach Syrien, Mesopotamien und Anatolien entsandt wurden. Hier untersuchte er nicht nur die ausgegrabenen Kulturpflanzenreste, sondern widmete sich auch der rezenten Flora der Umgebung der Ausgrabungsstätten, um die Ökologie aufzuklären.

Hans Helbaeks Arbeiten haben ausnahmslos die Abstammung und Geschichte der Kulturpflanzen zum Gegenstand. Dabei hat er sich nie mit der bloßen Bestimmung der archäologischen Kulturpflanzenreste begnügt, sondern war stets bestrebt, zu einer Auswertung der Befunde zu gelangen und größere Zusammenhänge zu entdecken. In seiner Arbeit „La Recherche Paletnobotanique" (1955) schuf er den Begriff Paläoethnobotanik, der seitdem für diese zwischen Botanik und Archäologie liegenden Forschungen international verwendet wird. Natürlich hat es derartige Untersuchungen schon früher gegeben, aber Helbaeks Verdienst besteht darin, diese Untersuchungen sehr folgerichtig und konzentriert durchgeführt zu haben und sie als ein Mittel zu sehen, die Entstehung des Ackerbaus zu ergründen, der die Grundlage der menschlichen Zivilisation darstellt. Sein Name wird für immer mit diesem Forschungsgebiet verknüpft sein.

Hans Helbaek wurde am 8. 4. 1907 in Roskilde, Dänemark, geboren. Nach langer Krankheit starb er am 10. 2. 1981 in Helsinge.

## Verzeichnis der Arbeiten von Hans Helbaek

1938: Planteavl. (pp. 165–166, 216–226, 249–250). In: G. Hatt, Jernalders Bopladser i Himmerland. – Aarbøger for Nordisk Oldkyndighed og Historie 1938, 119–266, franz. Res. XXXIII–XLVII, 12 Tavler.

1940: Studies on prehistoric and Anglo-Saxon cultivated plants in England. – Proc. Prehist. Soc. for 1940, N. S. **VI**, Part 1, 176–178.

1944: (mit K. Jessen) Cereals in Great Britain and Ireland in prehistoric and early historic times. – Det Kongelige Danske Videnskabernes Selskab, Biologiske Skrifter, Bind **III**, Nr. 2, 68 pp. København.

1948: Appendice I. Les empreintes de céréales. (pp. 205–207). In: P. J. Riis, Hama. Fouilles et Recherches de la Fondation Carlsberg 1931–1938, II, 3. Les Cimetières à Crémation. (Nationalmuseets Skrifter, Større Beretninger, I). – København.

1951: Tollund mandens sidste maaltid. Et botanisk bidrag til belsyning af oldtidens kost. Botanical study of the stomach contents of the Tollund man. – Aarbøger for Nordisk Oldkyndighed og Historie 1950, 311–341, T. I, II.

1951: Et vegetarisk maaltid fra Jernalderen. – Naturhistorisk Tidende **15**, 75–78.

1951: Ukrudtsfrø som naeringsmiddel i førromersk Jernalder. – Kuml, Årbog for Jysk arkaeologisk selskab 1951, 65–74.

1952: Early crops in Southern England. – Proc. Prehist. Soc. for 1952, N. S. **18**, No. 12, 194–233, Pl. XX–XXIV.

1952: Spelt (*Triticum spelta* L.) in Bronze Age Denmark. – Acta Archaeologica **23**, 97–107, Pl. I.

1952: Preserved apples and *Panicum* in the prehistoric site at Nørre Sandegaard in Bornholm. – Acta Archaeologica **23**, 107–115, Pl. II, III.

1953: Report on the cereals from the burnt corn level. (p. 91). In: M. A. Cotton and R. E. M. Wheeler, Verulamium, 1949. – St. Albans and Herts Architectural and Archaeological Society, Transactions, 13–97.

1953: Appendix I. [Plant remains from Forum Romanum] (pp. 155–157). In: E. Gjerstad, Early Rome. I. Stratigraphical researches in the Forum Romanum and along the Sacra Via. – Skrifter utgivna av Svenska Institutet i Rom, Acta Instituti Romani Regni Sueciae, Series in 4°, XVII:1.

1953: Agerbrugsrevolutionen. De første spor af kulturplanternes foraeldre. – Vor Viden **4**, 303–308.

1953: Saeddodder og Korn paa Bornholm i det 13. Aarhundrede. – Bornholmske Samlinger **33**, 9–15.

1953: Queen Ichetis' Wheat. A contribution to the study of Early Dynastic emmer of Egypt. – Det Kongelige Danske Videnskabernes Selskab, Biologiske Meddelelser **21**, no. 8, 1–17, Pl. I–IV.

1953: Did man once live by beer alone? – American Anthropologist **55**, 517–519.

1953: Archaeology and agricultural botany. – University of London, Institute of Archaeology, Ninth Annual Report, 44–59, Pl. I–IV.

1954: Prehistoric food plants and weeds in Denmark. A survey of archaeobotanical research 1923–1954. – Danmarks Geologiske Undersøgelse, II. Raekke, Nr. **80**, 250–261, Pl. IX.

1955: Store Valby–Kornavl i Danmarks første Neolitiske Fase. Store Valby, introduction of cereal husbandry in Denmark. – Aarbøger for nordisk Oldkyndighed og Historie 1954, 198–204.

1955: La Recherche Paletnobotanique. Une science née de la découverte des palafittes. – Sibrium **2**, 225–232.

1955: The Botany of the Vallhagar Iron Age Field. (pp. 653–699, 1206–1207). In: Vallhagar. A Migration Period Settlement on Gotland/Sweden. Part II. Editor M. Stenberger in collaboration with O. Klindt-Jensen. – Copenhagen.

1955: Ancient Egyptian Wheats. – Proc. Prehist. Soc. for 1955, N. S. **21**, 93–95, Pl. V.

1956: Appendix I. Vegetables in the funeral meals of Pre-Urban Rome. (pp. 287–294). In: E. Gjerstad, Early Rome. II. The Tombs. – Skrifter utgivna av Svenska Institutet i Rom, Acta Instituti Romani Regni Sueciae, Series in 4°, XVII:2.

1957: Bornholm plant economy in the first half of the first millennium A. D. (pp. 259–277, Pl. I–III). In: O. KLINDT-JENSEN, Bornholm i Folkevandringstiden og forudsaetningerne i tidlig Jernalder. 323 pp., Pl. I–III, Pl. A–F. – København.
1958: Carbonized Cereals. (pp. 206–209). In: G. P. BURSTOW and G. A. HOLLEYMAN, Late Bronze Age Settlement on Itford Hill, Sussex. – Proc. Prehist. Soc. for 1957, N. S. **23**, 167–212, Pl. 16–26.
1958: Korndyrkningens Udvikling paa Bornholm. – Bornholmske Samlinger **36**, 117–129.
1958: Appendix A. Plant Economy in Ancient Lachish. (pp. 309–317, Pl. 15). – In: O. TUFNELL, Lachish IV (Tell Ed-Duweir). The Bronze Age. – London, New York Toronto.
1958: The Oxbøl grain. – Acta Archaeologica **29**, 155–157.
1959: Notes on the evolution and history of *Linum*. – Kuml, Årbog for Jysk arkaeologisk selskab 1959, 103–129, Pl. 1–3.
1959: Grauballemandens sidste måltid. The Last Meal of Grauballe Man: An analysis of the food remains in the stomach. – Kuml, Årbog for Jysk Arkaeologisk Selskab 1958, 83–116, Pl. I–VII.
1959: Die Paläoethnobotanik des Nahen Ostens und Europas. – Opuscula Ethnologica Memoriae Ludovici Biró Sacra. – Budapest, 265–289.
1959: Domestication of food plants in the Old World. – Science **130**, 365–372.
1959: På Markvandring gennem Oldtidens Agre. – Nationalmuseets Arbejdsmark 1959, 69–82.
1959: How farming began in the Old World. – Archaeology **12**, 183–189.
1960: Appendix II. Cereals and weed grasses in Phase A. (pp. 540–543). In: R. J. BRAIDWOOD and L. S. BRAIDWOOD, Excavations in the Plain of Antioch. I. The Earlier Assemblages, Phases A–J. – The University of Chicago, Oriental Institute Publications **61**.
1960: The paleoethnobotany of the Near East and Europe. (pp. 99–118, Pl. 27–28). In: R. J. BRAIDWOOD and B. HOWE, Prehistoric Investigations in Iraqi Kurdistan. – The Oriental Institute of the University of Chicago, Studies in Ancient Oriental Civilization No. **31**, XXVIII +184 pp., 29 pl.
1960: Ancient crops in the Shahrzoor Valley in Iraqi Kurdistan. – Sumer **16**, 79–81.
1960: Ecological effects of irrigation in ancient Mesopotamia. – Iraq **22**, 186–196.
1960: Preliminary identifications in the Qantara deposit (A. D. c. 900). Appendix III. (pp. 10–19). In: T. JACOBSEN, Report on essential results of the Diyala Basin Archaeological Project 1957–1958. – Directorate General of Agriculture, Baghdad. [non vidi]
1960: Comment on *Chenopodium album* as a food plant in prehistory. – Ber. Geobot. Inst. Eidg. Techn. Hochschule, Stiftung Rübel, 31. Heft, Bericht über das Jahr 1959, 16–19.
1961: Late Bronze age and Byzantine crops at Beycesultan in Anatolia. – Anatolian Studies **11**, 77–97, Pl. XVI–XVIII.
1961: Studying the diet of ancient man. – Archaeology **14**, 95–101.
1962: Les grains carbonisés de la 48ème couche des fouilles de Tell Soukas. – Les Annales Archéologiques de Syrie **XI–XII**, 185–186.
1963: Late Cypriote vegetable diet at Apliki. – Opuscula Atheniensia 4, Skrifter utgivna av Svenska Institutet i Athen **8**, 171–186.
1963: Isin Larsan and Horian food remains at Tell Bazmosian in the Dokan Valley. – Sumer **19**, 27–35, Pl. I.
1963: Textiles from Catal Huyuk. – Archaeology **16**, 39–46.
1963: Palaeo-Ethnobotany. (pp. 177–185, Pl. IX). In: Science in Archaeology. A Comprehensive Survey of Progress and Research. Ed. by D. BROTHWELL and E. HIGGS. – Thames and Hudson, Bristol, 595 pp.
1964: First impressions of the Çatal Hüyük plant husbandry. – Anatolian Studies **14**, 121–123.
1964: Archaeological evidence for genetical changes in wheat and barley. – Tenth International Botanical Congress Edinburgh, Abstracts of Papers, p. 126.

1964: The Isca grain, a Roman plant introduction in Britain. – The New Phytologist **63**, 158–164.
1964: Early Hassunan vegetable food at Tell es-Sawwan near Samarra. – Sumer **20**, 45–48.
1965: Archaeological evidence for genetic changes in wheat and barley. – Plant Introd. Rev. (CSIRO) Canberra **2**, 10–16.
1966: Vendeltime farming products at Eketorp on Öland, Sweden. – Acta Archaeologica **37**, 216–221.
1966: 1966–Commentary on the phylogenesis of *Triticum* and *Hordeum*. – Econ. Bot. **20**, 350–360.
1966: Pre-pottery neolithic farming at Beidha – a preliminary report. – Palestine Exploration Quarterly **98**, 61–66.
1966: Appendix I. The plant remains from Nimrud. (pp. 613–620). In: M. E. L. MALLOWAN, Nimrud and its remains. Vol. **II**. Collins, London; Vol. **I**, p. 123 u. 138.
1966: What farming produced at Cypriote Kalopsidha. – Studies in Mediterranean Archaeology **II**, 115–126.
1966: Appendix IV. Report on carbonized grain from AF5 (GHD. Phase). p. 53. In: D. H. TRUMP, Skorba. Reports of the Research Committee of the Society of Antiquaries of London No. **XXII**.
1967: Agricoltura preistorica a Luni sul Mignone in Etruria. – Acta Instituti Romani Regni Sueciae, Series in 4°, **25**, 279–282.
1969: Appendix I. Plant collecting, dry-farming, and irrigation agriculture in prehistoric Deh Luran. (pp. 383–426, Pl. 40–41). In: F. HOLE, K. V. FLANNERY, J. A. NEELY, Prehistory and human ecology of the Deh Luran Plain. An early village sequence from Khuzistan, Iran. – Memoirs of the Museum of Anthropology, University of Michigan, Number 1, XVI + 440 pp., + plates.
1970: The Plant Husbandry of Hacilar. A study of cultivation and domestication. (pp. 189–244, Pl. I–V). In: J. MELLAART, Excavations at Hacilar (I). – Occasional Publications of the British Institute of Archaeology at Ankara Number 9, XIII + 249 pp., Edinburgh.
1971: Plant remains from the Tarxien cemetery. (p. 235). In: J. D. EVANS, Prehistoric Antiquities of the Maltese Islands. – London.
1971: Da rugen kom til Danmark. The arrival of rye in Denmark. – Kuml, Årbog for Jysk Arkaeologisk Selskab 1970, 279–296.
1971: The origin and migration of rye, *Secale cereale* L.; a palaeo-ethnobotanical study. (pp. 265–280). In: P. H. DAVIS, P. C. HARPER & I. C. HEDGE (eds.), Plant Life of South-West Asia. – Aberdeen, X + 335 pp.
1972: Plant anatomy in culture-historical research. (pp. 19–32, Pl. I–II). In: A. K. M. GHOUSE and M. YUNUS (eds.), Research Trends in Plant Anatomy. – Bombay/New Delhi.
1972: Traces of plants in the early ceramic site of Umm Dabaghiyah. – Iraq **34**, 17–19.
1972: Samarran irrigation agriculture at Choga Mami in Iraq. – Iraq **34**, 35–48.
1973: Eketorp Grain 1971. (pp. 19–20). In: H. JANKUHN, W. SCHLESINGER und H. STEUER (Hrsg.), Vor- und Frühformen der europäischen Stadt im Mittelalter. Bericht über ein Symposium in Reinhausen bei Göttingen vom 18. bis 24. April 1972, Teil II. – Abh. Akad. Wiss. Göttingen, Phil.-Hist. Kl. III/84.
1977: The Fyrkat Grain. A geographical and chronological study of rye. (pp. [1]–[41]). In: Fyrkat. En jysk vikingeborg. I. Borgen og Bebyggelsen af O. OLSEN og H. SCHMIDT med en bidrag af H. ØDUM og en ekskurs af HANS HELBAEK. – København. (Nordiske Fortidsminder, Serie B – in quarto. Bind 3. Udgivet af Det kgl. nordiske Oldskriftselskab), 241 + [41] pp. (preprint 1974).

Dr. JÜRGEN SCHULTZE-MOTEL
Zentralinstitut für Genetik
und Kulturpflanzenforschung
der Akademie der Wissenschaften der DDR
4325 Gatersleben

Kulturpflanze XXIX · 1981 · S. 447–463

# Literatur über archäologische Kulturpflanzenreste (1979/1980)

JÜRGEN SCHULTZE-MOTEL

(Eingegangen am 12. Januar 1981)


## Zusammenfassung

In Fortsetzung von Bibliographien über archäologische Kulturpflanzenreste wurde die einschlägige Literatur der Jahre 1979/1980 zusammengestellt und der Inhalt der erfaßten 95 Arbeiten kurz kommentiert.


## Einleitung

Hiermit wird die Bibliographie über archäologische Kulturpflanzenreste fortgesetzt (vgl. SCHULTZE-MOTEL 1980a). Es ist versucht worden, die entsprechende Literatur der Jahre 1979 und 1980 zu erfassen. Bisher übersehene Titel aus früheren Jahren oder solche, die nicht eher beschafft werden konnten, sind ebenfalls aufgenommen worden.

Auch in diesem Jahr haben mich Frau Bibliothekarin V. SCHNEIDER, Herr Direktor Dr. H. BEHRENS und Herr W. MATTHIAS, wissenschaftlicher Mitarbeiter (alle am Landesmuseum für Vorgeschichte Halle/S.), bei der Ermittlung und Bereitstellung der Literatur tatkräftig unterstützt. Auch Frau Dr. M. HOPF (Römisch-Germanisches Zentralmuseum, Mainz), Herr Prof. Dr. W. VAN ZEIST (Biologisch-Archäologisch Instituut, Groningen) und Herr Dr. H.-H. MÜLLER (Zentralinstitut für Alte Geschichte und Archäologie, Berlin) halfen durch Übermittlung von Zitaten und schwer beschaffbarer Literaturstellen. Allen Genannten sage ich meinen herzlichen Dank.

Es folgen die Einzelangaben zu den alphabetisch angeordneten Arten. Hierbei sind folgende Abkürzungen benutzt worden:
N = Neolithikum, Bz = Bronzezeit, Ez = Eisenzeit, Rz = Römerzeit, Ma = Mittelalter.

*Allium cepa* L., Zwiebel
Kozí Hrádek bei Tábor, ČSSR, 15. Jh., TEMPÍR 1974a.

*Amaranthus lividus* L.
Welzheim, Rz, KÖRBER-GROHNE 1979b; Wawel, Kraków, Ma, WASYLIKOWA 1979.

*Anethum graveolens* L., Dill
Aachen-Burtscheid, Rz, KNÖRZER 1980d; Welzheim, Rz, KÖRBER-GROHNE 1979b; Butzbach, Rz, BAAS 1979; Gasselte, Niederlande, Ma, VAN ZEIST und

PALFENIER-VEGTER 1979; Svendborg, Dänemark, Ma, JØRGENSEN 1980; Eschelbronn bei Heidelberg, 13. Jh., KÖRBER-GROHNE 1979a; Burg Brüggen, Kr. Viersen, spätes Ma, KNÖRZER 1979b.

*Apium graveolens* L., Sellerie
Aachen-Burtscheid, Rz, KNÖRZER 1980d; Welzheim, Rz, KÖRBER-GROHNE 1979b; Butzbach, Rz, BAAS 1979; Tornow, Kr. Calau, Kaiser-/Völkerwanderungszeit, LANGE 1979.

*Arachis hypogaea* L., Erdnuß
Guitarrero Cave, Peru, 500 v. u. Z.–500 u. Z., SMITH 1980a.

*Atriplex hortensis* L., Gartenmelde
Welzheim, Rz, KÖRBER-GROHNE 1979b.

*Avena* L., Hafer
Anza, Jugoslawien, N, RENFREW 1978; Shahr-i Sokhta, Iran, ca. 2000 v. u. Z., COSTANTINI 1977; Pirak, Pakistan, ca. 1600 v. u. Z., COSTANTINI 1979a; Hala Sultan Tekke, Zypern, ca. 1200 v. u. Z., HJELMQVIST 1979; Rajhrad, Bez. Brno-venkov, ČSSR, Hallstattzeit, KÜHN 1980a; Gussage All Saints, Dorset, England, Ez, EVANS und JONES 1979; Aldenhoven-Niedermerz, Kr. Düren, Ez, KNÖRZER 1980c; Neuss-Selikum, Ez, KNÖRZER 1980c; Malain-Mediolanum, Frankreich, 3. Jh. v. u. Z., ERROUX 1980; Aachen-Burtscheid, Rz, KNÖRZER 1980d; Welzheim, Rz, KÖRBER-GROHNE 1979b; Butzbach, Rz, BAAS 1979; Portchester Castle, England, Rz, MURPHY und RENFREW 1975; Svendborg, Dänemark, Ma, JØRGENSEN 1980; Gasselte, Niederlande, Ma, VAN ZEIST und PALFENIER-VEGTER 1979; Timmel, Wiegboldsbur und Etzel, Ostfriesland, Ma, KUČAN 1979; Wawel, Kraków, Ma, WASYLIKOWA 1979; Hauptmarkt in Kraków, Ma, WIESEROWA 1979; Eschelbronn bei Heidelberg, 13. Jh., KÖRBER-GROHNE 1979a.

*Beta vulgaris* L.
Burg Brüggen, Kr. Viersen, spätes Ma, KNÖRZER 1979b.

*Brassica napus* L.
Hauptmarkt in Kraków, Ma, WIESEROWA 1979.

*Brassica nigra* (L.) KOCH, Schwarzer Senf
Wawel, Kraków, Ma, WASYLIKOWA 1979; Hauptmarkt in Kraków, Ma, WIESEROWA 1979.

*Brassica oleracea* L., Kohl
Hauptmarkt in Kraków, Ma, WIESEROWA 1979.

*Brassica rapa* L., Rübsen
Wawel, Kraków, Ma, WASYLIKOWA 1979; Hauptmarkt in Kraków, Ma, WIESEROWA 1979; Burg Brüggen, Kr. Viersen, spätes Ma, KNÖRZER 1979b.

*Camelina sativa* (L.) CRANTZ, Leindotter
Hascherkeller, Bayern, frühe Ez, QUILLIAN 1980; Aldenhoven-Niedermerz, Kr. Düren, Ez, KNÖRZER 1980c.

*Cannabis sativa* L., Hanf
Wawel, Kraków, Ma, WASYLIKOWA 1979; Hauptmarkt in Kraków, Ma, WIESEROWA 1979.

*Capsicum sinense* JACQ.
Guitarrero Cave, Peru, 8600–8000 v. u. Z., SMITH 1980a.

*Carum carvi* L., Kümmel
Butzbach, Rz, BAAS 1979.

*Chenopodium* L.
ASCH und ASCH (1977) untersuchten die Frage, ob Vertreter dieser Gattung im östlichen Nord-Amerika in urgeschichtlicher Zeit kultiviert wurden.

*Cicer arietinum* L., Kichererbse
Tell-el-Hesi, Israel, Hellenistische Zeit, STEWART 1979.

*Citrullus colocynthis* (L.) SCHRAD.
Shahr-i Sokhta, Iran, ca. 2000 v. u. Z., COSTANTINI 1977; Pirak, Pakistan, ca. 1600 v. u. Z., COSTANTINI 1979a.

*Citrus medica* L.
Hala Sultan Tekke, Zypern, ca. 1200 v. u. Z., HJELMQVIST 1979.

*Coriandrum sativum* L., Koriander
Welzheim, Rz, KÖRBER-GROHNE 1979b; Butzbach, Rz, BAAS 1979; Bearsden, Schottland, Rz, DICKSON et al. 1979, DICKSON 1979; Svendborg, Dänemark, Ma, JØRGENSEN 1980; Burg Brüggen, Kr. Viersen, spätes Ma, KNÖRZER 1979b.

*Cucumis melo* L., Melone
Shahr-i Sokhta, Iran, ca. 2000 v. u. Z., COSTANTINI 1977; Butzbach, Rz, BAAS 1979.

*Cucumis sativus* L., Gurke
Shahr-i Sokhta, Iran, ca. 2000 v. u. Z., COSTANTINI 1977.

*Cucurbita maxima* DUCH., Riesenkürbis
Viru Valley, Peru, ca. 1800 v. u. Z., bis 100 u. Z., WEST und WHITAKER 1980.

*Cucurbita moschata* DUCH., Moschuskürbis
Viru Valley, Peru, ca. 600–ca. 1100 u. Z., WEST und WHITAKER 1980.

*Daucus carota* L., Mohrrübe
Aldenhoven-Niedermerz, Kr. Düren, Ez, KNÖRZER 1980c (ob kultiviert?); Aachen-Burtscheid, Rz, KNÖRZER 1980d; Butzbach, Rz, BAAS 1979.

*Fagopyrum sagittatum* GILIB., Buchweizen
Burg Brüggen, Kr. Viersen, spätes Ma, KNÖRZER 1979b.

*Ficus carica* L., Feigenbaum
Tel Ṣaf, Israel, Chalkolithikum, GOPHNA und KISLEV 1979; Hala Sultan Tekke, Zypern, ca. 1200 v. u. Z., HJELMQVIST 1979; Pompeji, Rz, JASHEMSKI 1977; Aachen-Burtscheid, Rz, KNÖRZER 1980d; Welzheim, Rz, KÖRBER-GROHNE

1979b; Butzbach, Rz, Baas 1979; Bearsden, Schottland, Rz, Dickson et al. 1979, Dickson 1979; Hauptmarkt in Kraków, Ma, Wieserowa 1979; Gamlebyen, Oslo, Norwegen, Ma, Griffin 1979; Neuss, Quirinusstraße, Ma: 14./15. Jh., Knörzer 1979c; Burg Brüggen, Kr. Viersen, spätes Ma, Knörzer 1979b.

*Foeniculum vulgare* Mill., Fenchel
Butzbach, Rz., Baas 1979; Burg Brüggen, Kr. Viersen, spätes Ma, Knörzer 1979b.

*Helianthus annuus* L., Sonnenblume
In seiner Darstellung der Entstehung der kultivierten Sonnenblume behandelt Heiser (1978) auch archäologische Reste. Yarnell (1978) bespricht die Domestikation dieser Art.

*Hordeum vulgare* L. s. l., Saatgerste
Yarym-Tepe, Nord-Irak, 6. und 5. Jt. v. u. Z., Bakhteyev und Yanushevich 1980; Balakot, Pakistan, Mitte des 4. Jt. v. u. Z., Dales 1979; Shahr-i Sokhta, Iran, ca. 3200–3000 v. u. Z., und ca. 2300–2200 v. u. Z., Costantini 1977; Kakanj, Jugoslawien, frühes N, Renfrew 1974; Obre II, Jugoslawien, spätes N, Renfrew 1974; Anza, Jugoslawien, N, Renfrew 1978; Cova de l'Or, Beniarrés-Alicante, Spanien, N, López García 1980; Tel Ṣaf, Israel, Chalkolithikum, Gophna und Kislev 1979; Kephala, Kykladen, spätes N, Renfrew 1977; Zürich, Schweiz, N, Jacomet-Engel 1980; Flögeln, Kr. Cuxhaven, Trichterbecherkultur, Zimmermann 1979; Broichweiden, Kr. Aachen, N, Knörzer 1979a; Lindebjerg, Dänemark, ältere Bz, Rowley-Conwy 1979; Pirak, Pakistan, ca. 1600 v. u. Z., Costantini 1979a; Swat Valley, Pakistan, Ende des 2. Jt. v. u. Z., Costantini 1979b; Hala Sultan Tekke, Zypern, ca. 1200 v. u. Z., Hjelmqvist 1979; Mönchengladbach-Rheydt, Bz, Knörzer 1980c; Atapuerca, Spanien, Jungbronzezeit, Apellániz und Hopf 1979; Hascherkeller, Bayern, frühe Ez, Quillian 1980; Berndorf, Ldkr. Kulmbach, Hallstattzeit, Hopf 1977; Jüchsen, Kr. Meiningen, Hallstattzeit, Gall 1980; Rajhrad, Bez. Brno-venkov, ČSSR, Hallstattzeit, Kühn 1980a; Aldenhoven-Niedermerz, Kr. Düren, Ez, Knörzer 1980c; Gussage All Saints, Dorset, England, Ez, Evans und Jones 1979; Malain-Mediolanum, Frankreich, 3. Jh. v. u. Z., Erroux 1980; Manching, Ldkr. Ingolstadt, Latène, Hopf 1979b; Tell-el-Hesi, Israel, Hellenistische Zeit, Stewart 1979; Welzheim, Rz, Körber-Grohne 1979b; Butzbach, Rz, Baas 1979; Portchester Castle, England, Rz, Murphy und Renfrew 1975; Bosau, Ostholstein, 1. Jh. u. Z., Kroll 1977; Krefeld-Gellep, römisch-fränkisch, Hopf 1979a; Wiegboldsbur, Timmel und Etzel, Ostfriesland, Ma, Kučan 1979; Gasselte, Niederlande, Ma, van Zeist und Palfenier-Vegter 1979; Wawel, Kraków, Ma, Wasylikowa 1979; Eschelbronn bei Heidelberg, 13. Jh., Körber-Grohne 1979a; Stillfried/March, Niederösterreich, 13./14. Jh., Felgenhauer et al.1978; Burg Brüggen, Kr. Viersen, spätes Ma, Knörzer 1979b.

Die im späten Paläolithikum (18300–17000 Jahre vor heute) von Wadi Kubbaniya bei Aswan, Ägypten, gefundenen Gerste-Reste müssen mit großer Skepsis aufgenommen werden (Wendorf et al. 1979).

*Humulus lupulus* L., Hopfen
Svendborg, Dänemark, Ma, Jørgensen 1980.

*Iva annua* L.
Die Domestikation dieser Art im nördlichen Nord-Amerika wird von YARNELL (1978) und von ASCH und ASCH (1978) behandelt.

*Juglans regia* L., Walnußbaum
Wawel, Kraków, Ma, WASYLIKOWA 1979; Hauptmarkt in Kraków, Ma, WIESEROWA 1979; Burg Brüggen, Kr. Viersen, spätes Ma, KNÖRZER 1979b; Funde im Gebiet der Westslawen, OPRAVIL 1979.

*Lagenaria siceraria* (MOL.) STANDL., Flaschenkürbis
Ho-mu-tu, Chekiang, China, N, NATURAL HISTORY SECTION ... 1978; Guitarrero Cave, Peru, 500 v. u. Z.–500 u. Z., SMITH 1980a; Viru Valley, Peru, 410 v. u. Z. bis ca. 1100 u. Z., WEST und WHITAKER 1980.

*Lathyrus cicera* L., Rote Platterbse
Tell-el-Hesi, Israel, Hellenistische Zeit, STEWART 1979.

*Lathyrus sativus* L., Saat-Platterbse
Kephala, Kykladen, spätes N, RENFREW 1977.

*Lens culinaris* MEDIK., Linse
Obre I, Jugoslawien, frühes N, RENFREW 1974; Obre II, Jugoslawien, spätes N, RENFREW 1974; Anza, Jugoslawien, N, RENFREW 1978; Bergheim-Zieverich, Bandkeramik, KNÖRZER 1980b; Tel Ṣaf, Israel, Chalkolithikum, GOPHNA und KISLEV 1979; Shahr-i Sokhta, Iran, ca. 2000 v. u. Z., COSTANTINI 1977; Swat Valley, Pakistan, Ende des 2. Jt. v. u. Z., COSTANTINI 1979b; Hala Sultan Tekke, Zypern, ca. 1200 v. u. Z., HJELMQVIST 1979; Hascherkeller, Bayern, frühe Ez, QUILLIAN 1980; Jüchsen, Kr. Meiningen, Hallstattzeit, GALL 1980; Tell-el-Hesi, Israel, Hellenistische Zeit, STEWART 1979; Malain-Mediolanum, Frankreich, 3. Jh v. u. Z., ERROUX 1980; Gussage All Saints, Dorset, England, Ez, EVANS und JONES 1979; Aachen-Burtscheid, Rz, KNÖRZER 1980d; Welzheim, Rz, KÖRBER-GROHNE 1979b; Wawel, Kraków, Ma, WASYLIKOWA 1979.

*Linum usitatissimum* L., Lein
Shahr-i Sokhta, Iran, ca. 2000 v. u. Z., COSTANTINI 1977; Pirak, Pakistan, ca. 1600 v. u. Z., COSTANTINI 1979a; Zürich, Schweiz, N, JACOMET-ENGEL 1980; Wanlo (Mönchengladbach), Bandkeramik, KNÖRZER 1980a; Eilsleben, Kr. Wanzleben, Bernburger Gruppe, SCHULTZE-MOTEL 1980b; Tell-el-Hesi, Israel, Hellenistische Zeit, STEWART 1979; Welzheim, Rz, KÖRBER-GROHNE 1979b; Gasselte, Niederlande, Ma, VAN ZEIST und PALFENIER-VEGTER 1979; Gamlebyen, Oslo, Norwegen, Ma, GRIFFIN 1979; Svendborg, Dänemark, Ma, JØRGENSEN 1980; Wawel, Kraków, Ma, WASYLIKOWA 1979; Hauptmarkt in Kraków, Ma, WIESEROWA 1979; Timmel, Wiegboldsbur und Etzel, Ostfriesland, Ma, KUČAN 1979; Eschelbronn bei Heidelberg, 13. Jh., KÖRBER-GROHNE 1979a; Burg Brüggen, Kr. Viersen, spätes Ma, KNÖRZER 1979b.

*Lupinus albus* L., Weiße Lupine
Hala Sultan Tekke, Zypern, ca. 1200 v. u. Z., HJELMQVIST 1979.

*Malus sylvestris* MILL., Apfel
Anza, Jugoslawien, N, RENFREW 1978; Wanlo (Mönchengladbach), Bandkera-

mik, Knörzer 1980a; Pompeji, Rz, Jashemski 1977; Butzbach, Rz, Baas 1979; Welzheim, Rz, Körber-Grohne 1979b; Svendborg, Dänemark, Ma, Jørgensen 1980; Wawel, Kraków, Ma, Wasylikowa 1979; Eschelbronn bei Heidelberg, 13. Jh., Körber-Grohne 1979a; Neuss, Quirinusstraße, Ma: 14./15. Jh., Knörzer 1979c; Kozí Hrádek bei Tábor, ČSSR, 15. Jh., Tempír 1974a; Burg Brüggen, Kr. Viersen, spätes Ma, Knörzer 1979b; Funde im Gebiet der Westslawen, Opravil 1979.

*Mespilus germanica* L., Mispel
Burg Brüggen, Kr. Viersen, spätes Ma, Knörzer 1979b.

*Nigella sativa* L., Schwarzkümmel
Wawel, Kraków, Ma, Wasylikowa 1979.

*Olea europaea* L., Ölbaum
Tel Şaf, Israel, Chalkolithikum, Gophna und Kislev 1979; Hala Sultan Tekke, Zypern, ca. 1200 v. u. Z., Hjelmqvist 1979; Tell-el-Hesi, Israel, Hellenistische Zeit, Stewart 1979; Pompeji, Rz, Jashemski 1977; Aachen-Burtscheid, Rz, Knörzer 1980d; Butzbach, Rz, Baas 1979.

*Oryza sativa* L., Reis
In Hemudu (Prov. Zhejiang, China) wurde Reis in Schichten nachgewiesen, die aus der Zeit 4900–4700 v. u. Z. stammen. Der Fund gehört einer Kulturform an und stützt die Vorstellung, daß der Reis in China entstanden sei (Natural History Section . . . 1978, You 1979). Glover (1979) diskutiert archäologische Reisfunde in Südost-Asien.
Weitere Funde: Ulu-Leang 1, Sulawesi, spätes N, ca. 4000 v. u. Z., Glover 1977, 1979; Pirak, Pakistan, ca. 1600–ca. 1000 v. u. Z., Costantini 1979a; Swat Valley, Pakistan, Ende des 2. Jt. v. u. Z., Costantini 1979b.

*Panicum miliaceum* L., Rispenhirse
Pirak, Pakistan, ca. 1600 v. u. Z., Costantini 1979a; Mönchengladbach-Rheydt, Bz, Knörzer 1980c; Eschweiler-Lohn, Kr. Aachen, Urnenfelderzeit, Knörzer 1980c; Hascherkeller, Bayern, frühe Ez, Quillian 1980; Rajhrad, Bez. Brno-venkov, ČSSR, Hallstattzeit, Kühn 1980a; Aldenhoven-Niedermerz, Kr. Düren, Ez, Knörzer 1980c; Butzbach, Rz, Baas 1979; Gasselte, Niederlande, Ma, van Zeist und Palfenier-Vegter 1979; Wawel, Kraków, Ma, Wasylikowa 1979; Hauptmark in Kraków, Ma, Wieserowa 1979.

*Papaver setigerum* DC., Borstenmohn
Wanlo (Mönchengladbach), Bandkeramik, Knörzer 1980a; Butzbach, Rz, Baas 1979.

*Papaver somniferum* L., Schlafmohn
Zürich, Schweiz, N, Jacomet-Engel 1980; Welzheim, Rz, Körber-Grohne 1979b; Bearsden, Schottland, Rz, Dickson et al. 1979, Dickson 1979; Wawel, Kraków, Ma, Wasylikowa 1979; Hauptmarkt in Kraków, Ma, Wieserowa 1979; Svendborg, Dänemark, Ma, Jørgensen 1980; Burg Brüggen, Kr. Viersen, spätes Ma, Knörzer 1979b.

*Petroselinum crispum* (MILL.) NYM., Petersilie
Burg Brüggen, Kr. Viersen, spätes Ma, KNÖRZER 1979b.

*Phaseolus lunatus* L., Limabohne
Guitarrero Cave, Peru, Schichten verschiedener Zeitstellung, SMITH 1980a.

*Phaseolus retusus* BENTH. (*Ph. metcalfei* WOOT. et STANDL.)
Einziger bisheriger archäologischer Fund: Gila Bend, Arizona, 1200–1300 u. Z., NABHAN et al. 1980.

*Phaseolus vulgaris* L., Gartenbohne
Guitarrero Cave, Peru, 8600–8000 v. u. Z. sowie in jüngeren Schichten, SMITH 1980a.

*Phoenix dactylifera* L., Dattelpalme
Siesbach, Kr. Birkenfeld, 174 u. Z., HOLSTEIN 1978.

*Pistacia terebinthus* L.
Hala Sultan Tekke, Zypern, ca. 1200 v. u. Z., HJELMQVIST 1979.

*Pisum sativum* L., Erbse
Obre I und Kakanj, Jugoslawien, frühes N, RENFREW 1974; Anza, Jugoslawien, N, RENFREW 1978; Eilsleben, Kr. Wanzleben, Linienbandkeramik, SCHULTZE-MOTEL 1980b; Wanlo (Mönchengladbach), Bandkeramik, KNÖRZER 1980a; Broichweiden, Kr. Aachen, N, KNÖRZER 1979a; Tel Şaf, Israel, Chalkolithikum, GOPHNA und KISLEV 1979; Swat Valley, Pakistan, Ende des 2. Jt. v. u. Z., COSTANTINI 1979b; Neuss-Selikum, Ez, KNÖRZER 1980c; Aldenhoven-Niedermerz, Kr. Düren, Ez, KNÖRZER 1980c; Jüchsen, Kr. Meiningen, Hallstattzeit, GALL 1980; Malain-Mediolanum, Frankreich, 3. Jh. v. u. Z., ERROUX 1980; Tell-el-Hesi, Israel, Hellenistische Zeit, STEWART 1979; BUTzbach, Rz, BAAS 1979; Welzheim, Rz, KÖRBER-GROHNE 1979b; Wawel, Kraków, Ma, WASYLIKOWA 1979; Gasselte, Niederlande, Ma, VAN ZEIST und PALFENIER-VEGTER 1979; Timmel, Ostfriesland, Ma, KUČAN; Kozí Hrádek bei Tábor, ČSSR, 15. Jh., TEMPÍR 1974a; Burg Brüggen, Kr. Viersen, spätes Ma, KNÖRZER 1979b.

*Portulaca oleracea* L., Portulak
Hauptmarkt in Kraków, Ma, WIESEROWA 1979.

*Prunus amygdalus* BATSCH, Mandelbaum
Pompeji, Rz, JASHEMSKI 1977; Lübeck, Ma/frühe Neuzeit, KROLL 1980b.

*Prunus avium* L., Süßkirsche
Pompeji, Rz, JASHEMSKI 1977; Butzbach, Rz, BAAS 1979; Aachen-Burtscheid, Rz, KNÖRZER 1980d; Wawel, Kraków, Ma, WASYLIKOWA 1979; Kozí Hrádek bei Tábor, ČSSR, 15. Jh., TEMPÍR 1974a; Burg Brüggen, Kr. Viersen, spätes Ma, KNÖRZER 1979b; Lübeck, Ma/frühe Neuzeit, KROLL 1980b; Funde im Gebiet der Westslawen, OPRAVIL 1979.

*Prunus cerasifera* EHRH., Kirschpflaume
Kozí Hrádek bei Tábor, ČSSR, 15. Jh., TEMPÍR 1974a; Funde im Gebiet der Westslawen, OPRAVIL 1979.

*Prunus cerasus* L., Sauerkirsche
Svendborg, Dänemark, Ma, JØRGENSEN 1980; Burg Brüggen, Kr. Viersen, spätes Ma, KNÖRZER 1979b; Lübeck, Ma/frühe Neuzeit, KROLL 1980b.

*Prunus domestica* L., Pflaume
Hala Sultan Tekke, Zypern, ca. 1200 v. u. Z., HJELMQVIST 1979; Aachen-Burtscheid, Rz, KNÖRZER 1980d; Welzheim, Rz, KÖRBER-GROHNE 1979b; Butzbach, Rz, BAAS 1979; Svendborg, Dänemark, Ma, JØRGENSEN 1980; Wawel, Kraków, Ma, WASYLIKOWA 1979; Eschelbronn bei Heidelberg, 13. Jh., KÖRBER-GROHNE 1979a; Kozí Hrádek bei Tábor, ČSSR, 15. Jh., TEMPÍR 1974a; Burg Brüggen, Kr. Viersen, spätes Ma, KNÖRZER 1979b; Lübeck, Ma bis frühe Neuzeit, KROLL 1980a, b; Funde im Gebiet der Westslawen, OPRAVIL 1979.

*Prunus persica* (L.) BATSCH, Pfirsich
Butzbach, Rz, BAAS 1979; Wawel, Kraków, Ma, WASYLIKOWA 1979; Kozí Hrádek bei Tábor, ČSSR, 15. Jh., TEMPÍR 1974a; Funde im Gebiet der Westslawen, OPRAVIL 1979.

*Punica granatum* L., Granatapfelbaum
Hala Sultan Tekke, Zypern, ca. 1200 v. u. Z., HJELMQVIST 1979.

*Pyrus domestica* MEDIK., Birnbaum
Butzbach, Rz, BAAS 1979; Svendborg, Dänemark, Ma, JØRGENSEN 1980; Wawel, Kraków, Ma, WASYLIKOWA 1979; Eschelbronn bei Heidelberg, 13. Jh., KÖRBER-GROHNE 1979a; Burg Brüggen, Kr. Viersen, spätes Ma, KNÖRZER 1979b; Funde im Gebiet der Westslawen, OPRAVIL 1979.

*Ribes uva-crispa* L., Stachelbeere
Butzbach, Rz, BAAS 1979.

*Rubus fruticosus* L., Brombeere
Aachen-Burtscheid, Rz, KNÖRZER 1980d; Butzbach, Rz, BAAS 1979; Svendborg, Dänemark, Ma, JØRGENSEN 1980; Neuss, Quirinusstraße, Ma: 14./15. Jh., KNÖRZER 1979c; Burg Brüggen, Kr, Viersen, spätes Ma, KNÖRZER 1979b.

*Rubus idaeus* L., Himbeere
Aachen-Burtscheid, Rz, KNÖRZER 1980d; Butzbach, Rz, BAAS 1979; Wawel, Kraków, Ma, WASYLIKOWA 1979; Svendborg, Dänemark, Ma, JØRGENSEN 1980; Neuss, Quirinusstraße, Ma: 14./15. Jh., KNÖRZER 1979c; Burg Brüggen, Kr. Viersen, spätes Ma, KNÖRZER 1979b.

*Rumex scutatus* L., Schildblättriger Ampfer
Welzheim, Rz, KÖRBER-GROHNE 1979b.

*Satureja hortensis* L., Bohnenkraut
Gasselte, Niederlande, Ma, VAN ZEIST und PALFENIER-VEGTER 1979.

*Secale cereale* L., Roggen
Shahr-i Sokhta, Iran, ca. 2000 v. u. Z., COSTANTINI 1977; Welzheim, Rz, KÖRBER-GROHNE 1979b; Krefeld-Gellep, römisch-fränkisch, HOPF 1979a; Timmel, Wiegboldsbur und Etzel, Ostfriesland, Ma, KUČAN 1979; Gasselte,

Niederlande, Ma, VAN ZEIST und PALFENIER-VEGTER 1979; Svendborg, Dänemark, Ma, JØRGENSEN 1980; Wawel, Kraków, Ma, WASYLIKOWA 1979; Hauptmarkt in Kraków, Ma, WIESEROWA 1979; Eschelbronn bei Heidelberg, 13. Jh., KÖRBER-GROHNE 1979a; Kozí Hrádek bei Tábor, ČSSR, 15. Jh., TEMPÍR 1974a; Burg Brüggen, Kr. Viersen, spätes Ma, KNÖRZER 1979b.

Der Ursprung des Roggens ist von SENGER und HAWKES (1980) dargestellt worden. Dabei werden auch archäologische Reste erwähnt.

*Setaria italica* (L.) PAL. BEAUV., Borstenhirse

Neuss-Selikum, Ez, KNÖRZER 1980c; Aldenhoven-Niedermerz, Kr. Düren, Ez, KNÖRZER 1980c.

Die Entstehung und Evolution der Borstenhirse ist von DE WET et al. (1979) behandelt worden.

*Sorbus domestica* L., Speierling

Butzbach, Rz, BAAS 1979.

*Triticum aestivum* L., Saatweizen (incl. *T. compactum*)

Yarym-Tepe, Nord-Irak, 6. und 5. Jt. v. u. Z., BAKHTEYEV und YANUSHEVICH (1980); Obre I und Kakanj, Jugoslawien, frühes N, RENFREW 1974; Obre II, Jugoslawien, spätes N, RENFREW 1974; Anza, Jugoslawien, N, RENFREW 1978; Zürich, Schweiz, N, JACOMET-ENGEL 1980; Cova de l'Or, Beniarrés-Alicante, Spanien, N, LÓPEZ GARCÍA 1980; Tel Ṣaf, Israel, Chalkolithikum, GOPHNA und KISLEV 1979; Shahr-i Sokhta, Iran, ca. 3000 v. u. Z., COSTANTINI 1977; Pirak, Pakistan, ca. 1600 v. u. Z., COSTANTINI 1979a; Swat Valley, Pakistan, Ende des 2. Jt. v. u. Z., COSTANTINI 1979b; Hala Sultan Tekke, Zypern, ca. 1200 v. u. Z., HJELMQVIST 1979; Atapuerca, Spanien, Jungbronzezeit, APELLÁNIZ und HOPF 1979; Rajhrad, Bez. Brno-venkov, ČSSR, Hallstattzeit, KÜHN 1980a; Gussage All Saints, Dorset, England, Ez, EVANS und JONES 1979; Malain-Mediolanum, Frankreich, 3. Jh. v. u. Z., ERROUX 1980; Tell-el-Hesi, Israel, Hellenistische Zeit, STEWART 1979; Butzbach, Rz, BAAS 1979; Welzheim, Rz, KÖRBER-GROHNE 1979b; Timmel, Ostfriesland, Ma, KUČAN 1979; Gasselte, Niederlande, Ma, VAN ZEIST und PALFENIER-VEGTER 1979; Wawel, Kraków, Ma, WASYLIKOWA 1979; Hauptmarkt in Kraków, Ma, WIESEROWA 1979.

Die vorgeschichtliche Ausbreitung des Saatweizens nach Asien ist von ZEVEN (1979) untersucht worden.

*Triticum dicoccon* SCHRANK, Emmer

Yarym-Tepe, Nord-Irak, 6. und 5. Jt. v. u. Z., BAKHTEYEV und YANUSHEVICH 1980; Shahr-i Sokhta, Iran, ca. 3000 v. u. Z., COSTANTINI 1977; Obre I und Kakanj, Jugoslawien, frühes N, RENFREW 1974; Obre II, Jugoslawien, spätes N, RENFREW 1974; Anza, Jugoslawien, N, RENFREW 1978; Zürich, Schweiz, N, JACOMET-ENGEL 1980; Eilsleben, Kr. Wanzleben, Linienbandkeramik, Stichbandkeramik und Bernburger Gruppe, SCHULTZE-MOTEL 1980b; Bergheim-Zieverich, Bandkeramik, KNÖRZER 1980b; Wanlo (Mönchengladbach), Bandkeramik, KNÖRZER 1980a; Broichweiden, Kr. Aachen, N, KNÖRZER 1979a; Flögeln, Kr. Cuxhaven, Trichterbecherkultur, ZIMMERMANN 1979; Cova de l'Or, Beniarrés-Alicante, Spanien, N, LÓPEZGARCÍA 1980; Atapuerca, Spanien, Jungbronzezeit, APELLÁNIZ und HOPF 1979; Lindebjerg, Dänemark, ältere Bz, ROWLEY-CONWY 1979; Mönchengladbach-Rheydt, Bz, KNÖRZER 1980c; Tel Şaf, Israel, Chalkolithikum, GOPHNA und KISLEV 1979; Hala Sultan Tekke, Zypern, ca. 1200 v. u. Z., HJELMQVIST 1979; Neuss-Selikum, Ez, KNÖRZER 1980c; Aldenhoven-Niedermerz, Kr. Düren, Ez, KNÖRZER 1980c; Portchester Castle, England, Rz, MURPHY und RENFREW 1975; Rajhrad, Bez. Brno-venkov, ČSSR, Hallstattzeit, KÜHN 1980a; Jüchsen, Kr. Meiningen, Hallstattzeit, GALL 1980; Manching, Ldkr. Ingolstadt, Latène, HOPF 1979b; Welzheim, Rz, KÖRBER-GROHNE 1979b; Tell-el-Hesi, Israel, Hellenistische Zeit, STEWART 1979; Gasselte, Niederlande, Ma, VAN ZEIST und PALFENIER-VEGTER 1979.

*Triticum durum* DESF., Hartweizen
Yarym-Tepe, Nord-Irak, 6. und 5. Jt. v. u. Z., BAKHTEYEV und YANUSHEVICH 1980.

*Triticum monococcum* L., Einkorn
Obre I und Kakanj, Jugoslawien, frühes N, RENFREW 1974; Obre II, Jugoslawien, spätes N, RENFREW 1974; Anza, Jugoslawien, N, RENFREW 1978; Zürich, Schweiz, N, JACOMET-ENGEL 1980; Bergheim-Zieverich, Bandkeramik, KNÖRZER 1980b; Wanlo (Mönchengladbach), Bandkeramik, KNÖRZER 1980a; Cova de l'Or, Beniarrés-Alicante, Spanien, N, LÓPEZ GARCÍA 1980; Atapuerca, Spanien, Jungbronzezeit, APELLÁNIZ und HOPF 1979; Shahr-i Sokhta, Iran, ca. 2300–2200 v. u. Z., COSTANTINI 1977; Mönchengladbach-Rheydt, Bz, KNÖRZER 1980c; Eschweiler-Lohn, Kr. Aachen, Urnenfelderzeit, KNÖRZER 1980c; Rajhrad, Bez. Brno-venkov, ČSSR, Hallstattzeit, KÜHN 1980a; Manching, Ldkr. Ingolstadt, Latène, HOPF 1979b; Welzheim, Rz, KÖRBER-GROHNE 1979b.

*Triticum parvicoccum* KISLEV
Anhand archäologischen Materials hat KISLEV (1979/80) diesen Weizen beschrieben, der im Nahen Osten und im Balkangebiet vorkommt. Bekannt sind nur Ährenspindel-Fragmente und kurze Körner (<5 mm). Dieser älteste Nacktweizen soll kurze, dichte Ähren haben und tetraploid sein (Genom AB). Ob mit der Aufstellung dieser Art einige Schwierigkeiten beseitigt sind, die sich bei der Rekonstruktion der Evolution des Weizens ergeben, sei dahingestellt. Die Art ist auf jeden Fall ungültig veröffentlicht, da eine lateinische Diagnose fehlt.

*Triticum spelta* L., Spelz
Yarym-Tepe, Nord-Irak, 6. und 5. Jt. v. u. Z., BAKHTEYEV und YANUSHEVICH 1980; Gussage All Saints, Dorset, England, Ez, EVANS und JONES 1979; Aldenhoven-Niedermerz, Kr. Düren, Ez, KNÖRZER 1980c; Malain-Mediolanum, Frankreich, 3. Jh. v. u. Z., ERROUX 1980; Portchester Castle, England, Rz, MURPHY und RENFREW 1975; Welzheim, Rz, KÖRBER-GROHNE 1979b.

*Triticum sphaeroccoccum* PERCIVAL, Indischer Kugelweizen
Shahr-i Sokhta, Iran, ca. 2500–2300 v. u. Z., COSTANTINI 1977; Pirak, Pakistan, ca. 1600 v. u. Z., COSTANTINI 1979a; Swat Valley, Pakistan, Ende des 2. Jt. v. u. Z., COSTANTINI 1979b.

*Vicia ervilia* (L.) WILLD., Erve
Tell-el-Hesi, Israel, Hellenistische Zeit, STEWART 1979.

*Vicia faba* L., Ackerbohne
Gussage All Saints, Dorset, England, Ez, EVANS und JONES 1979; Jüchsen, Kr. Meiningen, Hallstattzeit, GALL 1980; Tell-el-Hesi, Israel, Hellenistische Zeit, STEWART 1979; Pompeji, Rz, JASHEMSKI 1977; Welzheim, Rz, KÖRBER-GROHNE 1979b; Gasselte, Niederlande, Ma, VAN ZEIST und PALFENIER-VEGTER 1979; Timmel, Ostfriesland, Ma, KUČAN 1979.

*Vicia sativa* L., Saatwicke
Anza, Jugoslawien, N, RENFREW 1978; Timmel, Ostfriesland, Ma, KUČAN 1979.

*Vitis vinifera* L., Weinstock
Anza, Jugoslawien, N, Wildrebe, RENFREW 1978; Shahr-i Sokhta, Iran, ca. 2000 v. u. Z., COSTANTINI 1977; Pirak, Pakistan, ca. 1600 v. u. Z., COSTANTINI 1979a; Hala Sultan Tekke, Zypern, ca. 1200 v. u. Z., HJELMQVIST 1979; Swat Valley, Pakistan, Ende des 2. Jt. v. u. Z., COSTANTINI 1979b; Tell-el-Hesi, Israel, Hellenistische Zeit, STEWART 1979; Pompeji, Rz, JASHEMSKI 1977; Butzbach, Rz, BAAS 1979; Welzheim, Rz, KÖRBER-GROHNE 1979b; Gamblebyen, Oslo, Norwegen, Ma, GRIFFIN 1979; Wawel, Kraków, Ma, WASYLIKOWA 1979; Hauptmarkt in Kraków, Ma, WIESEROWA 1979; Eschelbronn bei Heidelberg, 13. Jh., KÖRBER-GROHNE 1979a; Neuss, Quirinusstraße, Ma: 14./15. Jh., KNÖRZER 1979c; Burg Brüggen, Kr. Viersen, spätes Ma, KNÖRZER 1979b; Funde im Gebiet der Westslawen, OPRAVIL 1979.

*Zea mays* L., Mais
Archäologische Funde des Mais liegen aus Peru vor (BIRD 1978; BIRD und BIRD 1980; SMITH 1980a, b). Mit der Evolution dieser Art setzte sich BIRD (1979, 1980) auseinander.

*Zizyphus jujuba* MILL., Gemeine Jujube
Pirak, Pakistan, ca. 1600 v. u. Z., COSTANTINI 1979a.

Zum Schluß seien zusammenfassende Arbeiten bzw. solche allgemeinen Inhaltes besprochen.

Die Rasterelektronenmikroskopie ist von FEINDT und MESTERHARM (1980) sowie von KÖRBER-GROHNE und PIENING (1980) zur Bestimmung paläoethnobotanischen Materials herangezogen worden.

Am Beispiel einer Ausgrabung in New Mexico (Salmon Ruin) setzten sich BOHRER und ADAMS (1977) ausführlich mit den paläoethnobotanischen Techniken im Gelände und im Labor auseinander. Die Pflanzenreste dieser Fundstelle sind von ADAMS (1980) eingehend beschrieben worden.

Mit der Identifizierung von Nahrungsmittelresten in archäologischen Funden haben sich ROTTLÄNDER und SCHLICHTHERLE (1979, 1980) beschäftigt.

Anhand quantitativer Analysen hat HUBBARD (1980) die Entwicklung des Ackerbaus in Europa und dem Nahen Osten untersucht. Auch STEWART (1979) berücksichtigt bei seinen Studien der Pflanzenreste von Tell-el-Hesi (Israel) quantitative Methoden. COSTANTINI und TOSI (1975) machen methodologische Vorschläge für paläobiologische Untersuchungen im Iran.

WILLERDING (1979a) hat den ersten Teil einer Bibliographie zur Paläo-Ethnobotanik des Mittelalters in Mittel-Europa erarbeitet. Jensen (1979) beschreibt Samen und andere Diasporen aus mittelalterlichen Schichten von Svendborg, Dänemark.

Berg von Linde (1979) untersuchte die Anfänge der Kultur und Domestikation von Weizen und Gerste.

Mit dem Beginn des Ackerbaus in Mittel-Europa hat sich Tempír (1974b) auseinandergesetzt. Kulturpflanzenfunde der Hallstatt-, Latène- und römischen Zeit aus dem Gebiet von Liptov/Slowakei (ČSSR) hat Hajnalová (1979) beschrieben. Kühn (1980b) behandelt die Getreide und Unkräuter der großmährischen Fundorte und Opravil (1979) stellt den Ostbau der Westslawen anhand archäobotanischer Funde dar.

Bei seiner Schilderung der natürlichen Umwelt der Trichterbecherkultur erwähnt Behre (1979) auch die bisher nachgewiesenen Kulturpflanzen. Willerding (1979b) beschreibt Vegetation und Ackerbau im Hannoverschen Wendland während des Mittelalters. Derselbe Autor (Willerding 1980) faßte das Wissen über den Ackerbau der Bandkeramiker zusammen. Januševič (1980) und Paškevič (1980) beschreiben Kulturpflanzen und Ackerbau in der Tripol'e-Kultur.

Funde von *Capsicum sinense* und *Phaseolus vulgaris* mit Kulturpflanzenmerkmalen lassen erkennen, daß diese Arten im 8000 v. u. Z. in Peru bereits voll domestiziert waren (Smith 1980a). Pearsall (1978) behandelt Fortschritte und Probleme der Paläoethnobotanik im westlichen Süd-Amerika.

Glover (1979) diskutiert archäologische Pflanzenreste aus Südost-Asien, darunter auch die Funde aus der Spirit Cave von Nordwest-Thailand.

## Summary

Literature on archaeological remains of cultivated plants (1979/1980)

The publications of 1979 and 1980 on archaeological remains of cultivated plants were put together and the contents of the 95 references briefly discussed.

## Краткое содержание

Литература об археологических находках остатков культурных растений (1979/1980 гг.)

Продолжение библиографии работ об археологических находках остатков культурных растений. Дается список соответствующей литературы за 1979 и 1980 годы. Кратко комментируется содержание 95 работ.

## Literatur

ADAMS, K. R., 1980: Pollen, Parched Seeds and Prehistory: A Pilot Investigation of Prehistoric Plant Remains from Salmon Ruin, a Chacoan Pueblo in Northwestern New Mexico. – Eastern New Mexico University Contributions in Anthropology Vol. **9** [non vidi]

APELLÁNIZ, J. M. und M. HOPF, 1979: Bronzezeitliche Pflanzenreste aus Alt-Kastilien. Die Cueva Mayor de Atapuerca (Burgos). – Madrider Mitteilungen **20**, 101–106, Taf. 8 u. 9.

ASCH, D. L., and N. B. ASCH, 1977: Chenopod as cultigen: A re-evaluation of some prehistoric collections from eastern North America. – Mid-Continental Journal of Archaeology **2**, 3–45.

–, – 1978: The economic potential of *Iva annua* and its prehistoric importance in the lower Illinois Valley. (pp. 300–341). In: R. I. FORD (ed.), The Nature and Status of Ethnobotany. – Anthropological Papers, Museum of Anthropology, University of Michigan No. **67**, Ann Arbor, Michigan.

BAAS, J., 1979: Kultur- und Nutzpflanzen aus einer römischen Grube in Butzbach und ihr Zusammenhang mit Pflanzenfunden aus anderen römischen Fundstätten. Ein Beitrag zur Geschichte unserer Kulturpflanzen. – Saalburg Jahrbuch, Bericht des Saalburg Museums **36**, 45–82.

BAKHTEYEV, F. KH., and Z. V. YANUSHEVICH, 1980: Discoveries of cultivated plants in the early farming settlements of Yarym-Tepe I and Yarym-Tepe II in Northern Iraq. – J. Archaeol. Sci. **7**, 167–178.

BEHRE, K.-E., 1979: Die natürliche Umwelt der Trichterbecherkultur. (pp. 199–202). In: H. SCHIRNIG (Hrsg.), Großsteingräber in Niedersachsen. – Veröffentlichungen der urgeschichtlichen Sammlungen des Landesmuseums Hannover **24**.

BERG VON LINDE, A., 1979: Man, wheat and barley. Studies on the beginnings of cultivation and domestication. Genetics in archaeology. – Doctoral Dissertation, Department of Classics, Lund University, Sweden, 137 pp.

BIRD, R. McK., 1978: Archaeological maize from Peru. – Maize Genetics Cooperation News Letter No. **52**, 90–92.

–, 1979: The evolution of maize: a new model for the early stages. – Maize Genetics Cooperation News Letter No. **53**, 53–54.

–, 1980: Maize evolution from 500 B. C. to the present. – Biotropica **12**, 30–41. [non vidi]

–, and J. B. BIRD, 1980: Gallinazo maize from the Chicama valley, Peru. – American Antiquity **45**, 325–332.

BOHRER, V. L. and K. R. ADAMS, 1977: Ethnobotanical techniques and approaches at Salmon Ruin, New Mexico. – San Juan Valley Archaeological Project, Technical Series No. 2, Eastern New Mexico University Contributions in Anthropology Vol. **8**, No. 1, XX, 220 pp. [non vidi]

COSTANTINI, L., 1977: Le piante. (pp. 159–171). In: G. TUCCI et al., La città bruciata del deserto salato. The burnt city in the Salt desert. – Ed. ERIZZO, Mestre (Venezia).

–, 1979a: Plant remains at Pirak, Pakistan. (pp. 326–333, planche LV–LVI). In: J.-F. JARRIGE et M. SANTONI, Fouilles de Pirak, Vol. **I**: Texte. – Publications de la Commission des Fouilles Archéologiques, Fouilles du Pakistan no. **2**, Paris.

–, 1979b: Notes on the palaeoethnobotany of protohistorical Swat. – South Asian Archaeology 1977, Papers from the Fourth International Conference of the Association of South Asian Archaeologists in Western Europe, Naples, ed. by M. TADDEI, Vol. **2**, Istituto Universitario Orientale, Seminario di Studi Asiatici, Series Minor VI, 703–708.

–, and M. TOSI, 1975: Methodological proposals for palaeobiological investigations in Iran. – Proceedings of the IIIrd Annual Symposium on Archaeological Research in Iran, 23 Oct.–1 Nov. 1974, 311–331, Tehran.

DALES, G. F., 1979: The Balakot project: Summary of four years excavations in Pakistan. – South Asian Archaeology 1977, Papers from the Fourth International Conference of the Association of South Asian Archaeologists in Western Europe, Naples, ed. by M. TADDEI, Vol. **1**, Istituto Universitario Orientale, Seminario di Studi Asiatici, Series Minor VI, 241–274.

DICKSON, J. H., 1979: Exotic food and drink in ancient Scotland. – Glasg. Nat. **19** (6), 437–442.

–, C. A. DICKSON, and D. J. BREEZE, 1979: Flour or bread in a Roman military ditch at Bearsden, Scotland. – Antiquity **53**, 47–51, pl. VIb.

ERROUX, J., 1980: Fouilles de Malain-Mediolanum. Étude des graines carbonisées de la fin du IIIe siècle après Jesus-Christ. – Revue Archéologique de l'Est et du Centre-Est 31, 100–104.

EVANS, A. M., and M. K. JONES, 1979: The Plant Remains. (Chapter XII: pp. 172–175). In: G. J. WAINWRIGHT (ed.), Gussage All Saints, An Iron Age Settlement in Dorset. – Department of the Environment, Archaeological Reports No. **10**, London HMSO, 202 pp.

FEINDT, F. und G. MESTERHARM, 1980: Rasterelektronenmikroskopie in der Paläoethnobotanik. – Mitt. Inst. Allg. Bot. Hamburg **17**, 135–143.

FELGENHAUER, S., P. RUCKENBAUER, O. STEINECK und F. SWATONEK, 1978: Bestimmung und Beschreibung von Gerstenkörnern von einem Fund aus dem 13./14. Jahrhundert in Stillfried/March, Niederösterreich. – Forschungen in Stillfried 3, 1978, Veröffentlichungen der Österreichischen Arbeitsgemeinschaft für Ur- und Frühgeschichte **10**, 62–66, Taf. 36–38.

GALL, W., 1980: Neue pflanzliche Großreste aus Thüringen. – Ausgrabungen und Funde **25**, 240–241, Taf. 34.

GLOVER, I. C., 1977: The Late Stone Age in eastern Indonesia. – World Archaeology **9**, 42–61, plates 5–6.

–, 1979: Prehistoric plant remains from Southeast Asia, with special reference to rice. – South Asian Archaeology 1977, Papers from the Fourth International Conference of the Association of South Asian Archaeologists in Western Europe, Naples, ed. by M. TADDEI, Vol. **1**, Istituto Universitario Orientale, Seminario di Studi Asiatici, Series Minor VI, 7–37.

GOPHNA, R. et M. KISLEV, 1979: Tel Ṣaf (1977–1978). – Revue Biblique **86**, 112–114, planche III.

GRIFFIN, K., 1979: Plant remains from „Oslogate 7". – De arkeologiske utgravninger i Gamblebyen, Oslo **2**, 124–133.

HAJNALOVÁ, E., 1979: Paleobotanické doklady pol'nohospodárstva a používania drevín v oblasti Liptova v dobe Halštatskej, Laténskej a Rímskej. Paläobotanische Belege über Landwirtschaft und Nutzung von Gehölzen im Liptov-Gebiet in der Hallstatt-, Latène- und römischen Zeit. – Slovenská Archeológia **27**-2, 437–474.

HEISER, C. B., Jr., 1978: Taxonomy of *Helianthus* and origin of domesticated sunflower. (Chapter 2: pp. 31–53). In: J. F. CARTER (ed.), Sunflower Science and Technology. – Agronomy **19**, Madison.

HJELMQVIST, H., 1979: Some economic plants and weeds from the Bronze Age of Cyprus. – Studies in Mediterranean Archaeology **45**: 5, 110–133.

HOLLSTEIN, E., 1978: Dattelimport vor 1800 Jahren. – Kurtrierisches Jahrbuch **18**, 177–179.

HOPF, M., 1977: Untersuchungsbericht der Getreidereste. (pp. 52–53, Taf. 8). In: D. SCHMUDLACH, Die Ausgrabungen im hallstattzeitlichen Hügelgräberfeld bei Berndorf, Landkreis Kulmbach. – Archivband **56** des Historischen Vereins von Oberfranken.

–, 1979a: Organische Reste aus dem römisch-fränkischen Gräberfeld von Krefeld-Gellep. (pp. 208–215, Taf. 121). In: R. PIRLING, Das römisch-fränkische Gräberfeld von Krefeld-Gellep 1964–1965. – Germanische Denkmäler der Völkerwanderungszeit, Serie B, Die fränkischen Altertümer des Rheinlandes Bd. **10**, Berlin.

–, 1979b: Getreideabdrücke aus Grobkeramik in Manching. (pp. 201–204). In: W. E. STÖCKLI, Die Grob- und Importkeramik von Manching. – Röm-German. Komm. d. Dt. Archäol. Inst. zu Frankfurt am Main. (Die Ausgrabungen in Manching Bd. **8**). FRANZ STEINER Verlag GmbH, Wiesbaden.

HUBBARD, R. N. L. B., 1980: Development of agriculture in Europe and the Near East: Evidence from quantitative studies. – Econ. Bot. **34**, 51–67.

JACOMET-ENGEL, S., 1980: Botanische Makroreste aus den neolithischen Seeufersiedlungen des Areals „Pressehaus Ringier" in Zürich (Schweiz). Stratigraphische und vegetationskundliche Auswertung. – Vierteljahrsschr. Naturforsch. Ges. Zürich **125**/2, 73–175.

JANUŠEVIČ, Z. V., 1980: Zemledelie v rannem Tripol'e (po dannym paleobotaniki). (pp. 225–234). In: Pervobytnaja Archeologija – Poiski i Nachodki. Sbornik Naučnych Trudov. – „Naukova Dumka", Kiev.

JASHEMSKI, W. F., 1977: Forschungen und Entdeckungen in den Gärten Pompejis. – Antike Welt **8** (14), 2–16.

JENSEN, H. A., 1979: Seeds and other diaspores in Medieval layers from Svendborg. – The Archaeology of Svendborg, Denmark, No. **2**, Odense University Press, Odense, Denmark, 101 pp.

JØRGENSEN, G., 1980: Om kostvaner i det middelalderlige Svendborg. – Naturens Verden **5–6**, 203–209.

KISLEV, M. E., 1979/1980: *Triticum parvicoccum* sp. nov., the oldest naked wheat. – Israel J. Bot. **28**, 95–107.

KNÖRZER, K.-H., 1979a: [Verkohlte Pflanzenreste aus 2 jungneolithischen Gruben innerhalb des Fundplatzes Broichweiden 6 B (Kr. Aachen)]. – Bonner Jahrb. **179**, 394–395.

–, 1979b: Spätmittelalterliche Pflanzenreste aus der Burg Brüggen, Kr. Viersen. (pp. 595–611). In: M. RECH, Ergebnisse der Ausgrabungen in Burg Brüggen, Kr. Viersen. – Bonner Jahrb. **179**, 565–611.

–, 1979c: [Pflanzenfunde des 14./15. Jh. aus einer mittelalterlichen Grube in Neuss, Quirinusstraße]. – Bonner Jahrb. **179**, 720–721.

–, 1980a: Pflanzliche Großreste des bandkeramischen Siedlungsplatzes Wanlo (Stadt Mönchengladbach). In: Naturwissenschaftliche Beiträge zur Archäologie. – Archaeo-Physika **7**, 7–20, Tabelle 1.

–, 1980b: Bandkeramische Pflanzenfunde aus Bergheim-Zieverich. In: Naturwissenschaftliche Beiträge zur Archäologie. – Archaeo-Physika **7**, 21–23.

–, 1980c: Neue metallzeitliche Pflanzenfunde im Rheinland. In: Naturwissenschaftliche Beiträge zur Archäologie. – Archaeo-Physika **7**, 25–34.

–, 1980d: Römerzeitliche Pflanzenfunde aus Aachen-Burtscheid. In: Naturwissenschaftliche Beiträge zur Archäologie. – Archaeo-Physika **7**, 35–60.

KÖRBER-GROHNE, U., 1979a: Samen, Fruchtsteine und Druschreste aus der Wasserburg Eschelbronn bei Heidelberg (13. Jahrhundert). – Forschungen und Berichte der Archäologie des Mittelalters in Baden-Württemberg **6**, 113–127.

–, 1979b: Botanische Untersuchungen an den römerzeitlichen Brunnenfüllungen des Ostkastells von Welzheim (Vorbericht). – Archäologisches Korrespondenzblatt **9**, 419–422, Taf. 75.

–, and U. PIENING, 1980: Microstructure of the surfaces of carbonized and non-carbonized grains of cereals as observed in scanning electron and light microscopes as an additional aid in determining prehistoric findings. – Flora **170**, 189–228.

KROLL, H., 1977: Getreideabdrücke in Gefäßscherben. (p. 41). In: W. WEBERS, H. HINZ, F. KONERDING und G. LÖFFLER, Bosau. Untersuchung einer Siedlungskammer in Ostholstein, unter Leitung von H. HINZ. II, Die Ausgrabungen und Untersuchungen im Dorf 1971–1975. – Offa-Bücher **37**, Neumünster.

–, 1980a: Botanische Funde aus der Lübecker Innenstadt. – Archäologie in Lübeck, Hefte zur Kunst und Kulturgeschichte der Hansestadt Lübeck **3**, 122–124.

–, 1980b: Mittelalterlich/frühneuzeitliches Steinobst aus Lübeck. – Lübecker Schriften zur Archäologie und Kulturgeschichte **3**, 167–173, Abb. 51–53, Tafel 31.

KUČAN, D., 1979: Mittelalterliche Kulturpflanzen und Unkräuter aus ostfriesischen Kirchen. – Probleme der Küstenforschung im südlichen Nordseegebiet **13**, 23–38, Taf. 1–5.

KÜHN, F., 1980a: Botanický rozbor obilí z halštatské doby z Rajhradu (okr. Brno-venkov). Botanische Getreideanalyse aus der Hallstattzeit in Rajhrad (Bez. Brno-venkov). – Přehled výzkumů 1977, ed. Archeologický ustav ČSAV v Brno, Brno, 43–46.

–, 1980b: Getreide und Unkräuter der großmährischen Fundorte in botanischer Hinsicht. – Rapports du IIIe Congrès International d'Archéologie Slave, Bratislava 7–14 septembre 1975, Tome **2**, Bratislava, 231–235.

Lange, E., 1979: Zum Stand und einigen Ergebnissen paläo-ethnobotanischer Arbeiten in Brandenburg. – Gleditschia **7**, 187–196.

López García, P., 1980: Los cereales. (pp. 175–192). In: B. Marti Oliver, V. Pascual Perez, M. D. Gallart Marti, P. López García, M. Perez Ripoll, J. D. Acuña Hernandez y F. Robles Cuenca, Cova de l'Or (Beniarrés-Alicante), Vol. II. – Servicio de Investigacion Prehistorica, Diputacion Provincial de Valencia, Serie de Trabajos Varios, Núm. **65**, Valencia.

Murphy, P., and J. M. Renfrew, 1975: The carbonized cereals. (pp. 416–418). In: B. Cunliffe, Excavations at Portchester Castle. Vol. **1**. Roman. – Soc. of Antiquaries of London.

Nabhan, G. P., J. W. Berry and C. W. Weber, 1980: Wild beans of the Greater Southwest: *Phaseolus metcalfei* and *P. ritensis*. – Econ. Bot. **34**, 68–85.

Natural History Section, Chekiang Provincial Museum, 1978: A study of the animal and plant remains unearthed at Ho-mu-tu. – Kaogu Xuebao (1), 106–107.

Opravil, E., 1979: Der Obstbau der Westslawen im Lichte archäobotanischer Funde. – Rapports du IIIe Congrès International d'Archéologie Slave, Bratislava 7–14 septembre 1975, Tome **1**, Bratislava, 587–595.

Paškevič, G. A., 1980: Kul'turnye rastenija Tripol'skich poselenij Podneprov'ja. (pp. 234–242). In: Pervobytnaja Archeologija–Poiski i Nachodki. Sbornik Naučnych Trudov. – „Naukova Dumka", Kiev.

Pearsall, D. M., 1978: Paleoethnobotany in western South America: Progress and problems. (pp. 389–416). In: R. I. Ford (ed.), The Nature and Status of Ethnobotany. – Anthropological Papers, Museum of Anthropology, University of Michigan No. **67**, Ann Arbor, Michigan.

Quillian, C. C., 1980: The plant remains at Hascherkeller: an Early Iron Age settlement in Bavaria. – A thesis presented to the Department of Anthropology. Harvard College, 114 pp.

Renfrew, J. M., 1974: Report on the carbonized cereal grains and seeds from Obre I, Kakanj and Obre II. – Wissenschaftliche Mitteilungen des Bosnisch-Herzegowinischen Landesmuseums **4**, Heft A, Archäologie, 47–53.

–, 1977: Seeds from Area K. (Appendix 3: p. 127–128, Pl. 92). In: J. Coleman, Keos. 1. Kephala: A Late Neolithic Settlement and Cemetery. – Am. School Classical Studies.

–, 1978: Carbonized seeds from Anza. (pp. 300–312). In: M. Gimbutas, Neolithic Macedonia – the excavations at Anza. – Monumenta Archaeologica **1**, U. C. L. A.

Rottländer, R. C. A., and H. Schlichtherle, 1979: Food identification of samples from archaeological sites. – Archaeo-Physika **10**, 260–267.

–, –, 1980: Gefäßinhalte. Eine kurz kommentierte Bibliographie. – In: Naturwissenschaftliche Beiträge zur Archäologie. Archaeo-Physika **7**, 61–70.

Rowley-Conwy, P., 1979: Forkullet korn fra Lindebjerg. En boplads fra aeldre bronzealder. The carbonized grain from Lindebjerg. – Kuml **1978**, 159–171.

Schultze-Motel, J., 1980a: Literatur über archäologische Kulturpflanzenreste (1978/1979). – Kulturpflanze **28**, 361–378.

–, 1980b: Neolithische Kulturpflanzenreste von Eilsleben, Kr. Wanzleben. – Z. Archäol. **14**, 213–216.

Sencer, H. A., and J. G. Hawkes, 1980: On the origin of cultivated rye. – Biol. J. Linn. Soc. **13**, 299–313.

Smith, C. E., Jr., 1980a: Plant Remains from Guitarrero Cave. (pp. 87–119). In: T. F. Lynch (ed.), Guitarrero Cave. Early Man in the Andes. – Academic Press, Inc., New York and London.

–, 1980b: Ancient Peruvian Highland Maize. (pp. 121–143). In: T. F. Lynch (ed.), Guitarrero Cave. Early Man in the Andes. – Academic Press, Inc., New York and London.

Stewart, R. B., 1979: Archeobotanic studies at Tell-el-Hesi. – Econ. Bot. **32**, 1978, 379–386.

TEMPÍR, Z., 1974a: Polní a zahradní plodiny v archeologických nálezech z Kozího Hrádku u Tábora (15. stol. n. l.). Field and garden plants in archeological findings from Kozí Hrádek near Tábor (15th century A. D.). – Vědecké Práce Zemědělského Muzea **14**, 5–15.

–, 1974b: Příspěvek k počátku pěstování rostlin ve střední Evropě. A contribution to the study of the beginning of plant growing in Central Europe. – Vědecké Práce Zemědělského Muzea **14**, 17–26.

WASYLIKOWA, K., 1979: Plant remains from early and late Medieval time found on the Wawel Hill in Cracow. Szczątki roślinne z wczesnogo i późnego średniowiecza znalezione na Wzgórzu Wawelskim w Krakowie. – Acta Palaeobotanica **19** (2), 1978, 115–200, plates I–IX, 15 folders.

WENDORF, F., R. SCHILD, N. E. HADIDI, A. E. CLOSE, M. KOBUSIEWICZ, H. WIECKOWSKA, B. ISSAWI, H. HAAS, 1979: Use of barley in the Egyptian Late Paleolithic. – Science **205** (4413), 1341–1347, cover.

WEST, M., and T. W. WHITAKER, 1980: Prehistoric cultivated cucurbits from the Viru Valley, Peru. – Econ. Bot. **33**, 1979, 275–279.

WET, J. M. J. DE, L. L. OESTRY-STIDD, and J. I. CUBERO, 1979: Origins and evolution of foxtail millets (*Setaria italica*). – Journ. d'Agric. Trad. et de Bota. Appl. **26**, 53–64.

WIESEROWA, A., 1979: Plant remains from the early and late Middle Ages found in the settlement layers of the Main Market Square in Cracow. – Acta Palaeobotanica **20** (2), 137–212, 5 folders.

WILLERDING, U., 1979a: Bibliographie zur Paläo-Ethnobotanik des Mittelalters in Mitteleuropa 1945–1977 (Teil 1). – Zeitschrift für Archäologie des Mittelalters **6**/1978, 173–223.

–, 1979b: Vegetation und Ackerbau im Hannoverschen Wendland während des Mittelalters. (pp. 905–915). – Rapports du IIIe Congrès International d'Archéologie Slave, Bratislava 7–14 septembre 1975, Tome **1**, 958 pp., Bratislava.

–, 1980: Zum Ackerbau der Bandkeramiker. In: Beiträge zur Archäologie Nordwestdeutschlands und Mitteleuropas. – Materialhefte zur Ur- und Frühgeschichte Niedersachsens **16**, 421–456.

YARNELL, R. A., 1978: Domestication of sunflower and sumpweed in eastern North America. (pp. 289–299). In: R. I. FORD (ed.), The Nature and Status of Ethnobotany. – Anthropological Papers, Museum of Anthropology, University of Michigan No. **67**, Ann Arbor, Michigan.

YOU, X.-L., 1979: On the origin, differentiation and spread of the cultivated rice in China – an investigation with the carbonized rice found at the Hemudu excavation. – Acta Agronomica Sinica **5** (3), 1–10, pl. I (chin., Engl. summ.).

ZEIST, W. VAN and R. M. PALFENIER-VEGTER, 1979: Agriculture in Medieval Gasselte. – Palaeohistoria **21**, 267–299.

ZEVEN, A. C., 1979: The prehistoric spread of bread wheat into Asia. (pp. 103–107). In: S. RAMANUJAM (ed.), Proceedings of the fifth International Wheat Genetics Symposium. Vol. **1**, Session 1, Evolution of wheat–changing concepts. New Delhi 1978.

ZIMMERMANN, W. H., 1979: Ein Hausgrundriß der Trichterbecherkultur von Flögeln – Im Örtjen, Kreis Cuxhaven. – Veröff. urgesch. Samml. Landesmus. Hannover **24**, 247–253.

Dr. J. SCHULTZE-MOTEL
Zentralinstitut für Genetik und Kulturpflanzenforschung
der Akademie der Wissenschaften der DDR
4325 Gatersleben

Kulturpflanze XXIX • 1981 • S. 465–483

# Taxonomy and evolution of cultivated plants: Literature review 1979/1980

JÜRGEN SCHULTZE-MOTEL, REINHARD FRITSCH, KARL HAMMER, PETER HANELT, JOACHIM KRUSE, HELGA I. MAASS, HARALD OHLE and KLAUS PISTRICK

(Eingegangen am 23. Januar 1981)

## Summary

Important papers on taxonomy and evolution of cultivated plants published in 1979 and 1980 were put together and briefly discussed.

## Introduction

In continuation of our literature reviews on taxonomy and evolution of cultivated plants (SCHULTZE-MOTEL et al. 1978, 1979, 1980) we present here the papers on this topic published in 1979 and 1980 as well as some publications of former years being overlooked in our earlier papers.

### *Araceae*

*Acorus* L.
RÖST (1979) has classified the genus by using micromorphological and chemotaxonomic characters mainly. The genus comprises the diploid *A. gramineus* SOLAND. and *A. calamus* L., which forms a supposedly autopolyploid series (2x, 3x, 4x). Keys are given for the determination of species as well as infraspecific races.

### *Compositae*

*Anacyclus* L.
A detailed revision of the genus has been given by HUMPHRIES (1979). Formerly the species *A. pyrethrum* (L.) LINK und *A. officinarum* HAYNE have been cultivated for medicinal uses in western Europe to a limited extent. The hypothesis most compatible with the results suggests that the latter 'species' is an unusual annual derivative of *A. pyrethrum*.

*Helianthus* L.
On the basis of morphological, geographical, genetical and immunochemical results ANAŠČENKO (1979) has proposed a phylogenetic scheme of the North American representatives of the genus. A lot of crosses have been produced using more than 100 races of wild and also cultivated species. Some of the crosses with *H. tuberosus* L. resulted in useful new lines for breeding purposes with *H. annuus* L.

## *Cruciferae*

*Brassica* L.

GUSTAFSSON et al. (1976) studied meiosis in *Brassica cretica* LAM. ssp. *cretica* and ssp. *nivea* (BOISS. et SPR.) ONNO. Almost all plants grown from spontaneous seeds and experimental hybrids had 2n=18 chromosomes. At MI quadrivalents and at AI cells with bridges and/or laggards were observed. The interpretation of these disturbances and the presumed nature of breeding barriers are discussed.

Karyotypes of *Brassica rapa* L., *B. juncea* (L.) CZERN. and *B. carinata* A. BRAUN have been worked out by DASTIDAR et al. (1979).

## *Cucurbitaceae*

*Cucumis* L.

With the aid of analyses of esterase, peroxidase and leucine-amino-peptidase patterns PUCHALSKI et al. (1978) have tried to discover relationships between 24 species of *Cucumis*. The zymogrammes of these species are compared with cytogenetical facts, and consequences for new infraspecific classifications are mentioned.

For breeding work in *Cucumis* there is a need of exact data 1) about the genetic variation in the Asiatic group of species (without *C. sativus* L.); 2) about the position of a species on the evolutionary tree for successful crossings with *C. sativus*; 3) about facts of the flower and pollination biology and 4) of the genetic embryo-endosperm-relation besides of problems of bridge-crosses. Previous results on species crossings enabled proposals for further research (KROON et al. 1979).

In the genus *Cucumis* there is a group of cross-compatible species from Africa and a group of incompatible ones. All of these species have been characterised by cytogenetic methods. In interspecific hybrids DANE (1977) has observed anomalies in the meiosis and reduced pollen fertility and he has studied the genetics of different alleles of peroxidase.

## *Ebenaceae*

*Diospyros lotus* L.

NG (1979) discussed the origin of the date-plum and presented a map of the distribution of its two varieties.

## *Ericaceae*

*Vaccinium* L.

Numerical, cytological, and experimental analyses of VANDER KLOET (1980) suggest that the highbush group belonging to *Vaccinium* L. subgen. *Cyanococcus* GRAY should be recognized as a single polymorphic species, namely *V. corymbosum* L. The author postulates that the group originated in northern Florida from a hybrid swarm which resulted when *V. tenellum* AITON came into contact with *V. darrowii* CAMP during the Pleistocene.

## *Gramineae*

*Aegilops* L. (see also *Triticum*!)

A monographic description has been given of the *Aegilops*-collection of the Gatersleben gene bank (HAMMER 1980b). Cultivation requirements and techniques, seed storage, and documentation of the collection are the main topics. Considering the characters of pollination ecology conclusions about the evolution within the genus are draw and the participation of *Aegilops* species in the evolution of *Triticum* is discussed. Thus the Gatersleben *Aegilops*-collection provides good possibilities for botanical research. Furtheron it is an important source of material for basic research as well as for practical plant breeding. Another paper has been published about the detailed description of the taxonomy and nomenclature of the genus (HAMMER 1980a).

In his book GANDILJAN (1980) presents determination keys for the species of *Aegilops*.

GENDEL's (1980) investigated the leaf anatomy of 22 *Aegilops* species considering different moisture conditions. On the basis of the anatomical characters corrections in the system of the genus *Aegilops* L. are proposed.

*Avena* L.

On the basis of electrophoretical studies of seed prolamine of di-, tetra- and hexaploid *Avena* species, KIM and MOSSÉ (1979) were able to detect the bands specific for genomes A, C and D. Like the diploids, the tetraploids can be divided into two groups: *A. barbata* POTT. ex LINK (derived from an A genome species only) and *A. magna* MURPHY et TERRELL–*A. murphyi* LAD. (derived from both A and C genome species). The hexaploids were derived from the *A. magna*–*A. murphyi* group and a not yet discovered D genome ancestor. JAIN and SINGH (1979) analyzed genetic similarity among 218 populations of 15 *Avena* species by scoring alloenzyme variation and were able to confirm largely genome relationships postulated from ontogenetic and morphological-ecogeographical studies.

*Eleusine* GAERTN.

A survey of leaf flavonoids of 7 *Eleusine* taxa was conducted by HILU et al. (1978). The morphologically well defined group *E. coracan* (L.) GAERTN.–*E. africana* KENNEDY–O'BYRNE–*E. indica* (L.) GAERTN. also forms a unit with respect to flavonoids. Race 'dezda' of *E. africana* has an identical flavonoid pattern with *E. coracan*. This suggests either an origin of the crop directly from this race, or extensive introgression from the crop into *E. africana*. Flavonoid patterns also support earlier conclusions that the African highland race is the progenitor of the other races of finger millet.

*Eragrostis* HOST

The relationships of *Eragrostis tef* (ZUCC.) TROTTER have been investigated by JONES et al. (1978) using morphological characters and chromosome number. *Eragrostis pilosa* (L.) P. BEAUV. and some taxa of the *E. cilianensis* (ALL.) HUBB. complex have been found to be most closely related to *E. tef*.

COSTANZA et al. (1980) studied 39 collections of *E. tef* in biosystematic detail and compared them with *E. aethiopica* CHIOV. and *E. pilosa*. The latter species is morphologically variable but closely allied to *E. tef*. Three morphologically distinct complexes are recognized within the cereal, but the ethnological significance of these complexes could not be explained.

*Hordeum* L.

Trofimovskaja and Gudkova (1979) examined the glumes, lemmas and caryopses (aleurone layer) of 24 *Hordeum* species anatomically. The anatomical as well as morphological characters have been discussed in respect to the classification to the genus *Hordeum* L. Bachteev (1978) discussed some theories about the origin of the six-rowed barley. A historical review about the discovery of *H. agriocrithon* Åberg and *H. lagunculiforme* Bacht. has been given by Bachteev (1977). The author regards *H. agriocrithon* as the ancestor of cultivated barley.

Nevo et al. (1979) investigated the genetic diversity of 28 populations of *Hordeum spontaneum* C. Koch, collected in Israel, in connection with some environmental factors. They found a very high and partly adaptive variability in wild barley.

The variation pattern of the storage protein hordein has been examined in barley by Doll and Brown (1979). A very high polymorphism has been found in hordein polypeptides of wild and cultivated barley. Immunoelectrophoretical studies of barley protein (Djagileva 1979) have shown that the relationship of *H. murinum* L. to *H. spontaneum* C. Koch and to the cultivated barley is closer than that of *H. bulbosum* L. and of other small-grained wild species to *H. spontaneum* and to the cultivated barley.

In his book Gandiljan (1980) presents determination keys for the species and botanical varieties of *Hordeum*.

*Oryza* L.

Katayama and Ogawa (1977) crossed diploid *Oryza punctata* Kotschy et Steud. with species of the C genome in order to clarify whether *O. punctata* was the donor of the B genome or not. They produced amphiploids with colchicine treatment of $F_1$ plants. Morphologically, the synthesized amphiploids, with the exception of grain size, were similar to tetraploid *O. punctata*. In meiosis, the number of bivalents ranged from 21–24. Comparing the isoenzyme variations of several population of *Oryza perennis* Moench from Thailand (annual, ratooned and mixture types, resp.), Oka and Sano (1978) draw the conclusion, that polymorphous and heterozygous intermediate populations of *O. perennis* could be the wild progenitor of *O. sativa* L.

*Pennisetum* L. C. Rich.

Basing on morphometric analyses of spike region, Brunken (1979) divides the taxa of *Pennisetum* sect. *Brevivalvula* into three species: *P. pedicellatum* Trin. (including a subspecies *unispiculum*), *P. hordeoides* and *P. polystachion* (L.) Schult. (with subsp. *setosum* and subsp. *atrichum*) and gives a comprehensive characterization of these species.

*Secale* L.

Sencer and Hawkes (1980) investigated the problem of the origin of cultivated rye, taking into consideration the arguments of numerous aspects. According to these results wild types of *S. cereale* L. emend. Sencer have evolved from *S. montanum* Guss. emend. Sencer and invaded fields of wheat and barley as a weed at an early time. Under these conditions rye types with characters of cultivated plants have been selected by man. First growing as a weed, the rye at last had become an autonomous cultivated plant in areas unfavourable for wheat and barley. Eastern Turkey has been regarded as the area of origin of cultivated rye.

KOBYLJANSKIJ (1979) discussed problems of the origin and the migration routes of rye cultivation, accordingly rye cultivation was distributed in three ways: from the Caucasus area towards the north and to western Europe, from Asia Minor to southern and western Europe and from a further diversity centre to central Asia.

In his book GANDILJAN (1980) presents determination keys for the species and botanical varieties of *Secale*.

*Setaria* P. BEAUV.

GUPTA and SINGH (1977) studied meiosis and distribution of flavonoids in several species. The base number of the genus was found to be $x=9$. The many polyploid collections are demonstrating the important role of polyploidy in the evolution of the genus. The thin layer chromatographic data were computed into a numerical chemotaxonomical study.

DE WET et al. (1979) were dealing with morphology, domestication and grain storage proteins of foxtail millets, *Setaria italica* (L.) P. BEAUV. They found it to be highly unlikely that this species was domesticated in China and introduced into Europe as a crop, or vice versa. Wild foxtail millets occur across Eurasia and could have been taken into cultivation at several places.

*Sorghum* MOENCH

Using the example of *Sorghum bicolor* (L.) MOENCH, DE WET and SHECHTER (1977) were dealing with the problem of domestication process of grasses. Harvesting alone does not necessarily lead to domestication, and the plant must be able of being adapted to a habitat that is permanently and continously being disturbed by man.

IVANJUKOVIČ and DORONINA (1978) give a dichotomous key to 31 cultivated species (in the sense of SNOWDEN), basing on SNOWDEN's polytomous key. IVANJUKOVIČ (1980) provides a review on the many theories on history and origin of the cultivated sorghums, especially emphasizing the role of hybridization in the evolutionary process.

*Triticum* L. (and *Aegilops* L.)

Based on the idea of a diphyletic origin of tetraploid wheat DOROFEEV and MIGUŠOVA (1978, 1979) developed a new system of this genus. It comprises subgen. *Triticum* (with the sections *Urartu*, *Dicoccoides* and *Triticum*) and subgen. *Boeoticum* (with the sections *Monococcon*, *Timopheevii* and *Kiharae*). KONAREV et al. (1978) discussed this system in respect to genome composition of the species.

UDAČIN (1978) discussed the role of Central Asia in the evolution of the genus *Triticum* (also on the base of this theory). The hexaploid naked wheats are autochthonous in this region.

A cytogenetic and cytoembryological investigation of $F_1$ hybrids from reciprocal crossings of *Triticum monococcum* L. var. *flavescens* KÖRN. with seven tetraploid wheat species was carried out by ČELAK (1980). The genome incongruity of the parental forms was the cause of the $F_1$ hybrid sterility. It was also found that the genome of *T. monococcum* var. *flavescens* did not participate in the origin of the genome A of the tetraploid wheat species.

The origin of the B genome in *T. dicoccon* SCHRANK was investigated by PENEVA (1979), in *T. carthlicum* NEVSKI by BARISAŠVILI (1979).

SEARS (1977) gave a short review on the origin of wheat as well as an outlook to its future (*Triticale*).

BUJUKLI (1977) described five new botanical varieties of *T. durum* DESF. in Moldavia which were selected from various crosses.

DOROFEEV and MIGUŠOVA (1977) described the new subspecies *kurdistanicum* of *T. araraticum* JAKUBC. from Iraq which comprises three botanical varieties.

In Xizang (Tibet) SHAO et al. (1980) detected a semi-wild wheat which SHAO described as the new subspecies *tibetanum* of the common wheat *T. aestivum* L. em. FIORI et PAOL. This wheat is characterized by mature spikelets which freely separate and is in this character similar to *T. aestivum* ssp. *yunnanensis* but the former is more primitive.

The volume I, Wheat, of the Flora of Cultivated Plants of the USSR appeared as a new edition (DOROFEEV and KOROVINA 1979). The treatment–following the above mentioned system–is very comprehensive and elaborate. Many new combinations have been made and many new botanical varieties have been described.

There is another book dealing with wheat (GANDILJAN 1980). It comprises determination keys for the species and botanical varieties of *Triticum* using mainly the author's formula-like descriptions which are handled as names likewise.

NISHIKAWA and FURUTA (1979) investigated the DNA content of different species of the genera *Triticum* and *Aegilops* and discussed the results in respect to the evolution of the genus *Triticum*.

The problem of the origin of B genome in polyploid wheats has been investigated by means of hybridized DNA (PARODA 1977). According to the results *Ae. sharonensis* EIG seems to be most likely the donor of the B genome.

DNA hybridization studies of the distribution patterns of *Triticum* and *Aegilops* species have been carried out by GERLACH et al. (1979). The results indicate only little changes of the B genome during the evolution of polyploid wheats and the existence of the same B genome type but a different A genome type in *Triticum timopheevii* ZHUK. and *T. dicoccon* SCHRANK. Resemblance between the B genome of the polyploid *Triticum* species and the genome of *Ae. longissima* SCHWEINF. has been found.

TANAKA et al. (1979) analysed cytological characters of a natural hybrid of *T. araraticum* JAKUBZ. and *T. dicoccoides* KÖRN. as well as an induced amphidiploid of *Ae. speltoides* TAUSCH and *T. boeoticum* BOISS. em. SCHIEM. The authors suppose that *T. araraticum* JAKUBZ. and *T. dicoccoides* KÖRN. have a common origin from a natural hybrid *Ae. speltoides* × *T. boeoticum*, originated in the area of sympatric distribution of both the parental species in Iran.

NAKAI (1979a) examined the esterase isozymes of diploid and tetraploid

*Triticum* species and *Ae. speltoides* as well as artificial amphiploid hybrids between both genera. The author concludes that the tetraploid species of *Triticum* originated from a hybrid between *T. monococcum* L. and one species of the *Sitopsis* group of the genus *Aegilops*. *Aegilops speltoides* is regarded as that species from which all other species if the *Sitopsis*-group originated. Turkey and Iraq are supposed to be the area of origin of *T. monococcum*.

JAASKA (1978) discussed the origin and phylogeny of polyploid wheats on the basis of the different types of NADP-dependent aromatic alcohol dehydrogenase. These investigations confirm the opinion that *Ae. speltoides* s.l. is the genetically most closely related diploid species to the ancestor form connected with the origin of tetraploid wheats. The examined isoenzymes indicate differences between the tetraploid wheats of the emmer group and the *timopheevi* group as well as within the emmer group. Further distinctions have been found between the West-European and the Asian spelt.

The variation patterns of six seedling esterases have been investigated with respect to their distribution in polyploid wheats and their diploid relatives in the genera *Triticum* and *Aegilops* (JAASKA 1980). The results refer to a monophyletic origin of tetraploid and hexaploid wheats. *Triticum boeoticum* BOISS., *Tr. urartu* TUM. ex GANDIL., *Ae. speltoides* TAUSCH and *Ae. tauschii* COSS. ssp. *strangulata* (EIG) TZVEL. or closely related forms are most likely the diploid ancestor of the polyploid wheats.

NAKAI (1979b) analyzed the esterase isozymes of synthesized hexaploid wheats and of different parental strains of tetraploid wheat species and *Ae. tauschii* COSS. The cymograms of special strains of *Ae. tauschii*, in connection with the findings of other authors, support the suggestion that the birthplace of common wheat is the area enclosing the southern coast of Caspian Sea and some parts of Transcausasus.

The pollen of the genera *Triticum* and *Aegilops* has been tested by means of SEM (KRUSE 1980). The wheat species show differences in exine sculpturing according to the genome and the ploidy level.

*Zea* L.

The origin of cultivated maize in South America, the subsequent history of its cultivation in Europe and the development of hybrid varieties in America and Europe are traced by BUNTING (1978).

Several papers dealing with the evolution of *Zea mays* L. are given in chapter 2 of a book edited by WALDEN (1978): A hypothesis is proposed by GALINAT, that maize has evolved from teosinte under the influence of man following the establishment of ancestral populations of teosinte near camp sites as a result of seed spillage from the harvest of wild populations. The possible process of the domestication of a maize ear from a teosinte-like spike is pictorially summarized. BEADLE proposed the hypothesis, that a dominant tunicate mutation occurred as first step in the transition from teosinte to maize. DE WET and HARLAN discussed the cytogenetic affinities between *Zea* and *Tripsacum* excluing *Tripsacum* as progenitor of maize. The pod-corn, teosinte, common-ancestry and tripartite theories for the origin of maize are outlined by BROWN. He also presents three anomalies regarding the teosinte theory which remain unsolved. Two further papers are dealing with the origins, main characters and migration of races of maize.

The pattern of genome organization of *Z. mays* and the relationship of maize to possible progenitor species as assessed by DNA: DNA hybridization have been analyzed by HAKE und WALBOT (1980). Teosinte and popcorn are approximately equally divergent from maize 'Midwestern Dent', whereas *Tripsacum laxum* NASH is much more divergent. The divergence times indicate that the divergence within *Z. mays* and between maize and its near relatives is at least an order of magnitude greater than expected.

Chloroplast and mitochondrial DNAs from 4 races of annual teosinte (*Zea mays* ssp. *mexicana* (SCHRAD.) ILTIS) from Mexico, 2 races of annual teosinte (*Z. mays* subsp. *luxurians* (DURIEU) ILTIS) from Guatemala, *Zea perennis* (HITCHC.) REEVES et MANGELSD., and *Z. mays* were compared and grouped by TIMOTHY et al. (1979) using restriction endonuclease fragment analysis. Only the race from southern Guatemala and perennial teosinte differed markedly from maize in both DNA fragment patterns.

### Labiatae

*Stachys* L.

For the 154 species of *Stachys* from Europe, Africa and Asia a new infrageneric classification is constructed. The species are distributed among 2 subgenera and 21 sections. Keys of all supraspecific ranks and groups are proposed and chorological facts with isoflor maps are demonstrated (BHATTACHARJEE 1980).

*Thymus* L.

A hybrid plant from *Thymus serpyllum* L. em. MILL. × *T. vulgaris* L., which the author found in a garden in Finland, has been described morphologically. Chromosome-number and the composition of the essential oils of the hybrid are intermediate between its parents (WIDÉN et al. 1977).

### Leguminosae

SMARTT (1980a) has reviewed the progress of biosystematic, karyological, phytogeographical, chemotaxonomical and archaeological studies in important food legumes referring to problems of their evolution. Most of the cultigens have been found to have conspecific wild relatives; the taxonomic rank of them should be generally the variety (which is of course not in agreement with other recent proposals). Two exceptions are *Vicia faba* L. and *Cajanus cajan* (L.) MILLSP. for which no obviously related wild forms are known. Tables with the cultigens and closest known relatives and the suggested nomenclature have been given. Special attention has been paid to *Phaseolus* L., *Arachis* L., *Cajanus* ADANS. and *Vigna* SAVI and the differences of evolutionary advancement under domestication between different crops have been stressed.

Gel-electrophoretic studies of seed albumins of different genera of the family, especially of *Pisum* and *Vicia*, have shown that by these means relationships could be verified only on a low taxonomic level, e.g. within one species. The banding patterns for different species or genera mostly don't make any taxonomical sense (WOLF 1980). The pattern for *Pisum* is partly in disagreement with the data for cross-fertility and/or gross morphology and need to be re-interpreted in a broader context.

LADIZINSKY (1979a) considered mutations that prevented dehiscence, dropping, and shattering of pods as one of the most important steps towards domestication of lentil, pea, grass pea, chick pea, and fenugreek.

*Arachis* L.
Branching pattern, intrageneric classification, breeding systems and the evolution of the genus in dependence of the geological history of the South-American Planalto and the development of its different drainage basins are the topics of an important paper by GREGORY et al. (1980); *A. hypogaea* L. is thought to have evolved from a wild allotetraploid along the eastern slopes of the Andes, the annual *A. batizocoi* KRAP. et GREG. and a still undescribed (but already named!) perennial species (both diploid) could be the ancestral taxa.

*Cajanus cajan* (L.) MILLSP.
Floristic, linguistic and cytological evidence points to an Indian origin of the pigeonpea (VAN DER MAESEN 1980).

*Cyamopsis* DC.
WHISTLER and HYMOWITZ (1979) dealt with the history, taxonomy, agronomy, production, and use of guar, *C. tetragonoloba* (L.) TAUB. They also presented a hypothesis for the origin of this annual crop that was never found in the wild. Biosystematic studies are needed to analyse the relationship between the species of this small genus.

*Glycine* WILLD.
HYMOWITZ and NEWELL (1977, 1980) have summarized the recent state of knowledge regarding taxonomic classification, distribution and variability of this genus and the domestication of *Gl. max* (L.) MERR. around the 11th century B. C. in the eastern half of North China, the migration of the crop to Korea (1st century AD), Japan (7th century AD) and other Asian countries and about the history of its establishment in North America.

By a numerical-taxonomic study of the phenotypic variation BROICH and PALMER (1980) have found a clear differentiation between *Gl. max* and the wild progenitor *Gl. soja* SIEB. et ZUCC.; the semi-wild or weedy races known as *Gl. gracilis* SKVORTZ. could have been shown as conspecific with *Gl. max*.

A designation of these plants (supposed to be of hybrid origin from the above mentioned species) as forma has been proposed.

*Lens* MILL.
LADIZINSKY (1979b) analysed the seed proteins for evolutionary studies of 36 lines of *L. culinaris* MEDIK., *L. ervoides* (BRIGN.) GRANDE, *L. nigricans* (BIEB.) GODR., and *L. orientalis* (BOISS.) SCHMALH. The protein profiles of the cultivated lentil and the latter two wild species were similar, thus corroborating the close morphological and cytogenetic affinities of these three taxa. *Lens ervoides* was distinct from the other species both in number and position of bands.

*Leucaena* BENTH.
Populations of *Leucaena* species, collected from Texas to Peru, were examined by BREWBAKER and OAKES (1978). Ten species were recognized as valid.

*Medicago* L.
LESINS and LESINS (1979) have presented a comprehensive monograph of this genus based on field and herbarium studies, analyses of karyology, breeding systems, crossing experiments, resistance pattern and biochemical evidences. The concept of the most important cultivated species *M. sativa* L. is a broad one including many taxa earlier described as species mostly by Soviet botanists. Likewise other taxa also known hitherto as species are interpreted as hybrids between *M. sativa* and *M. falcata* L., *M. glutinosa* M. BIEB. or *M. glomerata* BALB. Altogether 56 species (from 4 subgenera) are accepted, some being transferred from *Trigonella* L.

The adopted classification of an alfalfa monograph mainly concerned with breeding problems (IVANOV 1980) has an intermediate position between earlier Soviet classifications and the above mentioned one; 13 diploid and 6 tetraploid perennial species of the *sativa*-complex have been accepted and their variability and evolutionary relationships especially from taxa of the USSR have been recorded (compare also IVANOV 1977).

Data on the origin of Australian cultivars of annual fodder species of this genus have been collected by COCKS et al. (1980).

*Phaseolus* L., *Vigna* SAVI and related genera
This group is still attractive for taxonomists: BAUDET (1978) has reviewed the whole tribe *Phaseoleae* and reduced the number of its subtribes to three (*Glycininae Phaseolinae, Cajaninae*) by means of plant geographic, palynological, micromorphological and biochemical evidences.

Likewise LACKEY (1980) has used the chromosome numbers in this tribe for taxonomic conclusions and has summarized the distribution of this character within the genera of the different subtribes.

VERDCOURT (1980) has summarized shortly the recent state of classification of the genera *Dolichos* L. em. VERDC., *Lablab* ADANS., *Vigna* SAVI, *Phaseolus* L. and some allies stressing some typification problems and other nomenclatural ones not yet solved (conservation of *Vigna* and *Macrotyloma* (WIGHT et ARN.) VERDC. against *Voandzeia* THOU. and *Kerstingiella* HARMS).

The variability of growth habits of *V. unguiculata* (L.) WALP. in regard to photosensitivity and (in) determinate growth and the incorporation of the growth types into different farming systems in Africa and India have been analysed; the evolution from a West-African centre and the further diversification of two vegetable and fodder subspecies after migration to Southern Asia are other topics of a paper by STEELE and MEHRA (1980) who have reported also the occurrence of present-day cultivars intermediate between different cultivated subspecies.

The floral morphology and biology of different taxa of *V. unguiculata* have been analysed, the wild forest ssp. *mensensis* (SCHWEINF.) VERD. being more adapted to outbreeding than the regularly inbreeding savanna ssp. *dekindtiana* (HARMS) VERD. and the cultivated ssp. *unguiculata* (LUSH 1980).

JAIN and MEHRA (1980) have reviewed the evolution and relationships between the Indian *Vigna* species which have been transferred recently as part of the subgen. *Ceratotropis* (PIP.) VERDC. from *Phaseolus* to this genus. *Vigna radiata* (L.) WILCZ. and *V. mungo* (L.) HEPP. seem to have been domesticated independently from different races of the wild *sublobata*-taxon while *V. aconitifolia* (JACQ.) MARECH. and *V. trilobata* (L.) VERD. resp. *V. angularis* (WILLD.) OHWI et OHASHI and *V. umbellata* (THUNB.) OHWI et OHASHI represent other closely related pairs of species. However the data regarding karyology, crossing behaviour and hybrid fertility don't allow always the same interpretations.

The variability of local races of *Phaseolus vulgaris* from Kenya has been studied and classified into informal groups according to different colourations, sizes and shapes of the seeds; the most frequent groups have been described in detail (VAN RHEENEN 1980).

*Pisum* L.

MAKAŠEVA (1978) described 11 new agroecological groups of *P. sativum* L. comprising now altogether 29 such groups. The same author has treated the genus also for a new edition of the Flora of Cultivated Plants, vol. IV (1) (1979). Only two species are accepted: *P. fulvum* SIBTH. et SM. (with seven varieties) and *P. sativum* with the six subspecies *elatius* (BIEB) SCHMALH., *syriacus* (BOISS. et NOË) BERG., *abyssinicum* (A. BR.) BERG., *transcaucasicum* MAKASH., *asiaticum* GOV. and *sativum* which again is subdivided into 27 convarieties, 109 varieties, and 156 subvarieties. Many taxa were newly described.

HADAČOVÁ et al. (1980) and TURKOVÁ et al. (1980) examined identical seed samples of ten taxa of *Pisum*. Their results showed that *P. abyssinicum* A. BR. and *P. fulvum* have a rather isolated position biochemically, whereas *P. elatius* (BIEB.) STEV., *P. syriacum* (BERG.) LEHM. and *P. sativum* are more closely related.

The influence of different similarity-coefficients to the dendrogram formation has been studied in *P. sativum* L. (MEYER 1980).

*Psophocarpus* NECK ex. DC.

The evolution of the nowadays famous *Ps. tetragonolobus* (L.) DC. has been interpreted according to the transdomestication principle. Plant geographic evidence point to an African origin, the probable ancestral species being *Ps. grandiflorus* WILCZ. (SMARTT 1980b).

*Trifolium* L.

Australian cultivars of taxa of the *Tr. subterraneum* L.-group have been almost exclusively selected from ecotypes naturalised in Australia since the last century which have been originally introduced from Western Europe and Western Mediterranean areas into the continent. Enhanced by sporadic cross-pollinations evolutionary processes are now going on within these populations (COCKS et al. 1980).

*Trigonella* L.

The seed proteins in section *Foenum-graecum* SER. of *Trigonella* were studied by LADIZINSKY (1979c). None of the examined species can be considered as the immediate wild ancestor of the cultivated *T. foenum-graecum* L. According to morphological resemblance and similarity index of seed protein profile *T. berythaea* BOISS. et BL. is closer to fenugreek than *T. gladiata* STEV. ex BIEB. and *T. macrorrhyncha* BOISS., but the former two species are genetically isolated from one another.

*Vicia* L.

LEOKENE (1978) analysed morphological ynd physiological characters of 1500 strains of the common vetch collection at VAVILOV-Institute of Plant Industry

(Vir). Eight ecogeographical groups and 22 ecotypes of *Vicia sativa* L. ssp. *sativa* were separated (without key).

Abdalla and Günzel (1979) investigated crude protein content and seed protein profiles of 15 strains belonging to *V. faba* L. and one from *V. narbonensis* L. The electrophoresis showed *V. narbonensis* to be different from all the taxa of *V. faba* and, therefore, it can not be considered as immediate ancestor of the cultivated field bean.

The infrageneric system of *Vicia* has been reorganized many times, chiefly by writers of regional floras. Now Cvelev (1980) offers his taxonomical system of 40 species of this genus growing in the European part of URSS. Four subgenera and 12 sections are recognized, two of the latter are described respectively combined newly.

### Linaceae

*Linum* L.

Chržanovskij et al. (1979) have studied the microstructure of seeds from the sect. *Protolinum* Planch. (including the cultivated *L. usitatissimum* L.) and from the sect. *Adenolinum* Juz. (*L. perenne* L.). Derived conditions have been shown within the very closely related representatives of the sect. *Protolinum* whereas the characters of *L. perenne* are considered to be more primitive.

### Malvaceae

*Gossypium* L.

Descriptions have been given of the C genome group in the subgenus *Sturtia* (R. Br.) Tod. (Valíček 1978a), the D genome group in the subgenus *Houzingenia* Fryx. (Valíček 1978b), and the A, B, E and F genome groups in the subgenera *Gossypium* and *Karpas* Raf. (Valíček 1979) as well as some notes on the distribution and evolution of the genus *Gossypium*.

### Moraceae

*Cannabis* L.

A controlled cross between *C. sativa* L. and *C. ruderalis* Janisch. gave progeny intermediate in both cannabinoid content and morphology. The results strenghten the evidence that the cannabinoid content of *Cannabis* species is genetically controlled. The experiments also indicate that the cannabinoid profiles show no novel cannabinoids. This is one piece of evidence supporting the view that the genus *Cannabis* is monotypic (Beutler and der Marderosian 1978).

*Ficus carica* L.

Rivals (1978) has discussed the special pollination ecology of this species which he calls a triple symbiose (Caprificus – Blastophage – *Ficus*) and concluded that this fruit tree originated in the Near and Middle East.

### Rosaceae

*Fragaria* L.

Katsumata (1978a) has tried to describe the history of the discovery and introduktion of the two fruit-species *Fragaria virginiana* Duch. and *F. chiloensis* (L.) Duch.

The same author (KATSUMATA 1978b) has given new information about the species *Fragaria×ananassa*. Its origin and spread to America and Japan are described. Beside this our knowledge about the early history of cultivation of this fruit-species (originated from a hybridization of *F. chiloensis×F. virginiana*) is summarized.

*Malus* MILL.
In its morphological features a Siberian crab-apple from Transbaikalia is corresponding to the Linnéan type of *Pyrus baccata* from 1767. The Siberian crabs are very variable but in the author's opinion should be united into a single species, *M. baccata* (L.) BORKH. (synonyms are *M. sibirica* KOM. and *M. pallasiana* JUZ.). The large group of taxa (forms or cultivars) in Europe originating from hybridization with other *Malus* taxa should be classified under the name *M. cerasifera* SPACH. (PONOMARENKO 1978).

## *Solanaceae*

*Capsicum* L.
14 taxa of this genus have been examined by using starch-gel electrophoresis and grouped into 5 biological species (MCLEOD 1978): *C. cardenasii* HEISER et SMITH/*C. eximium* HUNZIKER/*C. pubescens* RUIZ et PAVON; *C. baccatum* L.; *C. praetermissum* HEISER et SMITH; *C. annuum* L./*C. chinense* JACQ.; *C. chacoense* HUNZIKER/an unnamed species. In another paper (MCLEOD et al. 1979) some of these ideas are further elaborated for the purple flowered taxa of the genus. *C. tovari* SMITH, nomen, was found to be a good species. The domesticated *C. pubescens* should be maintained as a separate species apart from *C. cardenasii*/*C. eximium*.

*Solanum* L. (tuber bearing species)
A potato-monograph has been edited by HARRIS (1978) as the scientific basis for the improvement of this crop.

BUKASOV (1978) has dealt with potato systematics and presents the system of the section *Tuberarium* (DUN.) BUK. which has 25 subsections and many series. The cultivated indigenous potato species from South America form three subsections: subsect. *Andigenum* BUK., subsect. *Subacaule* BUK., and subsect. *Chilozanum* BUK. (with altogether 17 species). The infraspecific classification of *Solanum chilotanum* HAWK. has been worked out by KOSTINA (1978) on the basis of 49 cv.-groups.

*Solanum* L. (fruit bearing species)
475 cultivars of *S. melongena* L. have been classified into 11 groups (MARTIN and RHODES 1979) by using numerical taxonomic methods. The results show that infraspecific classifications of this species can be done according to similarity groups and geographical origin.

## *Sterculiaceae*

*Theobroma cacao* L.
The origin of the main cv.-group of *Th. cacao*, viz. criollos and forasteros, has been discussed by BARRAU (1980). Apparently both of them originated from the area of the Upper Amazon and then spread to the north (Central American-Mexican area; criollo) vs. east (Basin of Amazon; forastero).

## General problems

A posthumous paper of SINSKAJA (1979) deals with the theory of botanical species and their evolution. Some examples concerning cultivated plants are given. BORKOVSKAJA and FILATENKO (1979) discussed the use of infraspecific taxa in the taxonomy of cultivated plants.

HEISER (1979) reviewed recent papers on origins of some cultivated New World plants (*Zea mays* L., *Phaseolus* spp., *Persea americana* MILL., *Capsicum* spp., *Solanum* spp., *Chenopodium* spp., *Ipomoea batatas* (L.) POIR., *Lupinus mutabilis* SWEET, *Scirpus californicus* BRITT. and *Lagenaria siceraria* (MOL.) STANDL.).

Certain elements of the theory of plant domestication are formulated by KUPZOW (1980). He illustrates economic and biological prerequisites, the different adaptability of plant ecotypes to cultivation and problems of naturalization and acclimatization in culture.

DE WET (1977) discussed increasing cereal yields as "evolution under domestication". The author emphasizes the importance of a sufficiently large gene pool for plant breeding.

## Zusammenfassung

Taxonomie und Evolution der Kulturpflanzen:
Literaturübersicht 1979/1980

Wichtige Arbeiten über Taxonomie und Evolution der Kulturpflanzen aus den Jahren 1979 und 1980 wurden zusammengestellt und kurz kommentiert.

### Краткое содержание

Таксономия и эволюция культурных растений: обзор литературы за 1979–1980 гг.

Даетса список работ по таксономии и эволюции культурных растений, вышедших в 1979–1980 годах. Работы кратко комментируются.

## Literature

ABDALLA, M. M. F. and G. GÜNZEL, 1979: Protein content and electrophoresis of seed proteins of certain *Vicia faba* L. stocks and their assumed ancestors. – Z. Pflanzenzüchtg. **83**, 148–154.

ANAŠČENKO, A. V., 1979: Filogenetičeskie svjazi v rode *Helianthus* L. – Trudy prikl. bot., genet. selekc. **64** (2), 146–156.

BACHTEEV, F. CH., 1977: [The problem of the origin and phylogeny of barley]. – Sb. nauk. tr. Saratov s-ch. in-t **84**, 107–113. (Plant Breed. Abstr. **49**, 1979, 749).

–, 1978: Sovremennye problemy proischoždenija jačmenja. – Dokl. MOIP. Obšč. biol. 1976 g. Sekc. antropol., biofiz., vitaminol., genet., gerontol., gistol. i èmbriol., istorii estestvozn., komis. po primeneniju mat. v biol. Mosk. o-va ispyt. prirody. M., 74–76. (Ref. Ž., 1979 (7), 7.55.133).

BARISAŠVILI, M. A., 1979: O proischoždenii genoma B u pšenicy *Triticum persicum* VAV. (T. *carthlicum* NEVSKI). – Bjull. vses. inst. rast. im. N. I. VAVILOVA **92**, 76–78.
BARRAU, J., 1979: Sur l'origine du cacaoyer, *Theobroma cacao* LINNÉ, Sterculiacées. – Journ. d'Agric. Trad. et de Bota. Appl. **26**, 171–180.
BAUDET, J. C., 1978: Prodrome d'une classification générique des Papilionaceae, *Phaseoleae*. – Bull. Jard. Bot. Natl. Belg. **48**, 183–220.
BEUTLER, J. A., and ARA H. DER MARDEROSIAN, 1978: Chemotaxonomy of *Cannabis*. I. Crossbreeding between *Cannabis sativa* and *C. ruderalis*, with analysis of cannabinoid content. – Econ. Bot. **32**, 387–394.
BHATTACHARJEE, R., 1980: Taxonomic studies in *Stachys*: II. A new infrageneric classification of *Stachys* L. – Notes Royal Bot. Gard. Edinburgh **38**, 65–90.
BORKOVSKAJA, V. A., i A. A. FILATENKO, 1979: O vnutrividovych taksonach kul'turnych rastenij. – Bjull. vses. inst. rast. im. N. I. VAVILOVA **91**, 60–64.
BREWBAKER, J. L. and A. J. OAKES, 1978: Taxonomic and genetic variation in the leguminous genus *Leucaena*. – Agronomy Abstracts, Madison (1978) 40.
BROICH, S. L. and R. G. PALMER, 1980: A cluster analysis of wild and domesticated soybean phenotypes. – Euphytica **29**, 23–32.
BRUNKEN, J. N., 1979: Morphometric variation and the classification of *Pennisetum* section *Brevivalvula* (Gramineae) in tropical Africa. – Bot. J. Linn. Soc. **79**, 51–64.
BUJUKLI, P. I., 1977: [New botanical varieties of winter durum wheat in Moldavia]. – Izv. AN MoldSSR, ser. biol. i chim. n. No. **6**, 82–83.
BUKASOV, S. M., 1978: Principi sistematiki kartofelja. – Trudy prikl. bot., genet. selekc. **62** (1), 3–35.
BUNTING, E. S., 1978: Maize in Europe. (pp. 1–13). In: BUNTING, E. S. et al., Forage Maize. Production and Utilisation. – London UK, Agric. Res. Council, XV+346 pp.
ČELAK, V. R., 1980: Citogenetičeskoe issledovanie gibridov nekotorych tetraploidnych vidov pšenicy s kul'turnoj odnozernjankoj (*T. monococcum* L. v. *flavescens*). – Citologija i Genetika **14** (2), 11–14.
CHRŽANOVSKIJ, V. G., S. F. PONOMARENKO i V. A. DOGUZAŠVILI, 1979: K voprosu proischoždenija i ėvoljucii roda *Linum* L. (sem. Linaceae). – Izvest. Akad. Nauk SSSR (Biol.) 1979 (5), 696–713.
COCKS, P. S., M. J. MATHISON, and E. J. CRAWFORD, 1980: From wild plants to pasture cultivars: Annual medics and subterranean clover in Southern Australia. – Advances in Legume Science, ed. by R. J. SUMMERFIELD and A. H. BUNTING, Kew, 569–596.
COSTANZA, S. H., J. M. J. DE WET and J. R. HARLAN, 1980: Literature review and numerical taxonomy of *Eragrostis tef* (T'ef). – Econ. Bot. **33**, 1979, 413–424.
CVELEV, N. N. (TZVELEV, N.), 1980: Sistema vidov roda *Vicia* L. evropejskoj časti SSSR (Systema specierum generis *Vicia* L. in parte europaea URSS). – Novosti sistematiki vysš. rast. **17**, 200–208.
DANE, F. K., 1977: Evolutionary studies in the genus *Cucumis*. – Dissertation Abstr. International, B. **37**, 3766B–3767B. (Plant Breeding Abstr. **50** (1980) 81, No. 876).
DASTIDAR, K. K., G. M. CHOWDHURY, and R. M. DATTA, 1979: Cytotaxonomic analysis of a few *Brassica* species with their cultivated varieties. – Broteria **48** (3–4), 59–70.
DJAGILEVA, G. E., 1979: Immunochimičeskij i ėlektroforetičeskij analiz belkov kul'turnogo i dikich vidov jačmenja v svjazi s voprosami sistematiki i filogenii roda *Hordeum*. – Bjull. vses. inst. rast. im. N. I. VAVILOVA **92**, 25–30.
DOLL, H., and A. H. D. BROWN, 1979: Hordein variation in wild (*Hordeum spontaneum*) and cultivated (*H. vulgare*) barley. – Can. J. Genet. Cytol. **21**, 391–404.
DOROFEEV, V. F., i O. N. KOROVINA (eds.), 1979: Kul'turnaja flora SSSR. I. Pšenica. – „Kolos", Leningrad, 347 pp.
–, i Ė. F. MIGUŠOVA, 1977: Utočnenija v sistematike *Triticum araraticum* JAKUBZ. – Bjull. vses. inst. rast. im. N. I. VAVILOVA **72**, 76.
DOROFEEV, V. F., i Ė. F. MIGUŠOVA, 1978: Novoe v sistematike pšenic. – Trudy prikl. bot., genet. selekc. **62** (2), 151.
–, –, 1979: Ideja difiletičeskogo proischoždenija tetraploidnoj pšenicy v trudach E. N. Sinskoj i sovremennoe ponimanie sistemy roda *Triticum* L. – Bjull. vses. inst. rast. im. N. I. VAVILOVA **91**, 24–26.

GANDILJAN, P. A., 1980: Opredelitel pšenicy, ėgilopsa, rži i jačmenja (opredelitel'nye i spravočnye tablicy vidov i raznovidnostej). – Izdatel'stvo AN Armjanskoj SSR, Erevan, 287 pp.
GENDEL'S, T. V., 1980: Ispol'zovanie anatomičeskich priznakov lista v sistematike roda *Aegilops* (Poaceae). The use of leaf anatomical characters in the systematics of the genus *Aegilops* (Poaceae). – Bot. Ž. **65**, 863–866.
GERLACH, W. L., R. APPELS, E. S. DENNIS and W. J. PEACOCK, 1979: [Evolution and analysis of wheat genomes using highly repeated DNA sequences]. – In: RAMANUJAM, S. (Ed.), Proceedings of the fifth international wheat genetics symposium. Vol. 1. Session 1. Evolution of wheat–changing concepts. New Delhi, 81–91. (Plant Breed. Abstr. **49**, 1979, 652).
GREGORY, W. C., A. KRAPOVICKAS, and M. P. GREGORY, 1980: Structure, variation, evolution, and classification in *Arachis*. – Advances in Legume Science, ed. by R. J. SUMMERFIELD and A. H. BUNTING, Kew, 469–481.
GUPTA, P. K. and R. V. SINGH, 1977: Variations in chromosomes and flavonoids in *Setaria* BEAUV. – Nucleus **20**, 167–171.
GUSTAFSSON, M., B. BENTZER, R. v. BOTHMER, and S. SNOGERUP, 1976: Meiosis in Greek *Brassica* of the *oleracea* group. – Bot. Notiser **129**, 73–84.
HADAČOVÁ, V., V. TURKOVÁ, E. HADAČ, and E. KLOZOVÁ, 1980: Comparison of seed proteins of some representatives of the genus *Pisum* from the point of view of their relationship. Comparison by disc electrophoresis. – Biol. plant. **22**, 7–16.
HAKE, S., and V. WALBOT, 1980: The genome of *Zea mays*, its organization and homology to related grasses. – Chromosoma **79**, 251–270.
HAMMER, K., 1980a: Zur Taxonomie und Nomenklatur der Gattung *Aegilops* L. – Feddes Rep. **91**, 225–258.
–, 1980b: Vorarbeiten zur monographischen Darstellung von Wildpflanzensortimenten: *Aegilops* L. – Kulturpflanze **28**, 33–180.
HARRIS, P. M. (Ed.), 1978: The potato crop. The scientific basis for improvement. – London, 730 pp.
HEISER, C. B., 1979: Origins of some cultivated New World plants. – Ann. Rev. Ecol. Syst. **10**, 309–326.
HILU, K. W., J. M. DE WET, and D. SEIGLER, 1978: Flavonoid patterns and systematics in *Eleusine*. – Biochem. Syst. Ecol. **6**, 247–249.
HUMPHRIES, C. J., 1979: A revision of the genus *Anacyclus* L. (Compositae: Anthemideae). – Bull. Br. Mus. nat. Hist. (Bot.) **7**, 83–142.
HYMOWITZ, T., and C. A. NEWELL, 1977: Current thoughts on origins, present status and future of soybeans. (pp. 197–209). In: Crop Resources, ed. by D. S. SEIGLER, Academic Press, New York, San Francisco, London, XI + 233 pp.
–, –, 1980: Taxonomy, speciation, domestication, dissemination, germplasm resources and variation in the genus *Glycine*. – Advances in Legume Science, ed. by R. J. SUMMERFIELD and A. H. BUNTING, Kew, 251–264.
IVANJUKOVIČ, L. K., 1980: K istorii i proischoždenii kul'tury sorgo. On the history and origin of the genus *Sorghum*. – Bot. Ž. **65**, 1014–1020.
– i J. A. DORONINA, 1978: Ključ dlja opredelenija kul'turnych vidov roda *Sorghum* MOENCH. A key for the determination of cultivated species of the genus *Sorghum* MOENCH. – Trudy prikl. bot., genet. selekc. **63** (1), 48–53.
IVANOV, A. I., 1977: Istorija, proischoždenie i ėvoljucija roda *Medicago* L. podroda *Falcago* (REICHB.) GROSSH. (History, origin and evolution of the genus *Medicago* of the subgenus *Falcago*). – Trudy prikl. bot., genet. selekc. **59** (1), 3–40.
–, 1980: Ljucerna. – Izd. ,Kolos', Moskva, 349 pp.
JAASKA, V., 1978: NADP-dependent aromatic alcohol. dehydrogenase in polyploid wheats and their diploid relatives. On the origin and phylogeny of polyploid wheats. – Theor. Appl. Genet. **53**, 209–217.
–, 1980: Electrophoretic survey of seedling esterases in wheats in relation to their phylogeny. – Theor. Appl. Genet. **56**, 273–284.
JAIN, H. K. and, K. L. MEHRA, 1980: Evolution, adaptation, relationships, and uses of the species of *Vigna* cultivated in India. – Advances in Legume Science, ed. by R. J. SUMMERFIELD and A. H. BUNTING, Kew, 459–468.

JAIN, S. K., and R. S. SINGH, 1979: Population Biology of *Avena*. VII. Allozyme variation in relation to the genome analysis. – Bot. Gazette **140**, 356–362.

JONES, B. M. G., J. PONTI, A. TAVASSOLI and P. A. DIXON, 1978: Relationships of the Ethiopian cereal T'ef (*Eragrostis tef* (ZUCC.) TROTTER: evidence from morphology and chromosome number. – Ann. Bot. **42**, 1369–1373.

KATAYAMA, T., and T. OGAWA, 1977: Cytogenetical studies on synthesized amphiploids (BBCC) in wild rice. – Sci. Bull. Fac. Agric., Kyushu Univ. Japan **32**, 113–117.

KATSUMATA, H., 1978a: Origin and development of the strawberry. – Agriculture and Horticulture **53**, 993–996. (Plant Breeding Abstr. **50**, 1980, 145, No. 1496).

–, 1978b: Origin and development of the strawberry. II. – Agriculture and Horticulture **53**, 1113–1117. (Plant Breeding Abstr. **50**, 1980, 145, No. 1497).

KIM, S. K. and J. MOSSÉ, 1979: Electrophoretic patterns of oat prolamines and species relationships in *Avena*. – Can. J. Genet. Cytol. **21**, 309–318.

KOBYLJANSKIJ, V. D., 1979: Proischoždenie i rasprostranenie kul'tury rži. – Bjull. vses. inst. rast. im. N. I. VAVILOVA **91**, 26–32.

KONAREV, V. G., V. F. DOROFEEV, V. F. MIGUŠOVA, i A. A. FILATENKO, 1978: Genomnyj sostav i sistema roda *Triticum* L. Genome composition and the system of the genus *Triticum* L. – Trudy prikl. bot., genet. selekc. **63**, (2), 6–14.

KOSTINA, L. I., 1978: Aborigennye sorta *Solanum chilotanum* Hawk. – Trudy prikl. bot., genet. selekc. **62** (1), 55–82.

KROON, G. H., J. B. M. CUSTERS, Y. O. KHO, A. P. M. DEN NIJS and H. Q. VAREKAMP, 1979: Interspecific hybridization in *Cucumis* (L.). I. Need for genetic variation, biosystematic relations and possibilities to overcome crossability barriers. – Euphytica **28**, 723–728.

KRUSE, J., 1980: Skulpturuntersuchungen an Pollen der Gattungen *Triticum* L. and *Aegilops* L. – Kulturpflanze **28**, 341–359.

KUPZOW, A. J. 1980: Theoretical basis of the plant domestication. – Theor. Appl. Genet. **57**, 65–74.

LACKEY, J. A., 1980: Chromosome numbers in the Phaseoleae (Fabaceae: Faboideae) and their relation to taxonomy. – Amer. J. Bot. **67**, 595–602.

LADIZINSKY, G., 1979a: Seed dispersal in relation to the domestication of Middle East legumes. – Econ. Bot. **33**, 284–289.

–, 1979b: Species relationships in the genus *Lens* as indicated by seed-protein electrophoresis. – Bot. Gaz. **140**, 449–451.

–, 1979c: Seed protein electrophoresis in section *Foenum-graecum* of *Trigonella* (Fabaceae). – Pl. Syst. Evol. **133**, 87–94.

LEOKENE, L. V., 1978: Ėkologo-geografičeskaja klassifikacija vozdelyvaemych sortov viki posevnoj (*Vicia sativa* L. subsp. *sativa*). (Ecogeographical classification of common vetch varieties). – Trudy prikl. bot., genet. selekc. **63** (1), 108–122.

LESINS, K. A., and I. LESINS, 1979: Genus *Medicago* (Leguminosae). A taxogenetic study. – Dr. W. JUNK bv-Publ., The Hague–Boston–London, 228 pp.

LUSH, W. M., 1980: Floral morphology of wild and cultivated cowpeas. – Econ. Bot. **33**, 1979, 442–447.

MCLEOD, M. J., 1978: A systematic and evolutionary study of the genus *Capsicum*. – Dissertation Abstr. International, B **38** (9), 4036 B.

–, W. H. ESHBAUGH and S. I. GUTTMAN, 1979: An electrophoretic study of *Capsicum* (Solanaceae): The purple flowered taxa. – Bull. Torr. Bot. Club **106**, 326–333.

MAESEN, L. J. G. VAN DER, 1980: India is the native home of the pigeonpea. – Liber Gratulatorius in honorem H. C. D. DE WIT. Miscellaneous Papers **19**, Landbouwhogeschool, Wageningen, The Netherlands, Wageningen, 257–262.

MAKAŠEVA, R. CH., 1978: Agroėkologičeskaja klassifikacija gorocha. (Agroecological classification of pea). – Trudy prikl. bot., genet. selekc. **63** (1), 123–135.

MAKAŠEVA, R. CH., 1979: Goroch. In: Kul'turnaja flora SSSR IV. Zernovye bobovye kul'tury, čast 1 (Edit. O. N. KOROVINA). – Leningrad (Kolos).

MARTIN, F. W., and A. M. RHODES, 1979: Subspecific grouping of eggplant cultivars. – Euphytica **28**, 367–383.

Meyer, D., 1980: Numerisch-taxonomische Untersuchungen an *Pisum sativum* L. – Kulturpflanze **28**, 285–340.

Nakai, Y., 1979a: [The origin of the tetraploid wheats revealed from the study on esterase isozymes]. – In: Ramanujam, S. (Ed.), Proceedings of the fifth international wheat genetics symposium. Vol. 1. Session 1. Evolution of wheat – changing concepts. New Delhi, 108–119. (Plant Breed. Abstr. **49**, 1979, 653).

–, 1979b: Isozyme variations in *Aegilops* and *Triticum*, IV. The origin of the common wheats revealed from the study on esterase isozymes in synthesized hexaploid wheats. – Japan. J. Genetics **54** (3), 175–189.

Nevo, E., D. Zohary, A. H. D. Brown, and M. Haber, 1979: Genetic diversity and environmental associations of wild barley, *Hordeum spontaneum*, in Israel. – Evolution **33**, 815–833.

Ng, F. S. P., 1979: The origin of *Diospyros lotus* and other notes on the genus *Diospyros* (Ebenaceae). – Malays. Forest. **42** (2), 165–170. (ref.: Ref. Ž. Biol. Bot. (Moskva) 1980 (2), 68, 2V431)

Nishikawa, K., and Y. Furuta, 1979: [DNA content of nucleus and individual chromosomes and its evolutionary significance]. – In: Ramanujam, S. (Ed.), Proceedings of the fifth international wheat genetics symposium. Vol. 1. Session 1. Evolution of wheat–changing concepts. New Delhi, 133–138. (Plant Breed. Abstr. **49**, 1979, 653).

Oka, H. I., and Y. Sano, 1978: The breeding behaviour of *Oryza perennis* populations observed in Thailand. – Nation. Inst. Genetics, Japan, Annual Report No. **28** (I) 1977, 104–105.

Paroda, C. M., 1977: Studies on B-genome in polyploid wheat by nucleic acid hybridization. – Diss. Abstr. Intern., B **37** (12), 5969 B. (Plant Breed. Abstr. **50**, 1980, 114).

Peneva, T. I., 1979: K voprosu o proischoždenii genoma B u *T. dicoccum* (Schrank) Schuebl. – Bjull. vses. inst. rast. im. N. I. Vavilova **92**, 10–13.

Ponomarenko, V. V., 1978: K sistematike *Malus baccata* (L.) Borkh. (On the systematics of *Malus baccata* (L.) Borkh.) – Trudy prikl. bot., genet. selekc. **62** (3), 24–35.

Puchalski, J. T., R. W. Robinson, and J. W. Shail, 1978: Comparative electrophoresis of isozymes of *Cucumis* species. – Cucurbit Genetics Cooperative No. 1, 39. (Plant Breeding Abstr. **49**, 1979, 700, No. 8508).

Rheenen, H. A. Van, 1980: Diversity of food beans in Kenya. – Econ. Bot. **33**, 1979, 448–454.

Rivals, P., 1978: La triple symbiose caprifiguier–blastophage–figuier et le problème des origines du *Ficus carica* L. – Journ. d'Agric. Trad. et de Bota. Appl. **25**, 287–290.

Röst, L. C. M., 1979: Biosystematic investigations with *Acorus*. 4. Communication. A synthetic approach to the classification of the genus. – Planta Med. **37**, 289–307.

Schultze-Motel, J., R. Fritsch, K. Hammer, P. Hanelt, J. Kruse, H. Ohle und H. I. Schäfer, 1978: Taxonomie und Evolution der Kulturpflanzen: Literaturübersicht 1975–1977. – Kulturpflanze **26**, 363–381.

–, –, –, –, –, –, and H. I. Maass, 1979: Taxonomy and evolution of cultivated plants: Literature review 1977/1978. – Kulturpflanze **27**, 247–264.

Schultze-Motel, J., R. Fritsch, K. Hammer, P. Hanelt, J. Kruse, H. I. Maass, H. Ohle and K. Pistrick, 1980: Taxonomy and evolution of cultivated plants: Literature review 1978/1979. – Kulturpflanze **28**, 379–397.

Sears, E. R., 1977: The origin and future of wheat. (pp. 193–196). In: Crop Resources, ed. by D. S. Seigler, Academic Press, Inc., New York, San Francisco, London, XI + 233 pp.

Sencer, H. A., and J. G. Hawkes, 1980: On the origin of cultivated rye. – Biol. J. Linn. Soc. **13**, 299–313.

Shao, Q., C. Li, C. Basang, 1980: Semi-wild wheat from Xizang (Tibet). – Acta Genetica Sinica **7**, 149–156 [chin., Engl. summ.].

Sinskaja, E. N., 1979: Vid i ego strukturnye časti na rasličnych urovnjach organičeskogo mira. – Bjull. vses. inst. rast. im. N. I. Vavilova **91**, 7–24.

Smartt, J., 1980a: Evolution and evolutionary problems in food legumes. – Econ. Bot. **34**, 219–235.

—, 1980b: Some observations on the origin and evolution of the winged bean (*Psophocarpus tetragonolobus*). – Euphytica **29**, 121–123.

STEELE, W. M., and K. L. MEHRA, 1980: Structure, evolution, and adaptation to farming systems and environments in *Vigna*. – Advances in Legume Science, ed. by R. J. SUMMERFIELD and A. H. BUNTING, Kew, 393–404.

TANAKA, M., T. KAWAHARA and J. SANO, 1979: [The evolution of wild tetraploid wheats]. – In: RAMANUJAM, S. (Ed.), Proceedings of the fifth international wheat genetics symposium. Vol. 1. Session 1. Evolution of wheat–changing concepts. New Delhi, 73–80. (Plant Breed. Abstr. **49**, 1979, 652).

TIMOTHY, D. H., C. S. LEVINGS III, D. R. PRING, M. F. CONDE, and J. L. KERMICLE, 1979: Organelle DNA variation and systematic relationships in the genus *Zea*: Teosinte. – Proc. Natl. Acad. Sci. USA **76**, 4220–4224.

TROFIMOVSKAJA, A. JA. i G. N. GUDKOVA, 1979: Anatomo-morfologičeskie issledovanija v svjazi s klassifikaciej roda *Hordeum* L. – Bjull. vses. inst. rast. im. N. I. VAVILOVA **91**, 71–79.

TURKOVÁ, V., E. KLOZOVÁ, and V. HADAČOVÁ, 1980: The comparison of seed proteins of several representatives of the genus *Pisum* with respect to their relationships. An immunological comparison. – Biol. plant. **22**, 17–24.

UDAČIN, R. A., 1978: Rol Srednej Azii v ėvoljucii roda *Triticum*. The role of Central Asia in the evolution of the genus *Triticum* L. – Trudy prikl. bot., genet. selekc. **63**, (2), 15–25.

VALÍČEK, P., 1978a: Wild and cultivated cottons (I). – Coton et Fibres Tropicales **33**, 363–387.

—, 1978b: Wild and cultivated cottons (II). – Coton et Fibres Tropicales **33**, 431–448.

—, 1979: Wild and cultivated cottons (III). – Coton et Fibres Tropicales **34**, 239–264.

VANDER KLOET, S. P., 1980: The taxonomy of the highbush blueberry, *Vaccinium corymbosum*. – Can. J. Bot. **58**, 1187–1201.

VERDCOURT, B., 1980: The classification of *Dolichos* L. emend. VERDC., *Lablab* ADANS., *Phaseolus* L., *Vigna* SAVI and their allies. – Advances in Legume Science, ed. by R. J. SUMMERFIELD and A. H. BUNTING, Kew, 45–48.

WALDEN, D. B. (Ed.), 1978: Maize breeding and genetics. Section 2. Evolution, 85–184. – JOHN WILEY & Sons, New York.

WET, J. M. J. DE, 1977: Increasing cereal yields: Evolution under Domestication. (pp. 111–118). In: Crop resources, ed. by D. S. SEIGLER, Academic Press, Inc., New York, San Francisco, London, XI + 233 pp.

—, L. L. OESTRY-STIDD and J. I. CUBERO, 1979: Origins and evolution of foxtail millets (*Setaria italica*). – Journ. d'Agric. Trad. et de Bota. Appl. **26**, 53–64.

WET, J. M. J. DE, and Y. SHECHTER, 1977: Evolutionary dynamics of *Sorghum* domestication. (pp. 179–191). In: Crop Resources, ed. by S. SEIGLER, Academic Press, Inc., New York, San Francisco, London, XI + 233 pp.

WHISTLER, R. L., and T. HYMOWITZ, 1979: Guar: Agronomy, production, industrial use, and nutrition. – West Lafayette, Indiana (Purdue University Press).

WIDÉN, K.-G., P. ALANKO, and M. UOTILA, 1977: *Thymus serpyllum* L. × *vulgaris* L., morphology, chromosome number and chemical composition. – Ann. bot. fenn. **14**, 29–34. (Ber. Biochem. Biol. **491**, 1979, 211, No. 34528).

WOLFF, G., 1980: Investigations on the relations within the family Papilionaceae on the basis of electrophoretic banding patterns. – Theor. Appl. Genet. **57**, 225–232.

Dr. J. SCHULTZE-MOTEL, Dr. R. FRITSCH, Dr. K. HAMMER, Dr. P. HANELT, Dr. J. KRUSE, Dr. H. I. MAASS, Dr. H. OHLE und K. PISTRICK
Zentralinstitut für Genetik und
Kulturpflanzenforschung der Akademie
der Wissenschaften der DDR
4325 Gatersleben

Kulturpflanze XXIX · 1981 · S. 485–496

# Variabilität der Stellungsverhältnisse im Sexualorganbereich bei verschiedenen Populationen von *Vaucheria sessilis* (Vauch.) D. C. 1805

Alfred Rieth

(Eingegangen am 12. Dezember 1980)

## Zusammenfassung

Es werden spontan aufgetretene Abweichungen von der als typisch angesehenen Ausbildung in Anordnung und Form der Sexualorgane bei Populationen von *Vaucheria sessilis* fünf verschiedener Herkünfte beschrieben und daraus einige Schlüsse über die Variabilitätsspanne der Art abgeleitet.

## Einleitung

Im Laufe vieler Jahre hatte ich Gelegenheit Populationen aus dem Formenkreis der weltweit verbreiteten *Vaucheria sessilis* an Freiland- und Kulturmaterial zahlreicher Vorkommen zu beobachten. Dabei zeigte sich, daß die Neigung zur Variabilität der Morphologie im Sexualorganbereich bei verschiedenen Herkünften unterschiedlich stark ausgeprägt ist. Die große Mehrzahl der untersuchten Populationen weist streng die für den Formenkreis typische Stellung der Fortpflanzungsorgane auf. Bei Material einiger Fundorte finden sich jedoch, in verschiedenem Grade gehäuft, Abweichungen, „Mißbildungen". Dabei sind jeweils bestimmte Bildungstendenzen bei einzelnen Populationen vorherrschend und für sie mehr oder minder typisch, treten aber, wie Kulturen zeigen, nicht in jeder Fruktifikationsperiode auf, sondern nur unter einstweilen nicht näher bekannten Bedingungen. Sporadische Beobachtungen anormaler Bildungen habe ich gelegentlich bereits mitgeteilt (z. B. Rieth 1969, 1980) während Angaben anderer Autoren darüber in der Literatur, soweit ich sehe, kaum vorliegen. Im folgenden sollen jetzt die Verhältnisse an umfangreicherem Material, bei fünf, durch Vorkommen eines relativ breiten Spektrums von Anomalien ausgezeichneten Populationen dargestellt und analysiert werden.

## Material

Den Untersuchungen liegen folgende Aufsammlungen, sämtliche von feuchter Erde, zugrunde:

I. Bei Trautenstein, Harzgebirge 20. 8. 1978. Rohkulturen in CHU 10 in Standortwasser. Sexualorgane beobachtet Februar–April 1979.

II. Bei Lence südl. Krakau (VR Polen) auf Gartenerde 8. 9. 75. Kulturen in Regenwasser mit Standorterde. Sexualorgane beobachtet Mitte Oktober–Mitte November 1975 und erneut April 1976.

III. Bei Nessebâr nördl. Burgas (VR Bulgarien), Schwarzmeerküste 21. 6. 1976. Rohkulturen in Regenwasser mit Standorterde. Sexualorgane beobachtet Juli 1976.

IV. Bei Alexisbad/Harzgebirge, am Ufer der Selke 31. 12. 1975. Rohkulturen in Selkewasser mit Standorterde. Sexualorgane beobachtet Ende April 1976.

V. Im Klostergrund bei Michaelstein/Harzgebirge 1. 6. 1979. Rohkulturen in Teichwasser sowie in CHU 10 in Bachwasser. Sexualorgane beaobachtet September 1979.

Die Pflanzen wurden stets lebend untersucht und dabei darauf geachtet, daß die räumliche Anordnung in den Präparaten nicht durch Deckglasdruck beeinträchtigt war. Bemerkenswerte Konfigurationen sind mit Hilfe eines Zeichenokulars (meist 430fach vergrößert) festgehalten.

## Beobachtungen

In den fünf Populationen wurden insgesamt 54 verschiedene Gruppierungen im Bereich der Sexualorgane gefunden. Sie sind in den Figuren 1–54 der Abbildungen 1a–1c schematisch dargestellt und größtenteils durch gleichlautend bezifferte, naturgetreue Figuren nach dem mikroskopischen Bild belegt (Abb. 2–6).

Am formenreichsten erweist sich das Material aus Trautenstein (I) mit 42 Konfigurationen (Abb. 1a; Abb. 2–4), die Herkunft II (Lence) zeigt im Vergleich dazu nur 7 (Abb. 1b, oben; Abb. 5, oben), die Aufsammlung III (Schwarzmeer-

Tabelle 1

Häufigkeit der einzelnen Sexualorgankonfigurationen bei *Vaucheria sessilis* Population I (Trautenstein)

| Konfiguration* | Anzahl | Prozent |
|---|---|---|
| 1 | 1211 | 73,71 |
| 2 | 149 | 9,07 |
| 3 | 114 | 6,94 |
| 4 | 51 | 3,10 |
| 5 | 24 | 1,46 |
| 6 | 14 | 0,85 |
| 7 | 6 | 0,35 |
| 8–11 und 15 | je 5 | je 0,30 |
| 12 | 4 | 0,24 |
| 13–14 | je 3 | je 0,18 |
| 16–18 | | |
| 19–23 und 36 | je 2 | je 0,12 |
| 24–35 } 36–42 } | je 1 | je 0,06 |

Gesamtzahl 1643

* Siehe Abb. 1a und 1b!

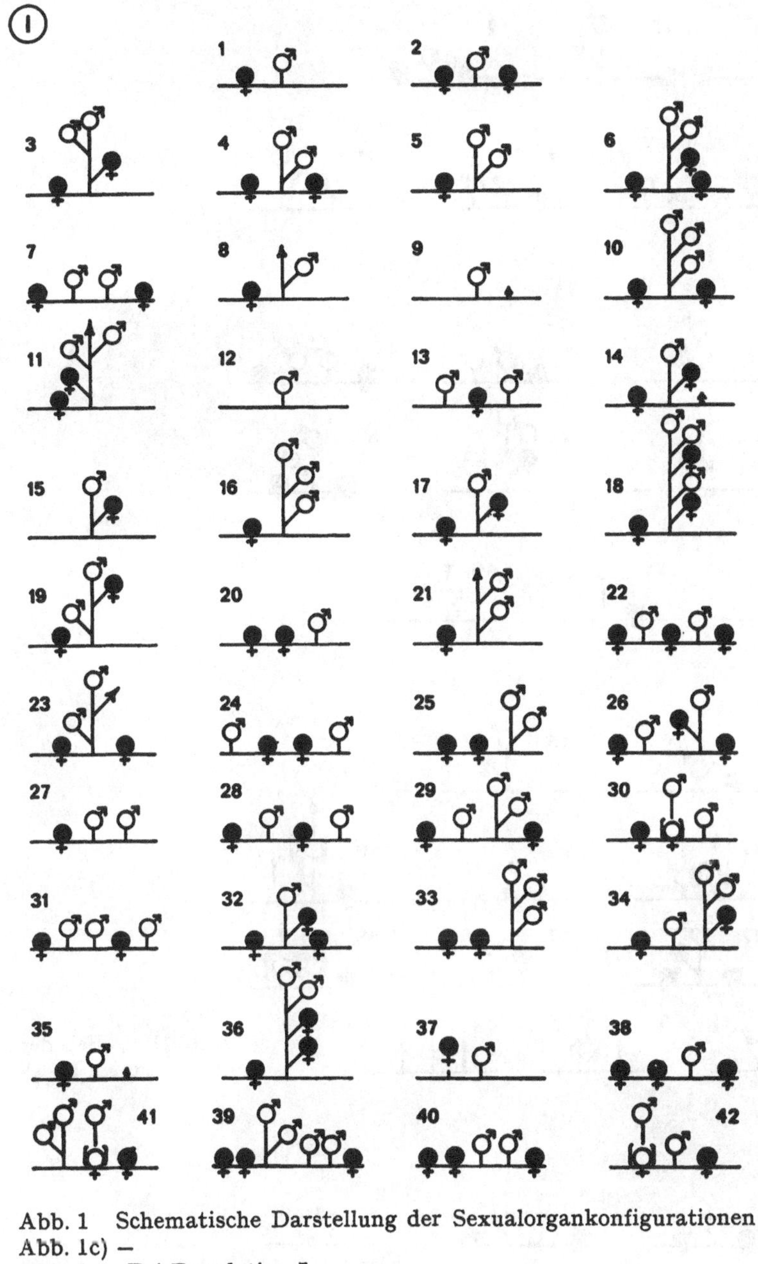

Abb. 1 Schematische Darstellung der Sexualorgankonfigurationen (Zeichenerklärung in Abb. 1c) –
Abb. 1a Bei Population I

küste) 11 (Abb. 1b, mitte; Abb. 5, unten), die Population aus Alexisbad (IV) 13 (Abb. 1b, unten; Abb. 6) und das Vorkommen im Klostergrund (V) 6 (Abb. 1c). Darunter sind bei II lediglich eine (Fig. 43 auf Abb. 1b und Abb. 5, oben), bei III drei (Fig. 44–46 auf Abb. 1b, mitte und Abb. 5, unten), bei IV sechs (Fig. 47–52 auf Abb. 1b, unten und Abb. 6) und bei V zwei (Fig. 53 und 54, Abb. 1c) in I nicht vertretene Konstellationen.

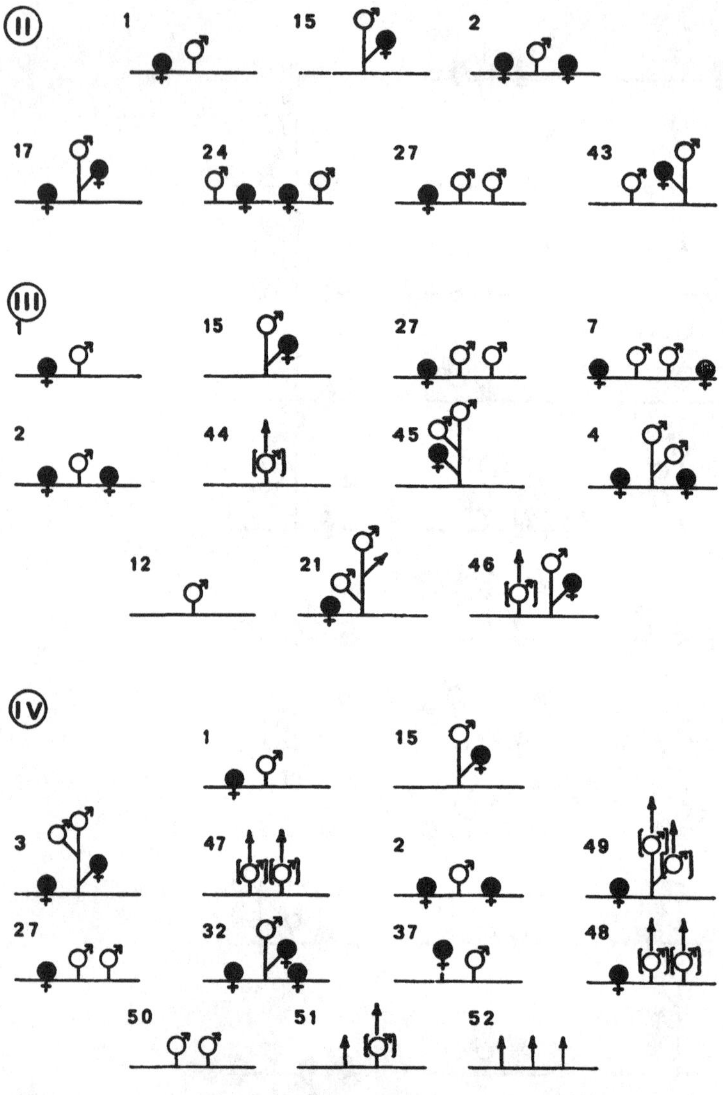


Abb. 1b Bei den Populationen II–IV

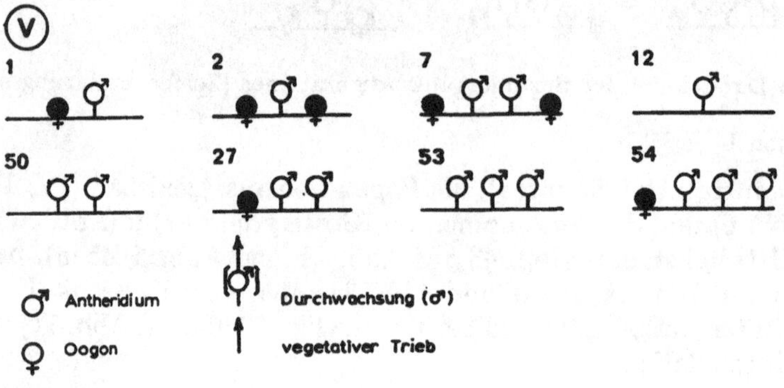


Abb. 1c Bei Population V

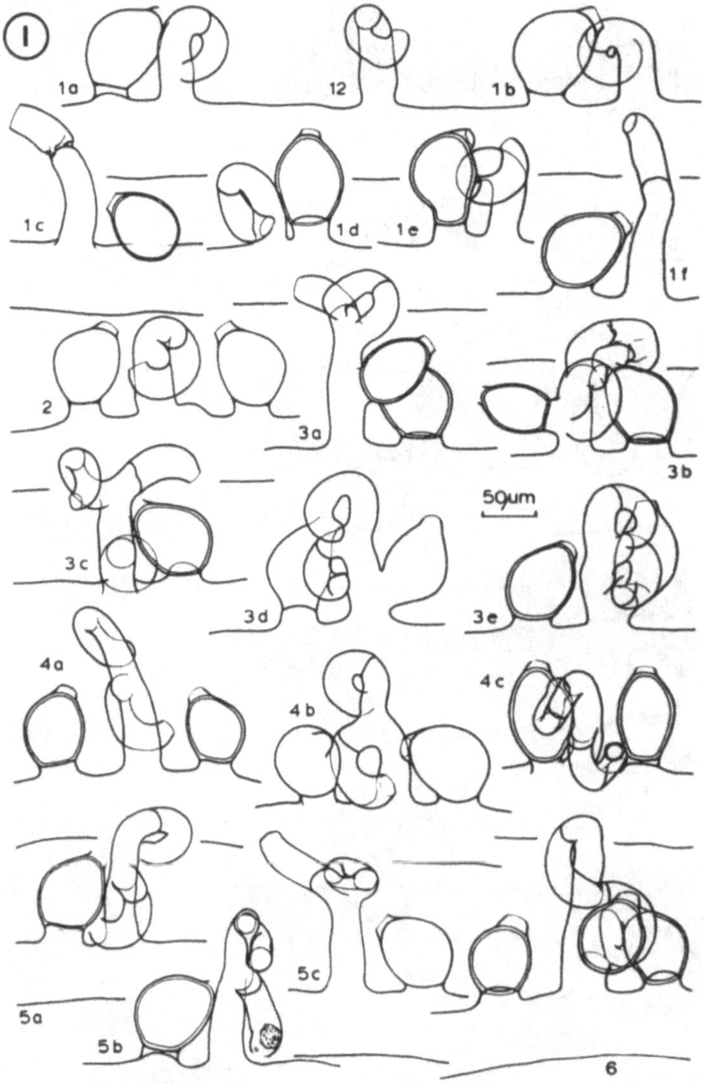


Abb. 2

Abb. 2–6 Die der schematischen Darstellung der Abb. 1 zugrunde liegenden Sexualorgankonfigurationen im mikroskopischen Bild –
Abb. 2–4 Aus Population I

Die durch größte Mannigfaltigkeit ausgezeichnete Population I (Trautenstein) wurde am eingehendsten untersucht und dabei über 1600 Sexualorgangruppierungen ausgewertet. Vorherrschend erwiesen sich, wie zu erwarten, die beiden als „Normalausbildung" anzusprechenden Konfigurationen Fig. 1 und 2 (Abb. 1a), die aus einem, samt Stiel mehr oder weniger schneckenartig gewundenen Antheridium mit zugeordnetem, dem Tragfaden ± ungestielt ansitzendem Oogon (Abb. 1a, Fig. 1; Abb. 2, Fig. 1a, 1b, 1d, 1e) oder aus einem solchen ♂ Kurztrieb zwischen zwei es symmetrisch einrahmenden und ihm den Schabel zukehrenden ± un-

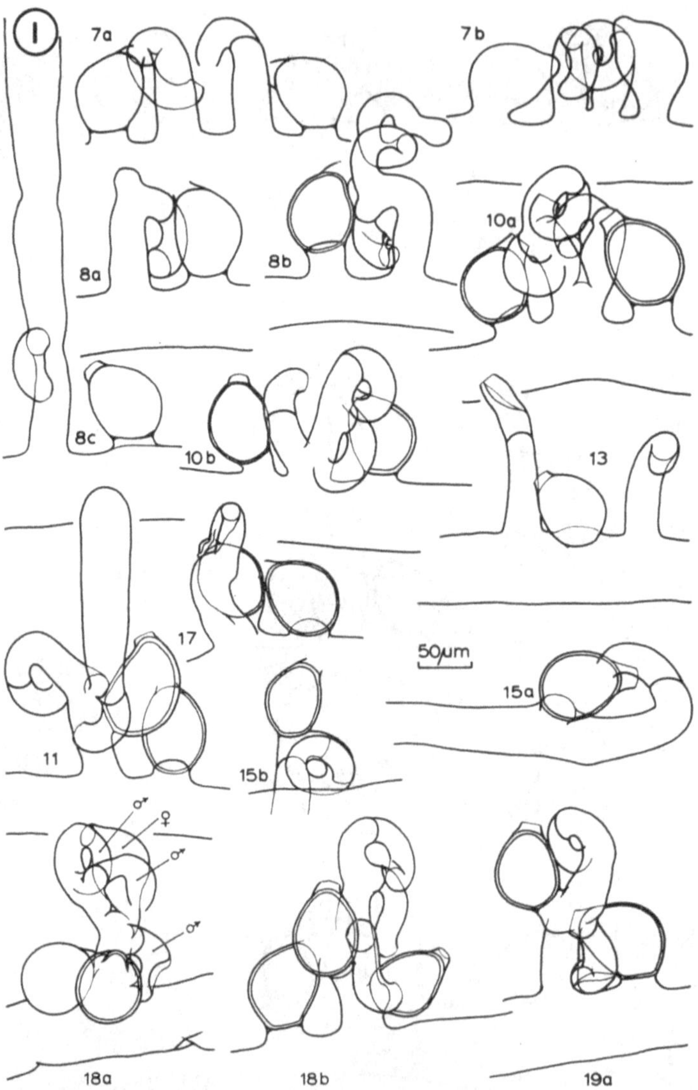


Abb. 3

gestielten Oogonien (Abb. 1a, Fig. 2, Abb. 2, Fig. 2), bestehen. Sie stellen mit 74%, bzw. 9%, 83% aller Gruppierungsvarianten dieser Population. Von den restlichen, als „abnorm" zu bezeichnenden 40 Bildungen sind mengenmäßig die Konfigurationen 3 mit 7%, 4 mit 3% und 5 mit 1,5% noch gut repräsentiert, die Konfiguration 6 kommt mit 14 Fällen knapp auf 1% Häufigkeit, alle übrigen 36 mit zusammen 4,5% erreichen jeweils 0,5% nicht, bleiben zum großen Teil weit darunter, treten in dem Material nur ein oder zweimal auf, sind also relativ selten (Tabelle 1).

Die verschiedenen, im Sexualorganbereich beobachteten Ausbildungsvarianten lassen sich in folgende Gruppen zusammenfassen:

a) Variationen in Zahl und Zuordnung normal ausgebildeter Antheridien und Oogonien am Tragfaden. Hierher gehören die Konfigurationen: 1, 2, 12, 13,

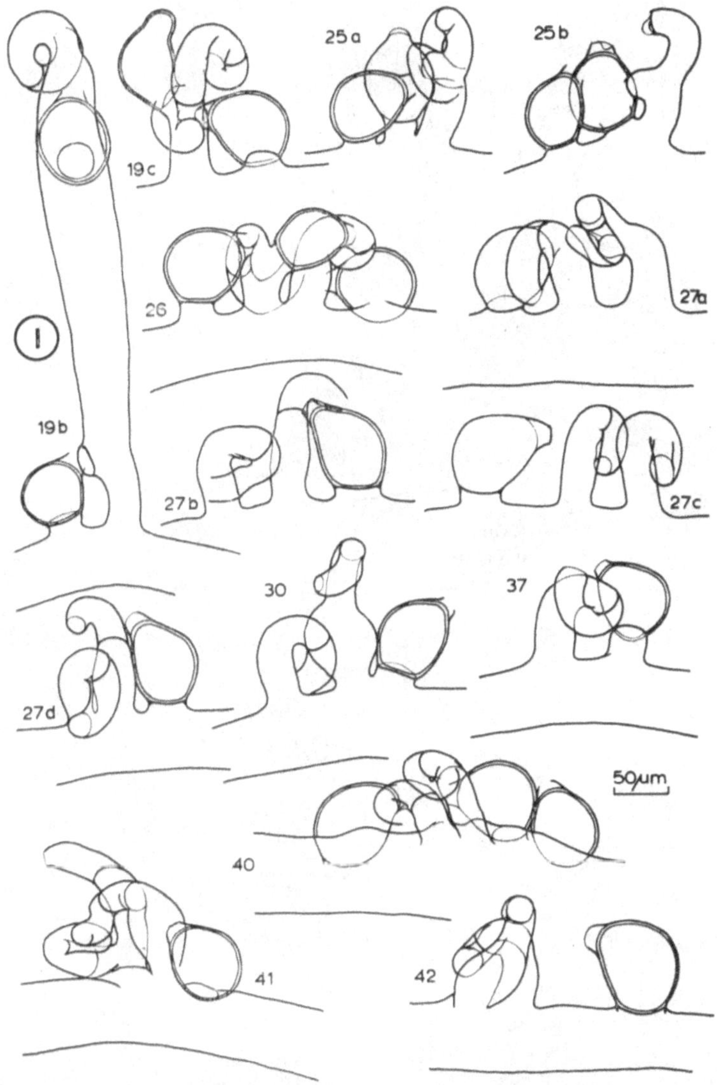


Abb. 4

20, 22, 24, 27, 28, 31, 37 (mit deutlich gestieltem ♀), 38, 40, 50, 53, 54 der Abbildungen 1a–1c.[1]

b) Mannigfaltige Verzweigungen der im Normalfall stets unverzweigten Antheridialäste.

α Zu mehrantheridigen, seitlich am Tragfaden inserierten, rein ♂ Ständen führend, gelegentlich auch vegetative Kurztriebe enthaltend: Fig. 4, 5, 8, 10, 16, 21, 23, 25, 29, 33, 39, 41 der Abb. 1a; Fig. 49, Abb. 1b.

1 Stellungsvarianten der Oogonien zum Tragfaden, wie sie sich im Winkel zwischen der Oogonlängsachse und der Tragfadenlängsachse und damit in der Richtung in der der Schnabel zeigt ausdrücken, blieben hier unberücksichtigt. Ich habe auf die Veränderlichkeit dieses Merkmals, das in der Vergangenheit zur Aufstellung eigener Arten oder Formen verwandt wurde (etwa einer *V. repens* Hassall, oder einer *V. orthocarpa* Reinsch, bereits 1965 anhand zahlreicher Messungen hingewiesen.

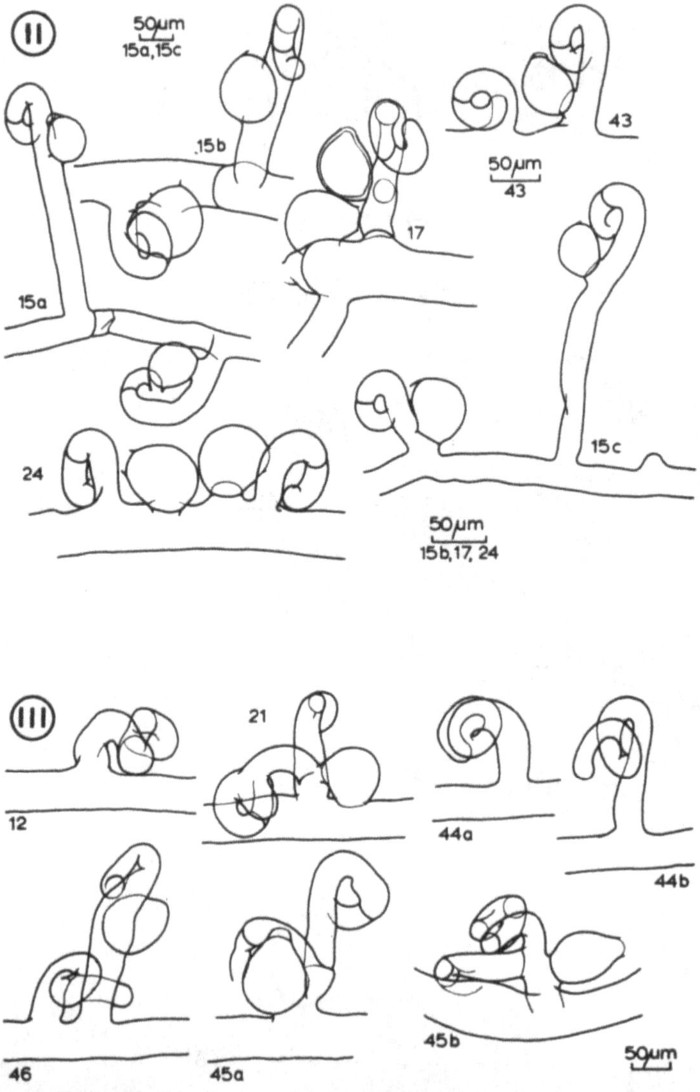


Abb. 5 Aus den Populationen II u. III

β Zu zwitterigen, seitlich am Tragfaden stehenden Ständen führend, gelegentlich auch mit vegetativen Kurztrieben: Fig. 6, 11, 14, 15, 17, 18, 19, 26, 32, 34, 36 der Abb. 1a; Fig. 43, 45, 46, Abb. 1b.

c) Durchwachsungen

α Mit Bildung eines neuen Sexualorgans (eines Antheridiums): Fig. 30, 41, 42 Abb, 1a und Abb. 4.

β Mit vegetativem Faden (oft begrenzten Wachstums): Fig. 8, 11, 21, 23 Abb. 1a; 44, 46, 47, 48, 49, 51, 52 Abb. 1b; Fig. 8 a–c, 11 Abb. 3; Fig. 44 a–b, 46 Abb. 5; Fig. 47 a–b, 48, 49, 51, 52 Abb. 6.

d) Formvarianten der Sexualorgane z. B. nicht spiralisierte (Abb. 2, Fig. 1c 1f,

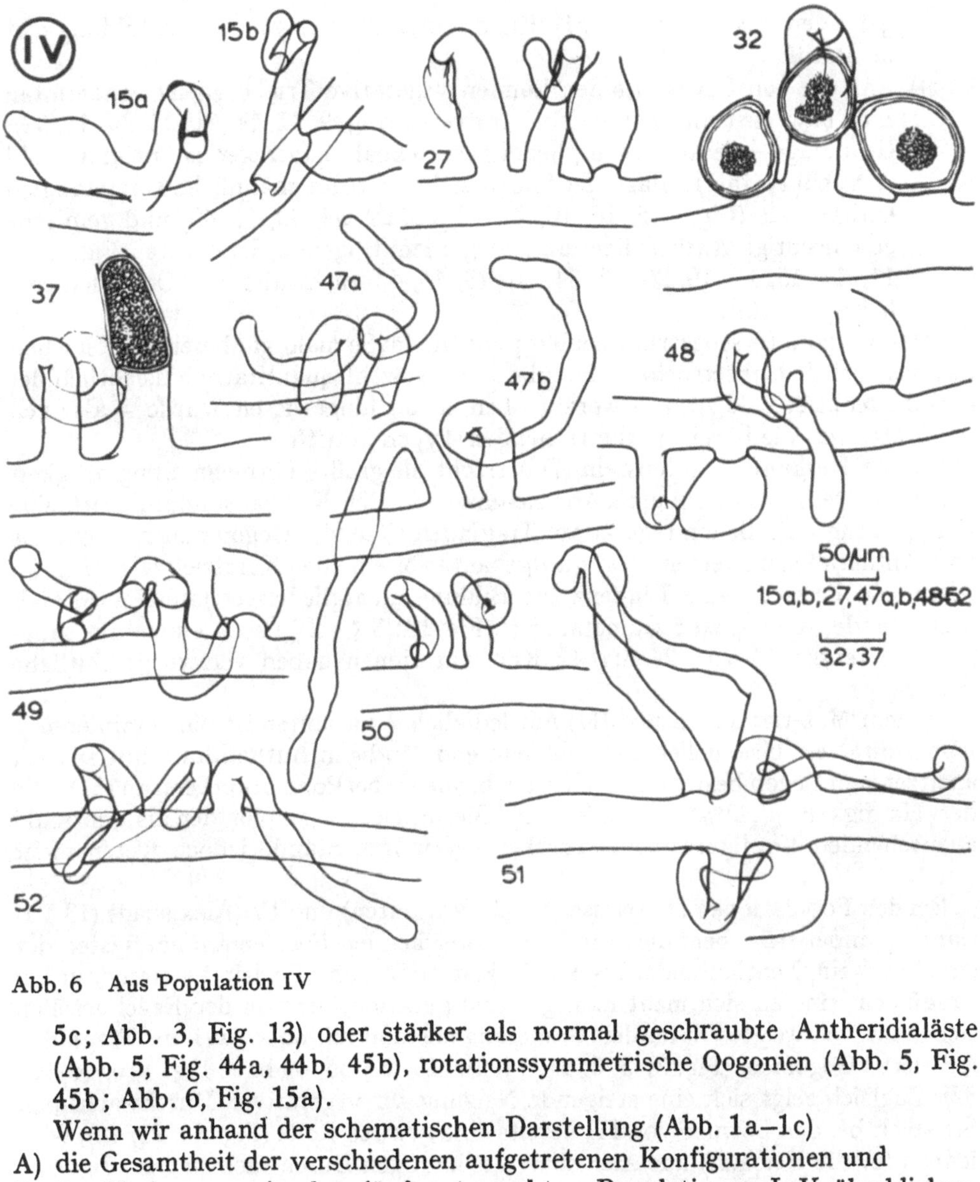


Abb. 6 Aus Population IV

5c; Abb. 3, Fig. 13) oder stärker als normal geschraubte Antheridialäste (Abb. 5, Fig. 44a, 44b, 45b), rotationssymmetrische Oogonien (Abb. 5, Fig. 45b; Abb. 6, Fig. 15a).

Wenn wir anhand der schematischen Darstellung (Abb. 1a–1c)

A) die Gesamtheit der verschiedenen aufgetretenen Konfigurationen und

B) ihr Vorkommen in den fünf untersuchten Populationen I–V überblicken, läßt sich allgemein folgendes feststellen:

A. 1) Es treten solitäre Antheridien (Fig. 9, 12, 44) und rein männliche Gruppen (Fig. 47, 50, 53) auf, jedoch nie einzelne Oogonien oder nur aus solchen bestehende Gruppen. Zur Bildung eines Oogons ist augenscheinlich – wie auch bei anderen monözischen Vaucheriaarten beobachtet (z. B. RIETH 1954) – stets die Induktion durch ein Antheridium erforderlich.

2) Durchwachsungen von Oogonien liefern in der Regel Antheridien (Fig. 30, 41, 42).

3) Bei Antheridien wurden nur Durchwachsungen mit vegetativen Fäden

gefunden (Fig. 44, 46–49, 51). Ein direkter Übergang zur Oogonbildung trat nicht auf.

4) Anstelle von Sexualorganen können vegetative Triebe, meist gehemmten Längenwachstums (Kurztriebe) auftreten (Fig. 9, 11, 14, 21, 23, 35, 51, 52).

5) Grundlage lateraler mehrgliederiger Sexualorganträger ist primär wohl stets ein Antheridialast. So finden sich zwar nur männliche Gametangien enthaltende (Fig. 4, 5, 10, 16, 21, 23, 25, 29, 33, 39, 41, 49) und gemischt geschlechtige Antheridien und Oogonien tragende Seitenäste (Fig. 3, 6, 11, 14, 15, 17–19, 26, 32, 34, 36, 43, 45, 46) nicht aber nur Oogonien enthaltende Stände.

B. Das Variabilitätsspektrum der Morphologie im Sexualorganbereich unterscheidet sich bei den untersuchten Populationen sowohl quantitativ hinsichtlich der Anzahl realisierter Typen – worauf oben schon hingewiesen wurde – als auch qualitativ was die jeweils vorherrschenden Typen betrifft.

Bei der Herkunft Trautenstein (I) herrscht die größte Formenmannigfaltigkeit (42 Varianten). Als häufigste Abweichung von der Normausbildung tritt eine Gruppierung auf, die ein regulär am Tragfaden sitzendes Oogon neben einem aus zwei Antheridien und einem Oogon aufgebauten bisexuellen Kurztrieb zeigt (Fig. 3). Allgemein herrscht eine Tendenz zur Bildung mehrgliederiger lateraler Sexualorganstände, mit bis zu 5 Gametangien (Fig. 18: 3♂+2♀ neben einem am Tragfaden sitzenden ♀) vor. 24 der 42 Konfigurationen haben verzweigte seitliche Stände.

In dem Material aus Lence (II) mit lediglich 7 Varianten ist die dominierende Abnormität ein bisexueller, nur aus dem endständigen Antheridium mit seitlich ansitzendem Oogon bestehender Kurztrieb, wie sie bei Population I erst an 15. Stelle der Häufigkeit (0,30%) zu finden ist. Sie rangiert sogar vor der als „normal" anzusehenden Konfiguration 2. Reicher verzweigte Stände jedoch treten nicht auf.

Bei den Populationen III (Nessebâr) (11 Varianten) und IV (Alexisbad) (13 Varianten) findet sich ebenfalls unter den abweichenden Formen am häufigsten der ein ♂ und ein ♀ enthaltende, bisexuelle Kurztrieb ohne Begleitgametangium am Tragfaden, eine an sich nicht häufige Konfiguration, denn in der Regel gesellen sich zum verzweigten lateralen Gametangienstand ein oder zwei am Tragfaden inserierte Oogonien (Fig. 3, 5, 8, 11, 14, 16–19, 21, 36, 49 bzw. Fig. 4, 6, 10, 23, 32). Zugleich zeigt sich eine steigende Neigung zur vegetativen Durchwachsung. Sie spielt bei den Formen von III bereits eine gewisse Rolle (Fig. 44, 21, 46), wird jedoch für die Herkunft aus Alexisbad mit häufigen Durchwachsungen (Fig. 47–49, 51) und Umwandlung von Sexualorganen in vegetative Triebe ± begrenzten Wachstums (Fig. 51, 52) charakteristisch.

Bei der Population (V) von Michaelstein mit 8 Varianten hingegen fehlen verzweigte Stände und vegetative Durchwachsungen völlig, Antheridialast und Oogonium sitzen dem Tragfaden an. Bemerkenswert ist jedoch ihre Anordnung. Neben den beiden „Normalkonfigurationen 1 u. 2 (Abb. 1a) deuten solitäre ♂ oder nur aus Antheridien bestehende Gruppen (Fig. 12, 50, 53), sowie gemischtgeschlechtige Gruppierungen mit überwiegender Antheridienzahl (Fig. 7, 27, 54) auf eine verstärkte ♂ Tendenz in diesem Material.

## Diskussion

Innerhalb der geschilderten Mannigfaltigkeit der Konfigurationen im Sexualorganbereich der *Vaucheria sessilis* lassen sich einige allgemeine Prinzipien, auch Grenzen der Variabilitätsmöglichkeit erkennen.

Auf die Tatsache, daß das Vorhandensein eines Antheridiums augenscheinlich unabdingbare Voraussetzungen für die Entstehung von Oogonien ist, wurde bereits hingewiesen und auch erwähnt, daß verzweigte seitliche Sexualorganstände immer von Antheridialästen ausgehen. Terminal an solchen lateralen, mehrgliederigen Ständen findet sich stets ein Antheridium, nie bildet ein Oogon den Abschluß. Die Induktion erfolgt danach in den Seitenästen wohl immer vom Antheridium nach basalwärts, entgegengesetzt der Wachstumsrichtung, nie nach vorwärts.

Die Oogonien sind dagegen durchweg sitzend und zeigen keine Neigung zur Stielbildung. Nur in einem einzigen Falle (Abb. 1a, Fig. 37; Abb. 4, Fig. 37) fand sich ein etwas vom Tragfaden abgehobenes Oogon.

Die für die Sektion Corniculatae in der Normalausbildung typische widderhornartige Windung von Antheridialast mit Antheridium kann fehlen, es treten, wenn auch verhältnismäßig selten, gestreckte, nicht eingerollte Antheridien bzw. Antheridialäste auf (Abb. 2, Fig. 1c, 1f, 5e; Abb. 3, Fig. 13). Die Determinierung des Merkmals „Einrollung des ♂ Astes ist danach in gewissem Grade variabel. Dies zeigt auch eine Beobachtung an *Vaucheria woroniniana*. Vor einigen Jahren fand ich in einer Population dieser Art relativ häufig gerade gestreckte, nicht eingerollte Antheridialäste, worauf hier nicht näher eingegangen werden soll.

Das relativ häufige Auftreten verzweigter, seitlicher ♂ oder bisexueller Sexualorganstände zeigt, daß das für die Subsektion Sessiles charakteristische Merkmal: Fehlen verzweigter seitlicher Sexualorganstände und dem Tragfaden direkt oder nur sehr kurz gestielt ansitzende Oogonien, ebenfalls nicht unabänderlich festgelegt ist, sondern vielmehr die Potenz zur Bildung solcher Stände existiert. Sie können rein ♂ (z. B. Abb. 1a, Fig. 4, 5, 10, 16, 21, 23, 25, 29, 33), dann aber immer von – dem gemeinsamen Tragfaden direkt ansitzenden – ♀ begleitet, sein, oder gemischtgeschlechtig (z. B. Abb. 1a, Fig. 3, 6, 11, 14, 15, 17, 18, 19, 26, 36) auftreten. In letzterem Falle findet sich der Sexualorganstand zuweilen solitär, d. h. ohne ihn am Tragfaden begleitendes Oogon (Abb. 1a, Fig. 15). Die Konfiguration gleicht dann denen wie sie für die Subsektion Racemosae typisch sind (siehe z. B. Abb. 5, Fig. 15a, 15b, 15c!). Die manchen Arten der Racemosae eigentümliche Anordnung der Oogonien in (meist dimeren) Wirteln unterhalb des Antheridiums wurde allerdings nie gefunden.

Die Fülle der beobachteten Konfigurationen deutet den Umfang der prinzipiell vorhandenen Bildungspotenz der Art an, deren „typische" Ausprägung durch Begrenzung dieser Möglichkeiten zustande kommt.

## Summary

Variation in the arrangement of Sexual Organs in Different Populations of *Vaucheria sessilis* (Vauch.) D. C. 1805

Description of spontaneous occuring anomalies in arrangement and shape of sexual organs in *Vaucheria sessilis* populations from five different localities and discussion of some consequences to the range of variation in the species.

## Краткое содержание

Изменчивость типа расположения и формы половых органов у различных популяций *Vaucheria sessilis* (Vauch.) D. C. 1805

Описываются спонтанно возникшие отклонения от (рассматриваемых как типичные) расположения и формы половых органов у популяций *Vaucheria sessilis*, полученных из пяти различных мест обитания. Делаются некоторые заключение о размахе изменчивости этого вида.

## Literatur

Rieth, A,. 1954: Beobachtungen zur Entwicklungsgeschichte einer *Vaucheria* der Sektion Woronina. Flora **142**, 156–182.

–, 1963: Die Algen der chinesisch-deutschen biologischen Sammelreise durch Nord- und Nordostchina 1956. I. Vaucheriaceae 1. Teil. Limnologica **1** (4), 287–313.

–, 1969: Beiträge zur Kenntnis der Vaucheriaceen XIV. Über das Auftreten von Sexualorganen an Vernarbungsmembranen im Thallusinneren. Kulturpflanze **XVII**, 103–107.

–, 1980: Xanthophyceae, 2. Teil: Vaucheriales. Süßwasserflora von Mitteleuropa. Begründet von A. Pascher, herausgegeben von H. Ettl, J. Gerloff, H. Heynig. Band 4. Jena und Stuttgart.

Prof. em. Dr. A. Rieth
Zentralinstitut für Genetik und Kulturpflanzenforschung
der Akademie der Wissenschaften der DDR
4325 Gatersleben

# III. Das Institut im Jahre 1980

Kulturpflanze XXIX · 1981 · S. 499–530

# A. Jahresberichte der Bereiche

## 1. Molekular- und Zellgenetik

Die Untersuchungen zur „Struktur, Mutabilität und Funktion des genetischen Materials" wurden weitergeführt. Folgende Themen standen dabei im Vordergrund:

1. Mutabilität und Reparaturprozesse
2. Mutabilität und Chromosomenaberrationen
3. Mutagenitätstestung
4. Analyse von Struktur und Expression eukaryotischer Gene
5. Genetik *in vitro* kultivierter Pflanzenzellen
6. Entwicklung von Methoden der Zell- und Gewebekultur pflanzlicher Zellen
7. Genetisch orientierte Entwicklungsbiologie der Säugetiere

**Bei der Charakterisierung spezifischer Komponenten der DNA-Reparatur und UV-Mutagenese bei Bakterien** (*Escherichia coli* und *Proteus mirabilis*) standen Fragen nach dem Mechanismus der DNA-Replikationsblockierung nach UV-Bestrahlung, der Interaktion induzierbarer Reparaturfunktionen mit der Plasmid-R 46-vermittelten UV-Mutagenese bei *P. mirabilis* sowie die Gewinnung und Funktionsanalyse von pBR322-Plasmidderivaten mit den Genen $recA^+$ aus *E. coli* bzw. *P. mirabilis* im Mittelpunkt des Interesses. Außerdem wurden die Arbeiten zur Genetik der Exoenzymbildung bei *Bacillus* weitergeführt. Die Blockung der DNA-Replikation durch UV-Schäden wurde an einer *E. coli*-Mutante mit Defekt im Initiationsmechanismus (dnaA 46) untersucht. UV-Bestrahlung verursacht eine „Suppression" des genetischen Defektes, d.h. führt zur Initiation von Replikationszyklen. Das führte zu der Schlußfolgerung, daß ein modifizierter Replikationskomplex (ohne dnaA-Protein?) wirksam wird und von Bedeutung für sogenannte Nach-UV-Reparaturprozesse sein könnte. In *P. mirabilis* wurde die induzierte Bildung von recA-Protein (dem Produkt des Strukturgens recA) autoradiographisch nach SDS-Acrylamid-Gelelektrophorese nachgewiesen. Im Vergleich zum *E. coli* recA-Protein hat das *P. mirabilis* Protein ein ca. 3000–5000 höheres relatives Molekulargewicht. Die Funktion der spezies-spezifischen recA-Proteine (aus *E. coli* bzw. *P. mirabilis*) konnte nach erfolgreicher Klonierung der entsprechenden chromosomalen DNA-Fragmente ($recA_{Ec}$ bzw. $recA_{Pm}$) auf dem Plasmidvektor pBR322 nach Transformation wechselseitig in beiden Wirtssystemen untersucht werden. Dabei wurde die Transformation von *P. mirabilis* mit pBR322-Plasmid-DNA erreicht. Trotz der unterschiedlichen Molekulargewichte sind die Genprodukte im jeweils heterologen Wirt funktionsfähig. Zur Erläuterung ist hervorzuheben, daß die Funktion des recA-Gens bei Bakterien eine zentrale Rolle bei DNA-Replikations-, Reparatur-, Mutagenese- und Zellteilungsprozessen spielt. Die noch nicht abgeschlossenen Restriktionskarten der entsprechenden recA-Gene lassen Unterschiede erkennen. Ziel der weiteren Untersuchungen ist die Analyse der $recA^+$-Genregulation. Sie wird als Schlüsselfunktion zur Aktivierung von DNA-Reparaturprozessen bei Bakterien angesehen.

In den Arbeiten an *Bacillus* wurden heterologe Transformationen (*Bacillus amyloliquefaciens* × *B. subtilis*) mit dem Ziel, Determinanten der alpha-Amylasebildung zu erfassen, ausgewertet. Außerdem wurde ein für geplante Arbeiten mit rekombinanten Plasmiden erforderlicher Wirtsstamm von *Bacillus subtilis* (u. a. mit den Markern recE, $r^-m^-$) hergestellt.

**Im Rahmen der Arbeiten zur experimentellen Induktion von chromatidalen Strukturumbauten** wurden in Kooperation mit dem Institut für Cytologie und Genetik der AdW der UdSSR in Novosibirsk (Prof. SALGANIK, Dr. DIANOV) weitere Experimente zur ortsgerichteten Induktion von Chromatidenaberrationen durchgeführt. Dabei wurden definierten RNA- bzw. DNA-Fraktionen von *Vicia faba* alkylierende und damit aberrationsinduzierende Chloräthylgruppen mit der Absicht angefügt, diese Kombinationsmoleküle durch Hybridisierung mit sequenzhomologen, auf den Chromosomen befindlichen DNA-Abschnitten innerhalb der Zelle zielgerichtet zu verbinden („Mutagenadressierung"). Die Auswertung der mit *Vicia faba* durchgeführten Versuche ergab, daß die neugeschaffenen Verbindungen in geringer Frequenz Chromosomenaberrationen auslösen. Die Zahl der Aberrationen war aber in den in Frage kommenden Regionen, in denen die Gene, die die entsprechenden RNA-Fraktionen kodieren (25S-, 18S- und 5S-RNA), lokalisiert sind, nicht signifikant erhöht. Versuche, die an die DNA der Zelle angefügten Chloräthylgruppen mit Hilfe von Natriumborhydrid zu aktivieren, zeigten erste positive Ergebnisse.

Zur molekularen und funktionellen Charakterisierung der *Vicia faba*-Chromosomen konnten nach unterschiedlicher Inkubationszeit in Bromdesoxycytidin (BrdC) spätreplizierende Regionen, asymmetrische Banden und Schwesterchromatid-Austausche (SCEs) in Analogie zur Inkubation in Bromdesoxyuridin) (BrdU) nachgewiesen werden. Für die beiden letztgenannten Phänomene ist dies der erste Befund dieser Art. Er besagt, daß mit BrdC kein ungleiches Guanin-Cyrosin (G-C)-Verhältnis zwischen den DNA-Strängen auf chromosomaler Ebene nachweisbar ist und stützt die Annahme, daß BrdC nicht anstelle von Cytidin (C), sondern für Thymidin (T) eingebaut wird.

Mit Hilfe der automatischen Bildauswertung wurden Schwesterchromatid-Austauschvorgänge (SCE) in BrdU-substituierten und fluorochromierten *Vicia faba*-Chromosomen mit dem selbstentwickelten Scanner BK1 und dem BVS 4200 (Institut für Züchtungsforschung Quedlinburg) densitometrisch erfaßt. Zur Bestimmung der SCE-Loci aus den erhaltenen Grauwertmatrizen wurden eine Reihe EDV-Verfahren am Kleinrechnersystem KRS 4200 erprobt. Die dabei gewonnenen Ergebnisse zeigten, daß die maschinelle Bildauswertung von Strukturen pflanzlicher Chromosomen die von Hand durchgeführten Messungen noch nicht ersetzen bzw. beschleunigen kann. Sie erbrachten aber Hinweise zur Verbesserung der Algorithmen.

In Zusammenarbeit mit dem Institut für Genetik der Bulgarischen Akademie der Wissenschaften in Sofia (Dr. NICOLOFF) und der FG „Angewandte Genetik" zu Fragen der Wirkungscharakteristika chemischer Mutagene bei rekombinanten Gerstenkaryotypen wurden der Standardkaryotyp, zwei rekombinante Karoytypen und vier Translokationslinien mit Maleinsäurehydrazid behandelt. Dabei traten besonders in einem Chromosomenabschnitt (Segment 44) bevorzugt Deletionen auf („Deletions-clustering"). Die unterschiedliche Lokalisation dieses Segments in den untersuchten Karyotypen erlaubte zu prüfen, ob die Positionsveränderung dieses Segmentes in der Lage ist, Einfluß auf die ortspezifische Häufung von Deletionen zu nehmen. In einem der Karyotypen (T505) blieb die Deletionshäufung in Segment 44 aus. Gleichzeitig fehlte in diesem Segment eine prominente Giemsa-Bande. Die Ergebnisse zeigen, daß Aberrationshäufung in bestimmten Chromosomen-Segmenten nach Mutageneinwirkung in bestimmten Fällen durch Karyotyp-„Umbauten" beeinflußbar ist. Gleiche Befunde wurden für rekonstruierte *Vicia faba* Karyotypen erhoben.

Im **Mutagenitäts-Testlaboratorium** konnten im Rahmen der vertraglich vereinbarten Routinetestung von Produkten der chemischen und pharmazeutischen Industrie u. a. folgende Probleme geklärt werden:

- In Versuchen zur Steigerung der Sensibilität des Ames-Testes (Mikrosomen-Assay) durch Zusatz von Nikotinamid wurde nachgewiesen, daß Nikotinamid nur dann eine Erhöhung der Biotransformationsrate bewirkt und so zur Verstärkung der Sensibilität des Testes führt, wenn NADP in unteroptimalen Konzentrationen vorliegt.
- Im Verlauf von Untersuchungen zur Struktur-Wirkungsbeziehungen der Organophosphorinsektizide konnte für weitere 13 untersuchte Verbindungen gezeigt werden, daß diese Stoffgruppe über Methylierung durch die O-Methylgruppen oder durch Halogenäthergruppen in den Seitenketten mutagen wirkt.
- Zur Entwicklung eines Genmutationstestes an Spermien über den histochemischen und immunologischen Nachweis mutativ veränderter Enzymproteine konnten in der ersten Phase Methoden für die histochemische Erfassung mutierter Spermien mit harnstoff-resistenter Succinatdehydrogenase und den immunologischen Nachweis von LDH-Mutanten in Spermienpopulationen erarbeitet werden.

Die Arbeiten zur Aufklärung der **Struktur und Expression eukaryotischer Gene** an den polytänen Riesenchromosomen der Dipteren-Art *Chironomus thummi* wurden weitergeführt. Nachdem im vorangegangenen Berichtszeitraum ein Teil der klonierten DNA-Sequenz des Balbiani-Ring c (BRc, vormals BR1) sequentiert werden konnte, wobei die tandemartige Anordnung einer 54–58 Basenpaare langen Repetitionseinheit nachgewiesen wurde, konnten weitere Erkenntnisse zum Gesamtaufbau der Balbiani-Ring-DNA erbracht werden. Mit dem Restriktionsenzym Alu I partiell gespaltene Gesamt-DNA ergab nach gelelektrophoretischer Auftrennung nicht – wie bei zufallsgemäßer

Anordnung der Spaltorte zu erwarten – eine statistische Verteilung der Spaltprodukte, sondern eine leiterartige Anordnung der Hybridisierungsbanden, die jeweils Multiple einer Grundeinheit darstellen. Für den Nachweis der Banden wurden eine radioaktiv markierte RNA-Fraktion, die in angereicherter Form Balbiani-Ring-RNA enthielt, als Probe verwendet. Wie eine genaue Analyse des Musters der Hybridisierungsbanden zeigte, beträgt die Länge der repetitiven Grundeinheit des BRb 200 (vormals BR 2), die des BRc 175 Basenpaare. Daraus folgt, daß ein großer Teil der rund 35000 Basenpaare langen Transkriptionseinheit der Balbiani-Ring-Gene aus repetitiven Sequenzen besteht. Weitere Einblicke in den feinstrukturellen Aufbau der Balbiani-Ring-DNA erbrachte die elektronenmikroskopische Analyse der mRNA beider Balbiani-Ringe, die affinitätschromatographisch durch Bindung an filtergebundene BR-DNA-Klone isoliert wurde. Es wurde nachgewiesen, daß unter Hybridisierungsbedingungen hoher Spezifität große Anteile der mRNA intra- und intermolekulare Doppelstrangstrukturen (Homoduplex, Heteroduplex) auszubilden vermögen, was auf die Existenz komplementärer Repetitionseinheiten als ein weiteres strukturbestimmendes Merkmal der Balbiani-Ring-DNA schließen läßt.

In Fortsetzung der Arbeiten zur weiteren Charakterisierung der Speicheldrüsensekretproteine wurden die Sekrethauptfraktionen mit der zweidimensionalen Fingerprinttechnik analysiert. Die Ergebnisse zeigen, daß alle Hauptfraktionen einen großen Anteil identischer Peptide enthalten und somit aller Wahrscheinlichkeit nach von sehr ähnlichen DNA-Sequenzen kodiert werden. Die Versuche zur *in vitro*-Translation wurden unter Verwendung von Balbiani-Ring-RNA weitergeführt. Die erhaltenen Translationsprodukte lassen nach elektrophoretischer Trennung leiterartige Polypeptidmuster erkennen, die auf die repetitive Natur der Translationsprodukte und der sie kodierenden RNA und DNA der Balbiani-Ringe hinweisen.

Im Rahmen der Arbeiten zur **„Genetik in vitro kultivierter Pflanzenzellen“** wurde das Ziel weiterverfolgt, vertiefte Einblicke in die Genetik und Biochemie der Stickstoffassimilation höherer Pflanzen am Modellobjekt Tabak zu gewinnen und Möglichkeiten zur genetischen Manipulation dieser Prozesse zu erkunden. Das Schwergewicht lag dabei im Berichtszeitraum auf der weiteren Charakterisierung der aus amphihaploiden Zellkulturen isolierten nitratreduktase-defekten Mutanten-Zellinien. Die im Vorjahr aus solchen Zellinien regenerierten diploiden, fertilen Pflanzen ermöglichten erstmalig eine genauere genetische Analyse durch sexuelle Kreuzung. Dabei ergab sich, daß bei der amphidiploiden Art *Nicotiana tabacum* der Wildtyp zwei aktive Strukturgene für das Nitratreduktase-Apoprotein enthält, die nicht gekoppelt sind, sich gegenseitig vertreten können und somit als homöolog anzusehen sind. Bei den bisher in dieser Hinsicht analysierten Mutanten (***nia***-128, ***nia***-130) beruht der Verlust der Nitratreduktase-Aktivität auf der gleichzeitigen Mutation beider homöologer Genloci. Es handelt sich also um Doppelmutanten. Die daraus durch Rückkreuzung mit dem Wildtyp hergestellten homozygoten einfach-mutierten Genotypen verfügen über ein zum normalen Wachstum auf Nitrat ausreichendes, wenn auch im Vergleich zum Wildtyp verringertes Niveau an aktiver Nitratreduktase. Enzymologische Untersuchungen ergaben, daß die im Kreuzungstest als allel erwiesenen Mutanten ***nia***-130 und ***nia***-128 sich phänotypisch unterscheiden. Während ***nia***-130-Zellen kein nachweisbares Nitratreduktase-Apoprotein enthalten, bilden ***nia***-128-Zellen ein nur partiell defektes Apoprotein, das zwar *in vivo* nicht funktionsfähig ist, nach Extraktion jedoch mit künstlichen Elektronendonatoren Nitrat zu reduzieren vermag.

Außer den Strukturgen-Mutanten konnten inzwischen auch die zum ***cnx***-Typ gehörenden nitratreduktasedefekten Mutanten genauer charakterisiert werden. Dabei ergab sich, daß alle vier bisher beim Tabak isolierten ***cnx***-Mutanten insofern übereinstimmen, als ihr Grunddefekt in der Aufnahme bzw. im intrazellulären Transport des Molybdats liegt. Unter Benutzung der Neurospora-Mutante ***nit***-1 als Indikator für den Molybdän-Kofaktor wurde gezeigt, daß ***cnx***-Zellen einen durch Molybdat-Zugabe reaktivierbaren und zur Komplementierung der Neurospora-Nitratreduktase befähigten Molybdän-Kofaktor enthalten. Die ***cnx***-Mutanten bilden dementsprechend ein komplettes Nitratreduktase-Multimeres (MG = 200000), dessen katalytische Inaktivität ausschließlich auf dem Fehlen von Molybdat beruht. Durch unphysiologisch hohe Molybdatkonzentrationen im Medium lassen sich *in vivo* sowohl Nitratreduktase als auch Xanthindehydrogenase partiell reaktivieren. – Da von den ***cnx***-Mutanten bisher keine fertilen Pflanzen regeneriert werden konnten und daher sexuelle Kreuzungen nicht möglich waren, wurde somatische Hybridisierung durch Protoplastenfusion für den Test auf genetische Komplementierung eingesetzt. Diese Versuche ergaben, daß die ***cnx***-Mutanten zwar mit verschiedenen ***nia***-Mutanten, nicht aber untereinander, komplementierende ($NR^+$) Hybriden ergeben. Die vier untersuchten ***cnx***-Mutanten sind daher als allel anzusehen.

Die **Untersuchungen zur Sproß- und Pflanzenregeneration bei Tomaten** wurden unter Verwendung zahlreicher Medienvarianten fortgesetzt. Es gelang, die Entwicklung der entstandenen Sproßknospen aus Langzeitkalluskulturen und protoplastenbürtigen Kallusgeweben hinsichtlich des Längenwachstums durch Gibberellinsäuregaben erheblich zu fördern. Die Zeiten, die ein Kallusgewebe auf einem sproßinduzierenden Medium verbringen muß, bis es zur Sproßinduktion kommt, ließen sich aber nicht wesentlich verringern. Aus Blattexplantaten und Sproßinternodien von Pflanzen, die keimfrei aus Samen der nur bedingt vitalen Tomaten-Mutante *chloronerva* stammten, konnten Sproßregenerate in größerer Auswahl aufgezogen werden, die sich spontan bewurzelten oder zur Bewurzelung angeregt werden konnten. Es scheint möglich zu sein, diese *chloronerva*-Pflanzen bei Zuführung entsprechender Nähr- und Wirkstoffe zu normalisieren. Ob der „Normalisierungsfaktor" Nicotianamin durch Biosynthese innerhalb des Pflanzenkörpers gebildet werden kann, soll geprüft werden. Im Zusammenhang mit den Untersuchungen zum „Normalisierungsfaktor" wurden größere Mengen an Kallusgeweben und suspendierten Zellmaterials für Aminosäurenanalysen produziert. Zum Einsatz gelangten verschiedene Tomaten-Herkünfte. Vorläufige Ergebnisse weisen darauf hin, daß sowohl in den suspendierten Zellen als auch in den benutzten Nährlösungen Nicotianamin vorhanden ist. Besonders deutlich ist die Existenz von Nicotianamin in den Kallusgeweben von *L. peruvianum* nachzuweisen. Aus den keimfrei angezogenen *chloronerva*-Pflanzen wurden Kalluskulturen angesetzt, die in Suspension übergeführt werden konnten. Es läßt sich bisher noch nicht sicher entscheiden, ob das Wachstum dieser Kulturen gegenüber dem von Kalluskulturen aus „Normalpflanzen" beeinträchtigt ist.

Aus Blattgeweben und Kalluskulturen der *chloronerva*-Mutante konnten Protoplasten isoliert und kultiviert werden, wobei nach bisherigen Erkenntnissen keine objektspezifischen Besonderheiten festgestellt wurden, die mit dem Charakter der Mutante verbunden sind.

Im Rahmen der Arbeiten am Objekt *Solanum* wurden Protoplasten aus dem Mesophyll von steril angezogenen Dihaploiden aus *Solanum tuberosum* Klon 7567 und Klon 7414 mit Protoplasten aus Suspensionskulturen von *S. phureja* und des dihaploiden Klons 7694 von *S. tuberosum* fusioniert. Um zu erreichen, daß sich nur Fusionsprodukte weiterentwickeln, wurden die Protoplasten aus der Suspensionskultur (S-Protoplasten) vor der Fusionierung mit Jodessigsäure oder deren Amid behandelt. Die weitere Entwicklung nichtfusionierter Mesophyllprotoplasten wurde durch ein cytokininfreies Medium verhindert. Kontrollversuche ohne Behandlung mit Polyäthylenglycol ergaben, daß alle S-Protoplasten abstarben, ohne eine Zellwand gebildet zu haben. Ein Teil der Mesophyllprotoplasten bildete zwar eine neue Zellwand, eine Zellteilung konnte jedoch in Abwesenheit von Cytokinin im Medium niemals beobachtet werden.

Aus den Fusionsprodukten aller erwähnten Kombinationen entwickelten sich Zellkolonien, die auf festes Medium übertragen werden konnten. Unter Bedingungen, bei denen auch aus Mesophyllprotoplasten Sprosse regeneriert werden, entwickelten sich aus den Fusionsprodukten 7567 + 7694 Sprosse, die bewurzelt und in Erde übergeführt werden konnten.

Die fehlende Zellteilung der dihaploiden Mesophyllprotoplasten in cytokininfreiem Medium erscheint als allgemeines Selektionskriterium nicht befriedigend, da sich Zellen aus Mesophyllprotoplasten monohaploider *S. chacoense* und einiger tetraploider Sorten von *S. tuberosum* auch ohne Cytokinin teilen. Deshalb wurde versucht, eine Komplementierung der Enzymsysteme durch irreversible Hemmung bestimmter Enzyme auch in den Mesophyllprotoplasten zu erreichen. Zum Einsatz gelangten Diäthylpyrocarbonat und Phenylmethansulfonylfluorid. Aus den bisherigen, noch unvollständigen Ergebnissen geht jedoch hervor, daß bei Einsatz letaler Dosen sowohl dieser Stoffe als auch von Jodessigsäureamid bei den Protoplasten auch die Fusionsprodukte nicht lebensfähig sind. Die Ursache hierfür liegt wahrscheinlich in der relativ unspezifischen Wirkung der verwendeten Hemmstoffe.

Die **„Entwicklungsbiologie der Säuger"** befaßte sich nach der erfolgreichen Etablierung der Methode zur Erzeugung von Aggregationschimären mit Versuchen zur Erzeugung von Chimären aus embryonalen Karzinomzellen (ECC) und frühen Embryonalstadien der Hausmaus. Das Ziel dieser Experimente besteht darin, *in vitro* kultivierte ECC in die Embryonalentwicklung zu integrieren und dadurch zur Normaldifferenzierung zu veranlassen.

In ersten Experimenten wurden frühe Furchungsstadien (8-Zeller) disaggregiert, um mit den gewonnenen Einzelblastomeren und ECC wieder Embryonen zu rekonstruieren. Nach diesen manipulativen Eingriffen entwickelten sich in der *in vitro*-Kultur morphologisch normale Blastozysten. Nach Transplantation dieser Blastozysten in pseudogravide Wirtsmütter wurden bisher aber noch keine Jungtiere geboren, bei denen sich eine Beteiligung der ECC an der Embryonalentwicklung nachweisen ließ.

Essentielle Voraussetzung für diese Experimente war die erfolgreiche *in vitro*-Kultivierung und Klonierung von ECC aus Teratokarzinomen. Bei diesen Experimenten ist es gelungen, eine nahezu euploide ECC-Kultur an das Wachstum auf Gelatine-beschichteten Plastekulturgefäßen zu adaptieren, ohne daß sie dabei morphologisch den ECC-Charakter verlor. Gleichzeitig wurde diese Zellinie einer umfangreichen biochemischen, karyologischen und zytologischen Charakterisierung unterzogen.

Die Arbeiten zur Entwicklung einer Methode zur immunologischen Geschlechtsbestimmung von präimplantativen Embryonen der Hausmaus durch Bestimmung des H-Y-Antigens wurden fortgesetzt. Die entscheidende Schwierigkeit besteht darin, daß das H-Y-Antigen ein sehr schwaches Antigen ist und die Erzeugung entsprechender Antikörper sehr erschwert ist. Wie Hauttransplantationsexperimente zeigten, reagieren die eingesetzten Inzuchtstämme jedoch mit einer spezifischen Immunantwort durch Abstoßung männlicher Transplantate bei weiblichen Tieren.

Aufbauend auf den bisherigen biochemischen Untersuchungsergebnissen zur Charakterisierung von Mukoviszidosezellen *in vitro* wurden biochemische Untersuchungen an Amnionzellen durchgeführt, um zu prüfen, ob diese Methoden für eine pränatale Diagnose geeignet sind. Deutliche Unterschiede in den $K_m$-Werten von 5 lysosomalen Enzymen und der alkalischen Phosphatase sowie der Thermostabilität der α-Fucosidase und der alkalischen Phosphatase zwischen Mukoviszidose- und Kontrollzellen ließen es gerechtfertigt erscheinen, erste pränatale Diagnosen an Risikoschwangerschaften durchzuführen (Geburten 1981).

## 2. Molekularbiologische Grundlagen der pflanzlichen Stoffproduktion

Die **Untersuchungen auf dem Gebiet des Proteinstoffwechsels pflanzlicher Samen** wurden auf der Ebene der Biosynthese-Regulation und der chemischen Charakterisierung von Reserveproteinen der Ackerbohne (*Vicia faba*) sowie auf der Ebene des Protein- und Aminosäure-Screenings an Getreiden und Leguminosen des Gaterslebener Kulturpflanzen-Weltsortiments fortgesetzt.

Ausgehend von hochgereinigter poly(A)-haltiger mRNA aus Kotyledonen der Ackerbohne konnte durch Umkehr-Transkription einsträngige und doppelsträngige cDNA synthetisiert und in Zusammenarbeit mit der FG Genwirkung kloniert werden. Mit dem Ziel, poly(A)-haltige RNA mit stark angereichertem Anteil an Legumin-spezifischen Sequenzen zu gewinnen und diese später für eine Selektion spezifischer cDNA-Klone zu verwenden, wurden mit Hilfe immunoaffinitätschromatographischer Methoden mit Legumin-Antikörpern reagierende Fraktionen aus den membrangebundenen Polysomen sich entwickelnder Ackerbohnenkotyledonen isoliert. *In vitro*-Translationsexperimente zur Bestimmung der Spezifität der mRNA aus derart selektierten membrangebundenen Polysomen und des erreichten Anreicherungsgrades spezifischer Sequenzen wurden begonnen.

Translationsversuche im zellfreien Weizenkeimlingssystem führten zum Nachweis eines gemeinsamen Präkursors für die alpha- und beta-Untereinheiten des Legumins. Dieses Polypeptid besitzt ein Molekulargewicht von ca. 60000. Ein Polypeptid von ähnlichem Molekulargewicht wurde nach Organellenfraktionierung im Immunopräzipitat (Legumin-Antikörper) der Mikrosomen-, nicht aber in der Aleuron-Fraktion gefunden. Mit der Identifizierung dieses Polypeptids wurde begonnen. Translationsversuche unter Einsatz von mRNA aus Kotyledonen verschiedener Altersstadien erhärteten die Vorstellung, daß die qualitative und quantitative Regulation der Reserveproteinsynthese vorwiegend auf der Stufe der mRNA-Synthese (Transkription) erfolgt.

Die Inkubation von Kotyledonen mit radioaktiven Aminosäuren ergab bei anschließenden mikroautoradiographischen Analysen eine kräftige Markierung des endoplasmatischen Reticulum (ER) und bestätigte dessen Funktion als Syntheseort der Reserveproteine. Zugabe von markierten Zuckern führte ebenfalls zur Markierung des ER, nicht aber des Golgi-Apparates, obwohl dieser neben dem ER als eines der innerzellulären Glykosylierungszentren angesehen wird. Ackerbohnenglobuline haben sich als relativ schwach glykosyliert erwiesen. Die radioaktive Markierung des ER nach Zuckerapplikation kann in Verbindung mit biochemischen Befunden anderer Laboratorien als Hinweis darauf interpretiert werden, daß eine Glykosylierung von Reserveproteinen bereits am ER, das ihren Syntheseort darstellt, stattfinden könnte. Ein endgültiger Beweis ist erst durch spezifischen Nachweis der Glykosylierung von Globulinen möglich.

Durch eine Kombination von Ammoniumsulfat-Fällung, Zonenpräzipitation und Hitzebehandlung konnte gelelektrophoretisch und serologisch reines Legumin gewonnen werden. Zu seiner Charakterisierung wurden das relative Molekulargewicht, die Aminosäure-Zusammensetzung, das Spektrum der N-terminalen Aminosäuren sowie der Absorptionskoeffizient bei 280 nm ermittelt. Es gelang, die ionenaustauscherchromatographisch isolierten Untereinheiten wieder zu einem Globulin zusammenzufügen, das von dem aus Kotyledonen gewonnenen Legumin weder gelelektrophoretisch noch serologisch unterschieden werden kann.

Reines Vicilin wurde durch Affinitätschromatographie an Concanavalin A-Sepharose gewonnen. Nach präparativer SDS-Gelelektrophorese wurden Polypeptide unterschiedlichen Molekulargewichts erhalten, deren N-terminale Aminosäuren mit der Hypothese in Einklang stehen, wonach das Vicilin nur einen Typ von Untereinheiten mit dem Molekulargewicht von ca. 50000 enthält und niedermolekularere Polypeptide aus diesem Typ durch vielleicht schon im Samen stattfindende Spaltungen hervorgehen. Eine genaue Analyse wird durch den wiederholt beschriebenen Polymorphismus des Vicilins sehr erschwert.

Das Protein- und Aminosäure-Screening an Formen der Gaterslebener Kulturpflanzen-Kollektion wurde durch Stickstoff- und Lysinanalysen an 1362 Gersten-, 254 Roggen-, 252 Weizen-, 80 Erbsen- und 65 Maisformen fortgesetzt. Außerdem wurde an 73 Sojaformen der Rohproteingehalt, an 72 Roggenformen der Tryptophangehalt und an 50 Erbsenformen der Gehalt an Cystein und Methionin ermittelt.

Es hat sich bestätigt, daß der im Vorjahr als Nicotianamin identifizierte ‚Normalisierungsfaktor' für die Tomatenmutante *chloronerva* nur in Gefäß-, nicht aber in Thalluspflanzen vorkommt. Er wird auch in Kallus- und Zellkulturen der Tomate synthetisiert. Die als ‚phänotypische Normalisierung' sich äußernde Überwindung des mutativ bedingten Syntheseblocks kommt auch zustande, wenn das Nicotianamin nicht über das Blatt, sondern über die Wurzel zugeführt wird. Damit ergibt sich eine zusätzliche Variante für den im Berichtszeitraum weiterentwickelten und standardisierten biologischen Test.

Vom 25. bis 27. Juni 1980 wurde das 2. Samenprotein-Symposium „Regulation of Protein Biosynthesis and Degradation During Embryogenesis and Germination of Plant Seeds" durchgeführt, an dem sich 72 Wissenschaftler aus 15 Ländern beteiligten (Programm s. S. 527 bis 529 dieses Bandes).

Im Rahmen der **Untersuchungen zur Ertragsphysiologie** bildete das Studium des Photosyntheseapparates von $C_3$-$C_4$-intermediären Arten einen gewissen Schwerpunkt. In der Gattung *Flaveria* (Asteraceae) wurden 2 neue $C_3$-$C_4$-intermediäre Arten (*F. anomala, F. pubescens*) gefunden. Wie bei anderen intermediären Arten liegt der $CO_2$-Kompensationspunkt bei $\sim$20 vpm $CO_2$ und seine Sauerstoffabhängigkeit ist nicht linear. Intermediär ist bei diesen Arten auch die Ausprägung der für $C_4$-Arten typischen Kranzanatomie. Von 6 Arten dieser Gattung wurden die Untereinheiten der Ribulose-1,5-biphosphat (RubP)-Carboxylase isoliert und isoelektrisch fokussiert. Anhand der isoelektrischen Punkte der großen Untereinheiten kann man zunächst 2 Entwicklungslinien innerhalb der Gattung unterscheiden. Die Gattung *Flaveria* ist für weiterführende Untersuchungen zur Evolution des $C_4$-pathways der Photosynthese insofern besonders geeignet, als hier $C_3$, $C_4$ und $C_3$-$C_4$-intermediäre Arten bekannt sind.

Untersuchungen über den Einfluß von Temperatur und Lichtintensität auf Ausbildung und Leistungsfähigkeit des Photosyntheseapparates bei Sommergerste ergaben:

Verminderung der Anzuchttemperatur um 6 °C (von 21 °C auf 15 °C) führte bei voller Lichtintensität (400 $\mu E \cdot m^{-2} \cdot s^{-1}$) zu keiner Verminderung von Menge und Aktivität der RubP-Carboxylase.

Verminderung der Lichtintensität (auf 100 $\mu E \cdot m^{-2} \cdot s^{-1}$) führt bei beiden Temperaturen zu einer Verminderung der Menge dieses Enzyms auf etwa die Hälfte und der Aktivität auf ein Viertel (bezogen auf die Blattfläche).

An Blättern von Gerstenjungpflanzen wurde der Einfluß von Sauerstoff auf den $CO_2$-Gaswechsel mit der RubP-Carboxylase-Aktivität verglichen. Die aus fruheren Versuchen bekannte Variabilität der $O_2$-Abhängigkeit des intrazellulären Widerstandes wurde bestätigt. Es ergab sich jedoch keine Beziehung zur RubP-Carboxylase-Aktivität.

Eine theoretische Analyse der kurzzeitigen Fixierung von $^{14}CO_2$ führte zu bisher noch fehlenden Vorstellungen über die Grundlagen dieser Methode zur Bestimmung der Lichtatmungsrate. Im Gegensatz zu den in der Literatur verbreiteten Ansichten erlaubt diese Methode keine zuverlässigeren Aussagen über die Lichtatmung als andere angewandte Verfahren.

Im Rahmen der **Forschungsarbeiten zur Aufklärung der molekularen Struktur des Photosyntheseapparates** wurden die Untersuchungen über die Chlorophyll-Proteine von *Vicia faba* und *Hordeum vulgare* fortgesetzt. Im Gegensatz zu den spektralen Eigenschaften vergleichbarer Chlorophyll-Holoproteine, die bei beiden Pflanzenarten weitgehend übereinstimmen, weisen die entsprechenden Apoproteine signifikante artspezifische Merkmale auf, was besonders in ihrem unterschiedlichen elektrophoretischen Verhalten zum Ausdruck kommt. Von den bisher charakterisierten drei $Chlorophyll_{a/b}$-Proteinen sind zwei Bestandteile einer als Lichtsammelkomplex (LHC) bezeichneten Struktureinheit. Das dritte $Chlorophyll_{a/b}$-Protein ($Chl_{a/b}$-P 1), das ebenfalls bei beiden Pflanzenarten vorkommt, fehlt in dieser Struktureinheit, was auf die Existenz weiterer Komplexe mit Lichtsammelfunktion hindeutet. Aus Untersuchungen an Gerstenmutanten kann abgeleitet werden, daß das Fehlen der beiden $Chlorophyll_{a/b}$-Proteine des LHC weder die Thylakoidmembranstruktur noch deren photosynthetische Leistung – und damit die Lebensfähigkeit der Mutanten – wesentlich beeinflußt. Dagegen scheint zumindest das Apoprotein von $Chl_{a/b}$-P 1 ein unentbehrlicher Strukturfaktor zu sein, da bislang keine photosynthetisch aktiven Mutanten bekannt sind, denen diese Komponente fehlt.

Der Lichtsammelkomplex LHC, dessen Isolierung gegenwärtig am besten beherrscht wird, wurde als Modellobjekt verwendet, um weitere Informationen über die Organisation der Chlorophyllmoleküle mittels optischer Methoden zu erlangen. Danach befinden sich nur die für die zwei Hauptspektralformen des Chlorophyll a verantwortlichen Pigmentmoleküle in einer stark asymmetrischen Umgebung. Durch die Weiterentwicklung der mathematischen Verfahren zur Zerlegung von Absorptionsspektren steht ein Instrumentarium zur Verfügung, das es gestattet, schwer auflösbare Spektralformen zu analysieren sowie die mit herkömmlichen Methoden erzielten Ergebnisse kritisch zu bewerten.

Nachdem in den vergangenen Jahren der als Kopplungsfaktor der Photophosphorylierung ($CF_1$) bezeichnete enzymatisch aktive Teil des ATPase-Komplexes ($CF_1$-$CF_0$) eingehend untersucht wurde, konzentrierten sich die 1980 durchgeführten Forschungsarbeiten auf die Analyse des Proteinkomplexes $CF_0$, der in die Photosynthesemembran eingebettet ist und die Funktion eines Protonenkanals hat. Am Aufbau des Teilkomplexes $CF_0$ sind vier unterschiedliche Polypeptidspezies beteiligt, von denen das anteilmäßig am stärksten vertretene Proteolipid im präparativen Maßstab isoliert wurde. Durch enzymatische Jodierung und kontrollierte Proteolyse des membrangebundenen ATPase-Komplexes konnten Rückschlüsse auf die räumliche Anordnung der Untereinheiten gezogen werden. Es gelang die Identifizierung derjenigen Polypeptide, die die Verbindung zwischen den Teilkomplexen $CF_1$ und $CF_0$ sowie zwischen $CF_1$ und der Thylakoidmembran bewirken. Die Untersuchungen liefern einen Beitrag zur Aufklärung der strukturellen Basis des ATPase-Komplexes, der die Energie eines Protonengradienten zur ATP-Synthese nutzt, und schaffen damit wichtige Voraussetzungen für das Verständnis der photosynthetischen Energieumwandlung.

Neben der Überarbeitung und Verbesserung vorhandener Programme wurde in der **Rechenstation** eine Reihe neuer Programme für den Kleinrechner KRS

4201 erarbeitet oder aus anderen Programmbibliotheken übernommen und an die Besonderheiten und Möglichkeiten des KRS 4201 angepaßt. Schwerpunkte lagen dabei auf dem Aufbau einer statistischen Programmbibliothek unter besonderer Berücksichtigung multivariater und parameterfreier Verfahren und auf der Erarbeitung eines Programmpakets zur Auswertung von Chromosomenbildern.

## 3. Angewandte Genetik, Taxonomie und Genbank der Kulturpflanzen

Auf dem **Gebiet der angewandten Genetik** konzentrierten sich die Untersuchungen auf die Entwicklung der apomiktischen Fortpflanzungsweise bei Futtergräsern sowie auf Probleme der Hybridforschung bei Gerste.

In der Apomixisforschung stand weiterhin die Selektion und Analyse von Art- und Gattungsbastarden mit apomeiotischer bzw. apomiktischer Samenausbildung im Mittelpunkt. Analysen von 277 Nachkommenschaften (etwa 3000 Pflanzen) nach Bestäubung mit dominanten Merkmalsträgern unterschiedlicher Ploidiestufen ergaben, daß (wie zu erwarten) in den meisten Fällen sexuelle Samenausbildung vorherrscht. Apomeiotische Bastarde konnten in erster Linie aus folgenden Kreuzungskombinationen erhalten werden: *Festuca pratensis* (2n = 28) × *F. arundinacea* (2n = 42), *F. pratensis* (2n = 28) × *F. gigantea* (2n = 42), *F. gigantea* (2n = 42) × *F. arundinacea* (2n = 42). Eine eingehende embryologische Analyse liegt bisher für Bastarde der beiden erstgenannten Kombinationen vor; bei ihnen werden apospore Embryosäcke nach dem *Polygonum*-Typ ausgebildet.

$M_2$-Material aus früheren Bestrahlungsversuchen mit sterilen $F_1$-Bastarden wurde erneut auf das Reproduktionsverhalten geprüft bzw. in größerem Umfang neu selektiert. Relativ hoher Samenansatz bei weitgehender morphologischer Homogenität wurde bisher bei einigen $M_3$-Nachkommenschaften aus der Kombination *Festuca gigantea* × *F. arundinacea* festgestellt. – Aus neu durchgeführten Art- und Gattungskreuzungen wurden 956 Nachkommen von 20 verschiedenen Kombinationen gewonnen, insbesondere unter Beteiligung von *Festuca pratensis*, *F. arundinacea* (2n = 28, 42, 56, 70), *F. gigantea* und *Lolium multiflorum*.

Die Versuche zur modifikativen Induktion der apomiktischen Samenausbildung haben bisher gezeigt, daß sowohl Phytoeffektoren als auch natürliche Wachstumsfaktoren (z. B. Tageslänge und Temperatur) den Samenansatz unterschiedlich beeinflussen können. Bei der fakultativ apomiktischen Art *Poa pratensis* führt z. B. Behandlung mit 2,4,5-Trichlorphenoxyessigsäure oder Gibberellinsäure ($GA_3$) zur signifikanten Erhöhung des Anteils polyembryoner Samen und damit höchstwahrscheinlich auch des Apomixiegrades.

Für Untersuchungen zur Entwicklung genetischer Systeme der Hybridsaatguterzeugung bei Gerste, insbesondere mit Hilfe balancierter tertiärer Trisomie (BTT), erfolgten u. a. weitere Prüfungen auf Kopplung zwischen Translokationsbruchpunkt und *ms*-Gen. In entsprechenden $F_2$- und $F_3$-Analysen ergab sich z. B. für 28 verschiedene Translokationen, an denen Chromosom 4 beteiligt ist, enge Kopplung für *ms24* in rund drei Viertel aller Fälle, während für *ms25* enge Kopplung kaum nachweisbar ist. Daher dürfte *ms24* sehr nahe am Zentromer lokalisiert sein. – In Nachkommenschaften translokationsheterozygoter Pflanzen aus derartigen Kreuzungen konnte durch cytologisches Screening eine größere Anzahl trisomer Pflanzen isoliert werden. Ihre Vitalität ist außerordentlich unterschiedlich, abhängig von der beteiligten Translokation bzw. der spezifischen Struktur des Zusatzchromosoms. In einigen Fällen entsprechen Vitalität und Wüchsigkeit nahezu derjenigen disomer Pflanzen.

In den Versuchen zur Möglichkeit der Entwicklung funktioneller Diploidie bei autotetraploider Gerste (Amphidiploidisierung) mittels multipler Translokationen wurden mathematische Modellvorstellungen erweitert, mit deren Hilfe das Ausmaß der präferentiellen Paarung strukturgleicher Chromosomen in der Meiose besser beurteilt werden kann. Nachdem bisher Karyotypen mit ein bis vier Translokationen untersucht wurden, erfolgten nun Arbeiten zur Herstellung neuer duplexheterozygoter Formen unter Einbeziehung von fünf und sechs Translokationen.

Fortgeführt wurden auch Arbeiten an den Sortimenten von Cucurbitaceen und Mais sowie bei Sojabohnen durch den Anbau des vorhandenen Welt- und Mutantensortiments (etwa 2700 Sippen) für die Selektion züchterisch geeigneter Formen. – Im Rahmen einer Forschungsvereinbarung mit der IAEA Wien wurden, in Ergänzung zu früheren Arbeiten, Untersuchungen zu den potentiellen Erträgen und der N-Bilanz proteinreicher Mutanten und anderer Formen bei Gerste durchgeführt.

In den Arbeiten zur **Taxonomie und Evolution** von Kulturpflanzen nahm die planmäßige Fortführung der Neubearbeitung des „Vorläufigen Verzeichnisses landwirtschaftlich oder gärtnerisch kultivierter Pflanzenarten" von R. MANSFELD einen wichtigen Platz ein. Die Zahl der zu berücksichtigenden Sippen hat wiederum zugenommen, in einer einschlägigen Kartei sind derzeit etwa 4350 Artnamen erfaßt. Für eine größere Anzahl von Familien (z. B. Malvaceae, Zingiberaceae, Chenopodiaceae, Umbelliferae, Asclepiadaceae), für umfangreichere Gruppen der Rosaceae, Leguminosae-Mimosoideae und Gramineae und viele weitere Gattungen und Arten wurden Manuskriptentwürfe abgefaßt, zu früher fertiggestellten Familien Ergänzungen und Materialsammlungen zu bisher nicht bearbeiteten Familien vorgenommen sowie die parallel zu diesem Vorhaben vorgesehenen Bibliographien zur Literatur über Taxonomie und Evolution von Kulturpflanzen und zur Archäologie von Kulturpflanzen zusammengestellt.

In Weiterführung der numerisch-taxonomischen Arbeiten bei Kulturpflanzen wurde das vorliegende Datenmaterial für die Arten der Gattungen *Triticum* und *Aegilops* mit Hilfe von 9 unterschiedlichen Ähnlichkeitskoeffizienten und 6 verschiedenen Clustermethoden berechnet, wobei sich deutliche Unterschiede in der Brauchbarkeit der Ähnlichkeitsmaße ergaben. Neben dem GOWER similarity-Koeffizient brachte die mean character difference für das bearbeitete Material die taxonomisch sinnvollsten Ergebnisse. Im Gegensatz dazu war der Einfluß unterschiedlicher Clustermethoden erstaunlich gering, so daß beliebige Kombinationen der beiden genannten Ähnlichkeitsmaße mit allen untersuchten Clusterverfahren gleichermaßen brauchbare Resultate lieferten. Die scharfe

Trennung der beiden Gattungen *Triticum* und *Aegilops* spiegelte sich in fast allen der erhaltenen Dendrogramme wider. Ferner ist die Sonderstellung der Art *Aegilops mutica* eindeutig belegbar, und die klaren taxonomischen Abstände der Arten *Triticum dicoccon* und *Tr. dicoccoides* liefern neue Aspekte für die Diskussion zur Taxonomie der behandelten Sippen. – Mit Hilfe der SUMRAT-Informationsanalyse wurde der Informationswert der ermittelten Merkmale für die Dendrogrammbildung berechnet und eine Gliederung in unterschiedlich informative Merkmalssätze aufgestellt.

Für die zum Abschluß gekommene numerische Bearbeitung einer *Pisum*-Kollektion wurden auf Grund unterschiedlicher Merkmalssätze ebenfalls verschiedene zusätzliche Clustermethoden geprüft, wobei diese in Übereinstimmung mit den genannten Befunden bei Gräsern eine im Vergleich zu anderen Einflußgrößen der Dendrogrammbildung geringfügige Wirkung zeigten. Verfeinerte Ermittlungen von SUMRAT- und SAMRAT-Informationsmaßen bestätigten die bereits früher gemachten Aussagen über die unterschiedliche Wertigkeit der verwendeten Ähnlichkeitskoeffizienten.

Die monographische Bearbeitung der Gattung *Aegilops* L. auf der Grundlage der *Aegilops*-Kollektion der Genbank kam zum Abschluß (Taxonomie und Evolution der Gattung, blütenbiologische Untersuchungen, Erhaltung des Materials).

Für die taxonomische Bearbeitung der Gattung *Raphanus* wurden Merkmalsanalysen an Lebendmaterial aller Entwicklungsstadien, u. a. auch UV-Muster der Petalen und anatomische Untersuchungen der Speicherorgane, fortgeführt und in die Analysen Populationen von Wildsippen der Gattung einbezogen. – Für eine Fortführung der rasterelektronenmikroskopischen Studien der Pollenskulpturen von *Triticum*- und *Aegilops*-Arten wurde Pollenmaterial aller zu prüfenden Sippen gewonnen und für die serienmäßigen Untersuchungen vorbereitet. – Cytologische Untersuchungen an Sortimentsmaterial von *Allium, Papaver* und *Hordeum* wurden vorgenommen.

Spezielle Untersuchungen an Freiland- und Kulturmaterial von Algen der heimischen Flora liefen weiter, eine Untersuchung der Variationsbreite von Merkmalen der Sexualorgane von *Vaucheria sessilis* konnte abgeschlossen werden.

Die Archive (Herbar, Ähren- und Samensammlung) wurden durch Material aus den Kulturpflanzenkollektionen und von den Sammelreisen ergänzt. Ordnungs-, Etikettierungs- und Präparationsarbeiten nahmen den vorgesehenen Fortgang, die Etikettierung der *Hordeum*- und *Secale*-Ährensammlung ist abgeschlossen; für andere Einrichtungen in der DDR erfolgten Determinationsarbeiten.

Ein Symposium mit dem Thema „Collection of European land-races of cultivated plants and their evaluation“ wurde in der Zeit vom 1. bis 4. Juli 1980 ausgerichtet. An ihm nahmen 73 Wissenschaftler aus 17 Ländern teil. In den 30 Vorträgen und 10 Postern wurden Fragen der Sammlung, Bearbeitung und züchterischen Nutzung europäischer Landsorten sowie Probleme der Taxonomie und Biosystematik der Kulturpflanzen dargestellt.

Der Bestand der **Genbank der Kulturpflanzen** erweiterte sich im Jahre 1980 um 1465 Muster und beträgt 48959 Sippen:

| | |
|---|---|
| Getreide und Gräser | 25069 |
| Hülsenfrüchte | 9769 |

| | |
|---|---|
| Gemüse | 6731 |
| Heil- und Gewürzpflanzen, Industriepflanzen | 4591 |
| Mutanten (*Antirrhinum, Glycine, Lycopersicon*) | 2799 |

Im nationalen und internationalen Samenaustausch kamen 9242 Proben zur Abgabe, davon erhielten Einrichtungen in der DDR 2781; für die Genbank und Forschungsgruppen im Institut gingen 1883 Proben ein.

Im Rahmen der Vereinbarung der Akademie der Wissenschaften der DDR mit dem Consiglio Nazionale delle Ricerche in Italien wurden mit Mitarbeitern des Istituto del Germoplasma in Bari zwei Reisen (Juni und August/September 1980) zur Sammlung von Material indigener Kulturpflanzen in den Regionen Puglia, Basilicata und Campania durchgeführt. Dabei konnten insgesamt 460 Saatgutproben, überwiegend von Gemüsen und Körnerleguminosen, gesammelt werden. Dieses Material ist eine wertvolle Bereicherung der Kollektionen der Genbanken in Bari und Gatersleben.

Auf der Grundlage langfristiger Vereinbarungen mit dem Instytut Hodowli i Aklimatyzacji Roślin (IHAR) in Radzików, VR Polen, erfolgte vom 26. August bis 5. September 1980 in der DDR eine Exkursion zur Sammlung von Samenproben von Wildformen wirtschaftlich wichtiger Futtergräser. Die beteiligten Mitarbeiter des Zentralinstituts, des IHAR und des Instituts für Futterpflanzenzüchtung in Malchow sammelten 365 Samenproben im Harz, im Mittel-Elbe- und Havel-Gebiet sowie in West-Mecklenburg.

Die Untersuchungen des Materials der Genbank richteten sich hauptsächlich auf die Charakterisierung der in den letzten Jahren gesammelten Landsorten, die Erfassung von Krankheitsresistenzmerkmalen an Gersten und Weizen und die Ermittlung der Variabilität des Gehaltes an Protein und limitierenden essentiellen Aminosäuren in ausgewählten Mustern der Gersten-, Weizen- und Roggenkollektion. Die Einlagerung von Samen in das Samenkühllagerhaus wurde in vorgesehener Weise fortgeführt.

## B. Kolloquien; Ausstellungen; Vortragsabende

08. 01. 1980 Dr. U. Wobus: Feinstrukturanalyse eines Insektengens: Interne Repetition als Bauprinzip der Balbianiring-DNA von *Chironomus thummi*.

22. 01. 1980 Dr. M. Hartmann, Zentralinstitut für Mikrobiologie und experimentelle Therapie, Jena: Restriktionsenzyme als Werkzeuge der Molekularbiologie.

08. 04. 1980 Dr. E. Vogel, Department of Radiation Genetics and Chemical Mutagenesis, State University of Leiden, Leiden, The Netherlands: Die Beziehungen zwischen chemischen Reaktionsmechanismen und mutagenen Eigenschaften alkylierender Verbindungen in *Drosophila*.

29. 04. 1980 Dr. R. Krauspe, Institut für Biochemie der Pflanzen, Halle: Gentransfer bei Pflanzen durch *Agrobacterium tumefaciens*.

06. 05. 1980 Dr. B. Kaina: Ortsspezifische Mutagenese – Methoden, Ergebnisse, Probleme.

20. 5. 1980 Dr. J. Cullum, Institut für Biologie III der Universität Freiburg i. Br./BRD: Insertions sequence-mediated DNA rearrangements in bacteria.

05. 09. 1980 Prof. Dr. F. Ruddle, Departments of Biology and Human Genetics, Yale

| | |
|---|---|
| | University, New Haven/USA: Gene mapping and gene transfer in mammalian cells. |
| 08. 09. 1980 | Prof. D.F.C. KAFATOS, Biological Laboratories, Harward University, Cambridge/USA: The *Drosophila* chorian genes: amplification and expression. |
| 08. 09. 1980 | Prof. J. J. GALL, Department of Biology, Yale University, New Haven/USA: Molecular biology of lampbrush chromosomes. |
| 16. 09. 1980 | P. STEPHAN: Untersuchungen zum Bevölkerungswachstum in einer Gemeinde des Nordharzvorlandes im 17. und 18. Jahrhundert. |
| 30. 09. 1980 | Prof. Dr. K. R. H. REPKE, Zentralinstitut für Molekularbiologie, Berlin-Buch: Wie arbeiten energiewandelnde ATPasen mit hohem energetischem Wirkungsgrad? |
| 14. 10. 1980 | Dr. K.-W. HAAKE, Bereich Medizin der Karl-Marx-Universität Leipzig: *In Vitro*-Befruchtung beim Menschen – Stand und Probleme. |
| 16. 10. 1980 | Dr. A. LAISK, Institut für Astrophysik und Physik der Atmosphäre der Estnischen Akademie der Wissenschaften, Tartu/UdSSR: Limiting factors in photosynthesis. |
| 21. 10. 1980 | Dipl.-Biol. I. SAALBACH: Untersuchungen zur Nitratreduktion und Nitrataufnahme bei Zellkulturen von *Nicotiana tabacum* unter Verwendung von nitratreduktasedefekten Mutanten. Zugleich Vortrag anläßlich der Verteidigung der Dissertation im Rahmen des Promotionsverfahrens A. |
| 28. 10. 1980 | Prof. Dr. H. THEILE, Bereich Medizin der Karl-Marx-Universität Leipzig: Ergebnisse und Probleme der Stoffwechselgenetik des Menschen. |
| 30. 10. 1980 | Prof. Dr. C. B. OSMOND, Australian National University, Canberra City/Australien: Photorespiration, photoinhibition and plants responses to stress. |
| 04. 11. 1980 | Dr. K. REGEL, Bereich Strahlenbiophysik des Zentralinstituts für Molekularbiologie, Abteilung Rossendorf: *In vivo*-Organisation der Eukaryoten-DNS: Schlußfolgerungen aus Sedimentationsuntersuchungen nach Einwirkung ionisierender Strahlen. |
| 18. 11. 1980 | Dr. R. BORRIS, VEB PROWIKO, Schönebeck/Elbe: Extrazelluläre Enzyme von Bakterien: Herstellung und Anwendung. |
| 25. 11. 1980 | Prof. Dr. U. WOLF, Institut für Humangenetik und Anthropologie der Universität Freiburg i. Br./BRD: Funktion und Genetik des H-Y-Antigens. |
| 09. 12. 1980 | Dr. P. KINZEL, Bezirkskrankenhaus Gera: Quantitative Chromosomenanalyse. |
| 16. 12. 1980 | Dr. K. HAMMER: Vorarbeiten zur monographischen Darstellung von Wildpflanzensortimenten: *Aegilops* L. Zugleich Vortrag anläßlich der Verteidigung der Dissertation im Rahmen des Promotionsverfahrens B. |

## Bereichskolloquien

| | |
|---|---|
| 23. 01. 1980 | Dr. CHR. O. LEHMANN: Über die Genbank für das Mittelmeergebiet in Bari. |
| 28. 01. 1980 | Dr. S. KLAUS, Zentralinstitut für Mikrobiologie und experimentelle Therapie, Jena: Die DNA von *Streptomyces*-Phagen. |
| 10. 03. 1980 | Dr. E. ANANIEV, Institut für Molekulargenetik, Moskva/UdSSR: Mobile dispersed genes in *Drosophila*. |
| 18. 03. 1980 | Dr. H. MAHLKE, Zentralinstitut für Mikrobiologie und experimentelle Therapie, Jena: Streptokokken-Plasmide: Struktur, Funktion und genetische Manipulierung. |
| 01. 04. 1980 | Dr. CH. HORSTMANN: Ergebnisse des Studienaufenthaltes am John Innes Institute, Norwich/Großbritannien. |
| 07. 04. 1980 | Dr. V. LANZOV, Institut für Kernphysik, Leningrad/UdSSR: Hot spots of recombination in the *E. coli* chromosome. |

| | |
|---|---|
| 08. 04. 1980 | Dr. U. RÄNDEL, Museum für Naturkunde der Humboldt-Universität zu Berlin: Die Wildflora von Ost- und Westkuba. |
| 21. 04. 1980 | Dipl.-Biol. G. BUSSE: Untersuchungen zur Morphogenese an *in vitro*-Kulturen von *Vicia faba*. |
| 16. 05. 1980 | Dr. K. D. SCHWENKE, Zentralinstitut für Ernährung, Bergholz-Rehbrücke: Technologisch-funktionelle Eigenschaften von Pflanzenproteinen. |
| 27. 05. 1980 | Dr. J. FOREJT, Institut für Molekulargenetik der Akademie der Wissenschaften der ČSSR, Praha/ČSSR: Relationships between the T-gene-complex of the mouse and terato carcinome differentiation. |
| 27. 05. 1980 | Dr. K. GRÖBER: Moderne Trends in Obstbau und Obstzüchtung. |
| 03. 06. 1980 | Dipl.-Math. A. BERGMANN: Entscheidungsregeln bei speziellen Wahrscheinlichkeitsverteilungen – Möglichkeiten und Grenzen der quadratischen Diskriminanzanalyse. |
| 17. 06. 1980 | Dr. R. CASEY, John Innes Institute, Norwich/Großbritannien: Variability in the structure of legumin from *Pisum*. |
| 26. 06. 1980 | Dr. N. N. KOLESNIKOV und Dr. E. J. KARAKIN, Institut für Cytologie und Genetik, Novosibirsk/UdSSR: Functional organization of polytene chromosomes. |
| 08. 07. 1980 | Dr. A. M. EVANS, Department of Applied Biology, University of Cambridge, Cambridge/Großbritannien: On taxonomy, evolution, and breeding of *Phaseolus* beans. |
| 21. 07. 1980 | Dr. CHR. O. LEHMANN, Dr. K. GRÖBER und Dr. F. SCHOLZ: Einige Schwerpunkte des Anbaus von Material der Genbank sowie genetischer Versuche mit Futtergräsern und Gerste. |
| 24. 07. 1980 | Prof. Dr. F. H. KASTEN, Med. Center, Department of Anatomy, Louisiana State University, New Orleans/USA: Proliferation und Differenzierung von Herzmuskelzellen. |
| 04. 08. 1980 | Prof. Dr. R. RIEGER und Dr. I. SCHUBERT: Molekulare und funktionelle Charakterisierung von differentiell mutagensensiblen Chromosomenbereichen und -beständen. |
| 08. 09. 1980 | Prof. D. F. C. KAFATOS, Biological Laboratories, Harvard University, Cambridge/USA: Organization and function of a multigene family: the chorion genes of insects. |
| 09. 09. 1980 | Prof. Dr. K. MÜNTZ: Bericht über den II. Kongreß der Föderation Europäischer Gesellschaften für Pflanzenphysiologie (FESPP) in Santiago de Compostela/Spanien. |
| 11. 09. 1980 | Dr. D. DUDITS: Institut für Genetik des Biologischen Forschungszentrums, Szeged/VR Ungarn: Premature chromosome condensation in plants. |
| 15. 09. 1980 | Dr. U. WOBUS: Genstruktur, mRNA-Synthese und Processing: Ein Bericht über den 4. Arolla-Workshop. |
| 19. 09. 1980 | Dr. G. BENCZE: Institut für Genetik des Biologischen Forschungszentrums, Szeged/VR Ungarn: Genetic analysis of the 87 C (heat shock gene) region of *Drosophila melanogaster*. |
| 15. 10. 1980 | Dr. A. LAISK, Institut für Astrophysik und Physik der Atmosphäre der Estnischen Akademie der Wissenschaften, Tartu/UdSSR: Ein Modell der Photosynthese und seine experimentelle Prüfung. |
| 27. 10. 1980 | Dr. CHR. O. LEHMANN: Bericht über den IX. EUCARPIA-Kongreß 1980 in Leningrad. |
| 27. 10. 1980 | Dr. F. SCHOLZ: Über ein IAEA Research Coordination Meeting in Nicosia und Eindrücke aus Zypern. |
| 27. 10. 1980 | Dipl.-Biol. H.-P. JÄKEL: Enzymologische Untersuchungen zur Mukoviszidose und Möglichkeiten der pränatalen Diagnose dieses Syndroms. |
| 30. 10. 1980 | Prof. Dr. C. B. OSMOND, Australian National University, Canberra City/Australien: Untersuchungen zum $O_2$-Gaswechsel und zu hierfür relevanten Stoffwechselprozessen. |
| 10. 11. 1980 | Dipl.-Biol. C. ANSORGE: Pflanzenregeneration aus Kartoffelexplantaten als Vorarbeiten zur Gewinnung von Resistenzmutanten. |

| | |
|---|---|
| 10. 11. 1980 | Prof. Dr. K. MÜNTZ und Prof. Dr. O. MACHOLD: Bericht über die Jahrestagung der EMBO „Molecular Biologists Look at Green Plants", Heidelberg/BRD sowie über den Internationalen Photosynthesekongreß in Kallithea-Kassandra, Halkidiki/Griechenland. |
| 24. 11. 1980 | Dipl.-Gärtner B. WOLTERSTORFF: Aktuelle Aspekte der Gemüsezüchtung. |
| 26. 11. 1980 | Dr. U. MÜLLER, Institut für Humangenetik und Anthropologie der Universität Freiburg i. Br./BRD: Zur immunologischen Charakterisierung des H-Y-Antigens. |
| 02. 12. 1980 | Dr. A. MEISTER: Formen von Chlorophyll, Identifikation und Interpretation. |
| 15. 12. 1980 | Dr. G. SCHOLZ und Dr. B. SCHLESIER: Quartärstruktur von Reserveproteinen. |
| 15. 12. 1980 | Dr. J. BOGAJEWSKI, Institut für Humangenetik der Polnischen Akademie der Wissenschaften, Poznan/VR Polen: Anaphase-Auswertung von Ascitestumoren als Methode der Mutagenitätstestung. |
| 15. 12. 1980 | Dr. K. HAMMER und Dr. P. HANELT: Bericht über die 1980 durchgeführten Sammelreisen nach Süditalien. |

## Ausstellungen

| | |
|---|---|
| 13. 02. bis 27. 02. 1980 | Ausstellung der Laienkünstler des Zentralinstituts Gatersleben: Malerei, Grafik, Kunsthandwerk. |
| 18. 06. bis 06. 07. 1980 | Ausstellung der Zirkel für Volkskunst des Zentralinstituts Gatersleben: Plastik, Textiles Gestalten und Kindermalen. |
| 10. 10. bis 24. 10. 1980 | Kunstausstellung HEIDI MANTHEY, Keramikerin. |

## Vortragsabende

| | |
|---|---|
| 23. 01. 1980 | Gemeinschaftskonzert junger Künstler der ČSSR und der DDR mit LADISLAV KOŽDERKA, Trompete, JINDRA KRAMPEROVA, Klavier, JOCHEN AMELN, Violoncello, und GÜNTHER ANGER, Klavier. |
| 24. 01. 1980 | Dr. D. KÜHLMANN, Museum für Naturkunde der Humboldt-Universität zu Berlin: Auf Entdeckungsfahrt in das Große Barriere-Riff (Australien). |
| 13. 02. 1980 | Kammermusikabend mit BRIGITTE FUNKE, Violine, CHRISTIAN FUNKE, Violine, und PETER WAAS, Klavier. |
| 13. 03. 1980 | Dr. U. WOBUS: 3 Monate in Cambridge/USA: Eindrücke aus den Vereinigten Staaten. |
| 26. 03. 1980 | Theaterabend. PETER HACKS: Ein Gespräch im Hause Stein über den abwesenden Herrn von Goethe. Frau von Stein: TRAUTE RICHTER. |
| 21. 05. 1980 | Konzert mit dem Berliner Harfentrio: RICHARD WAAGE, Flöte, MANFRED HOFMANN, Viola, und MARION HOFMANN, Harfe. |
| 05. 06. 1980 | Dr. E. BEUSCHOLD, VEB Fernwasserversorgung Elbaue-Ostharz, Blankenburg: Das Rapp-Bode-Talsperrensystem: Entstehung, Funktion und Bedeutung für unsere Trinkwasserversorgung. |
| 18. 06. 1980 | Konzert des Hausmusikkreises des Zentralinstituts Gatersleben. |
| 18. 09. 1980 | Dr. H. KNAPPE und Dr. H. SCHEFFLER, Harzmuseum Wernigerode: Vom Werden und Wert einer Landschaft – der Harz. |
| 24. 09. 1980 | Konzert mit dem „Improvisationstrio": MANFRED SCHULZE, Klarinette/Bariton-Saxophon, und HERMANN KELLER, Klavier. |
| 19. 11. 1980 | Klavierabend mit CHRISTOPH TAUBERT. |
| 27. 11. 1980 | Dr. F. EBEL: Botanischer Garten der Martin-Luther-Universität Halle-Wittenberg, Halle: Als Botaniker in Mexiko. |

03. 12. 1980 Konzert mit ANDREAS AIGMÜLLER, Schlagzeug, und CHRISTIAN KOZIK, Klavier.
11. 12. 1980 K.-G. EICKENJÄGER, Berlin: Mediziner und andere unnütze Membra.

# C. Veröffentlichungen aus dem Institut 1980


ANASTASSOVA-KRISTEVA, MARLEN, RIGOMAR RIEGER, GOTTFRIED KÜNZEL, HRISTO NICOLOFF and ARNE HAGBERG: Further evidence on "nucleolar dominance" in barley translocation lines. Barley Genetics Newsletters **10** (1980) S. 3–6.

APEL, PETER: $CO_2$ compensation concentration and its $O_2$ dependence in *Moricandia spinosa* and *Moricandia moricandioides* (Cruciferae). Biochemie u. Physiol. Pflanzen **175** (1980) S. 386–388.

– und MARTIN PEISKER: $CO_2$-Versorgung der photosynthetischen Carboxylierungszentren. Colloquia Pflanzenphysiol. Humboldt-Univ. Berlin (1980) Nr. 3, S. 181–204.

BÄUMLEIN, HELMUT, ULRICH WOBUS and REINHARD PANITZ: Analysis of Balbiani ring DNA in *Chironomus thummi* by cloning in *E. coli*. In: ZADRAZIL, S. and J. SPONAR (Eds.): DNA – Recombination, Interactions and Repair. Oxford u. New York (1980) S. 73–75.

BAUWE, HERMANN: Biochemische und gaswechselphysiologische Untersuchungen zur Charakterisierung von Pflanzenarten mit $C_3$-$C_4$-intermediären Eigenschaften sowie zur Prüfung mathematischer Modellvorstellungen des photosynthetischen $CO_2$-Stoffwechsels. Diss. A, Martin-Luther-Univ. Halle-Wittenberg, Halle (1980).

– and PETER APEL: Investigations on $C_3$ and $C_3$–$C_4$ intermediate species checking mathematical models of photosynthetic $CO_2$ metabolism. Abstract, 2nd Congr. Feder. Eur. Soc. Plant Physiol. (FESPP), Santiago de Compostela/Spanien, 27. Juli bis 1. August 1980, Abstracts of lectures and poster demonstrations (1980) S. 216–217.

– and PETER APEL: Investigations on $C_3$ and $C_3$–$C_4$ intermediate species checking mathematical models of photosynthetic $CO_2$ metabolism. Abstract, Adaptivnye reakcij v formirovanii i aktivnosti fotosintetičeskogo apparata. Tezisy dokladov, Puščino (1980) S. 4–5.

– und PETER APEL: $CO_2$-Austausch und Carboxylierungsreaktionen bei $C_3$–$C_4$-intermediären Pflanzenarten. Colloquia Pflanzenphysiol. Humboldt-Univ. Berlin (1980) Nr. 3, S. 118–133.

–, PETER APEL and MARTIN PEISKER: Ribulose 1,5-bisphosphate carboxylase/oxygenase and $CO_2$ exchange characteristics in $C_3$ and $C_3$–$C_4$ intermediate species checking mathematical models of carbon metabolism. Photosynthetica **14** (1980) S. 550–556.

BECKER, KLAUS: Einfluß von Steroidhormonen auf den Meioseablauf in den Oocyten der Hausmaus. Diss. A, Martin-Luther-Univ. Halle-Wittenberg, Halle (1980).

– and JÖRG SCHÖNEICH: Effects of estrogens on meiosis in female mice. Abstracts 10th Annual Meeting of the European Environmental Mutagen Society on Environmental Mutagenesis, Athen/Griechenland, 14. bis 19. September 1980 (1980) S. 31.

BÖRNER, THOMAS and ARMIN MEISTER: Chlorophyll and carotenoid content of ribosome-deficient plastids. Photosynthetica **14** (1980) S. 589–593.

BRAUN, ROLF, JÖRG SCHÖNEICH and DIETER ZIEBARTH: Influence of nicotinamide on mutagenic activity of N-nitrosomorpholine in the AMES/Salmonella microsome test. Exper. Pathol. **18** (1980) S. 314–319.

–, JÖRG SCHÖNEICH and DIETER ZIEBARTH: Testing of drugs for combined mutagenesis with sodium nitrite in the host-mediated assay. Arch. Toxicol. (1980) Suppl. 4, S. 49–53.

–, JÖRG SCHÖNEICH, PETER BREZANI, LUDWIG WEISSFLOG and WOLFGANG DEDEK: Activity of organophosphorus insecticides in bacterial tests for mutagenicity and DNA repair: Per se alkylation versus metabolic activation. Abstracts 10th Annual Meeting of the European Environmental Mutagen Society on Environmental Mutagenesis, Athen/Griechenland, 14. bis 19. September 1980 (1980) S. 105.

BUDĚŠINSKÝ, MILOŠ, HERBERT BUDZIKIEWICZ, ŽELIMÍR PROCHÁZKA, HELMUT RIPPERGER, AXEL RÖMER, GÜNTER SCHOLZ and KLAUS SCHREIBER: Nicotianamine, a possible phytosiderophore of general occurrence. Phytochemistry **19** (1980) S. 2295–2297.

Budĕšinský, Miloš, Želimír Procházka, Herbert Budzikiewicz, Axel Römer, Helmut Ripperger, Klaus Schreiber and Günter Scholz: On the 'normalizing factor' for the tomato mutant '*chloronerva*'. XI. Mass and NMR spectroscopic investigations of the phytosiderophore nicotianamine and some of its derivatives. Tetrahedron **37** (1981) S. 191–196.

Čajka, Marija Tichonovna, Galina Evseevna Savčenko und Armin Meister: Osobennosti obrazovanija protochlorofillogo pigmenta v zelenejuščich klubnjach kartofelja. Doklady Akad. nauk BSSR **24** (1980) S. 457–460.

Chroboczek, Jadwiga, Michail Witt, K. Ostrówka, Ronald Bassüner, Manfred Püchel and Wlodzimierz Zagórski: Seed transmissibility of plant viruses may be modulated by competition between viral and cellular messengers. A proposal. Plant Sci. Letters **19** (1980) S. 263–270.

Constabel, Friedrich, Helmut Koblitz, Jim W. Kirkpatrick and S. Rambold: Fusion of cell sap vacuoles subsequent to protoplast fusion. Canad. J. Bot. **58** (1980) S. 1032–1034.

Eitner, Gudrun, Barbara Adler and Petra Frank: High-level synthesis of the recA protein is not sufficient for the expression of SOS-functions in *Escherichia coli* K-12. Biol. Zbl. **99** (1980) S. 443–451.

Frank, Reinhard: Grain yield and ear development of spring barley as influenced by environmental conditions during early stages of plant development. Biol. Plantarum **22** (1980) S. 274–281.

Giller, Juri Efimovič und Armin Meister: Izučenie sostojanija chlorofilla a v iskusstvennom pigment-belkovom komplekse metodom proizvodnoj spektrofotometrii. Studia Biophys. **78** (1980) S. 13–20.

Grebenščikov, Igor: Zur Möglichkeit einer Schätzung von $F_2$-Leistungen (je Pflanze) bei einfachen Maishybriden auf Grund der entsprechenden Angaben über $F_1$ und $\overline{P}$. Kulturpflanze **28** (1980) S. 239–255.

Gröber, Kurt, Fritz Matzk and Martin Zacharias: Hybrid effect and fertility in the $F_1$ generation of interspecific and intergeneric hybrids of *Lolium* and *Festuca*. In: Proc. 13th Internat. Grassland Congr., Leipzig, 18. bis 27. Mai 1977, Berlin (1980), Vol. 1, S. 341–344.

Hammer, Karl: Vorarbeiten zur monographischen Darstellung von Wildpflanzensortimenten: *Aegilops* L. Kulturpflanze **28** (1980) S. 33–180.

– Zur Taxonomie und Nomenklatur der Gattung *Aegilops* L. Feddes Repert. **91** (1980) S. 225–258.

– Vorarbeiten zur monographischen Darstellung von Wildpflanzensortimenten: *Aegilops* L. Diss. B, Akad. Wiss. DDR, Berlin (1980).

– und Peter Hanelt: Variabilitäts-Indices von *Papaver rhoeas*-Populationen und ihre Beziehungen zum Entwicklungsstand der Landwirtschaft. Biol. Zbl. **99** (1980) S. 325–343.

Hofemeister, Jürgen: Blockage and recovery of DNA replication from UV-induced lesions in bacteria. Biol. Zbl. **99** (1980) S. 533–551.

Horstmann, Christian and Roderick Casey: Vergleich der Untereinheiten von Legumin aus *Pisum sativum* und *Vicia faba* durch „peptide mapping". Abstract, 12. Jahrestag. biochem. Gesellsch. DDR, Magdeburg, 8. bis 10. Oktober 1980, Kurzreferate der Vorträge und Posterbeiträge (1980) P 2/4.

Hüttner, Edith, Rolf Braun and Jörg Schöneich: Testing of isoniazid in the mouse for genetic activity using the mammalian spot test and the sperm abnormality assay. Abstract, Mutat. Res. **74** (1980) S. 228.

Kaina, Bernd: Ortsspezifische Mutagenese: Resultate, Probleme und Perspektiven. Biol. Zbl. **99** (1980) S. 513–531.

Koblitz, Helmut: The physiology of fusion in isolated higher plant protoplasts as a basis for somatic cell genetics. Proc. 5th internat. Protoplast Symposium, Szeged/VR Ungarn, 9. bis 14. Juli 1979, Budapest (1980) S. 383–387.

Kruse, Joachim: Skulpturuntersuchungen an Pollen der Gattungen *Triticum* L. und *Aegilops* L. Kulturpflanze **28** (1980) S. 341–359.

Kühn, František, Karl Hammer und Peter Hanelt: Botanische Ergebnisse einer Reise in die ČSSR 1977 zur Sammlung autochthoner Landsorten von Kulturpflanzen. Kulturpflanze **28** (1980) S. 183–226.

Lichtenfeld, Claudia, Renate Manteuffel, Klaus Müntz und Armin Rudolph: Physikochemische und strukturelle Veränderungen der Reserveproteine durch Proteasen im Keimungsprozeß von *Vicia faba* L. Abstract, 12. Jahrestag. biochem. Gesellsch. DDR, Magdeburg, 8. bis 10. Oktober 1980, Kurzreferate der Vorträge und Posterbeiträge (1980) P. 3/17.

Machold, Otto and Armin Meister: Chlorophyll-proteins of *Vicia faba*. Photobiochem. & Photobiophys. **1** (1980) S. 213–218.

Matzk, Fritz: Relationships between interspecific incompatibility and self-incompatibility in grasses. Incompatibility Newsletter **12** (1980) S. 88–94.

–, Kurt Gröber und Martin Zacharias: Ergebnisse von Art- und Gattungskreuzungen mit Gramineen im Zusammenhang mit den natürlichen Isolationsmechanismen zwischen den Arten. Kulturpflanze **28** (1980) S. 257–284.

Mendel, Ralf R.: Comparative affinity chromatography of nitrate reductase from wild-type and molybdenum cofactor-defective cell cultures of *Nicotiana tabacum*. Biochemie u. Physiol. Pflanzen **175** (1980) S. 216–227.

– and Andreas J. Müller: Comparative characterization of nitrate reductase from wild-type and molybdenum cofactor-defective cell cultures of *Nicotiana tabacum*. Plant Sci. Letters **18** (1980) S. 277–288.

Meyer, Doris: Numerisch-taxonomische Untersuchungen an *Pisum sativum* L. Kulturpflanze **28** (1980) S. 285–340.

Müntz, Klaus: Fortschritte in der Erforschung des pflanzlichen Aminosäurestoffwechsels. Biol. Rdsch. **18** (1980) S. 261–281.

– Regulation von Biosynthese und Akkumulation der Reserveproteine während der Entwicklung pflanzlicher Samen. Abstract, 12. Jahrestag. biochem. Gesellsch. DDR, Magdeburg, 8. bis 10. Oktober 1980, Kurzreferate der Vorträge und Posterbeiträge (1980) V 2.3.

– Stoffwechselphysiologische Grundlagen für die Bildung des Sameneiweißertrages bei Kulturpflanzen. Aus der Arb. Plenum u. Klassen Akad. Wiss. DDR. **5** (1980) Nr. 4, S. 30–57.

Opatrný, Zdenek, Ursula Schumann, Slavek Rakouský and Helmut Koblitz: The role of some exogenous and endogenous factors in the isolation of protoplasts from potato cell cultures and their recovery in cell colonies. Biol. Plantarum **22** (1980) S. 107–116.

Peisker, Martin: Differential short-time uptake of $^{14}CO_2$ and $^{12}CO_2$ as a method for determining photorespiration. A theoretical approach. Photosynthetica **14** (1980) S. 406–412.

– Opredelenie fotodychanija pri pomošči kratkovremennogo izmerenija pogloščenija $^{14}CO_2$. Kritičeskij analiz metoda. Abstract, Adaptivnye reakcij v formirovanii i aktivnosti fotosintetičeskogo apparata. Tezisy dokladov, Puščino (1980) S. 26–27.

– and Peter Apel: Dark respiration and the effect of oxygen on $CO_2$ compensation concentration in wheat leaves. Z. Pflanzenphysiol. **100** (1980) S. 389–395.

–, Ingrid Tichá und Jiři Čatský: Sootnošenie meždu uglekislotnym kompensacionnym punktom i temnovym dychaniem v tečenie ontogeneza list'ev fasoli. Abstract, Adaptivnye reakcij v formirovanii i aktivnosti fotosintetičeskogo apparata. Tezisy dokladov, Puščino (1980) S. 27.

– and J. Václavik: Relationship between transpiration and $CO_2$ uptake in leaves of *Zea mays* L. after excision. Photosynthetica **14** (1980) S. 545–549.

Püchel, Manfred, Helmut Bäumlein, Peter Silhengst und Ulrich Wobus: Klonierung von doppelsträngiger cDNA, gewonnen aus polysomaler mRNA der Ackerbohnen-Kotyledonen. Abstract, 12. Jahrestag. biochem. Gesellsch. DDR, Magdeburg, 8. bis 10. Oktober 1980, Kurzrefcrate der Vorträge und Posterbeiträge (1980) P 3/16.

Rieth, Alfred: Xanthophyceae, 2. Teil. In: Ettl, H., J. Gerloff u. H. Heynig (Hrsg.): Süßwasserflora von Mitteleuropa, Bd. 4, Jena (1980).

– Beiträge zur Kenntnis der Vaucheriaceae. XXII. Eine neue Sektion und 2 neue Formen. Arch. Protistenkd. **123** (1980) S. 446–449.

– Süßwasser-Algenarten in Einzeldarstellung. VIII. *Spirogyra pseudonodifera* O. et W. Bock 1956, nebst Bemerkungen über Spirogyren mit braunen Mesosporverdickungen. Kulturpflanze **28** (1980) S. 399–415.

Saalbach, Isolde: Untersuchungen zur Nitratreduktion und Nitrataufnahme bei Zellkulturen von *Nicotiana tabacum* unter Verwendung von nitratreduktasedefekten Mutanten. Diss. A, Akad. Wiss. DDR, Berlin (1980).

Schöneich, Jörg, Rolf Braun und Edith Hüttner: Ergebnisprotokoll über die Prüfung der Substanz STS 557 des VEB Jenapharm auf Mutagenität. Gatersleben (1980).

–, Rolf Braun und Edith Hüttner: Gutachten über die Prüfung der Verbindung Etephon (CKB 1108) des VEB Chemiekombinat Bitterfeld auf Mutagenität. Gatersleben (1980).

–, Rolf Braun, Edith Hüttner und Anna-M. Wobus: Gutachten über die Prüfung der Substanz P 4 des VEB Apogepha auf Mutagenität. Gatersleben (1980).

Scholz, Günter and Helmut Böhme: Biochemical mutants in higher plants as tools for chemical and physiological investigations – a survey. Kulturpflanze **28** (1980) S. 11–32.

Schreiber, Klaus, Helmut Ripperger und Günter Scholz: Eisen-Carrier in höheren Pflanzen. Spectrum **11** (1980) Nr. 8, S. 28–29.

Schubert, Ingo, Gottfried Künzel, Hermann Bretschneider, Rigomar Rieger and Hristo Nicoloff: Sister chromatid exchanges in barley. Theoret. & appl. Genetics **56** (1980) S. 1–4.

– and Rigomar Rieger: Cytochemical and cytogenetic features of the nucleolus organizing region (NOR) of *Vicia faba*. Biol. Zbl. **99** (1980) S. 65–72.

–, Rigomar Rieger, Arnd Michaelis, Konrad Heindorff and Bernd Kaina: Genomic and environmental influences on mutagen-induced recombination in plant chromosomes. In: Zadrazil, S. and J. Sponar (Eds.): DNA – Recombination, Interactions and Repair. Oxford u. New York (1980) S. 403–410.

Schubert, Rudolf, Klaus Helmecke, Joachim Kruse, Enrique del Risco, René Capote, Roberto Vandama, Ramona Oviedo und Leda Menéndez: Ergebnisse der Expedition von Wissenschaftlern der Akademie der Wissenschaften Kubas und der DDR zur Isla de Pinos (Kuba) 1975. III. Ergebnisse geobotanischer Untersuchungen in den Mangroven an der Playa de Herradura bei Bibijagua. Feddes Repert. **91** (1980) S. 11–23.

–, Klaus Helmecke, Joachim Kruse, Enrique del Risco, René Capote, Roberto Vandama, Ramona Oviedo, Daysi Vilamajo und Leda Menéndez: Ergebnisse der Expedition von Wissenschaftlern der Akademie der Wissenschaften Kubas und der DDR zur Isla de Pinos (Kuba) 1975. IV. Ergebnisse geobotanischer Untersuchungen in den Citrus-Plantagen der Schule Carlos Baliño in der Zone las Sacra. Feddes Repert. **91** (1980) S. 313–325.

Schultze-Motel, Jürgen: Literatur über archäologische Kulturpflanzenreste (1978/1979). Kulturpflanze **28** (1980) S. 361–378.

– Neolithische Kulturpflanzenreste von Eilsleben, Kr. Wanzleben. Z. Archäol. **14** (1980) S. 213–216.

– Reinhard Fritsch, Karl Hammer, Peter Hanelt, Joachim Kruse, Helga I. Maass, Harald Ohle and Klaus Pistrick: Taxonomy and evolution of cultivated plants: Literature review 1978/1979. Kulturpflanze **28** (1980) S. 379–397.

Schumann, Ursula, Helmut Koblitz and Zdenek Opatrný: Plant recovery from long-term callus cultures and from suspension culture derived protoplasts of *Solanum phureja*. Biochemie u. Physiol. Planzen **175** (1980) S. 670–675.

Serfling, Edgar: Struktur und Expression eukaryotischer 5S RNA-Gene – Eine Übersicht. Biol. Zbl. **99** (1980) S. 129–147.

Süss, Karl-Heinz: Identification of chloroplast thylakoid membrane polypeptides. ATPase complex ($CF_1$–$CF_0$) and light-harvesting chlorophyll a/b-protein (LHCP) complex. FEBS Letters **112** (1980) S. 255–259.

– and Eberhard Brecht: Polypeptide composition and spectral properties of light-harvesting chlorophyll a/b-protein complexes from intact and trypsin-treated chloroplast thylakoid membranes. Biochim. & Biophys. Acta **592** (1980) S. 369–374.

– and Alexander Malyan: Isolation, purification and properties of coupling factor ATPase from *Rhodopseudomonas palustris* chromatophores. Biochemie. u. Physiol. Pflanzen **175** (1980) S. 491–498.

TICHÁ, INGRID, J. ČATSKY and MARTIN PEISKER: Ontogenetic changes in the internal limitations to bean-leaf photosynthesis. 6. $CO_2$ dependence of net photosynthetic rate. Photosynthetica **14** (1980) S. 489–496.

TITTEL, CLAUS: Versuche zur Verlangsamung der Samenalterung. Kongress- u. Tag.-Ber. Martin-Luther-Univ. Halle-Wittenberg, Wiss. Beiträge 1980/20 (S 23), **3** (1980) S. 567–571.

WALTHER, URSULA und CHRISTIAN O. LEHMANN: Resistenzeigenschaften im Gersten- und Weizensortiment Gatersleben. 24. Prüfung von Sommer- und Wintergersten auf ihr Verhalten gegenüber Zwergrost (*Puccinia hordei* Otth.) Kulturpflanze **28** (1980) S. 227–238.

WEBER, ERNST, RENATE MANTEUFFEL und DIETER NEUMANN: Biosynthese und Akkumulation der Reserveproteine während der Samenentwicklung von *Vicia faba*, Abstract, 12. Jahrestag. biochem. Gesellsch. DDR, Magdeburg, 8. bis 10. Oktober 1980, Kurzreferate der Vorträge und Posterbeiträge (1980) P 3/4.

– and DIETER NEUMANN: BPP-Review. Protein bodies, storage organelles in plant seeds. Biochemie u. Physiol. Pflanzen **175** (1980) S. 279–306.

WOBUS, ULRICH, HELMUT BÄUMLEIN, REINHARD PANITZ, EDGAR SERFLING and F. C. KAFATOS: Periodicities and tandem repeats in a Balbiani ring gene. Cell **22** (1980) S. 127–135.

## Nachtrag zum Verzeichnis der Veröffentlichungen aus dem Institut 1979

BEYER, EDITH, ROLF BRAUN and JÖRG SCHÖNEICH: Genetic activity of isoniazid in the mammalian spot test. Abstract, Mutat. Res. **64** (1979) S. 152.

FRANKE, GUNTHER, KARL HAMMER, PETER HANELT, HANS-ALBRECHT KETZ, GÜNTHER NATHO und HORST REINBOTHE: Darove na zemjata. [Früchte der Erde]. Sofia (1979).

–, KARL HAMMER, PETER HANELT, HANS-ALBRECHT KETZ, GÜNTHER NATHO und HORST REINBOTHE: Plody zemli. [Früchte der Erde]. Moskva (1979).

HÜTTNER, EDITH, ROLF BRAUN and JÖRG SCHÖNEICH: Mutagenic activity of isoniazid in the mouse. Pol. J. Pharmacol. & Pharm. **31** (1979) S. 689–694.

KOBLITZ, HELMUT: Die Physiologie isolierter Protoplasten aus höheren Pflanzen als Voraussetzung für die somatische Zellgenetik. Ergebnisse exper. Med. **34** (1979) S. 63 bis 75.

PORSCHE, WOLFGANG, CHRISTIAN LEHMANN und KLAUS RICHTER: Ergebnisse von Untersuchungen an proteinreichen Herkünften des Weizens (*Triticum aestivum* L.). Tag.-Ber. Akad. Landwirtsch.-Wiss. DDR, Berlin (1979) Nr. 175, S. 81–86.

RIEGER, RIGOMAR: Struktur und Strukturumbauten der Chromosomen vielzelliger Organismen. Sitz.-Ber. Akad. Wiss. DDR, Math. - Naturwiss. - Technik (1979) Nr. 20/N, 40 S.

SAALBACH, GERHARD: Organredifferenzierung aus Kalluskulturen verschiedener Teile der Gerstenpflanze unter Benutzung eines Keimpflanzenextraktes. Ergebnisse exper. Med. **34** (1979) S. 135–141.

– und HELMUT KOBLITZ: Möglichkeiten und Probleme von *In-vitro*-Verfahren auf zellulärer Ebene bei Gerste. Tag.-Ber. Akad. Landwirtsch.-Wiss. DDR, Berlin (1979) Nr. 168, S. 241–254.

SCHEUNERT, ERNST-ULRICH: Einige zytologische und karyologische Untersuchungen an *in vitro* kultivierten Zellstämmen der Gerste. Ergebnisse exper. Med. **34** (1979) S. 121–133.

SCHUBERT, INGO and ARMIN MEISTER: Possible connections between the modes of origination of sister chromatid exchanges (SCEs) and chromatid aberrations. Studia Biophys. **76** (1979) S. 211–217.

SCHUBERT, RUDOLF, KLAUS HELMECKE, JOACHIM KRUSE, ENRIQUE DEL RISCO, RENÉ CAPOTE, ROBERTO VANDAMA, RAMONA OVIEDO, DAYSI VILAMAJO und LEDA MENÉNDEZ: Ergebnisse der Expedition von Wissenschaftlern der Akademie der Wissenschaften Kubas und der DDR zur Isla de Pinos (Kuba) 1975. I. Ergebnisse geobotanischer Untersuchungen in den Mogoten El Abra und Bibijagua. Feddes Repert. **89** (1979) S. 601–628.

SCHUBERT, RUDOLF, KLAUS HELMECKE, JOACHIM KRUSE, ENRIQUE DEL RISCO, RENÉ CAPOTE, ROBERTO VANDAMA, RAMONA OVIEDO, DAISY VILAMAJO und LEDA MENÉNDEZ: Ergebnisse der Expedition von Wissenschaftlern der Akademie der Wissenschaften Kubas und der DDR zur Isla de Pinos (Kuba) 1975. II. Ergebnisse geobotanischer Untersuchungen in den Pinares von Baños termales del Rosario und Loma la Cañada. Feddes Repert. **90** (1979) S. 19–50.

SÜSS, KARL-HEINZ and EBERHARD BRECHT: Isolation, purification and properties of light-harvesting chlorophyll a/b-protein and a P 700 chlorophyll a-protein containing complex from chloroplast thylakoid membranes. Abstracts 17th Symposium CMEA-problem I-18.2, Reinhardsbrunn, 22. bis 27. Oktober 1979, S. 39–40.

## D. Vorträge von Mitarbeitern bei wissenschaftlichen Veranstaltungen anderer Institutionen

Dr. APEL, PETER: Variabilität der Keimwurzelzahl bei *Hordeum vulgare* L. - Arbeitstagung zum RGW-Thema 10, Bernburg (3. September 1980).

– und Dr. MARTIN PEISKER (vorgetragen von Dr. P. APEL): $CO_2$-Versorgung der photosynthetischen Carboxylierungszentren. – Tagung der Arbeitsgemeinschaft Photosynthese der Biologischen Gesellschaft der DDR, Egsdorf, 31. März bis 3. April 1980 (2. April 1980).

Dipl.-Biochem. BAUWE, HERMANN: $CO_2$-Austausch und Carboxylierungsreaktionen bei $C_3$-$C_4$-intermediären Pflanzenarten. – Tagung der Arbeitsgemeinschaft Photosynthese der Biologischen Gesellschaft der DDR, Egsdorf, 31. März bis 3. April 1980 (1. April 1980).

– Biochemische und gaswechselphysiologische Untersuchungen zur Charakterisierung von Pflanzenarten mit $C_3$-$C_4$-intermediären Eigenschaften sowie zur Prüfung mathematischer Modellvorstellungen des photosynthetischen $CO_2$-Stoffwechsels. – Promotion A, Martin-Luther-Universität Halle-Wittenberg, Sektion Biowissenschaften, Vortrag anläßlich der Verteidigung, Halle (9. April 1980).

– und Dr. PETER APEL: Untersuchungen an $C_3$ und $C_3$-$C_4$-intermediären Arten zur Prüfung mathematischer Modelle der Photosynthese. – Poster anläßlich der XV. Koordinierungskonferenz zum RGW-Thema I-18.3., Puščino/UdSSR, 1. bis 6. Dezember 1980.

Dipl.-Biol. BECKER, KLAUS: Einfluß von Steroidhormonen auf den Meioseablauf in den Oozyten der Hausmaus. – Promotion A, Martin-Luther-Universität Halle-Wittenberg, Sektion Biowissenschaften, Vortrag anläßlich der Verteidigung, Halle (23. Juli 1980).

– Untersuchungen zur Abschätzung des genetischen Risikos oraler Kontrazeptiva. – Poster anläßlich des Tages des Jungen Wissenschaftlers, Berlin (5. Dezember 1980).

– and Dr. JÖRG SCHÖNEICH: Effects of oestrogens on meiosis in female mice. – Poster presented at the 10th Annual Meeting of the European Environmental Mutagen Society, Athen/Griechenland, 14. bis 19. September 1980 (16. September 1980).

Dr. BENEDIX, ERICH HEINZ: Der biologische Fortschritt bei Fruchtkörperformen höherer Pilze (von der Wurzellorchel zur Stinkmorchel). – Kulturbund der DDR, Fachgruppe Mykologie, Dresden (18. März 1980).

Prof. Dr. BÖHME, HELMUT und Dr. JÖRG SCHÖNEICH (vorgetragen von Prof. Dr. H. BÖHME): Legislative Aspekte der Mutagenitätstestung in der DDR. – Tagung der Gesellschaft für Umwelt-Mutationsforschung e. V. „Legislative Aspekte der Mutagenitätsprüfung“, München/BRD, 2. bis 5. Dezember 1980 (3. Dezember 1980).

Dr. BRAUN, ROLF: Mutagene Effekte des p-Phenylendiamin. – 2. Symposium der Gesellschaft für Dermatologie der DDR zu biologischen Effekten para-substituierter Kohlenwasserstoffe, Neubrandenburg, 10. bis 13. April 1980 (11. April 1980).

– Recent advances in the detection of gene mutations in the male gonad. – Seminar at the Chemical Industry Institute of Toxicology, Research Triangle Park, N. C./USA (12. November 1980)

Dr. BRAUN, ROLF: Mutagenicity testing in the German Democratic Republik. – Colloquium at the Biological Division of the Oak Ridge National Laboratory, Oak Ridge, Tenn./USA (13. Dezember 1980).

–, Dr. JÖRG SCHÖNEICH, Dr. LUDWIG WEISSFLOG und Dr. WOLFGANG DEDEK (vorgetragen von Dr. R. BRAUN): Activity of organophosphorus insecticides related to butonate and its metabolites in bacterial tests for mutagenicity and DNA repair. – Poster presented at the International Symposium on Chemical Carcinogenesis and Mutagenesis – Relationships and Practical Aspects, Warszawa/VR Polen, 20. bis 24. April 1980 ( 21. April. 1980).

–, Dr. JÖRG SCHÖNEICH, Dr. PETER BREZANI. Dr. LUDWIG WEISSFLOG and Dr. WOLFGANG DEDEK (vorgetragen von Dr. J. SCHÖNEICH): Activity of organophosphorus insecticides in bacterial tests for mutagenicity and DNA repair: Per se alkylation versus metabolic activation. – 10th Annual Meeting of the European Environmental Mutagen Society, Athen/Griechenland, 14. bis 19. September 1980 (16. September 1980).

Dipl.-Phys. BRECHT, EBERHARD and Dr. KARL-HEINZ SÜSS (vorgetragen von Dipl.-Phys. E. BRECHT): Isolation and characterization by spectroscopic methods of the light-harvesting chlorophyll a/b-protein complex. – Kolloquium am Institut für Pflanzenphysiologie der Ungarischen Akademie der Wissenschaften, Szeged/VR Ungarn (26. Februar 1980).

– und Dr. KARL-HEINZ SÜSS (vorgetragen von Dipl.-Phys. E. BRECHT): Spektroskopische Charakterisierung einer Präparation des lichtsammelnden Chlorophyll a/b-Protein-Komplexes. – Tag des Jungen Wissenschaftlers, Berlin (5. Dezember 1980).

Dr. BRETSCHNEIDER, HERMANN: Experiments on amphidiploidisation of autotetraploid barley by means of translocations. – Kolloquium am Institut für Cytologie und Genetik der Sibirischen Abteilung der Akademie der Wissenschaften der UdSSR, Novosibirsk/UdSSR (30. Mai 1980).

MR Prof. Dr. Dr. FRANZ, HARTMUT, Dr. CLAUS TITTEL, Dr. PETER ZISKA und CARLA FRIEDRICH (vorgetragen von MR Prof. Dr. Dr. H. FRANZ): Besteht ein Zusammenhang zwischen Keimfähigkeit und Lektingehalt bei Leguminosen? – 4. Arbeitstagung „Lektine" des Staatlichen Instituts für Immunpräparate und Nährmedien, Berlin, 5. bis 6. Juni 1980 (6. Juni 1980).

Dr. FRITSCH, REINHARD: Methoden der Chromosomenzählung bei Bryophyten. – 2nd Workshop of Bryologists of Socialistic Countries, Poznan/VR Polen, 26. bis 29. Juni 1980 (26. Juni 1980).

Dr. GRAFE, REINHARD: Genetic analysis of nitrate-reductase defective mutants in higher plants. – Kolloquium am Institute of Physiological Botany, Uppsala/Schweden (15. Oktober 1980).

Dr. HAMMER, KARL: Eine botanische Reise durch Spanien. – Gemeinsame Veranstaltung des Kulturbunds der DDR und des Museums Heineanum, Halberstadt (23. Januar 1980).

– Spanische Impressionen. – Staatliches Museum für Naturkunde, Görlitz (14. November 1980).

Dr. HANELT, PETER: Warum Änderung der wissenschaftlichen Pflanzennamen? – Bezirkstagung des Bezirksfachausschusses Botanik und des Floristischen Arbeitskreises Nordharz und Vorland des Kulturbunds der DDR, Halberstadt (23. Februar 1980).

– Taxonomy of cultivated plants – activities in Gatersleben in past and present. – Kolloquium am Istituto del Germoplasma, C. N. R., Bari/Italien (6. Juni 1980).

– Zur Geschichte einheimischer Kulturpflanzen. – Naturwissenschaftliche Vorträge des Staatlichen Museums für Naturkunde, Görlitz (21. Oktober 1980).

– und Dr. KARL HAMMER (vorgetragen von Dr. K. HAMMER): Zur Kulturpflanzenflora Perus. – Symposium „Flora und Vegetation von Peru" der Sektion Biowissenschaften der Karl-Marx-Universität Leipzig und der Sektion Geobotanik und Phytotaxonomie der Biologischen Gesellschaft der DDR, Leipzig (19. November 1980).

Dr. HOFEMEISTER, JÜRGEN: Kontrolle der DNA-Replikation bei Prokaryoten. – Arbeitstagung „Physiologie mikrobieller Wachstums- und Vermehrungsprozesse" der

Arbeitsgemeinschaft Physiologie und Biochemie der Mikroorganismen der Gesellschaft für Allgemeine und Technische Mikrobiologie der Biologischen Gesellschaft der DDR, Kloster/Hiddensee, 8. bis 11. Oktober 1980 (8. Oktober 1980).

– Mutagenese und DNA-Reparatur bei Prokaryoten. – Vorlesung am Wissenschaftsbereich Genetik der Sektion Biowissenschaften der Martin-Luther-Universität Halle-Wittenberg, Halle (29. Oktober 1980).

Dr. Horstmann, Christian: A phytohemagglutinin from seeds of *Vicia faba* L. – Kolloquium am John Innes Institute, Norwich/Grossbritannien (22. Februar 1980).

– and Roderick Casey: Vergleich der Untereinheiten von Legumin aus *Pisum sativum* und *Vicia faba* durch „peptide mapping". – Poster anläßlich der 12. Jahrestagung der Biochemischen Gesellschaft der DDR, Magdeburg, 8. bis 10. Oktober 1980 (9. Oktober 1980).

Dipl.-Biol. Hüttner, Edith and Dr. Jörg Schöneich: The effect of smoking on the frequencies of chromosomal aberrations and SCE in man. – Poster presented at the 10th Annual Meeting of the European Environmental Mutagen Society, Athen/Griechenland, 14. bis 19. September 1980 (16. September 1980).

Dipl.-Biol. Jäkel, Hans-Peter: Die Pathogenese der Mukoviszidose – der aktuelle Stand. – 12. Arbeitstagung der Arbeitsgemeinschaft zur Bekämpfung der Mukoviszidose in der Gesellschaft für Pädiatrie der DDR, Dresden (21. Januar 1980).

– Möglichkeiten der pränatalen Diagnose der Mukoviszidose durch Induktion der alkalischen Phosphatase und die Bestimmung der physikochemischen Eigenschaften verschiedener Enzyme. – 1. Sitzung der Gesellschaft für Humangenetik der DDR, Neubrandenburg, 9. bis 13. November 1980 (12. November 1980).

Dipl.-Math. Knüpffer, Helmut: Nutzung großer Datenbestände mit Hilfe des KRS 4201 – Konzeption. – Frühjahrsschulung 1980 der Nutzergemeinschaft KRS 4200/4201 der Akademie der Wissenschaften der DDR, Biesenthal, 14. bis 18. April 1980 (16. April 1980).

– Übernahme eines Datenspeichersystems von der BESM-6 auf einen Kleinrechner KRS 4201 für die Dokumentation von genetischen Kulturpflanzenressourcen. – Poster anläßlich der Kreis-MMM der Akademie der Wissenschaften der DDR, Berlin, 19. bis 23. Mai 1980 und der Neuerer-Angebotsmesse des Bezirksvorstands des FDGB, Halle, 8. bis 9. September 1980.

– Ispol'zovanie ĖVM v Gaterslebenskom genbanke [Die Anwendung der EDV in der Gaterslebener Genbank]. – Arbeitsberatung im Rahmen der Zusammenarbeit des RGW zum Thema „Anwendung der EDV in Genbanken", Praha–Ruzyně/ČSSR, 24. bis 28. November 1980 (25. November 1980).

Prof. Dr. Koblitz, Helmut: Probleme der Protoplastenkultur und einige Anwendungsmöglichkeiten. – Kolloquium am Institut für Züchtungsforschung, Quedlinburg (29. Mai 1980).

Dr. Künzel, Gottfried: Nutzung mutativ veränderter Gerstenkaryotypen in der Nukleolusforschung. – Kolloquium am Institut für Züchtungsforschung, Quedlinburg (5. November 1980).

Dr. Lehmann, Christian O.: Eiweißscreening an der Gaterslebener Erbsen-Kollektion. – Arbeitsberatung in der Saatzuchtstation, Biendorf (22. Januar 1980).

– Eiweißscreening an der Gaterslebener Hafer-Kollektion. – Arbeitsberatung im Zentralinstitut für Genetik und Kulturpflanzenforschung, Gatersleben (7. Mai 1980).

– Screening for disease resistance within the material of the gene bank of the GDR for their utilization of its resistant strains in breeding – 9th EUCARPIA Congress, Leningrad/UdSSR, 16. bis 20. September 1980 (17. September 1980).

Dipl.-Chem. Lichtenfeld, Claudia, Dr. Renate Manteuffel, Prof. Dr. Klaus Müntz und Dr. Armin Rudolph: Physikochemische und strukturelle Veränderungen der Reserveproteine durch Proteasen im Keimungsprozeß von *Vicia faba* L. – Poster anläßlich der 12. Jahrestagung der Biochemischen Gesellschaft der DDR, Magdeburg, 8. bis 10. Oktober 1980 (10. Oktober 1980).

Prof. Dr. Machold. Otto: Aufbau und Funktion der Photosynthesemembran. – VI. Tagung der Gesellschaft für Physikalische und Mathematische Biologie der DDR in Zusammenarbeit mit der Friedrich-Schiller-Universität Jena und der HFR Biomembranforschung „Probleme der Modell- und Biomembranen", Eisenach, 16. bis 21. März 1980 (19. März 1980).

–, Dr. DAVID SIMPSON und Dr. BIRGER LINDBERG MØLLER: Chlorophyll-proteins of *Hordeum vulgare* and *Vicia faba*. – Poster presented at the 5th International Congress on Photosynthesis, Kallithea-Kassandra, Halkidiki/Griechenland, 7. bis 13. September 1980 (12. September 1980).

Dr. MANTEUFFEL, RENATE: Der Einsatz immunologischer Methoden in der Pflanzenphysiologie. – 10. Generalversammlung und wissenschaftliche Vortragstagung der Biologischen Gesellschaft der DDR, Weimar, 2. bis 4. Mai 1980 (3. Mai 1980).

Dr. MEISTER, ARMIN: Forms of chlorophyll – interpretation and identification. – 18. Koordinierungskonferenz zum RGW-Thema I-18.2 und Symposium „Photosynthetic Solar Energy Conversion and Storages", Warszawa/VR Polen, 17. bis 24. September 1980 (18. September 1980).

– und Dipl.-Math. HANS-WOLFGANG JANK (vorgetragen von Dr. A. MEISTER): Spektrenzerlegung und Kurvenanpassung mit den FORTRAN-Programmen RESO und NNNG. – Frühjahrsschulung 1980 der Nutzergemeinschaft KRS 4200/4201 der Akademie der Wissenschaften der DDR, Biesenthal, 14. bis 18. April 1980 (16. April 1980).

Dr. MENDEL, RALF-RAINER: Moralische und ethische Konsequenzen der modernen Molekulargenetik. – URANIA-Referentenkonferenz des Bezirks Halle „Der Mensch und die moderne Wissenschaft", Halle (18. Juni 1980).

– Genetic engineering bei höheren Pflanzen – gegenwärtiger Stand und Perspektive. – Jahressitzung der Sektion Biologie der URANIA, Kreisvorstand Aschersleben, Aschersleben (22. Oktober 1980).

– Eigenschaften, Nachweismethoden und Reinigung der Nitratreduktase. – Bereichskolloquium an der Sektion Biowissenschaften der Martin-Luther-Universität Halle-Wittenberg, Halle (14. November 1980).

Dr. MÜLLER, ANDREAS J.: Studies on nitrate-reductase mutants of *Nicotiana tabacum*. – Kolloquium am Friedrich-Miescher-Institut, Basel/Schweiz (24. September 1980).

Prof. Dr. MÜNTZ, KLAUS: Synthese und Speicherung von Proteinen in Samen von Cerealien und Leguminosen. – Kolloquium an der Sektion Biologie der Friedrich-Schiller-Universität, Jena (5. Mai 1980).

– Reserveproteine – ihre Biosynthese und Akkumulation in Kulturpflanzensamen. – – Kolloquium an der Sektion Biowissenschaften der Karl-Marx-Universität, Leipzig (15. Mai 1980).

– Biosynthesis and deposition of globulins in the cotyledons of developing *Vicia faba* seeds. – 6th Annual EMBO-Meeting „Molecular Biologists Look at Green Plants", Heidelberg/BRD, 29. September bis 2. Oktober 1980 (29. September 1980).

– Analyse der Reserveproteine, ihrer Synthese und Akkumulation in Pflanzensamen. – Kolloquium am Institut für Biochemie der Biologischen Bundesanstalt für Land- und Forstwirtschaft, Braunschweig/BRD, (6. Oktober 1980).

– Regulation von Biosynthese und Akkumulation der Reserveproteine während der Entwicklung pflanzlicher Samen. – 12. Jahrestagung der Biochemischen Gesellschaft der DDR, Magdeburg, 8. bis 10. Oktober 1980 (8. Oktober 1980).

– Regulation der Reserveproteinbiosynthese in pflanzlichen Samen. – Festkolloquium der Klasse Biowissenschaften der Akademie der Wissenschaften der DDR anläßlich des 80. Geburtstags von Akademie-Mitglied KURT MOTHES, Berlin (13. November 1980).

–, Dr. HELMUT BÄUMLEIN, Dipl.-Biol. RONALD BASSÜNER, Dr. RENATE MANTEUFFEL, Dr. MANFRED PÜCHEL, Dr. PETER SCHMIDT and Dr. ULRICH WOBUS: Poly (A)-containing RNA from field bean cotyledons (*Vicia faba* L. var. minor), its reserve transcription and cloning of double-stranded cDNA. – Poster presented at the 2nd Congress of the Federation of European Societies of Plant Physiology, Santiago de Compostela/Spanien, 27. Juli bis 1. August 1980.

Dr. PEISKER, MARTIN: Determination of photorespiration by measuring short-time $^{14}CO_2$ uptake. A critical analysis of the method. – Poster anläßlich der XV. Koordinierungskonferenz zum RGW-Thema I-18.3, Puščino/UdSSR, 1. bis 6. Dezember 1980.

Dr. PEISKER, MARTIN, Dr. INGRID TICHÁ und Dr. JIŘI ČATSKÝ: Sootnošenie meždu uglekislotnym kompensacionnym punktom i temnovym dychaniem v tečenie ontogeneza list'ev fasoli. [Die Beziehung zwischen $CO_2$-Kompensationskonzentration und Dunkelatmung während der Ontogenese von Bohnenblättern]. – Poster anläßlich der XV. Koordinierungskonferenz zum RGW-Thema I-18.3, Puščino/UdSSR, 1. bis 6. Dezember 1980.

Dipl.-Biol. PISTRICK, KLAUS: Anthropogene Veränderungen der Moosflora. – Bezirkstagung des Bezirksfachausschusses Botanik und des Floristischen Arbeitskreises Nordharz und Vorland des Kulturbunds der DDR, Halberstadt (23. Februar 1980).

Dr. PÜCHEL, MANFRED: Nukleinsäuremetabolismus und Regulation der Reservestoffbiosynthese während der Samenentwicklung. – Nachwuchswissenschaftler-Symposium der Biologischen Gesellschaft der DDR, Almsfeld (17. März 1980).

– Klonierung von doppelsträngiger cDNA. – Tagung der Sektion Biologie der Ernst-Moritz-Arndt-Universität Greifswald „Die Regulation von RNA-Processing, Translation und Posttranslation bei Mikroorganismen, Pflanzen und Tieren", Kloster/Hiddensee, 28. bis 31. Mai 1980 (29. Mai 1980).

–, Dr. HELMUT BÄUMLEIN, Dipl.-Biol. PETER SILHENGST und Dr. ULRICH WOBUS: Klonierung von doppelsträngiger cDNA, gewonnen aus polysomaler mRNA der Ackerbohnen-Kotyledonen. – Poster anläßlich der 12. Jahrestagung der Biochemischen Gesellschaft der DDR, Magdeburg, 8. bis 10. Oktober 1980 (10. Oktober 1980).

Prof. Dr. RIEGER, RIGOMAR: Über die Entstehung und biologische Bedeutung von chromosomalen Strukturumbauten (Chromosomenmutationen). – Sitzung der Leopoldina, Halle (20. Mai 1980).

– Rekonstruierte Karyotypen von *Vicia faba* und ihre Nutzung zur Untersuchung von Mutageneseprozessen. – Vortrag anläßlich der Emeritierung von Prof. Dr. U. NÜRNBERG, Sektion Biologie der Humboldt-Universität zu Berlin, Berlin (24. September 1980).

Dr. SCHIEMANN, JOACHIM: Nitratreduktase-defekte Mutanten des Tabaks als Marker für Gentransferexperimente mit höheren Pflanzen. – Bereichskolloquium am Institut für Biochemie der Pflanzen, Halle (4. Juni 1980).

Dipl.-Biol. SCHMIDT, OSKAR: Der Abbau der Thylakoidmembranproteine der Chloroplasten von *Vicia faba* während der Vergilbung der Blätter und in Modellversuchen. – Kolloquium am Institut für Photobiologie der Akademie der Wissenschaften der BSSR, Minsk/UdSSR (13. Februar 1980).

Dr. SCHÖNEICH, JÖRG: Aufgaben und Probleme der Mutagenitätstestung und der Entwicklungsbiologie der Säuger. – Weiterbildungsveranstaltung des Bezirkskabinetts für Lehrerweiterbildung des Bezirks Magdeburg, Magdeburg (14. Februar 1980).

– Einleitende Bemerkungen zu Aufgaben und Zielstellungen des Forschungsprojekts Humangenetik. – Sitzung der Klassen Biowissenschaften und Medizin der Akademie der Wissenschaften der DDR, Berlin (12. Juni 1980).

– Isonikotinsäurehydrazid (INH). Ein Beispiel für die Mutagenitätstestung und Risikoabschätzung von Arzneimitteln. – Kolloquium am Institut für Genetik der Gesellschaft für Strahlen- und Umweltforschung, Neuherberg/BRD, (7. Oktober 1980).

– Prüfung von Pharmaka auf Mutagenität. – 3. Arbeitstagung der Arbeitsgemeinschaft Antivirale Testungen der Gesellschaft für Mikrobiologie und Epidemiologie der DDR, Berlin (25. November 1980).

Dr. SCHOLZ, FRIEDRICH: Problems and possibilities in genetic improvement of protein yields in cereals. – 1st Research Coordination Meeting of the FAO/IAEA/GSF/SIDA "Coordinated Research Programme on the Use of Nuclear Techniques for Cereal Grain Protein Improvement", Nicosia/Zypern, 21. bis 25. April 1980 (22. April 1980).

– Genetisch-züchterische Probleme der Hybridforschung bei Gerste. – 3. Erwin-Baur-Kolloquium des Instituts für Züchtungsforschung Quedlinburg und des Zentralinstituts für Genetik und Kulturpflanzenforschung Gatersleben, Quedlinburg (20. Juni 1980).

Dr. SCHOLZ, GÜNTER: Untersuchungen am Vicilin, einem Reserveprotein der Ackerbohne. – Kolloquium an der Pädagogischen Hochschule „Liselotte Herrmann", Güstrow (12. März 1980).

– Der 'Normalisierungsfaktor' der auxotrophen Tomatenmutante 'chloronerva' – ein

Eisen-Carrier. – Kolloquium der Interdisziplinären Arbeitsgemeinschaft Genetik der Martin-Luther-Universität Halle-Wittenberg, Halle (15. April 1980).

Dr. Schubert, Ingo: Die menschlichen Chromosomen – Methoden ihrer Darstellung und Identifikation. – Weiterbildungsveranstaltung der Medizinischen Akademie, Magdeburg (11. Februar 1980).

– Molekulare Aspekte der Mutagenese. – Arbeitstagung der Gesellschaft für Geschwulstforschung „Molekulare Aspekte der Mutagenese und Kanzerogenese", Berlin-Buch (5. März 1980).

– Moderne Einsatzgebiete und Resultate der differenziellen Chromosomen- und Chromatidenmarkierung. – Vortragsreihe „Probleme der modernen Genetik" des Wissenschaftsbereichs Genetik der Martin-Luther-Universität Halle-Wittenberg, Halle (8. Oktober 1980).

Dr. Serfling, Edgar: BR-RNA und Sekretproteinsynthese in den larvalen Speicheldrüsen von *Chironomus thummi*. – Tagung der Sektion Biologie der Ernst-Moritz-Arndt-Universität Greifswald „Die Regulation von RNA-Processing, Translation und Posttranslation bei Mikroorganismen, Pflanzen und Tieren", Kloster/Hiddensee, 28. bis 31. Mai 1980 (28. Mai 1980).

– Struktur ribosomaler RNA-Gene, Synthese und Processing ribosomaler RNA eukaryotischer Organismen. – Tagung der Sektion Biologie der Ernst-Moritz-Arndt-Universität Greifswald „Die Regulation von RNA-Processing, Translation und Posttranslation bei Mikroorganismen, Pflanzen und Tieren", Kloster/Hiddensee, 28. bis 31. Mai 1980 (28. Mai 1980).

Dr. Süss, Karl-Heinz: Der ATPase Komplex und andere Proteine der Photosynthesemembran höherer Pflanzen. – Kolloquium am Zentralinstitut für Molekularbiologie, Berlin-Buch (27. Juni 1980).

Dr. Tittel, Claus: Versuche zur Verlangsamung der Samenalterung. – 2. Saatgutsymposium der Sektion Pflanzenproduktion der Martin-Luther-Universität Halle-Wittenberg, „Saatgutvitalität und Pflanzenertrag", Halle, 9. bis 12. September 1980 (11. September 1980).

Dr. Walter, Gaby und Dr. Armin Meister (vorgetragen von Dr. G. Walter): Biogenese des Chlorophylls. – Tagung der Arbeitsgemeinschaft Photosynthese der Biologischen Gesellschaft der DDR, Egsdorf, 31. März bis 3. April 1980 (31. März 1980).

Dipl.-Biol. Walter, Gerald: Entwicklungsbiologie und zellbiologische Manipulationen beim Säuger. – Tagung der Sektion Genetik der Biologischen Gesellschaft der DDR „Genetische Experimente im Biologieunterricht", Halle, 24. bis 25. Oktober 1980 (25. Oktober 1980).

Dr. Weber, Ernst: Protein-processing bei Pflanzen. – Tagung der Sektion Biologie der Ernst-Moritz-Arndt-Universität Greifswald „Die Regulation von RNA-Processing, Translation und Posttranslation bei Mikroorganismen, Pflanzen und Tieren", Kloster/Hiddensee, 28. bis 31. Mai 1980 (29. Mai 1980).

–, Dr. Renate Manteuffel und Dr. Dieter Neumann: Biosynthese und Akkumulation der Reserveproteine während der Samenentwicklung von *Vicia faba*. – Poster anlässlich der 12. Jahrestagung der Biochemischen Gesellschaft der DDR, Magdeburg, 8. bis 10. Oktober 1980 (10. Oktober 1980).

Dr. Wobus, Ulrich: Feinstrukturanalyse eines Insekten-Gens: Interne Repetition als Bauprinzip der Balbiani Ring-DNA von *Chironomus thummi*. – Kolloquium der Problemkommission Nukleinsäuren und Eiweiße des Forschungszentrums für Molekularbiologie und Medizin, Berlin-Buch (18. Januar 1980).

– Die Klonierung von eukaryotischen Genen – ein neuer Weg zur Analyse von Struktur und Funktion der Gene. – 10. Generalversammlung und wissenschaftliche Vortragstagung der Biologischen Gesellschaft der DDR, Weimar, 2. bis 4. Mai 1980 (3. Mai 1980).

– Periodicities and internal repeats in Balbiani ring DNA. – 4th Arolla Workshop „Formation of Messenger RNA in Eukaryotic Cells", Arolla/Schweiz, 18. bis 22. August 1980 (19. August 1980).

– Verfahren und neue Resultate der Gen-Manipulation. – Vortragsreihe „Probleme der modernen Genetik" des Wissenschaftsbereichs Genetik der Martin-Luther-Universität Halle-Wittenberg, Halle (10. Dezember 1980).

Dr. WOBUS, ULRICH, Dr. HELMUT BÄUMLEIN, Dr. REINHARD PANITZ, Dr. EDGAR SERFLING and Prof. F. C. KAFATOS: Internal repeats in Balbiani ring transcription units of *Chironomus thummi*. – Poster presented at the 4th Arolla Workshop "Formation of Messenger RNA in Eukaryotic Cells", Arolla/Schweiz, 18. bis 22. August 1980.

# E. Tagungen im Institut

## Tag des Jungen Wissenschaftlers (21. Mai 1980)

Dipl.-Agraring. J. MATZEL, Gatersleben: Untersuchungen zur modifikativen Induktion der Apomixis bei Futtergräsern.

Dipl.-Biol. A. HEIDL, Gatersleben: Markertransfer im Bazillus durch heterologe Transformation.

Dipl.-Biol. K. HEINDORFF, Gatersleben: Untersuchungen über den Mechanismus und die Modifizierbarkeit der aberrationsinduzierenden Wirkung des Maleinsäurehydrazids an rekonstruierten Karyotypen von *Vicia faba*.

Dipl.-Phys. E. BRECHT und Dr. K.-H. SÜSS, Gatersleben (vorgetragen von Dipl.-Phys. E. BRECHT): Spektroskopische Charakterisierung einer Präparation des lichtsammelnden Chlorophyll a/b-Protein-Komplexes.

Dipl.-Biol. J. SCHOTTE, Gatersleben: Die Beziehung zwischen der Struktur der Photosynthesemembran und deren photochemischer Aktivität.

Dipl.-Biol. K. BECKER, Gatersleben: Untersuchungen zur Abschätzung des genetischen Risikos oraler Kontrazeptiva.

2nd Symposium on Seed Proteins des Zentralinstituts für Genetik und Kulturpflanzenforschung Gatersleben und der Sektion Physiologie und Biochemie der Pflanzen der Biologischen Gesellschaft der DDR: "Regulation of Protein Biosynthesis and Degradation During Embryogenesis and Germination of Plant Seeds" in Gatersleben vom 25. bis 27. Juni 1980

### Plenarvorträge

## Section 1:
## Regulation of Storage Protein Biosynthesis During Seed Development (25. Juni 1980)

ZAGÓRSKI, W., J. CHROBOCZEK and W. RYCHLIK, Warszawa/VR Polen: Wheat germ cell-free system for protein synthesis.

PEDERSEN, K. and B. A. LARKINS, West Lafayette/USA: Factors regulating zein biosynthesis during maize endosperm development.

INGVERSEN, J., A. BRANDT and V. CAMERON-MILLS, København/Dänemark: The structure, biosynthesis and intracellular transport of barley endosperm reserve protein.

MÜNTZ, K., R. BASSÜNER, H. BÄUMLEIN, R. MANTEUFFEL, M. PÜCHEL, P. SCHMIDT and U. WOBUS, Gatersleben/DDR: The biosynthesis of storage proteins in developing field bean seeds.

NIELSEN, N. C., M. MOREIRA, P. STASWICK, M. A. HERMODSON, N. TUMER and V. H. THANH, West Lafayette/USA: The structure of glycinin from soybeans.

## Section 2: Accumulation of Storage Proteins and the Formation of Protein Bodies (26. Juni 1980)

Boulter, D., Durham/Großbritannien: Ultrastructure and composition of protein bodies.

Weber, E., D. Neumann and R. Manteuffel, Gatersleben und Halle/DDR: Storage protein accumulation and formation of protein bodies in *Vicia faba* seeds.

## Section 3: Regulation of Synthesis and Degradation of Proteins During Seed Germination (27. Juni 1980)

Barth, A., Halle/DDR: The function of proteases.

Lichtenfeld, C., R. Manteuffel, K. Müntz and G. Scholz, Gatersleben/DDR: Protein degradation in germinating legume seeds.

Mikola, J., Jyväskylä/Finnland: Three stages of reserve protein mobilization in germinating barley grain.

Peumans, W. J., A. R. Carlier, A. Manickam and B. M. Delay, Leuven/Belgien: Protein biosynthesis at the beginning of embryo development in germinating seeds.

Ajtchožin, M. A., Alma Ata/UdSSR: Informosomes of plant cells and some problems of regulation of protein synthesis in higher plants.

# Poster Abstracts

## 1 Isolation, Purification and Structure of Storage Proteins

Buschbeck, R., and R. Wilp, Güstrow/DDR: Differentiation of rye cultivars and species by their multiple forms of seed proteins and enzymes.

Casey, R., J. F. March and E. V. Sanger, Norwich/Großbritannien: Heterogeneity of legumin from *Pisum*.

Casey, R. and Ch. Horstmann, Norwich/Großbritannien und Gatersleben/DDR: Comparison of legumin subunits from *Pisum sativum* and *Vicia faba* by means of peptide mapping.

Guldager, P., L. Busch and K. Henriksen, København/Dänemark: Vicilin from two lines of pea (*Pisum sativum* L.).

Manickam, A., A. R. Carlier and W. J. Peumans, Leuven/Belgien: Storage albumin in mung bean embryonic axes.

Schlesier, B. and R. Manteuffel, Gatersleben/DDR: Preparation of legumin subunits and the reconstitution of the globulin.

Schwenke, K. D., K.-J. Linow, B. Raab, M. Schultz, K. Gast, D. Zirwer and R. Reichelt, Bergholz-Rehbrücke, Berlin-Buch und Teltow-Seehof/DDR: Isolation and characterization of a 12 S globulin from rapeseed (*Brassica napus* L.).

Plietz, P., G. Damaschun and K.-D. Schwenke, Berlin-Buch und Bergholz-Rehbrücke/DDR: Structure of 11 S globulins from different plants.

Švachulová, J., V. Turková and E. Klozová, Praha/ČSSR: Comparison of seed proteins in 3 cultivars of *Phaseolus vulgaris* L. with and without phytohemagglutinating activity.

## 2 Analysis of the Regulation of Storage Protein Synthesis and Accumulation *in vivo* and *in vitro*

Chroboczek, J., K. Ostrówka, M. Witt, R. Bassüner, M. Püchel and W. Zagórski, Warszawa und Poznań/VR Polen und Gatersleben/DDR: Accumulation of messengers for storage proteins may be responsible for embryo immunity against viral infection.

FUKAZAWA, C., K. UDAKA and K. KAINUMA, Ibaraki und Tokyo/Japan: Characterization of glycinin mRNA and reiteration frequency of glycinin DNA.

GARCIA-OLMEDO, F., P. CARBONERO, C. HERNANDEZ-LUCAS, F. PONZ, O. VICENTE and J. M. STERRA, Madrid/Spanien: Synthesis and location of thionins in developing endosperms of barley and wheat.

MERBACH, W., Halle/DDR: Seed protein yield and composition of leguminous plants as influenced by N-fertilizing in the period of flowering.

OTTAVIANO, E. and C. SOAVE, Milano/Italien: Chromosomal location and interaction effects of gene controlling zein components in Maize.

POPOVIĆ, V., K. KONSTANTINOV and M. DENIĆ, Beograd/SFR Jugoslavien: *In vitro* RNA and protein synthesis by RNA polymerase from $Opaque_2$ and normal *Zea mays* endosperm.

PÜCHEL, M., H. BÄUMLEIN, P. SILHENGST and U. WOBUS, Gatersleben/DDR: Cloning of double stranded cDNA derived from polysomal poly(A)-containing RNA of field bean cotyledons (*Vicia faba* L.).

SCHMIDT, P., R. BASSÜNER, H.-W. JANK, R. MANTEUFFEL and M. PÜCHEL, Gatersleben/DDR: Polysomes and poly(A)-containing RNA of membrane-bound polysomes from developing broad bean seeds (*Vicia faba* L. var. minor).

SOBOLEV, A., M. AZARKOVIČ and M. DMITRIEVA, Moskva/UdSSR: The use of isolated endosperm culture for investigation of protein synthesis and accumulation in castor bean seeds.

## 3 Compartmentation of Storage Protein Biosynthesis and Accumulation

BERGFELD, R. and P. SCHOPFER, Freiburg i. Br./BRD: Fine structural aspects of storage protein formation in developing cotyledons of *Sinapis alba* L.

WEBER, E., R. MANTEUFFEL, M. JACUBEK and D. NEUMANN, Gatersleben und Halle/DDR und Poznań/VR Polen: Comparative studies on protein body contents of *Vicia* and *Pisum* seeds.

## 4 Protein Biosynthesis and Degradation During Seed Germination

BELOZERSKIJ, M. and JA. DUNAEVSKIJ, Moskva/UdSSR: Degradation of the main storage globulin in germinating seeds of buckwheat *Fagopyrum esculentum* Moench.

BERNHARDT, D., Greifswald/DDR: The early protein biosynthesis in imbibing *Agrostemma githago* seeds.

CROCOMO, O. J., J. A. CURY and N. SACCHI, Piracicaba/Brasilien: Degradation of storage protein and synthesis of isoenzymes during *Phaseolus* embryo development in culture.

GUMILEVSKAJA, N., M. SKAŽENNIK, A. ACHMATOVA, L. ČUMIKINA and V. KRETOVIČ, Moskva/UdSSR: Polypeptides and RNAs synthesised in pea cotyledons at early imbibition.

HADAČOVÁ, V. and K. BENEŠ, Praha/ČSSR: Isoenzymes of o-mannosidase and β-xylosidase in root growth zones of *Vicia faba* L. and *Zea mays* L.

KONOPSKA, L., Łódź/VR Polen: Legumins in aleurone grains from pea seeds during germination.

LICHTENFELD, C., R. MANTEUFFEL and K. MÜNTZ, Gatersleben/DDR: *In vivo*- and *in vitro*-proteolysis of storage proteins by proteases from germinating field bean seeds (*Vicia faba* L.).

MIKOLA, J. and M. VIRTANEN, Jyväskylä/Finnland: Regulation of pH in the starchy endosperm of barley grain during reserve protein mobilization.

PEUMANS, W. J., B. M. DELAEY and A. R. CARLIER, Leuven/Belgien: Protein synthesis in a pea cell-free system.

SOPANEN, T. and S. NYMAN, Espoo/Finnland: Development and regulation of the uptake systems for peptides and amino acids in the scutellum of germinating barley grain.

JURINA, N. P., N. A. KARPOVA and M. S. ODINČOVA, Moskva/UdSSR: Investigation of the acidic proteins of 80 S ribosomes from pea seedlings.

**Symposium on Collection of European Land-Races of Cultivated Plants and their Evaluation** (1. bis 4. Juli 1980).

Die Vorträge und Poster des Symposiums sind in dem vorliegenden Band der Schriftenreihe „Die Kulturpflanze", S. 19 bis 399, enthalten.

## Berichtigung

Im Band XXVIII (1980) der ‚Kulturpflanze' muß es auf S. 426, 7. Zeile von unten richtig heißen: „. . . im zellfreien Weizenkeimlingssystem . . ." anstatt „. . . in verschiedenen zellfreien Systemen . . .".

Der Schriftleiter

# Sachregister

Table 7

Matrix of genetic distances (below diagonal) and genetic similarities (above diagonal) between cultivated melons in 35 Spanish prov[illegible] based on 15 loci. Distances and similarities calculated using the method of Nei (1972).

| | Albacete | Alicante | Almeria | Badajoz | Baleares | Barcelona | Caceres | Cadiz | Castellon | Ciudad-Real | Cordoba | Cuenca | Gerona | Granada | Guadalajara |
|---|---|---|---|---|---|---|---|---|---|---|---|---|---|---|---|
| Albacete | – | .96 | .79 | .98 | .90 | .99 | .94 | .93 | .87 | .96 | .96 | >.99 | .92 | .98 | .9 |
| Alicante | .05 | – | .89 | .98 | .95 | .96 | .93 | .94 | .97 | .99 | .99 | .96 | .95 | .98 | .9 |
| Almeria | .24 | .11 | – | .86 | .91 | .82 | .86 | .85 | .91 | .89 | .85 | .80 | .90 | .86 | .8 |
| Badajoz | .02 | .02 | .16 | – | .95 | .97 | .91 | .93 | .91 | .98 | .96 | .97 | .90 | .99 | .9 |
| Baleares | .10 | .05 | .09 | .05 | – | .94 | .84 | .97 | .97 | .98 | .90 | .90 | .89 | .92 | .8 |
| Barcelona | .01 | .04 | .20 | .03 | .06 | – | .93 | .97 | .91 | .98 | .95 | .99 | .95 | .97 | .9 |
| Caceres | .06 | .08 | .15 | .09 | .17 | .07 | – | .87 | .84 | .91 | .94 | .96 | .98 | .96 | .9 |
| Cadiz | .07 | .07 | .16 | .07 | .03 | .03 | .14 | – | .94 | .97 | .90 | .93 | .92 | .91 | .8 |
| Castellon | .14 | .03 | .09 | .09 | .04 | .09 | .17 | .06 | – | .97 | .94 | .88 | .92 | .91 | .9 |
| Ciudad-Real | .04 | .01 | .12 | .02 | .02 | .02 | .10 | .03 | .03 | – | .97 | .96 | .94 | .97 | .9 |
| Cordoba | .04 | .01 | .16 | .04 | .10 | .05 | .06 | .10 | .07 | .03 | – | .97 | .95 | .99 | .9 |
| Cuenca | <.01 | .04 | .22 | .03 | .11 | .01 | .04 | .07 | .12 | .04 | .03 | – | .94 | .99 | .9 |
| Gerona | .08 | .06 | .10 | .10 | .12 | .07 | .02 | .08 | .09 | .06 | .05 | .06 | – | .94 | .9 |
| Granada | .02 | .02 | .16 | .01 | .09 | .03 | .04 | .09 | .10 | .03 | .01 | <.01 | .06 | – | .9 |
| Guadalajara | .08 | .02 | .19 | .08 | .13 | .08 | .10 | .12 | .06 | .05 | .10 | .06 | .06 | .05 | – |
| Huelva | .07 | .07 | .16 | .07 | .03 | .03 | .14 | 0 | .06 | .03 | .10 | .07 | .08 | .09 | .1 |
| Huesca | .10 | .03 | .07 | .06 | .01 | .06 | .14 | .03 | .01 | .02 | .07 | .10 | .07 | .08 | .0 |
| Jaen | .04 | .03 | .12 | .06 | .06 | .02 | .05 | .02 | .05 | .02 | .04 | .03 | .02 | .04 | .0 |
| Leon | .06 | .04 | .26 | .09 | .17 | .07 | .08 | .14 | .10 | .07 | .02 | .05 | .07 | .05 | .0 |
| Lerida | .14 | .07 | .08 | .07 | .01 | .10 | .22 | .07 | .05 | .04 | .13 | .15 | .16 | .11 | .1 |
| Logrono | .07 | .06 | .12 | .06 | .02 | .03 | .13 | <.01 | .05 | .02 | .09 | .07 | .08 | .08 | .1 |
| Madrid | .05 | .01 | .13 | .04 | .03 | .02 | .10 | .02 | .02 | <.01 | .03 | .04 | .05 | .04 | .0 |
| Malaga | .11 | .03 | .10 | .07 | .02 | .06 | .15 | .03 | <.01 | .02 | .06 | .10 | .08 | .08 | .0 |
| Murcia | .12 | .06 | .10 | .12 | .07 | .07 | .09 | .04 | .04 | .05 | .08 | .10 | .02 | .10 | .0 |
| Navarra | .05 | .03 | .24 | .08 | .16 | .06 | .07 | .12 | .10 | .06 | .01 | .04 | .06 | .04 | .0 |
| Salamanca | .02 | .03 | .14 | .02 | .04 | <.01 | .07 | .02 | .07 | .01 | .04 | .02 | .05 | .02 | .0 |
| Segovia | .22 | .06 | .10 | .14 | .08 | .17 | .24 | .14 | .01 | .07 | .09 | .20 | .14 | .14 | .0 |
| Sevilla | .12 | .07 | .09 | .13 | .08 | .08 | .08 | .04 | .05 | .06 | .08 | .10 | .02 | .10 | .0 |
| Tarragona | .15 | .06 | .05 | .07 | .01 | .10 | .19 | .07 | .04 | .04 | .11 | .15 | .13 | .10 | .1 |
| Teruel | .16 | .05 | .12 | .13 | .07 | .12 | .19 | .09 | .01 | .05 | .07 | .15 | .09 | .12 | .0 |
| Toledo | .02 | .04 | .18 | .04 | .12 | .03 | .02 | .09 | .12 | .05 | .02 | .01 | .03 | .01 | .0 |
| Valencia | .13 | .05 | .07 | .06 | .01 | .09 | .19 | .06 | .02 | .03 | .10 | .13 | .13 | .10 | .1 |
| Valladolid | <.01 | .04 | .21 | .02 | .07 | <.01 | .07 | .04 | .11 | .03 | .04 | .01 | .07 | .02 | .0 |
| Zamora | .13 | .05 | .03 | .07 | .02 | .08 | .13 | .05 | .02 | .03 | .09 | .12 | .07 | .09 | .1 |
| Zaragoza | .14 | .06 | .18 | .15 | .10 | .10 | .16 | .07 | .03 | .06 | .07 | .13 | .07 | .13 | .0 |

| Huesca | Jaen | Leon | Lerida | Logrono | Madrid | Malaga | Murcia | Navarra | Salamanca | Segovia | Sevilla | Tarragona | Teruel | Toledo | Valencia | Valladolid | Zamora | Zaragoza |
|---|---|---|---|---|---|---|---|---|---|---|---|---|---|---|---|---|---|---|
| .90 | .96 | .94 | .87 | .93 | .95 | .90 | .89 | .95 | .98 | .81 | .89 | .86 | .85 | .98 | .88 | .99 | .88 | .87 |
| .97 | .97 | .96 | .94 | .94 | .99 | .97 | .94 | .97 | .97 | .94 | .94 | .94 | .95 | .96 | .95 | .96 | .96 | .94 |
| .93 | .88 | .77 | .92 | .88 | .88 | .91 | .91 | .78 | .87 | .90 | .91 | .95 | .88 | .83 | .93 | .81 | .97 | .84 |
| .94 | .95 | .91 | .94 | .94 | .96 | .93 | .88 | .93 | .98 | .87 | .88 | .93 | .88 | .86 | .94 | .98 | .93 | .86 |
| .99 | .95 | .84 | .99 | .90 | .97 | .98 | .93 | .85 | .96 | .93 | .93 | .99 | .93 | .88 | .99 | .93 | .98 | .90 |
| .94 | .98 | .93 | .90 | .97 | .98 | .94 | .93 | .94 | >.99 | .84 | .93 | .90 | .89 | .97 | .91 | >.99 | .92 | .90 |
| .87 | .95 | .92 | .80 | .88 | .91 | .86 | .91 | .93 | .93 | .79 | .92 | .83 | .83 | .98 | .82 | .93 | .88 | .85 |
| .97 | .98 | .87 | .93 | >.99 | .98 | .97 | .96 | .88 | .98 | .87 | .96 | .93 | .92 | .91 | .94 | .96 | .95 | .93 |
| .99 | .95 | .90 | .95 | .95 | .98 | >.99 | .96 | .90 | .93 | .99 | .95 | .96 | .99 | .89 | .97 | .90 | .98 | .97 |
| .98 | .98 | .93 | .96 | .98 | >.99 | .99 | .95 | .94 | .99 | .93 | .95 | .96 | .95 | .95 | .97 | .97 | .97 | .94 |
| .93 | .96 | .98 | .88 | .91 | .97 | .94 | .93 | .99 | .96 | .91 | .92 | .89 | .93 | .98 | .91 | .96 | .92 | .93 |
| .91 | .97 | .95 | .86 | .93 | .96 | .90 | .91 | .96 | .98 | .82 | .90 | .86 | .86 | .99 | .88 | .99 | .89 | .88 |
| .93 | .98 | .93 | .85 | .93 | .95 | .93 | .98 | .94 | .95 | .87 | .98 | .88 | .91 | .97 | .88 | .93 | .94 | .93 |
| .92 | .96 | .95 | .90 | .92 | .96 | .92 | .90 | .96 | .98 | .87 | .90 | .90 | .89 | .99 | .91 | .98 | .92 | .88 |
| .92 | .95 | .99 | .86 | .89 | .96 | .94 | .93 | .99 | .92 | .93 | .92 | .87 | .96 | .95 | .89 | .93 | .90 | .96 |
| .97 | .98 | .87 | .93 | >.99 | .98 | .97 | .96 | .88 | .98 | .87 | .96 | .93 | .92 | .91 | .94 | .96 | .95 | .93 |
| – | .97 | .88 | .98 | .98 | .98 | >.99 | .97 | .89 | .96 | .96 | .96 | .98 | .97 | .90 | .99 | .93 | .99 | .95 |
| .03 | – | .94 | .91 | .98 | .99 | .97 | .98 | .95 | .99 | .89 | .98 | .92 | .94 | .97 | .93 | .97 | .96 | .95 |
| .12 | .06 | – | .80 | .87 | .94 | .90 | .91 | >.99 | .92 | .87 | .90 | .81 | .92 | .96 | .84 | .93 | .85 | .93 |
| .02 | .10 | .22 | – | .95 | .94 | .96 | .89 | .82 | .94 | .93 | .89 | >.99 | .91 | .85 | >.99 | .89 | .98 | .87 |
| .02 | .02 | .14 | .05 | – | .98 | .97 | .96 | .88 | .98 | .89 | .96 | .95 | .92 | .91 | .96 | 96 | .97 | .93 |
| .02 | .01 | .06 | .06 | .02 | – | .99 | .97 | .95 | .98 | .94 | .96 | .94 | .97 | .95 | .96 | .97 | .96 | .97 |
| <.01 | .04 | .10 | .04 | .03 | .01 | – | .97 | .91 | .95 | .97 | .96 | .97 | .99 | .90 | .98 | .93 | .98 | .97 |
| .03 | .02 | .10 | .12 | .04 | .03 | .03 | – | .91 | .94 | .91 | >.99 | .91 | .96 | .92 | .92 | .92 | .96 | .98 |
| .11 | .05 | <.01 | .20 | .13 | .05 | .10 | .09 | – | .93 | .87 | .91 | .83 | .91 | .97 | .85 | .95 | .87 | .93 |
| .04 | .01 | .08 | .07 | .02 | .02 | .05 | .06 | .07 | – | .87 | .94 | .94 | .90 | .97 | .94 | .99 | .95 | .91 |
| .04 | .12 | .14 | .07 | .12 | .07 | .03 | .09 | .14 | .14 | – | .90 | .94 | .99 | .83 | .96 | .83 | .95 | .94 |
| .04 | .02 | .11 | .12 | .04 | .04 | .04 | <.01 | .10 | .06 | .10 | – | .90 | .95 | .92 | .92 | .91 | .96 | .96 |
| .02 | .08 | .21 | <.01 | .05 | .06 | .03 | .09 | .19 | .07 | .06 | .09 | – | .92 | .86 | >.99 | .89 | .99 | .88 |
| .03 | .07 | .09 | .09 | .08 | .03 | .02 | .04 | .09 | .10 | .01 | .04 | .08 | – | .87 | .94 | .87 | .95 | .98 |
| .10 | .03 | .04 | .17 | .09 | .05 | .10 | .08 | .03 | .03 | .19 | .08 | .15 | .14 | – | .86 | .98 | .90 | .89 |
| .01 | .07 | .18 | <.01 | .04 | .04 | .02 | .08 | .17 | .06 | .04 | .09 | <.01 | .06 | .15 | – | .90 | .99 | .90 |
| .07 | .03 | .07 | .11 | .04 | .03 | .08 | .09 | .06 | .01 | .18 | .09 | .12 | .13 | .02 | .10 | – | .91 | .89 |
| <.01 | .04 | .16 | .03 | .03 | .04 | .02 | .04 | .14 | .05 | .05 | .04 | .01 | .05 | .11 | .01 | .10 | – | .92 |
| .05 | .05 | .07 | .14 | .07 | .03 | .03 | .03 | .07 | .10 | .06 | .04 | .13 | .02 | .12 | .10 | .11 | .08 | – |

Kolibri 360 GmbH
Ferdinand-Jühlke-Straße 7
99095 Erfurt, Deutschland
produktsicherheit@kolibri360.de